Selected Titles in This Series

19 **William G. Dwyer, Stephen Halperin, Richard Kane, Stanley O. Kochman, Mark E. Mahowald, and Paul S. Selick (Editor-in-Chief), Editors,** Stable and unstable homotopy, 1998

18 **Panos M. Pardalos and Henry Wolkowicz, Editors,** Topics in semidefinite and interior-point methods, 1998

17 **Joachim J. R. Cuntz and Masoud Khalkhali, Editors,** Cyclic cohomology and noncommutative geometry, 1997

16 **Victor P. Snaith, Editor,** Algebraic K-theory, 1997

15 **Stephen P. Braham, Jack D. Gegenberg, and Robert J. McKellar, Editors,** Sixth Canadian conference on general relativity and relativistic astrophysics, 1997

14 **Mourad E. H. Ismail, David R. Masson, and Mizan Rahman, Editors,** Special functions, q-series and related topics, 1997

13 **Peter A. Fillmore and James A. Mingo, Editors,** Operator algebras and their applications, 1997

12 **Dan-Virgil Voiculescu, Editor,** Free probability theory, 1997

11 **Colleen D. Cutler and Daniel T. Kaplan, Editors,** Nonlinear dynamics and time series: Building a bridge between the natural and statistical sciences, 1997

10 **Jerrold E. Marsden, George W. Patrick, and William F. Shadwick, Editors,** Integration algorithms and classical mechanics, 1996

9 **W. H. Kliemann, W. F. Langford, and N. S. Namachchivaya, Editors,** Nonlinear dynamics and stochastic mechanics, 1996

8 **Larry M. Bates and David L. Rod, Editors,** Conservative systems and quantum chaos, 1996

7 **William F. Shadwick, Perinkulam Sambamurthy Krishnaprasad, and Tudor Stefan Ratiu, Editors,** Mechanics day, 1996

6 **Anna T. Lawniczak and Raymond Kapral, Editors,** Pattern formation and lattice gas automata, 1996

5 **John Chadam, Martin Golubitsky, William Langford, and Brian Wetton, Editors,** Pattern formation: Symmetry methods and applications, 1996

4 **William F. Langford and Wayne Nagata, Editors,** Normal forms and homoclinic chaos, 1995

3 **Anthony Bloch, Editor,** Hamiltonian and gradient flows, algorithms and control, 1994

2 **K. A. Morris, Editor,** Control of flexible structures, 1993

1 **Michael J. Enos, Editor,** Dynamics and control of mechanical systems: The falling cat and related problems, 1993

FIELDS INSTITUTE COMMUNICATIONS

THE FIELDS INSTITUTE FOR RESEARCH IN MATHEMATICAL SCIENCES

Stable and Unstable Homotopy

William G. Dwyer
Steven Halperin
Richard Kane
Stanley O. Kochman
Mark E. Mahowald
Paul S. Selick (Editor-in-Chief)
Editors

American Mathematical Society
Providence, Rhode Island

The Fields Institute
for Research in Mathematical Sciences

The Fields Institute is named in honour of the Canadian mathematician John Charles Fields (1863–1932). Fields was a visionary who received many honours for his scientific work, including election to the Royal Society of Canada in 1909 and to the Royal Society of London in 1913. Among other accomplishments in the service of the international mathematics community, Fields was responsible for establishing the world's most prestigious prize for mathematics research—the Fields Medal.

The Fields Institute for Research in Mathematical Sciences is supported by grants from the Ontario Ministry of Education and Training and the Natural Sciences and Engineering Research Council of Canada. The Institute is sponsored by McMaster University, the University of Toronto, the University of Waterloo, and York University and has affiliated universities in Ontario and across Canada.

1991 *Mathematics Subject Classification.* Primary 55–06; Secondary 55Pxx.

Library of Congress Cataloging-in-Publication Data

Stable and unstable homotopy / William G. Dwyer... [et al.], editors.
p. cm. — (Fields Institute Communications, ISSN 1069-5265 ; 19)
Includes bibliographical references.
ISBN 0-8218-0824-9 (hardcover ; alk. paper)
1. Homotopy theory. I. Dwyer, William G., 1947– . II. Series: Fields Institute communications ; v. 19.
QA612.7.S73 1998
514′.24—dc21 98-11194
CIP

♾ The paper used in this book is acid-free and falls within the guidelines
established to ensure permanence and durability.
This publication was prepared by The Fields Institute.
Visit the AMS home page at URL: `http://www.ams.org/`

10 9 8 7 6 5 4 3 2 1 02 01 00 99 98

Contents

Preface

During 1995-96 an emphasis program in Homotopy Theory was held at The Fields Institute for Research in the Mathematical Sciences, constituting the Institute's major program that year. The organizers of the program were: Bill Dwyer, Steve Halperin, Richard Kane (Chief Organizer), Stan Kochman, Doug Ravenel, and Paul Selick.

The program had several workshops associated with it including one on Stable Homotopy Theory held in January (organized by Stan Kochman, Mark Mahowald, and Doug Ravenel), and programs in Unstable and Rational Homotopy Theory (organized by Bill Dwyer, Steve Halperin, Richard Kane, and Paul Selick) held in May and June. This volume contains the proceedings of those workshops. Another workshop in the program, The Second Great Lakes Algebraic K-Theory Conference (organized by Rick Jardine, Manfred Kolster, and Vic Snaith) was held in April and its proceedings can be found in The Fields Institute Communication Series, Volume 16. A conference in Applications of Homotopy Theory to Physics and Geometry (organized by Jacques Hurturbise and Richard Kane) was held in March. The program also included a one day conference held at Simon Fraser University in December.

The organizers are grateful to The Fields Institute and NSERC for financial support and to The Fields Institute and its Director, John Chadam, for providing excellent facilities in which to conduct our research. We would especially like to thank the staff of The Fields Institute, Sue Embro, Pauline Grant, Becky Sappong, Karen Walker, and Alesia Zuccala, for helping to create the atmosphere that made the stay of all of the participants so enjoyable. Finally, we would like to thank Erna Unrau for her many hours of work in the preparation of this volume.

The Editors

October 1997

Fields Institute Communications
Volume **19**, 1998

A Functorial Model for Iterated Snaith Splitting with Applications to Calculus of Functors

Greg Arone
The University of Chicago
Department of Mathematics
Chicago, IL 60637 USA
arone@math.uchicago.edu

Marja Kankaanrinta
Department of Mathematics
P.O. Box 4 (Yliopistonkatu 5)
University of Helsinki
FIN-00014, Helsinki, Finland
mkankaan@cc.helsinki.fi

Abstract. We study the Goodwillie tower of the identity functor from based spaces to based spaces by evaluating the Goodwillie tower of a certain cosimplicial functor that approximates the identity. We provide fairly concrete descriptions for the "Taylor polynomials" of the identity and recover B. Johnson's description of the Goodwillie derivatives of the identity. The main ingredient of the proof is a new (in a sense more functorial) model for the "Snaith filtration" and its iterates.

Introduction

The stable homotopy functor $QX = \Omega^\infty \Sigma^\infty X$, together with the obvious natural transformations $\mathrm{Id} \to Q$ and $QQ \to Q$, gives rise to a monad in the category of spaces. Therefore, there exists (a coaugmented) cosimplicial space

$$[k] \mapsto Q^{k+1}X. \tag{0.1}$$

This cosimplicial space can be thought of as a space level version of the R-localization construction due to Bousfield and Kan [**4**], with the role of the ring R played by the "ring up to homotopy" $Q(S^0)$. For the purposes of this paper, we refer to (0.1) as the Bousfield-Kan cosimplicial space and we call the associated tower of fibrations the Bousfield-Kan tower. The total space of this cosimplicial space is the **S**-localization of X, written $L_{\mathbf{S}}X$. There is a natural transformation $X \to L_{\mathbf{S}}X$, which is a weak homotopy equivalence if X is "**S**-good", i.e., if $\pi_1(X)$ is nilpotent and acts nilpotently on the stable homotopy of X. In particular, it is an equivalence

---

1991 *Mathematics Subject Classification.* 55P99.

MK was supported by the Academy of Finland and the Emil Aaltonen Foundation.

if X is simply connected. Thus for many practical purposes one can think of the Bousfield-Kan tower as converging to the identity functor.

Our goal in this note is to make precise the idea that one can study the Goodwillie tower of the identity functor by first evaluating the Goodwillie tower of the Bousfield-Kan cosimplicial space. As an illustration for our approach, we recover a reformulated form of B. Johnson's description of the Goodwillie derivatives of the identity [**12**] and provide fairly concrete formulas for the Taylor polynomials of the identity. In particular, our methods provide ways to study the "k-invariants" in the Goodwillie tower of the identity rather than just the fibers in the tower.

We will assume some familiarity with the basic ideas of Goodwillie's calculus of functors. In particular, at one point we will use the result on the relation between cross-effects and differentials. We refer the reader to [**9, 10, 11**], as well as [**2, 12**], for background on Goodwillie's calculus in general and on the Goodwillie tower of the identity in particular. We use some standard notation from there. Thus, given a homotopy functor F from spaces to spaces, we let P_nF denote its n-th Taylor approximation. By a slight abuse of notation, we may use P_nF to denote any functor with the right homotopy type. In other words, we may use P_nF to denote a functor which is only naturally weakly equivalent to what is denoted P_nF in [**11**]. Similarly, D_nF denotes the homotopy fiber of the map $P_nF \to P_{n-1}F$, or any functor naturally equivalent to it. Also, if $F = \mathrm{Id}$, the identity functor, then we often abbreviate $P_n\mathrm{Id}(X)$ as P_nX and $D_n\mathrm{Id}(X)$ as D_nX.

The basic ideas of the paper are the following: It is easier to identify P_n and D_n of higher iterates of $Q(-)$ (by means of the Snaith splitting) than to identify P_n and D_n of the identity directly. More explicitly, the following is true:

$$P_nQQ(X) \simeq \prod_{i=1}^{n} Q(X^{\wedge i}_{h\Sigma_i})$$

and

$$D_nQQ(X) \simeq Q(X^{\wedge n}_{h\Sigma_n}).$$

As for higher iterates of $Q(-)$, combinatorial analysis of the functor $Q^{k+1}X$ by means of the standard splitting shows that there are natural weak equivalences

$$D_nQ^{k+1}X \simeq \Omega^\infty \left(\mathrm{Map}_* \left(T^k_n, \Sigma^\infty X^{\wedge n}\right) \wedge_{\Sigma_n} E\Sigma_{n+}\right), \tag{0.2}$$

and

$$P_nQ^kX \simeq \prod_{m=1}^{n} D_mQ^kX.$$

Here T^k_n is the set of k-simplices in a certain simplicial set with an action of Σ_n (the simplicial set is essentially the nerve of the category of unordered partitions of the set $\underline{n}$ - see Definition 4.2), whose geometric realization we denote K_n. The next observation is that Goodwillie's operators P_n and D_n commute (more or less) with homotopy inverse limits of functors. Thus, applying P_n or D_n degree-wise to the BK cosimplicial space, one gets a cosimplicial space whose total space is $P_n\mathrm{Id}(X)$ or $D_n\mathrm{Id}(X)$, respectively (here $P_n(-)$ and $D_n(-)$ are meant strictly in the sense of [**11**]). Since P_nX and D_nX of $Q^{k+1}X$ are "known", at least up to homotopy, one should be able to recover information about $P_n\mathrm{Id}(X)$ or $D_n\mathrm{Id}(X)$, respectively. This would be analogous to a situation in which one is given an expansion of a function $f(x)$ as an infinite sum $\Sigma f_i(x)$ of other functions, and one recovers the Taylor expansion of f by expanding the functions f_i and rearranging terms.

There are a couple of technical problems with this plan. The first issue is convergence. P_n and D_n, whose definitions involve finite holim and filtered hocolim, commute (up to natural homotopy equivalence) in general only with finite holim, while the total space of a cosimplicial space is an infinite holim. So it is not entirely automatic that the total spaces of the cosimplicial spaces $k \mapsto P_n Q^{k+1} X$ and $k \mapsto D_n Q^{k+1} X$ are equivalent to $P_n X$ and $D_n X$, respectively. However, it is not difficult to show that for this case it is true. This follows because in the tower of fibrations associated with the BK cosimplicial space, the Taylor expansions of the m-th fiber begins with a term of degree m. In particular, the degree of the first term goes to infinity as one goes up in the tower. We make this precise in Section 5.

The second, far more serious, issue is functoriality. As indicated above, up to homotopy $P_n Q^k X$ and $D_n Q^k X$ are well-known - they are given by an iterated application of the standard stable splitting (Snaith splitting) of QX. Further analysis shows that the coface and codegeneracy maps in the cosimplicial space $k \mapsto D_n Q^{k+1} X$ are dual, at least in the homotopy category, to the face and degeneracy maps in the simplicial set $k \mapsto T_n^k$. This suggests that

$$D_n X \simeq \Omega^\infty \left(\mathrm{Map}_* \left(K_n, \Sigma^\infty X^{\wedge n}\right) \wedge_{\Sigma_n} E\Sigma_{n+}\right),$$

which is the form of Johnson's theorem that we want to prove. The problem is that identification (0.2) is only up to a weak homotopy equivalence, and is not functorial enough for our purposes. By this we mean that it does not behave well with respect to natural transformations. Thus given a coface map $Q^k X \to Q^{k+1} X$ it seems to be impossible, using the known models, to define maps

$$Q^k X \to \prod_{m=1}^{n} D_m Q^k X \; (\simeq P_n Q^k X)$$

and

$$\prod_{m=1}^{n} D_m Q^k X \to \prod_{m=1}^{n} D_m Q^{k+1} X$$

in such a way that there would be a strictly commutative diagram

$$\begin{array}{ccccc} Q^k X & \to & \prod_{m=1}^{n} D_m Q^k X & \leftarrow & D_n Q^k X \\ \downarrow & & \downarrow & & \downarrow \\ Q^{k+1} X & \to & \prod_{m=1}^{n} D_m Q^{k+1} X & \leftarrow & D_n Q^{k+1} X. \end{array}$$

Therefore, the known models for the Snaith splitting can not be used to write down a cosimplicial space homotopy equivalent to

$$P_n QX \leftrightarrows P_n QQX \cdots P_n Q^k X \cdots$$

or even to

$$D_n QX \leftrightarrows D_n QQX \cdots D_n Q^k X \cdots .$$

At best, the Snaith splitting can be pushed, using the methods of [**13**], to determine the homotopy type of the maps $D_n Q^k X \to D_n Q^{k+1} X$, thus producing a cosimplicial space in the homotopy category. But cosimplicial spaces in the homotopy category do not have realizations, and moreover two different cosimplicial spaces, with different total spaces, may become isomorphic when projected to the homotopy category. Most of the work in the paper is to overcome this difficulty.

This situation is somewhat akin to the foundational problem that there was with construction of THH before the work of Bökstedt and Elmendorf, Kriz, Mandell & May. However, their work alone does not solve our problem, because their constructions are entirely in the stable category, while our situation is unstable in an essential way. Even though all the spaces involved are infinite loop spaces, we have to consider maps that are not infinite loop maps. We will, however, make some use of Bökstedt's ideas for the construction of THH.

Thus we are looking for a model for the Snaith splitting (and its iterates) that is functorial with respect to the coface maps in the BK cosimplicial space. Before we indicate how we intend to obtain such model, we offer two heuristic explanations for the failure of the known models of the Snaith splitting to be functorial enough for our purposes. First, recall once more that up to homotopy the (iterated) Snaith splitting is nothing but the Goodwillie tower of the functor Q^kX. From calculus' point of view, the Goodwillie tower is a tower of fibrations that is associated to a functor in a way that is functorial with respect to natural transformations, while the fact that the tower splits is rather accidental. Thus, one might expect that in a functorial model for P_nQ^kX the fact that $P_nQ^kX \simeq \prod_{m=1}^{n} D_mQ^kX$ would not be obvious from the construction. Second, the Snaith splitting represents D_nQ^kX as a homotopy orbit spectrum, in particular a homotopy direct limit, while Goodwillie's construction of P_n and of D_n involves *inverse* homotopy limits rather than direct homotopy limits. Therefore it is reasonable to expect that a functorial description of P_n and D_n will be in terms of homotopy inverse limits, perhaps homotopy fixed points, rather than homotopy direct limits. There is in fact a canonical way to replace homotopy orbits of a spectrum by homotopy fixed points of another spectrum by means of the Adams isomorphism [**14, 15**], which is indeed a part of what we are going to do.

Guided by these general comments, we are looking for models defined in terms of inverse limits rather than direct limits for which the splitting is not explicit. This suggests the relevance of [**1**], where an explicit description of the Goodwillie tower of the functor $F(Y) = Q\mathrm{Map}_*(K,Y)$, (here K is a fixed finite CW-complex), is given. Namely, it is proved there that

$$P_nQ\mathrm{Map}_*(K,Y) \simeq \mathrm{Nat}_{M_n}(\Sigma^\infty K^{\wedge i}, \Sigma^\infty Y^{\wedge i}),$$

where Nat denotes the space of natural transformations, M_n denotes the category of standard finite sets $\emptyset, \underline{1}, \dots, \underline{n}$ with surjections and $\Sigma^\infty K^{\wedge i}, \Sigma^\infty Y^{\wedge i}$ are certain evident functors from M_n to spectra. The space of natural transformations is in fact an inverse limit [**8, 1**]. It follows from this formula that

$$D_nF(Y) \simeq \mathrm{Map}_*\left(K^{(n)}, QY^{\wedge n}\right)^{\Sigma_n},$$

where $K^{(n)}$ is $K^{\wedge n}$ with the fat diagonal collapsed to a point (notice that the formula is given in terms of fixed points, rather than homotopy orbits).

Taking $K = S^m$, $Y = S^mX$ and letting $m \to \infty$ one obtains a description for P_nQQX. It may be illuminating to observe, for instance, that for $n = 2$ our description says that P_2QQX is equivalent to $Q_{\Sigma_2}(X^{\wedge 2})^{\Sigma_2}$, which is equivalent to $QX \times Q(X^{\wedge 2}_{h\Sigma_2})$ by the classical tom Dieck splitting, just as one expects.

In Section 2, Theorem 2.21, we show that there is a similar description of P_nQ^kX as a space of natural transformations between functors whose domain is not M_n, but rather a category closely related to the category of simplices of M_n

(we hope that these models for polynomial approximations of Q^kX will be of some independent interest). In Section 3 we show that our formulas for P_nQ^kX are functorial enough to fit into a semi-cosimplicial space, i.e. a space with coface maps, but no codegeneracy maps. This is sufficient for our purposes since semi-cosimplicial spaces have "realizations" (total spaces). In Section 4 we use this semi-cosimplicial space to recover the main theorem of [**12**]. Other applications will be pursued elsewhere.

1 Preliminaries

For a natural number n, let $\underline{n}$ denote the set $\{1, \ldots, n\}$. In the course of the paper we will sometimes neglect to underline cardinalities. Thus we may use the symbol n to denote both the set $\underline{n}$ and the cardinality n.

We are now going to define several small categories, which will be used extensively throughout the paper.

Definition 1.1 Let M be the category whose objects are finite sets and whose morphisms are surjections of sets. For a finite set S, let M_S denote the full subcategory of M whose objects are subsets of S. Let M_n stand for $M_{\underline{n}}$.

We let $\mathcal{S}$ denote the category of finite sets with all maps. Thus M is a subcategory of $\mathcal{S}$. Notice that the empty set is an object of M, but the only morphism to or from the empty set is the identity $\emptyset \to \emptyset$. We think of the empty set as the basepoint of M. Clearly, the full subcategory of M_n containing the non-empty sets has a final object $\underline{1}$ and we identify M_n with the disjoint union $M_n/\underline{1} \coprod \{\emptyset\}$, where $M_n/\underline{1}$ is the over category. Thus we denote an object $\underline{i}$ of M_n by $\underline{i} \twoheadrightarrow \underline{1}$ whenever $i > 0$.

We will need to deal with categories closely related to the nerve category of M_n. In other words, we will need to deal with chains of surjections. As a matter of notation, throughout the paper we use $\boldsymbol{\alpha}$ to denote a generic chain of surjections of the form $\underline{i_k} \overset{\alpha_k}{\twoheadrightarrow} \underline{i_{k-1}} \overset{\alpha_{k-1}}{\twoheadrightarrow} \cdots \overset{\alpha_2}{\twoheadrightarrow} \underline{i_1} \twoheadrightarrow \underline{i_0} = \underline{1}$. We call k the length of $\boldsymbol{\alpha}$.

Definition 1.2 Let $n > 0$. We define N_0M_n to be the category with two objects: $\{\emptyset\}$ and $\underline{1}$ and identity morphisms only. For $k \geq 1$ let N_kM_n denote the category of $(k-1)$-simplices of M_n.

Thus the objects of N_kM_n are chains of surjections of the form

$$\underline{i_k} \overset{\alpha_k}{\twoheadrightarrow} \underline{i_{k-1}} \overset{\alpha_{k-1}}{\twoheadrightarrow} \cdots \overset{\alpha_2}{\twoheadrightarrow} \underline{i_1} \twoheadrightarrow \underline{i_0} = \underline{1}$$

and one more object corresponding to the empty set. A morphism $\boldsymbol{\alpha} \to \boldsymbol{\alpha}'$ is a collection of surjections $h_j \colon \underline{i_j} \twoheadrightarrow \underline{i'_j}$ such that the diagram

$$\begin{array}{ccccccccc}
\underline{i_k} & \overset{\alpha_k}{\twoheadrightarrow} & \underline{i_{k-1}} & \overset{\alpha_{k-1}}{\twoheadrightarrow} & \cdots & \overset{\alpha_2}{\twoheadrightarrow} & \underline{i_1} & \overset{\alpha_1}{\twoheadrightarrow} & \underline{1} \\
\downarrow_{h_k} & & \downarrow_{h_{k-1}} & & & & \downarrow_{h_1} & & \downarrow_{h_0=\mathrm{id}} \\
\underline{i'_k} & \overset{\alpha'_k}{\twoheadrightarrow} & \underline{i'_{k-1}} & \overset{\alpha'_{k-1}}{\twoheadrightarrow} & \cdots & \overset{\alpha'_2}{\twoheadrightarrow} & \underline{i'_1} & \overset{\alpha'_1}{\twoheadrightarrow} & \underline{1}
\end{array}$$

commutes.

Let $\boldsymbol{\alpha} = (\underline{i_2} \twoheadrightarrow \underline{i_1})$ and $\boldsymbol{\alpha'} = (\underline{i'_2} \twoheadrightarrow \underline{i'_1})$ be two surjections. Let $(h_2, h_1) : \boldsymbol{\alpha} \to \boldsymbol{\alpha'}$ be a morphism.

Definition 1.3 We say that (h_2, h_1) is *admissible* if for all $s \in \underline{i_1}$, h_2 maps the inverse image of s in $\underline{i_2}$ surjectively onto the inverse image of $h_1(s)$ in $\underline{i'_2}$

Admissible morphisms have the following useful property.

Proposition 1.4 *Let $\boldsymbol{\alpha} = (\underline{i_2} \twoheadrightarrow \underline{i_1})$ and $\boldsymbol{\alpha}' = (\underline{i_2'} \twoheadrightarrow \underline{i_1'})$ be two surjections. Let $(h_2, h_1)\colon \boldsymbol{\alpha} \to \boldsymbol{\alpha}'$ be an admissible morphism. Then if h_2 is an isomorphism of sets then so is h_1.*

Proof Easy. □

Let

$$\boldsymbol{\alpha} = (\underline{i_k} \overset{\alpha_k}{\twoheadrightarrow} \underline{i_{k-1}} \overset{\alpha_{k-2}}{\twoheadrightarrow} \cdots \overset{\alpha_{j+1}}{\twoheadrightarrow} \underline{i_j} \twoheadrightarrow \cdots \underline{i_1} \twoheadrightarrow \underline{1})$$

and

$$\boldsymbol{\alpha}' = (\underline{i_k'} \overset{\alpha_k'}{\twoheadrightarrow} \underline{i_{k-1}'} \overset{\alpha_{k-2}'}{\twoheadrightarrow} \cdots \overset{\alpha_{j+1}'}{\twoheadrightarrow} \underline{i_j'} \twoheadrightarrow \cdots \underline{i_1'} \twoheadrightarrow \underline{1})$$

be two chains of surjections. We say that a morphism $\boldsymbol{\alpha} \to \boldsymbol{\alpha}'$ is admissible if for all $k \geq j_2 \geq j_1 \geq 0$ the induced morphism

$$(\underline{i_{j_2}} \twoheadrightarrow \underline{i_{j_1}}) \to (\underline{i_{j_2}'} \twoheadrightarrow \underline{i_{j_1}'})$$

is admissible.

Definition 1.5 For $k \geq 0$ let $O_k M_n$ denote the following subcategory of $N_k M_n$: the objects of $O_k M_n$ are the same as of $N_k M_n$, the morphisms of $O_k M_n$ are the admissible morphisms of $N_k M_n$.

The disjoint union $\coprod_{k\geq 0} O_k M_n$ can be given the structure of a simplicial category $O_\bullet M_n$. The face maps are given by the following formulas:

$$d_0(\underline{i_k} \overset{\alpha_k}{\twoheadrightarrow} \underline{i_{k-1}} \overset{\alpha_{k-1}}{\twoheadrightarrow} \cdots \overset{\alpha_2}{\twoheadrightarrow} \underline{i_1} \twoheadrightarrow \underline{i_0} = \underline{1}) = \begin{cases} \underline{i_k} \overset{\alpha_k}{\twoheadrightarrow} \underline{i_{k-1}} \overset{\alpha_{k-1}}{\twoheadrightarrow} \cdots \overset{\alpha_2}{\twoheadrightarrow} \underline{i_1} & \text{if } i_1 = 1 \\ \emptyset & \text{otherwise} \end{cases}$$

and for $1 \leq j \leq k$

$$d_j(\underline{i_k} \overset{\alpha_k}{\twoheadrightarrow} \cdots \overset{\alpha_2}{\twoheadrightarrow} \underline{i_1} \twoheadrightarrow \underline{i_0} = \underline{1}) = (\underline{i_k} \overset{\alpha_k}{\twoheadrightarrow} \cdots \underline{i_{j+1}} \overset{\alpha_j \circ \alpha_{j+1}}{\twoheadrightarrow} \underline{i_{j-1}} \cdots \overset{\alpha_2}{\twoheadrightarrow} \underline{i_1} \twoheadrightarrow \underline{i_0} = \underline{1}).$$

Similarly, the degeneracy maps $s_j : O_k M_n \to O_{k+1} M_n$ are defined by inserting $\underline{i_j}$ in the appropriate place. It is easy to check that d_j and s_j induce well-defined maps on morphisms of $O_k M_n$.

Throughout the paper, X stands for a compactly generated space with a non-degenerate basepoint.

Definition 1.6 For a finite set T, $X^{\wedge T}$ stands for the total cofiber of the cubical diagram (see [**10**] for the relevant definitions) $U \mapsto X^U$, $U \subseteq T$, in which the maps are basepoint inclusions.

Basically, $X^{\wedge T}$ is just smash product of $|T|$ copies of X made functorial in T. Note that $X^{\emptyset} \cong X^{\wedge \emptyset} \cong *$.

A map of sets $\alpha\colon \underline{n} \to \underline{m}$ induces a map of spaces $X^{\underline{m}} \to X^{\underline{n}}$ and it is easy to see that if α is a surjection, then α induces a well-defined map

$$\begin{aligned} \hat{\alpha}\colon \quad & X^{\wedge m} && \to X^{\wedge n} \\ & (x_1 \wedge \cdots \wedge x_m) && \mapsto (x_{\alpha(1)} \wedge \cdots \wedge x_{\alpha(n)}). \end{aligned}$$

In fact, the correspondence $\underline{i} \mapsto X^{\wedge i}$ defines a contravariant functor from M to based spaces and, by restriction, a contravariant functor from M_n to based spaces.

Similarly, given a $(k+1)$-tuple of based spaces $(X_k, \ldots, X_0)$, the correspondence

$$F_{X_k,\ldots,X_0} \colon (\underline{i_k} \overset{\alpha_k}{\twoheadrightarrow} \underline{i_{k-1}} \overset{\alpha_{k-1}}{\twoheadrightarrow} \cdots \overset{\alpha_2}{\twoheadrightarrow} \underline{i_1} \twoheadrightarrow \underline{i_0} = \underline{1}) \mapsto \bigwedge_{j=0}^{k} X_j^{\wedge i_j}$$

defines a contravariant functor from $O_k M_n$ to based spaces. We will be interested in some special cases of this functor. We write them in a separate definition:

Definition 1.7 Let $\mathbf{l}= (l_k, \ldots, l_0)$ be a sequence of finite sets. Let $S^l = (S^1)^{\wedge l}$ denote the l-dimensional sphere. Define

$$S^{\mathbf{l}} = F_{S^{l_k},\ldots,S^{l_0}}.$$

For a based space X, define

$$S^{\mathbf{l}}_X = F_{S^{l_k}\wedge X, S^{l_{k-1}},\ldots,S^{l_0}}.$$

We also need a variation of this. Let $\mathbb{R}^l$ denote the l-dimensional euclidean space. Define the contravariant functor $\mathbb{R}^{\mathbf{l}}$ from $O_k M_n$ to unbased spaces by the following formula:

$$(\underline{i_k} \overset{\alpha_k}{\twoheadrightarrow} \underline{i_{k-1}} \overset{\alpha_{k-1}}{\twoheadrightarrow} \cdots \overset{\alpha_2}{\twoheadrightarrow} \underline{i_1} \twoheadrightarrow \underline{i_0}) \mapsto \prod_{j=0}^{k} \mathbb{R}^{l_j i_j}.$$

Let $\boldsymbol{\alpha} = (\underline{i_k} \twoheadrightarrow \underline{i_{k-1}} \twoheadrightarrow \cdots \twoheadrightarrow \underline{1}) \in O_k M_n$. We write the group of automorphisms of $\boldsymbol{\alpha}$ in $O_k M_n$ as $\Sigma_{\boldsymbol{\alpha}}$. Thus the elements of $\Sigma_{\boldsymbol{\alpha}}$ are $(k+1)$-tuples of set isomorphisms $h = (h_k, \ldots, h_0 = \mathrm{id})$ such that the diagram

$$\begin{array}{ccccccccc}
\underline{i_k} & \overset{\alpha_k}{\twoheadrightarrow} & \underline{i_{k-1}} & \overset{\alpha_{k-1}}{\twoheadrightarrow} & \cdots & \overset{\alpha_2}{\twoheadrightarrow} & \underline{i_1} & \overset{\alpha_1}{\twoheadrightarrow} & \underline{1} \\
\downarrow h_k & & \downarrow h_{k-1} & & & & \downarrow h_1 & & \downarrow h_0 \\
\underline{i_k} & \overset{\alpha_k}{\twoheadrightarrow} & \underline{i_{k-1}} & \overset{\alpha_{k-1}}{\twoheadrightarrow} & \cdots & \overset{\alpha_2}{\twoheadrightarrow} & \underline{i_1} & \overset{\alpha_1}{\twoheadrightarrow} & \underline{1}
\end{array}$$

commutes.

Let $X_0, X_1, \ldots, X_k$ be based topological spaces. Since the correspondence

$$\boldsymbol{\alpha} \mapsto F_{X_k,\ldots,X_0}(\boldsymbol{\alpha}) = \bigwedge_{j=0}^{k} X_j^{\wedge i_j}$$

is functorial, there is an action of $\Sigma_{\boldsymbol{\alpha}}$ on $\bigwedge_{j=0}^{k} X_j^{\wedge i_j}$. Let $\Delta^{\boldsymbol{\alpha}}$ denote the subspace of $\bigwedge_{j=0}^{k} X_j^{\wedge i_j}$ on which this action is not free. We will later provide an explicit description for $\Delta^{\boldsymbol{\alpha}}$.

We will need to consider the space of natural transformations between functors. Let us recall some easy generalities. Let $\mathcal{C}$ be a small category. Let F and G be two functors from $\mathcal{C}$ to based spaces, such that either both are covariant or both are contravariant (in the examples that will interest us most, the functors will be contravariant). Let $\mathrm{Nat}_{\mathcal{C}}(F, G)$ be the set of natural transformations from F to G. Obviously, $\mathrm{Nat}_{\mathcal{C}}(F, G)$ is a subset of the space

$$\prod_{c \in \mathrm{Ob}(\mathcal{C})} \mathrm{Map}_*(F(c), G(c))$$

and we endow it with the subspace topology. If F and G are functors from $\mathcal{C}$ to spectra, then, with a little care, $\mathrm{Nat}_{\mathcal{C}}(F, G)$ is still a topological (infinite loop) space.

Following [**8**], we define $a\mathcal{C}$, the twisted arrow category of $\mathcal{C}$, as follows: The objects of $a\mathcal{C}$ are maps $c_0 \to c_1$ in $\mathcal{C}$ and the morphisms $(c_0 \to c_1) \to (c_0' \to c_1')$ are commutative squares

$$\begin{array}{ccc} c_0 & \leftarrow & c_0' \\ \downarrow & & \downarrow \\ c_1 & \rightarrow & c_1' \end{array}$$

(note the twist in the direction of the horizontal arrows). Given two contravariant functors F and G as above, define the functor $\hom_a(F, G)$ from $a\mathcal{C}$ to based spaces by

$$(c_0 \to c_1) \mapsto \mathrm{Map}_*(F(c_1), G(c_0)).$$

Proposition 1.8 *Let F and G be two contravariant functors. There is a homeomorphism*

$$Nat_{\mathcal{C}}(F, G) \cong \underleftarrow{\lim}^{a\mathcal{C}} \hom_a(F, G).$$

Proof A trivial modification of Proposition 3.2 in [**8**]. The only difference is that in [**8**] the functors are covariant. □

Proposition 1.9 *Suppose $\mathcal{C}$ can be written as a disjoint union of subcategories $\mathcal{C} = \coprod_i \mathcal{C}_i$. Then*

$$Nat_{\mathcal{C}}(F, G) \cong \prod_i Nat_{\mathcal{C}_i}(F, G).$$

Proof Easy. □

Proposition 1.10 *Suppose $\mathcal{C}$ is a connected groupoid. Let c be an object of $\mathcal{C}$. Let $Aut(c)$ be the group of automorphisms of c. Then $Aut(c)$ acts on $F(c)$ and $G(c)$ and*

$$Nat_{\mathcal{C}}(F, G) \cong Map_*(F(c), G(c))^{Aut(c)}.$$

Proof Easy. □

Let H be a finite group acting on $\mathcal{C}$. By this we mean that there are maps $H \times \mathrm{Ob}(\mathcal{C}) \to \mathrm{Ob}(\mathcal{C})$ and $H \times \mathrm{Mor}(\mathcal{C}) \to \mathrm{Mor}(\mathcal{C})$ compatible with product in H and with identities and composition in $\mathcal{C}$. Let $H \rtimes \mathcal{C}$ denote the semidirect product category. Thus $\mathrm{Ob}(H \rtimes \mathcal{C}) = \mathrm{Ob}(\mathcal{C})$ and morphisms $c_1 \to c_2$ are pairs (h, f) where $h \in H$ and f is a morphism $hc_1 \to c_2$ in $\mathcal{C}$. The composition is given by $(h_1, f_1) \circ (h_2, f_2) = (h_1 \circ h_2, f_1 \circ h_1 f_2)$. The assignment $f \mapsto (1, f)$ embeds $\mathcal{C}$ as a subcategory of $H \rtimes \mathcal{C}$. Let F and G be two functors from $H \rtimes \mathcal{C}$ to based spaces. Then it is clear that H acts on $\mathrm{Nat}_{\mathcal{C}}(F, G)$ and

Proposition 1.11

$$Nat_{H \rtimes \mathcal{C}}(F, G) \cong (Nat_{\mathcal{C}}(F, G))^H.$$

Proof Easy. □

If, moreover, $\mathcal{C}$ has only identity morphisms, then $\mathcal{C}$ is just an H-set and $H \rtimes \mathcal{C}$ is a groupoid. If the action of H is transitive, then it is a connected groupoid. Let H_c be the stabilizer group of an element c of $\mathcal{C}$. The set of objects of $\mathcal{C}$ is isomorphic to H/H_c. Combining Propositions 1.10 and 1.11 we obtain the familiar identification

$$\mathrm{Map}_*(F(c), G(c))^{H_c} \cong (\mathrm{Nat}_{\mathcal{C}}(F, G))^H \cong \mathrm{Map}_*(H/H_c, \mathrm{Map}_*(F(c), G(c)))^H.$$

Here the second identification follows from the facts that $\mathcal{C}$ is discrete and F and G have to be constant functors.

Another issue that we need to discuss in this preliminary section is the precise definition of QX. We would like $QX = \Omega^\infty S^\infty X$ to be the infinite loop space associated with the suspension spectrum of X. We have to be precise about the model that we use for $\Omega^\infty S^\infty X$. For the application in Section 4, we will have to use the version introduced by Bökstedt in [**3**], in which the direct limit is taken over the category of finite sets with all injective maps rather than only with the canonical inclusions. Let I be the standard skeletal subcategory of finite sets with injections and let $\mathbb{N}$ be the subcategory of I generated by inclusions. More precisely:

Definition 1.12 I is the category whose objects are $\emptyset$, $\underline{1}$, $\underline{2}, \dots$ and whose morphisms are injective maps. $\hat{I}$ is the category I with a final object $\{\infty\}$ adjoined. Let $\mathbb{N}$ be the subcategory of I having the same objects as I, whose morphisms are generated by the obvious inclusions $\underline{n} \subset \underline{n+1}$.

We adopt the convention to underline the letters denoting finite sets when they are considered objects of M, but not when they are considered objects of I.

Remark 1.13 Whenever we have a functor $\Psi\colon I \to$ Spaces we extend its definition to a functor $\hat{I} \to$ Spaces by defining $\Psi(\infty) = \mathrm{hocolim}_{l\in I}\Psi(l)$.

The correspondence

$$l \mapsto \Omega^l S^l X$$

defines a functor $I \to \mathrm{Spaces}_*$.

Definition 1.14

$$QX = \Omega^\infty \Sigma^\infty X = \mathrm{hocolim}_I \Omega^l S^l X.$$

The more traditional definition of QX is $\mathrm{colim}_{\mathbb{N}} \Omega^l S^l X$, which is homotopy equivalent to $\mathrm{hocolim}_{\mathbb{N}} \Omega^l S^l X$. There is a natural (in X) map

$$\mathrm{hocolim}_{\mathbb{N}} \Omega^l S^l X \to \mathrm{hocolim}_I \Omega^l S^l X$$

induced by inclusion of categories, which is known [**3**] to be a homotopy equivalence. More generally, the following lemma is proved in [**3**]:

Lemma 1.15 *Let Ψ be a functor from I^k to spaces. Assume that the connectivity of the map $\Psi(n_1, \dots, n_k) \to \Psi(m_1, \dots, m_k)$ induced by a morphism $(n_1, \dots, n_k) \to (m_1, \dots, m_k)$ tends to ∞ uniformly with Σn_i. Then the inclusion $\mathrm{hocolim}_{\mathbb{N}^k}\Psi \to \mathrm{hocolim}_{I^k}\Psi$ is a weak homotopy equivalence.*

Unfortunately, a published proof of this lemma still seems lacking, although it is now "folklore". The reader should perhaps be warned that the category I^k is not filtered. Therefore homotopy direct limit over I^k need not in general commute up to homotopy with finite homotopy inverse limit. However, if the functor in question satisfies the conditions of Lemma 1.15, then there is no problem since $\mathbb{N}^k$ is filtered. The advantage of using I instead of $\mathbb{N}$ is that there is a natural map

$$\mu\colon \mathrm{hocolim}_{(l_1,l_2)\in I^2} \Omega^{l_1}\Omega^{l_2} S^{l_2} S^{l_1} X \to \mathrm{hocolim}_{l\in I} \Omega^l S^l X$$

that is strictly associative and unital. As mentioned above, we only need to use I in Section 4. Everything else can be done with $\mathbb{N}$.

As was indicated in the introduction, we will want to study the BK cosimplicial space, whose space of k-cosimplices is $Q^{k+1}X$. There is some difficulty in defining the codegeneracy maps in the BK cosimplicial space with our models for QX, and there is some more difficulty in defining the codegeneracy maps in our models for the polynomial approximations of the BK space. Therefore we dispose of codegeneracies entirely. Thus we deal only with semi-cosimplicial spaces, meaning they have only coface maps, but no codegeneracies. Semi-cosimplicial spaces are sufficient for our purposes, since they have "realizations" (total spaces).

It is clear how to define the BK semi-cosimplicial space with our model for QX. The indexing of the coface maps is such that d^0 corresponds to the map which is not an infinite loop map.

2 Higher iterates of $\Omega^\infty S^\infty$ and their polynomial approximations

We introduce the following

Definition 2.1 Let $\mathbf{l} = (l_k, l_{k-1}, \ldots, l_0) \in \hat{I}^{k+1}$. Define $\Omega S(\mathbf{l})X$ as follows: if $\mathbf{l} = (l_k, l_{k-1}, \ldots, l_0) \in I^{k+1}$, then

$$\Omega S(\mathbf{l})X = \Omega^{l_0} S^{l_0} \left(\cdots \Omega^{l_k} S^{l_k} X\right).$$

We extend the domain of the functor to $\hat{I}^{k+1}$ as in Remark 1.13.

Similarly, the correspondence

$$\mathbf{l} \mapsto \mathrm{Nat}_{O_k M_n}(S^{\mathbf{l}}, S^{\mathbf{l}}_X)$$

defines a functor $I^{k+1} \to \mathrm{Spaces}_*$ and we view it as a functor $\hat{I}^{k+1} \to \mathrm{Spaces}_*$ in our usual way.

We need to construct a transformation

$$p_n : \Omega S(\mathbf{l})X \to \mathrm{Nat}_{O_k M_n}(S^{\mathbf{l}}, S^{\mathbf{l}}_X)$$

that is natural in X and $\mathbf{l}$. To construct the natural transformation, it is enough to construct a map

$$\Omega S(\mathbf{l})X \to \prod_{\boldsymbol{\alpha} \in O_k M_n} \mathrm{Map}_*\left(S^{\mathbf{l}}(\boldsymbol{\alpha}), S^{\mathbf{l}}_X(\boldsymbol{\alpha})\right)$$

whose image is in $\mathrm{Nat}_{O_k M_n}(S^{\mathbf{l}}, S^{\mathbf{l}}_X)$.

The map is constructed by induction on k. If $k = 0$ then $\mathbf{l} = (l_0)$ and p_n is simply the identity map. Assume we have constructed p_n for all tuples $(l_{k-1}, \ldots, l_0)$. Let $\mathbf{l} = (l_k, \ldots, l_1, l_0)$. Let $d_k\mathbf{l} = (l_{k-1}, \ldots, l_1, l_0)$. Then

$$\Omega S(\mathbf{l})X = \Omega S(d_k\mathbf{l})\Omega^{l_k} S^{l_k} X.$$

By induction, we have a map

$$\Omega S(d_k\mathbf{l})\Omega^{l_k} S^{l_k} X \to \prod_{\boldsymbol{\alpha} \in O_{k-1} M_n} \mathrm{Map}_*\left(S^{d_k\mathbf{l}}(\boldsymbol{\alpha}), S^{d_k\mathbf{l}}_{\Omega^{l_k} S^{l_k} X}(\boldsymbol{\alpha})\right).$$

The face map $d_k : O_k M_n \to O_{k-1} M_n$ induces a map

$$\prod_{\boldsymbol{\alpha} \in O_{k-1} M_n} \mathrm{Map}_*\left(S^{d_k\mathbf{l}}(\boldsymbol{\alpha}), S^{d_k\mathbf{l}}_{\Omega^{l_k} S^{l_k} X}(\boldsymbol{\alpha})\right)$$

$$\to \prod_{\boldsymbol{\alpha} \in O_k M_n} \mathrm{Map}_*\left(S^{d_k\mathbf{l}}(d_k(\boldsymbol{\alpha})), S^{d_k\mathbf{l}}_{\Omega^{l_k} S^{l_k} X}(d_k(\boldsymbol{\alpha}))\right).$$

In more explicit terms, the right-hand side is

$$\prod_{\boldsymbol{\alpha} \in O_k M_n} \mathrm{Map}_*\left(\bigwedge_{j=0}^{k-1} S^{l_j i_j}, \left(\bigwedge_{j=0}^{k-1} S^{l_j i_j} \wedge \left(\Omega^{l_k} S^{l_k} X\right)^{\wedge i_{k-1}}\right)\right).$$

The surjections $\alpha_k : \underline{i_k} \to \underline{i_{k-1}}$ induce a map from this space to

$$\prod_{\boldsymbol{\alpha}\in O_k M_n} \mathrm{Map}_* \left(\bigwedge_{j=0}^{k-1} S^{l_j i_j}, \left(\bigwedge_{j=0}^{k-1} S^{l_j i_j} \wedge \left(\Omega^{l_k} S^{l_k} X\right)^{\wedge i_k} \right)\right).$$

Clearly, there is a tautological map from this space to

$$\prod_{\boldsymbol{\alpha}\in O_k M_n} \mathrm{Map}_* \left(\bigwedge_{j=0}^{k} S^{l_j i_j}, \left(\bigwedge_{j=0}^{k} S^{l_j i_j} \wedge X^{\wedge i_k} \right)\right),$$

which is nothing but

$$\prod_{\boldsymbol{\alpha}\in O_k M_n} \mathrm{Map}_* \left(S^{\mathbf{l}}(\boldsymbol{\alpha}), S^{\mathbf{l}}_X(\boldsymbol{\alpha})\right).$$

Composing, we get the desired map p_n. It is easy to check by induction that the image of the resulting map is indeed in $\mathrm{Nat}_{O_k M_n}(S^{\mathbf{l}}, S^{\mathbf{l}}_X)$. It is also easy to check that the map p_n is compatible with suspensions. In other words, it is a natural transformation of functors of I^{k+1}. Therefore, we may allow the possibility that $l_i = \infty$ for some values of i.

Remark 2.2 It is not obvious from the definition that $\mathrm{Nat}_{O_k M_n}(S^{\mathbf{l}}, S^{\mathbf{l}}_X)$ is a homotopy functor of X, or that it satisfies the conditions of Lemma 1.15 as a functor of $\mathbf{l}$. However, we will see below (Corollary 2.10) that in fact it does have these good properties.

The map p_n is probably of little significance in general. However, if $l_k = \cdots = l_0 = \infty$, then it is highly connected. In fact, it becomes a model for Goodwillie's n-th Taylor approximation map. We will formulate this precisely in Theorem 2.21. We need to analyze the functor $\mathrm{Nat}_{O_k M_n}(S^{\mathbf{l}}, S^{\mathbf{l}}_X)$ in order to prepare the ground for proving Theorem 2.21. There is a restriction map

$$\mathrm{Nat}_{O_k M_n}(S^{\mathbf{l}}, S^{\mathbf{l}}_X) \to \mathrm{Nat}_{O_k M_{n-1}}(S^{\mathbf{l}}, S^{\mathbf{l}}_X).$$

We need to understand the homotopy fiber of this map. Before we can do it, we have to introduce some more notation.

Let ${}^nO_{k-1}M_n$ be the full subcategory of $O_k M_n$ whose objects are chains of surjections of the form

$$\underline{n} \twoheadrightarrow \underline{i_{k-1}} \twoheadrightarrow \cdots \twoheadrightarrow \underline{1}$$

and the basepoint. We make $k \mapsto {}^nO_{k-1}M_n$ into a simplicial category by defining

$$d_k(\underline{n} \overset{\alpha_k}{\twoheadrightarrow} \underline{i_{k-1}} \overset{\alpha_{k-1}}{\twoheadrightarrow} \cdots \overset{\alpha_2}{\twoheadrightarrow} \underline{i_1} \overset{\alpha_1}{\twoheadrightarrow} \underline{i_0} = \underline{1}) = \begin{cases} \underline{i_{k-1}} \overset{\alpha_{k-1}}{\twoheadrightarrow} \underline{i_{k-2}} \overset{\alpha_{k-2}}{\twoheadrightarrow} \cdots \overset{\alpha_1}{\twoheadrightarrow} \underline{1} & \text{if } i_{k-1} = n \\ \{\emptyset\} & \text{otherwise,} \end{cases}$$

and defining d_j and s_j as in $O_k M_n$ for $0 \leq j \leq k-1$.

Proposition 2.3 *The category ${}^nO_{k-1}M_n$ is a groupoid.*

Proof Let

$$\boldsymbol{\alpha} = (\underline{n} \overset{\alpha_k}{\twoheadrightarrow} \underline{i_{k-1}} \overset{\alpha_{k-1}}{\twoheadrightarrow} \cdots \twoheadrightarrow \underline{i_1} \twoheadrightarrow \underline{1})$$

and

$$\boldsymbol{\alpha}' = (\underline{n} \overset{\alpha_k'}{\twoheadrightarrow} \underline{i_{k-1}'} \overset{\alpha_{k-1}'}{\twoheadrightarrow} \cdots \twoheadrightarrow \underline{i_1'} \twoheadrightarrow \underline{1})$$

be two objects of ${}^nO_{k-1}M_n$. Let $(h_k, h_{k-1}, \ldots, h_0)$ be a morphism from $\boldsymbol{\alpha}$ to $\boldsymbol{\alpha}'$. We claim that h_j is an isomorphism of sets for all $k \geq j \geq 0$. Indeed, it is clear that

h_k is an isomorphism since $i_k = i'_k = n$ and h_k is a surjection. The claim follows by downward induction on j using Proposition 1.4. It follows that $(h_k, h_{k-1}, \dots, h_0)$ has an inverse, which completes the proof. □

In fact, each connected component of ${}^nO_{k-1}M_n$ is a groupoid of a special kind, namely a set with an action of a group. To be more specific, we have the following proposition:

Proposition 2.4 *Let $\boldsymbol{\alpha} = (\underline{n} \twoheadrightarrow \underline{i_{k-1}} \twoheadrightarrow \cdots \twoheadrightarrow \underline{1})$. For the moment, let C be the connected component of ${}^nO_{k-1}M_n$ containing $\boldsymbol{\alpha}$. Let C_0 be the set of objects of C considered as a category with identity morphisms only. Then*

$$C \cong (\Sigma_n \times \Sigma_{i_{k-1}} \times \cdots \times \Sigma_{i_1}) \ltimes C_0.$$

Proof We define a functor

$$\eta : (\Sigma_n \times \Sigma_{i_{k-1}} \times \cdots \times \Sigma_{i_1}) \ltimes C_0 \to C$$

as follows: η is the identity on objects. To define η on morphisms, let

$$(g, c) = ((g_k, \dots, g_1), c) \in \Sigma_n \times \Sigma_{i_{k-1}} \times \cdots \times \Sigma_{i_1} \times \mathrm{Ob}(C_0).$$

Clearly, (g, c) can be identified canonically with a morphism in

$$(\Sigma_n \times \Sigma_{i_{k-1}} \times \cdots \times \Sigma_{i_1}) \ltimes C_0$$

from c to $g(c)$. The k-tuple $(g_k, \dots, g_1)$ gives rise to a k-tuple of set isomorphisms which clearly defines a morphism from c to $g(c)$ in C. We define $\eta(g, c)$ to be this morphism. It is easy to check that η is an isomorphism of categories. □

For $\boldsymbol{\alpha}$ an object of ${}^nO_{k-1}M_n$ let $\Sigma_{\boldsymbol{\alpha}}$ be the group of automorphisms of $\boldsymbol{\alpha}$ in ${}^nO_{k-1}M_n$. Thus

$$\Sigma_{\boldsymbol{\alpha}} \subseteq \Sigma_n \times \Sigma_{i_{k-1}} \times \cdots \times \Sigma_{i_1}.$$

In fact, it is easy to see that the composed map

$$\Sigma_{\boldsymbol{\alpha}} \to \Sigma_n \times \Sigma_{i_{k-1}} \times \cdots \times \Sigma_{i_1} \to \Sigma_n$$

is an injection. This is so essentially because given surjections $\underline{n} \twoheadrightarrow \underline{i_1}$, $\underline{n} \twoheadrightarrow \underline{i_2}$ and $\underline{n} \twoheadrightarrow \underline{n}$, there is at most one surjection $\underline{i_1} \twoheadrightarrow \underline{i_2}$ that makes the obvious square diagram commute. Thus we may think of $\Sigma_{\boldsymbol{\alpha}}$ as a subgroup of Σ_n. We have observed already that $\Sigma_{\boldsymbol{\alpha}}$ acts on $S^{\mathbf{l}}(\boldsymbol{\alpha}) = \bigwedge_{j=0}^k S^{l_j i_j}$. Let $\Delta^{\boldsymbol{\alpha}}(\mathbf{l})$ denote the subspace of $\bigwedge_{j=0}^k S^{l_j i_j}$ on which $\Sigma_{\boldsymbol{\alpha}}$ does not act freely.

For a category $\mathcal{C}$, let $[\mathcal{C}]$ be the set of connected components of $\mathcal{C}$. The following theorem provides an inductive description of $\mathrm{Nat}_{O_k M_n}(S^{\mathbf{l}}, S^{\mathbf{l}}_X)$.

Theorem 2.5 *There exists a commutative square (the products on the right hand side are both taken over the same set of representatives)*

$$\begin{array}{ccc}
Nat_{O_k M_n}(S^{\mathbf{l}}, S^{\mathbf{l}}_X) & \to & \prod_{\boldsymbol{\alpha} \in [{}^nO_{k-1}M_n]} Map_*\left(S^{\mathbf{l}}(\boldsymbol{\alpha}), S^{\mathbf{l}}_X(\boldsymbol{\alpha})\right)^{\Sigma_{\boldsymbol{\alpha}}} \\
\downarrow & & \downarrow \\
Nat_{O_k M_{n-1}}(S^{\mathbf{l}}, S^{\mathbf{l}}_X) & \to & \prod_{\boldsymbol{\alpha} \in [{}^nO_{k-1}M_n]} Map_*\left(\Delta^{\boldsymbol{\alpha}}(\mathbf{l}), S^{\mathbf{l}}_X(\boldsymbol{\alpha})\right)^{\Sigma_{\boldsymbol{\alpha}}}
\end{array}$$

which is a strict pullback. Moreover, the right vertical map is the product of restriction maps. In particular, it is a fibration and therefore the square is also a homotopy pullback.

Proof By Proposition 1.8, there is a homeomorphism

$$\mathrm{Nat}_{O_k M_n}(S^{\mathbf{l}}, S^{\mathbf{l}}_X) \cong \underleftarrow{\lim}^{aO_k M_n} \hom_a(S^{\mathbf{l}}, S^{\mathbf{l}}_X).$$

We write $aO_k M_n$ as a union of two subcategories $aO_k M_n^1$ and $aO_k M_n^2$ defined as follows: $aO_k M_n^1$ is the full subcategory of $aO_k M_n$ whose objects are arrows $\boldsymbol{\alpha} \twoheadrightarrow \boldsymbol{\alpha}'$ such that $\boldsymbol{\alpha}$ is an object of ${}^nO_{k-1}M_n$. $aO_k M_n^2$ is the full subcategory of $aO_k M_n$ whose objects are arrows $\boldsymbol{\alpha} \twoheadrightarrow \boldsymbol{\alpha}'$ such that $\boldsymbol{\alpha}'$ is not an object of ${}^nO_{k-1}M_n$. It is quite easy to see that the nerve of $aO_k M_n$ is the union of the nerves of $aO_k M_n^1$ and $aO_k M_n^2$. It follows, as in [**10**, Proposition 0.2], that there is a pullback square

$$\begin{array}{ccc} \underleftarrow{\lim}^{aO_k M_n} \hom_a(S^{\mathbf{l}}, S^{\mathbf{l}}_X) & \rightarrow & \underleftarrow{\lim}^{aO_k M_n^1} \hom_a(S^{\mathbf{l}}, S^{\mathbf{l}}_X) \\ \downarrow & & \downarrow \\ \underleftarrow{\lim}^{aO_k M_n^2} \hom_a(S^{\mathbf{l}}, S^{\mathbf{l}}_X) & \rightarrow & \underleftarrow{\lim}^{aO_k M_n^1 \cap aO_k M_n^2} \hom_a(S^{\mathbf{l}}, S^{\mathbf{l}}_X). \end{array}$$

We will prove that this pullback square is in fact of the form that is claimed in the theorem. To analyze

$$\underleftarrow{\lim}^{aO_k M_n^1} \hom_a(S^{\mathbf{l}}, S^{\mathbf{l}}_X),$$

let $\widetilde{aO_k M_n^1}$ be the subcategory of $aO_k M_n^1$ whose objects are isomorphism arrows.

Proposition 2.6 *$\widetilde{aO_k M_n^1}$ is an initial subcategory of $aO_k M_n^1$.*

Proof We need to show that for every object of $aO_k M_n^1$ there exists an object of $\widetilde{aO_k M_n^1}$ over it. Furthermore, we have to show that any two such objects are connected by a chain of morphisms in the appropriate "over category". Let $\boldsymbol{\alpha} \twoheadrightarrow \boldsymbol{\alpha}'$ be an object of $aO_k M_n^1$. Thus $\boldsymbol{\alpha}$ is an object of ${}^nO_{k-1}M_n$. Then the identity arrow $\boldsymbol{\alpha} \twoheadrightarrow \boldsymbol{\alpha}$ is an object of $\widetilde{aO_k M_n^1}$ over $\boldsymbol{\alpha} \twoheadrightarrow \boldsymbol{\alpha}'$. Moreover, it is easy to check that for any object $\boldsymbol{\beta} \twoheadrightarrow \boldsymbol{\beta}'$ of $\widetilde{aO_k M_n^1}$ over $\boldsymbol{\alpha} \twoheadrightarrow \boldsymbol{\alpha}'$, there is a morphism from the identity arrow $\boldsymbol{\alpha} \twoheadrightarrow \boldsymbol{\alpha}$ to $\boldsymbol{\beta} \twoheadrightarrow \boldsymbol{\beta}'$. In particular, the over category is connected. □

It follows (dual to Theorem IX 3.1 in [**16**]) that

$$\underleftarrow{\lim}^{aO_k M_n^1} \hom_a(S^{\mathbf{l}}, S^{\mathbf{l}}_X) \cong \underleftarrow{\lim}^{\widetilde{aO_k M_n^1}} \hom_a(S^{\mathbf{l}}, S^{\mathbf{l}}_X).$$

It is easy to see that $\widetilde{aO_k M_n^1}$ is nothing but $a^nO_{k-1}M_n$. Therefore

$$\underleftarrow{\lim}^{\widetilde{aO_k M_n^1}} \hom_a(S^{\mathbf{l}}, S^{\mathbf{l}}_X) \cong \mathrm{Nat}_{{}^nO_{k-1}M_n}(S^{\mathbf{l}}, S^{\mathbf{l}}_X).$$

It now follows from Propositions 1.9, 1.10 and 2.4 that

$$\underleftarrow{\lim}^{\widetilde{aO_k M_n^1}} \hom_a(S^{\mathbf{l}}, S^{\mathbf{l}}_X) \cong \prod_{\boldsymbol{\alpha} \in [{}^nO_{k-1}M_n]} \mathrm{Map}_*\left(S^{\mathbf{l}}(\boldsymbol{\alpha}), S^{\mathbf{l}}_X(\boldsymbol{\alpha})\right)^{\Sigma_{\boldsymbol{\alpha}}}.$$

To analyze

$$\underleftarrow{\lim}^{aO_k M_n^2} \hom_a(S^{\mathbf{l}}, S^{\mathbf{l}}_X),$$

let $\widetilde{aO_k M_n^2}$ be the subcategory of $aO_k M_n^2$ whose objects are arrows $\boldsymbol{\alpha} \twoheadrightarrow \boldsymbol{\alpha}'$ such that $\boldsymbol{\alpha}$ is not in ${}^nO_{k-1}M_n$. Then $\widetilde{aO_k M_n^2}$ is an initial subcategory of $aO_k M_n^2$ (the proof is the same level of difficulty as that of Proposition 2.6 and we omit it). It follows that

$$\underleftarrow{\lim}^{aO_kM_n^2}\hom_a(S^{\mathbf{l}},S^{\mathbf{l}}_X)\cong\underleftarrow{\lim}^{\widetilde{aO_kM_n^2}}\hom_a(S^{\mathbf{l}},S^{\mathbf{l}}_X).$$

On the other hand it is easy to see that $\widetilde{aO_kM_n^2}$ is nothing but aO_kM_{n-1} and therefore

$$\mathrm{Nat}_{O_kM_{n-1}}(S^{\mathbf{l}},S^{\mathbf{l}}_X)\cong\underleftarrow{\lim}^{aO_kM_n^2}\hom_a(S^{\mathbf{l}},S^{\mathbf{l}}_X).$$

It remains to analyze $\underleftarrow{\lim}^{aO_kM_n^1\cap aO_kM_n^2}\hom_a(S^{\mathbf{l}},S^{\mathbf{l}}_X)$. We want to prove that

$$\underleftarrow{\lim}^{aO_kM_n^1\cap aO_kM_n^2}\hom_a(S^{\mathbf{l}},S^{\mathbf{l}}_X)\cong\prod_{\boldsymbol{\alpha}\in[^nO_{k-1}M_n]}\mathrm{Map}_*\left(\Delta^{\boldsymbol{\alpha}}(\mathbf{l}),S^{\mathbf{l}}_X(\boldsymbol{\alpha})\right)^{\Sigma_{\boldsymbol{\alpha}}}.$$

Fix a connected component of ${}^nO_{k-1}M_n$. For the moment, let $\mathcal{D}$ be the full subcategory of $aO_kM_n^1\cap aO_kM_n^2$ whose objects are arrows $\boldsymbol{\alpha}\twoheadrightarrow\boldsymbol{\alpha}'$ such that $\boldsymbol{\alpha}$ is in our chosen connected component. Let $\mathcal{E}$ be the subcategory of $\mathcal{D}$ which has the same objects and whose morphisms are those which induce the identity on $\boldsymbol{\alpha}$. It is routine to check that $\mathcal{D}\cong(\Sigma_n\times\Sigma_{i_{k-1}}\times\cdots\times\Sigma_{i_1})\ltimes\mathcal{E}$. Therefore

$$\underleftarrow{\lim}^{\mathcal{D}}\hom_a(S^{\mathbf{l}},S^{\mathbf{l}}_X)\cong\underleftarrow{\lim}^{\mathcal{E}}\hom_a(S^{\mathbf{l}},S^{\mathbf{l}}_X)^{(\Sigma_n\times\Sigma_{i_{k-1}}\times\cdots\times\Sigma_{i_1})}.$$

Now fix $\boldsymbol{\alpha}$ in our connected component of ${}^nO_{k-1}M_n$. Let $\boldsymbol{\alpha}/O_kM_{n-1}$ be the category whose objects are arrows $\boldsymbol{\alpha}\twoheadrightarrow\boldsymbol{\alpha}'$, where $\boldsymbol{\alpha}'\in O_kM_{n-1}$ and whose morphisms are commuting triangles

$$\begin{array}{ccc}\boldsymbol{\alpha} & \rightarrow & \boldsymbol{\alpha}'\\ & \searrow & \downarrow\\ & & \boldsymbol{\alpha}''.\end{array}$$

This category is a connected component of $\mathcal{E}$ and it is easy to see that

$$\underleftarrow{\lim}^{\mathcal{E}}\hom_a(S^{\mathbf{l}},S^{\mathbf{l}}_X)^{(\Sigma_n\times\Sigma_{i_{k-1}}\times\cdots\times\Sigma_{i_1})}\cong\underleftarrow{\lim}^{\boldsymbol{\alpha}/O_kM_{n-1}}\hom_a(S^{\mathbf{l}},S^{\mathbf{l}}_X)^{\Sigma_{\boldsymbol{\alpha}}}.$$

The right-hand side is, in turn, homeomorphic to

$$\mathrm{Map}_*\left(\left(\operatorname*{co\,lim}_{\boldsymbol{\alpha}/O_kM_{n-1}}S^{\mathbf{l}}(\boldsymbol{\alpha}')\right),S^{\mathbf{l}}_X(\boldsymbol{\alpha})\right)^{\Sigma_{\boldsymbol{\alpha}}}.$$

We next prove that there is a homeomorphism

$$\operatorname*{co\,lim}_{\boldsymbol{\alpha}/O_kM_{n-1}}S^{\mathbf{l}}(\boldsymbol{\alpha}')\overset{\cong}{\rightarrow}\Delta^{\boldsymbol{\alpha}}(\mathbf{l}),$$

where the colimit is taken over elements $(\boldsymbol{\alpha}\rightarrow\boldsymbol{\alpha}')\in\boldsymbol{\alpha}/O_kM_{n-1}$. Clearly, there is a map

$$\operatorname*{co\,lim}_{\boldsymbol{\alpha}/O_kM_{n-1}}S^{\mathbf{l}}(\boldsymbol{\alpha}')\rightarrow S^{\mathbf{l}}(\boldsymbol{\alpha}).$$

The next two lemmas say that the image of this map is precisely $\Delta^{\boldsymbol{\alpha}}(\mathbf{l})$ and that the map is one-to-one.

Lemma 2.7

$$\Delta^{\boldsymbol{\alpha}}(\mathbf{l})=\bigcup_g\{Im(S^{\mathbf{l}}(g))\}.$$

Here the union is taken over the set of morphisms $g:\boldsymbol{\alpha}\twoheadrightarrow\boldsymbol{\beta}$ *where* $\boldsymbol{\beta}$ *is an object of* O_kM_{n-1}.

Proof Let $\boldsymbol{\beta} \in O_k M_{n-1}$ be a sequence of surjections $\underline{e_k} \overset{\beta_k}{\twoheadrightarrow} \underline{e_{k-1}} \overset{\beta_{k-1}}{\twoheadrightarrow} \cdots \overset{\beta_2}{\twoheadrightarrow} \underline{e_1} \twoheadrightarrow \underline{e_0} = \underline{1}$ so $e_k < n$. Let $g\colon \boldsymbol{\alpha} \twoheadrightarrow \boldsymbol{\beta}$ be a morphism. We claim that there exists $h \in \Sigma_{\boldsymbol{\alpha}}$, $h \neq 1$, such that $g \circ h = g$. Indeed, since $e_k < n$, there are $p, q \in \underline{n}$ with $g_k(p) = g_k(q)$. Let s be the largest j such that $\alpha_j \circ \cdots \circ \alpha_k(p) = \alpha_j \circ \cdots \circ \alpha_k(q)$. We can now choose h to be the element which induces the obvious action of Σ_2 on $\boldsymbol{\alpha}^{-1}(\alpha_s \circ \cdots \circ \alpha_k(q))$ and which equals to the identity elsewhere. Thus $g = g \circ h$, which implies that $\operatorname{Im}(S^{\mathbf{l}}(g)) \subset \Delta^{\boldsymbol{\alpha}}(\mathbf{l})$.

To prove the converse, let $1 \neq h \in \Sigma_{\boldsymbol{\alpha}}$. We define an equivalence relation in $\underline{i_j}$, $0 \leq j \leq k$, as follows: $p \sim q$ if and only if $p = h_j^m(q)$ for some $m \in \mathbb{Z}$. Let $\underline{i'_j}$ denote the set of equivalence classes in $\underline{i_j}$ and let $f_j\colon \underline{i_j} \to \underline{i'_j}$ be the natural projection. Since $\alpha_j \circ h_j = h_{j-1} \circ \alpha_j$, also $\alpha_j \circ h_j^m = h_{j-1}^m \circ \alpha_j$, for $m \geq 0$. Thus we get unique surjections $\alpha'_j\colon \underline{i'_j} \to \underline{i'_{j-1}}$ such that $\alpha'_j \circ f_j = f_{j-1} \circ \alpha_j$ for every j. Notice that if h_j is the identity, then also f_j is the identity. On the other hand, since $h \neq 1$, it follows that $h_s \neq 1$ for some s. Then $h_j \neq 1$ for every $j \geq s$. This implies that $i'_k < n$.

Finally, let $x \in S^{\mathbf{l}}(\boldsymbol{\alpha})$ be such that $h(x) = x$. We will show that $x \in \operatorname{Im}(S^{\mathbf{l}}(f))$. Write $x = x^0 \wedge \cdots \wedge x^{i_k}$ and $x^j = x_1^j \wedge \cdots \wedge x_{i_j}^j$. Let $p, q \in \underline{i_j}$. By induction one sees that if $p = h_j^m(q)$, for some $m \in \mathbb{Z}$, then $x_p^j = x_q^j$. This simply means that $x^j \in \operatorname{Im}(F_{S^{l_j}}(f_j))$, which completes the proof. □

Lemma 2.8 *Assume $\boldsymbol{\alpha} \in {}^nO_{k-1}M_n$ and assume $\boldsymbol{\alpha}', \boldsymbol{\alpha}'' \in O_k M_n$ are such that $i'_k < n$ and $i''_k < n$. Let $g'\colon \boldsymbol{\alpha} \twoheadrightarrow \boldsymbol{\alpha}'$ and $g''\colon \boldsymbol{\alpha} \twoheadrightarrow \boldsymbol{\alpha}''$ be morphisms. Then there exists $\boldsymbol{\beta} \in O_k M_n$ and $h'\colon \boldsymbol{\alpha}' \twoheadrightarrow \boldsymbol{\beta}$ and $h''\colon \boldsymbol{\alpha}'' \twoheadrightarrow \boldsymbol{\beta}$ such that the square*

$$\begin{array}{ccc} \boldsymbol{\alpha} & \xrightarrow{\;g'\;} & \boldsymbol{\alpha}' \\ {\scriptstyle g''}\big\downarrow & & \big\downarrow{\scriptstyle h'} \\ \boldsymbol{\alpha}'' & \xrightarrow{\;h''\;} & \boldsymbol{\beta} \end{array}$$

commutes and, for $p = h' \circ g'$, $\operatorname{Im} F_{X_k,\dots,X_0}(p) = \operatorname{Im} F_{X_k,\dots,X_0}(g') \cap \operatorname{Im} F_{X_k,\dots,X_0}(g'')$.

Proof We define an equivalence relation in $\underline{i_j}$ for all $0 \leq j \leq k$, as follows: $q_1 \sim q_2$ if and only if there exist $s_1, \dots, s_m$, for some $m > 0$, where $q_1 = s_1$, $q_2 = s_m$, such that for every $1 \leq t < m$ either $g'_j(s_t) = g'_j(s_{t+1})$ or $g''_j(s_t) = g''_j(s_{t+1})$. Let $\underline{i_j(\boldsymbol{\beta})}$ denote the set of equivalence classes and let $p_j\colon \underline{i_j} \to \underline{i_j(\boldsymbol{\beta})}$ be the natural projection. Then the mappings $\beta_j\colon \underline{i_j(\boldsymbol{\beta})} \to \underline{i_{j-1}(\boldsymbol{\beta})}$, $s \mapsto p_{j-1}(\alpha_j(p_j^{-1}(s)))$, are well-defined surjections making the diagram

$$\begin{array}{ccccccccc} \underline{i_k} & \overset{\alpha_k}{\twoheadrightarrow} & \underline{i_{k-1}} & \overset{\alpha_{k-1}}{\twoheadrightarrow} & \cdots & \overset{\alpha_2}{\twoheadrightarrow} & \underline{i_1} & \overset{\alpha_1}{\twoheadrightarrow} & \underline{1} \\ \downarrow_{p_k} & & \downarrow_{p_{k-1}} & & & & \downarrow_{p_1} & & \downarrow_{p_0 = \mathrm{id}} \\ \underline{i_k(\boldsymbol{\beta})} & \overset{\beta_k}{\twoheadrightarrow} & \underline{i_{k-1}(\boldsymbol{\beta})} & \overset{\beta_{k-1}}{\twoheadrightarrow} & \cdots & \overset{\beta_2}{\twoheadrightarrow} & \underline{i_1(\boldsymbol{\beta})} & \overset{\beta_1}{\twoheadrightarrow} & \underline{1} \end{array}$$

commute. We denote the bottom chain of surjections by $\boldsymbol{\beta}$. It is easy to see that $\boldsymbol{\beta}$ is the finest common coarsening of $\boldsymbol{\alpha}'$ and $\boldsymbol{\alpha}''$. It is easy to check that

$$h'_j\colon \underline{i'_j} \to \underline{i_j(\boldsymbol{\beta})}, \ s \mapsto p_j({g'_j}^{-1}(s)),$$

and

$$h''_j\colon \underline{i''_j} \to \underline{i_j(\boldsymbol{\beta})},\ s \mapsto p_j({g''_j}^{-1}(s)),$$

are well-defined and satisfy $p_j = h'_j \circ g'_j = h''_j \circ g''_j$.

Finally, let $y_1 \wedge \cdots \wedge y_{i_j} = F_{X_j}(g'_j)(x_1 \wedge \cdots \wedge x_{i'_j}) = x_{g'_j(1)} \wedge \cdots \wedge x_{g'_j(i_j)}$. Then $g'_j(p) = g'_j(q)$ implies $y_p = y_q$. If $y_1 \wedge \cdots \wedge y_{i_j} \in \operatorname{Im} F_{X_j}(g''_j)$, then similarly $y_p = y_q$ whenever $g''_j(p) = g''_j(q)$. Therefore, for $y_1 \wedge \cdots \wedge y_{i_j} \in \operatorname{Im} F_{X_j}(g'_j) \cap \operatorname{Im} F_{X_j}(g''_j)$, $y_p = y_q$ if $p \sim q$. But this is just a way to say that $y_1 \wedge \cdots \wedge y_{i_j} \in \operatorname{Im} F_{X_j}(p_j)$. Hence the second claim follows. □

We now return to the proof of Theorem 2.5. It follows that

$$\lim_{\leftarrow}{}^{aO_kM_n^1 \cap aO_kM_n^2} \hom_a(S^{\mathbf{l}}, S^{\mathbf{l}}_X)$$

is indeed as claimed. Therefore, a pullback square as claimed in the statement of the theorem exists. Moreover, it is easy to see that the right vertical map is a product of restriction maps and therefore is a fibration. Hence it is also a homotopy pullback square. □

Corollary 2.9 *The homotopy fiber of the restriction map*

$$Nat_{O_kM_n}(S^{\mathbf{l}}, S^{\mathbf{l}}_X) \to Nat_{O_kM_{n-1}}(S^{\mathbf{l}}, S^{\mathbf{l}}_X)$$

has the homotopy type of

$$\prod_{\boldsymbol{\alpha}\in[{}^nO_{k-1}M_n]} Map_*\left(S^{\mathbf{l}}(\boldsymbol{\alpha})/\Delta^{\boldsymbol{\alpha}}(\mathbf{l}), S^{\mathbf{l}}_X(\boldsymbol{\alpha})\right)^{\Sigma_{\boldsymbol{\alpha}}}.$$

Corollary 2.10 $Nat_{O_kM_n}(S^{\mathbf{l}}, S^{\mathbf{l}}_X)$ *is a homotopy functor of X and satisfies the conditions of Lemma 1.15 as a functor of* $\mathbf{l}$.

Proof Consider the mapping space $\operatorname{Map}_*\left(S^{\mathbf{l}}(\boldsymbol{\alpha})/\Delta^{\boldsymbol{\alpha}}(\mathbf{l}), S^{\mathbf{l}}_X(\boldsymbol{\alpha})\right)$. The action of $\Sigma_{\boldsymbol{\alpha}}$ on the source of this mapping space is free in the based sense. Moreover, the connectivity of the target is at least as large as the dimension of the source. It follows that the statement is true for $\operatorname{Map}_*\left(S^{\mathbf{l}}(\boldsymbol{\alpha})/\Delta^{\boldsymbol{\alpha}}(\mathbf{l}), S^{\mathbf{l}}_X(\boldsymbol{\alpha})\right)^{\Sigma_{\boldsymbol{\alpha}}}$. By Corollary 2.9, the statement is true for the homotopy fiber of the map

$$\operatorname{Nat}_{O_kM_n}(S^{\mathbf{l}}, S^{\mathbf{l}}_X) \to \operatorname{Nat}_{O_kM_{n-1}}(S^{\mathbf{l}}, S^{\mathbf{l}}_X).$$

The corollary follows by induction on n. □

To prove that the natural transformation

$$p_n\colon Q^{k+1}X \to \operatorname{hocolim}_{I^{k+1}} \operatorname{Nat}_{O_kM_n}(S^{\mathbf{l}}, S^{\mathbf{l}}_X)$$

is highly connected, we use the "little cubes" model for iterated loop spaces. The standard reference for this material is [**17**]. Let $C(\mathbb{R}^l, i)$ be the space of i-tuples of disjoint little cubes in $\mathbb{R}^l$. Define the space

$$C(l)(X) = (\coprod_{i=0}^{\infty} C(\mathbb{R}^l, i) \times X^i)_{\approx}$$

where the subscript $\approx$ stands for the usual identifications (see [**17**]). There is a natural mapping $C(l)X \to \Omega^l S^l X$ and one of the main theorems of [**17**] states that this map is a weak equivalence if X is connected. Notice that $C(\mathbb{R}^l, i)$ is Σ_i-homotopy equivalent to the space of ordered i-tuples of distinct points in $\mathbb{R}^l$. This configuration space can be identified with the complement of the "fat diagonal"

$\Delta^{\Sigma_i}(l)$ in $\mathbb{R}^{li}$, i.e., with the complement of the subspace of $\mathbb{R}^{li}$ on which Σ_i does not act freely. Thus, we may write

$$C(\mathbb{R}^l, i) \simeq_{\Sigma_i} \mathbb{R}^{li} \setminus \Delta^{\Sigma_i}(l).$$

The map in the $\rightarrow$ direction is given by sending each cube to its center.

Now let $\mathbf{l'} = (l_k, \ldots, l_1) \in \hat{I}^k$. Thus we omit l_0 till further notice. Iterating the little cubes construction k times one gets a model for $\Omega S(\mathbf{l'})X$

$$C(\mathbf{l'})(X) = \Big(\coprod_{i_1=0}^{\infty} C(\mathbb{R}^{l_1}, i_1) \times \Big(\coprod_{i_2=0}^{\infty} C(\mathbb{R}^{l_2}, i_2) \times \ldots \times \Big(\coprod_{i_k=0}^{\infty} C(\mathbb{R}^{l_k}, i_k) \times X^{i_k}\Big)_{\approx}^{i_{k-1}} \ldots\Big)_{\approx}^{i_1}\Big)_{\approx}.$$

To analyze the combinatorics of this construction, we need to generalize the notion of the little cubes space. Let $\boldsymbol{\alpha} = (\underline{i_k} \xrightarrow{\alpha_k} \cdots \twoheadrightarrow \underline{i_1} \twoheadrightarrow \underline{1})$ be a chain of surjections of length k. We define $C(\mathbb{R}^{\mathbf{l'}}, \boldsymbol{\alpha})$ by induction on k.

Definition 2.11 If $k = 1$ then

$$C(\mathbb{R}^{\mathbf{l'}}, \boldsymbol{\alpha}) = C(\mathbb{R}^{l_1}, i_1).$$

If $k > 1$ then

$$C(\mathbb{R}^{\mathbf{l'}}, \boldsymbol{\alpha}) = \Big(\prod_{s \in \underline{i_{k-1}}} C(\mathbb{R}^{l_k}, |\alpha_k^{-1}(s)|)\Big) \times C(\mathbb{R}^{(l_{k-1}, \ldots, l_1)}, d_k(\boldsymbol{\alpha})).$$

Observe that there is a free action of $\Sigma_{\boldsymbol{\alpha}}$ on $C(\mathbb{R}^{\mathbf{l'}}, \boldsymbol{\alpha})$. There is a canonical $\Sigma_{i_k} \times \cdots \times \Sigma_{i_1}$-equivariant map

$$C(\mathbb{R}^{\mathbf{l'}}, \boldsymbol{\alpha}) \rightarrow \prod_{j=1}^{k} \mathbb{R}^{l_j i_j}$$

sending each cube to its center. The image of this map is a $\Sigma_{\boldsymbol{\alpha}}$-invariant subspace of

$$\prod_{j=1}^{k} \mathbb{R}^{l_j i_j} \setminus \Delta^{\boldsymbol{\alpha}}(\mathbf{l'}).$$

Here $\Delta^{\boldsymbol{\alpha}}(\mathbf{l'})$ denotes the subspace of $\prod_{j=1}^{k} \mathbb{R}^{l_j i_j}$ where $\Sigma_{\boldsymbol{\alpha}}$ does not act freely. For a general $\mathbf{l'}$, the map

$$C(\mathbb{R}^{\mathbf{l'}}, \boldsymbol{\alpha}) \rightarrow \prod_{j=1}^{k} \mathbb{R}^{l_j i_j} \setminus \Delta^{\boldsymbol{\alpha}}(\mathbf{l'})$$

is not a homotopy equivalence. However, the next proposition says that if $l_1 = \cdots = l_k = \infty$ then it is.

Proposition 2.12 *If* $l_1 = \cdots = l_k = \infty$ *then* $C(\mathbb{R}^{\mathbf{l'}}, \boldsymbol{\alpha})$ *and* $\prod_{j=1}^{k} \mathbb{R}^{l_j i_j} \setminus \Delta^{\boldsymbol{\alpha}}(\mathbf{l'})$ *are both contractible (thus both are models for* $E\Sigma_{\boldsymbol{\alpha}}$*) and the map*

$$C(\mathbb{R}^{\mathbf{l'}}, \boldsymbol{\alpha}) \rightarrow \mathbb{R}^{\mathbf{l'}}(\boldsymbol{\alpha}) \setminus \Delta^{\boldsymbol{\alpha}}(\mathbf{l'})$$

is a $\Sigma_{\boldsymbol{\alpha}}$*-equivariant equivalence.*

Proof It is not difficult to see that $\operatorname{codim}(\Delta^{\boldsymbol{\alpha}}(\mathbf{l'})) \geq \min\{l_s | 1 \leq s \leq k\}$. Thus $\operatorname{codim}(\Delta^{\boldsymbol{\alpha}}(\mathbf{l'})) = \infty$ when each l_j equals infinity. This proves the contractibility of $\mathbb{R}^{\mathbf{l'}}(\boldsymbol{\alpha}) \setminus \Delta^{\boldsymbol{\alpha}}(\mathbf{l'})$. Let Δ denote the fat diagonal in $C(\mathbb{R}^{l_1}, i_1)$. Then $\dim(\Delta) = l_1(i_1 - 1)$ and $\operatorname{codim}(\Delta) = l_1$. Thus $C(\mathbb{R}^{l_1}, i_1)$ is contractible when $l_1 = \infty$. It now follows by induction on k that $C(\mathbb{R}^{\mathbf{l'}}, \boldsymbol{\alpha})$ is contractible since it is a product of contractible spaces when $l_k = \cdots = l_1 = \infty$. The proposition follows since any equivariant map of free contractible spaces is an equivariant equivalence. □

The space $C(\mathbf{l'})(X)$ admits a presentation as a coend that is more or less dual to our description of $P_n Q^k X$. First, we need to define a (somewhat strange) small category that will play the role of $O^k M_n$:

Definition 2.13 Let $I^k M_n$ be the following category: the objects of $I^k M_n$ are the same as of $N_k M_n$, and a morphism $\delta\colon \boldsymbol{\alpha} \to \boldsymbol{\alpha}'$ is a collection of *injections* $\delta_j\colon \underline{i_j} \to \underline{i'_j}$ such that the diagram

$$\begin{array}{ccccccccc} \underline{i_k} & \overset{\alpha_k}{\twoheadrightarrow} & \underline{i_{k-1}} & \overset{\alpha_{k-1}}{\twoheadrightarrow} & \cdots & \overset{\alpha_2}{\twoheadrightarrow} & \underline{i_1} & \overset{\alpha_1}{\twoheadrightarrow} & \underline{1} \\ \downarrow_{\delta_k} & & \downarrow_{\delta_{k-1}} & & & & \downarrow_{\delta_1} & & \downarrow_{\delta_0 = \mathrm{id}} \\ \underline{i'_k} & \overset{\alpha'_k}{\twoheadrightarrow} & \underline{i'_{k-1}} & \overset{\alpha'_{k-1}}{\twoheadrightarrow} & \cdots & \overset{\alpha'_2}{\twoheadrightarrow} & \underline{i'_1} & \overset{\alpha'_1}{\twoheadrightarrow} & \underline{1} \end{array}$$

commutes.

Define ${}^n I_{k-1} M_n$ to be the full subcategory of $I_k M_n$ whose objects are chains of surjections of the form

$$\underline{n} \twoheadrightarrow \underline{i_{k-1}} \twoheadrightarrow \cdots \twoheadrightarrow \underline{i_1} \twoheadrightarrow \underline{1}.$$

The following proposition is easily checked:

Proposition 2.14 *The category ${}^n I_{k-1} M_n$ is isomorphic to ${}^n O_{k-1} M_n$. In particular, it is a groupoid.*

In the next theorem we regard $C(\mathbb{R}^{\mathbf{l'}}, \boldsymbol{\alpha})$ and X^{i_k} as functors $(I^k M_n)^{op} \to$ Spaces and $I_k M_n \to$ Spaces, respectively. Indeed, consider a configuration space $C(\mathbb{R}^l, i)$ as a space of embeddings $\mathrm{Emb}(\underline{i}, \mathbb{R}^l)$. Then an injection $\delta\colon \underline{i} \to \underline{j}$ induces a map $\hat{\delta}\colon \mathrm{Emb}(\underline{j}, \mathbb{R}^l) \to \mathrm{Emb}(\underline{i}, \mathbb{R}^l)$, i.e., a map $C(\mathbb{R}^l, j) \to C(\mathbb{R}^l, i)$. Similarly one can consider $C(\mathbb{R}^{\mathbf{l'}}, \boldsymbol{\alpha})$ as a product of embedding spaces. Clearly, a morphism $\delta\colon \boldsymbol{\alpha} \to \boldsymbol{\alpha}'$ in $I^k M$ induces a map $\hat{\delta}\colon C(\mathbb{R}^{\mathbf{l'}}, \boldsymbol{\alpha}') \to C(\mathbb{R}^{\mathbf{l'}}, \boldsymbol{\alpha})$. It is easy to see that the correspondence $\boldsymbol{\alpha} \mapsto C(\mathbb{R}^{\mathbf{l'}}, \boldsymbol{\alpha})$ gives rise to a functor $(I^k M)^{op} \to$ Spaces. On the other hand, the correspondence $\boldsymbol{\alpha} \mapsto X^{i_k}$ gives rise to a functor $I^k M \to$ Spaces via basepoint inclusion.

Proposition 2.15

$$C(\mathbf{l'})(X) \cong \int^{I^k M} C(\mathbb{R}^{\mathbf{l'}}, \boldsymbol{\alpha}) \times X^{i_k}$$

where $\int$ stands for coend [**16**].

Proof The proof is by a tedious but straightforward counting argument. □

This suggests a natural filtration of $C(\mathbf{l'})(X)$.

Definition 2.16 We denote

$$C_n(\mathbf{l'})(X) = \int^{I^k M_n} C(\mathbb{R}^{\mathbf{l'}}, \boldsymbol{\alpha}) \times X^{i_k}.$$

If $l_1 = \cdots = l_k = \infty$ then we denote $C_n(\mathbf{l'})(X)$ by $C_n^k(X)$.

Next we introduce the "fat wedge" functor, which will play the role of the fat diagonal:

Definition 2.17 Let $W^n X \subset X^n$ be the space of all n-tuples $(x_1, \ldots, x_n) \in X^n$ such that $x_i = *$ for some $1 \le i \le n$.

Now we are ready to formulate the dual analogue of Theorem 2.5, that provides an inductive description of $C_n(\mathbf{l'})(X)$.

Theorem 2.18 *There exists a commutative square (the coproducts on the left-hand side are both taken over the same set of representatives)*

$$\begin{array}{ccc} \coprod\limits_{\boldsymbol{\alpha}\in[{}^n I_{k-1} M_n]} C(\mathbb{R}^{\mathbf{l'}}, \boldsymbol{\alpha}) \times_{\Sigma_{\boldsymbol{\alpha}}} X^n & \rightarrow & C_n(\mathbf{l'})(X) \\ \uparrow & & \uparrow \\ \coprod\limits_{\boldsymbol{\alpha}\in[{}^n I_{k-1} M_n]} C(\mathbb{R}^{\mathbf{l'}}, \boldsymbol{\alpha}) \times_{\Sigma_{\boldsymbol{\alpha}}} W^n X & \rightarrow & C_{n-1}(\mathbf{l'})(X) \end{array}$$

which is a strict pushout. Moreover, the left vertical map is the coproduct of inclusion maps. In particular, it is a cofibration and therefore the square is also a homotopy pushout.

Proof The proof is very much dual to the proof of Theorem 2.5. Details are straightforward and are left to the reader. □

We now have the dual analogue of Corollary 2.9:

Corollary 2.19 *There are cofibration sequences*

$$C_{n-1}(\mathbf{l'})(X) \rightarrow C_n(\mathbf{l'})(X) \rightarrow \bigvee_{\boldsymbol{\alpha}\in[{}^n I_{k-1} M_n]} C(\mathbb{R}^{\mathbf{l'}}, \boldsymbol{\alpha})_+ \wedge_{\Sigma_{\boldsymbol{\alpha}}} X^{\wedge n}.$$

As a functor of X,

$$Q\left(\bigvee_{\boldsymbol{\alpha}\in[{}^n O_{k-1} M_n]} C(\mathbb{R}^{\mathbf{l'}}, \boldsymbol{\alpha})_+ \wedge_{\Sigma_{\boldsymbol{\alpha}}} X^{\wedge n}\right)$$

is homogeneous of degree n. It follows, by induction on n, that QC_n^k is excisive of degree n. In fact, it is weakly equivalent to $P_n Q^{k+1}$, as we will show later.

Remark 2.20 Recall that $\mathbf{l} = (\mathbf{l'}, l_0) = (l_k, \ldots, l_0) \in \hat{I}^{k+1}$. Take $l_0 = \infty$. Observe that under the last assumption

$$\mathrm{Map}_*\left(S^{\mathbf{l}}(\boldsymbol{\alpha}) / \Delta^{\boldsymbol{\alpha}}(\mathbf{l}), S_X^{\mathbf{l}}(\boldsymbol{\alpha})\right)^{\Sigma_{\boldsymbol{\alpha}}}$$

is an infinite loop space. In fact,

$$\mathrm{Map}_*\left(S^{\mathbf{l}}(\boldsymbol{\alpha}) / \Delta^{\boldsymbol{\alpha}}(\mathbf{l}), S_X^{\mathbf{l}}(\boldsymbol{\alpha})\right)^{\Sigma_{\boldsymbol{\alpha}}} \simeq \Omega^\infty\left(\mathrm{Map}_*\left(S^{\mathbf{l'}}(\boldsymbol{\alpha}) / \Delta^{\boldsymbol{\alpha}}(\mathbf{l'}), \Sigma^\infty S_X^{\mathbf{l'}}(\boldsymbol{\alpha})\right)^{\Sigma_{\boldsymbol{\alpha}}}\right).$$

There exists a map N

$$N : Q\left(C\left(\mathbb{R}^{\mathbf{l}}, \boldsymbol{\alpha}\right)_+ \wedge_{\Sigma_{\boldsymbol{\alpha}}} X^{\wedge n}\right) \rightarrow \mathrm{Map}_*\left(S^{\mathbf{l}}(\boldsymbol{\alpha}) / \Delta^{\boldsymbol{\alpha}}(\mathbf{l}), S_X^{\mathbf{l}}(\boldsymbol{\alpha})\right)^{\Sigma_{\boldsymbol{\alpha}}}$$

defined as follows: a point in $C\left(\mathbb{R}^{\mathbf{l}}, \boldsymbol{\alpha}\right)$ can be identified with a cube in $\mathbb{R}^{i_k l_k + \cdots + i_1 l_1}$ (where $i_k = n$) that generates a free orbit under the action of $\Sigma_{\boldsymbol{\alpha}}$. Thus a point in

$\left(C\left(\mathbb{R}^{\mathbf{l}},\boldsymbol{\alpha}\right)_+\wedge_{\Sigma_{\boldsymbol{\alpha}}} X^{\wedge n}\right)$ is an unordered $|\Sigma_{\boldsymbol{\alpha}}|$-tuple of cubes labeled by appropriate permutations of an n-tuple $(x_1,\dots,x_n)$ of points in X. This data determines a $\Sigma_{\boldsymbol{\alpha}}$-equivariant map $S^{\mathbf{l}}(\boldsymbol{\alpha})\to S^{\mathbf{l}}_X(\boldsymbol{\alpha})$ given by first collapsing the complement of the union of the $|\Sigma_{\boldsymbol{\alpha}}|$ cubes to the basepoint, thus obtaining a space homeomorphic to $\Sigma_{\boldsymbol{\alpha}+}\wedge S^{\mathbf{l}}(\boldsymbol{\alpha})$, and then mapping this space to $S^{\mathbf{l}}_X(\boldsymbol{\alpha})$ by the folding map. Since the action of $\Sigma_{\boldsymbol{\alpha}}$ on the cubes is free, the map factors through a map from $S^{\mathbf{l}}(\boldsymbol{\alpha})/\Delta^{\boldsymbol{\alpha}}(\mathbf{l})$ to $S^{\mathbf{l}}_X(\boldsymbol{\alpha})$. The map N is homotopic to the composition of the following chain of maps:

$$
\begin{array}{c}
Q\left(C\left(\mathbb{R}^{\mathbf{l}},\boldsymbol{\alpha}\right)_+\wedge_{\Sigma_{\boldsymbol{\alpha}}} X^{\wedge n}\right)\\
\downarrow\\
Q\left(\left(\mathbb{R}^{\mathbf{l}}(\boldsymbol{\alpha})\setminus\Delta^{\boldsymbol{\alpha}}(\mathbf{l})\right)_+\wedge_{\Sigma_{\boldsymbol{\alpha}}} X^{\wedge n}\right)\\
\downarrow\\
\Omega^\infty\left(\mathrm{Map}_*\left(S^{\mathbf{l'}}(\boldsymbol{\alpha})/\Delta^{\boldsymbol{\alpha}}(\mathbf{l'}),\Sigma^\infty S^{\mathbf{l'}}_X(\boldsymbol{\alpha})\right)_{h\Sigma_{\boldsymbol{\alpha}}}\right)\\
\downarrow\\
\mathrm{Map}_*\left(S^{\mathbf{l}}(\boldsymbol{\alpha})/\Delta^{\boldsymbol{\alpha}}(\mathbf{l}),S^{\mathbf{l}}_X(\boldsymbol{\alpha})\right)^{\Sigma_{\boldsymbol{\alpha}}}.
\end{array}
$$

Here the first map is induced by the map $C(\mathbb{R}^{\mathbf{l}},\boldsymbol{\alpha})_+\to(\mathbb{R}^{\mathbf{l}}(\boldsymbol{\alpha})\setminus\Delta^{\boldsymbol{\alpha}}(\mathbf{l}))_+$ discussed in Proposition 2.12, the second map is the Spanier-Whitehead duality (strictly speaking, this is a weak map because we have to replace $\Sigma_{\boldsymbol{\alpha}}$-orbits by homotopy orbits, we can do it because the action of $\Sigma_{\boldsymbol{\alpha}}$ is free), and the third map is the "norm", or "Tate" map (essentially a version of the Adams transfer). The last two maps are equivalences for all values of $l_1,\dots,l_k$ (the implicit assumption is that $l_0=\infty$) and the first map is an equivalence if $l_0=\cdots=l_k=\infty$. Thus N is an equivalence if $l_0=\cdots=l_k=\infty$.

It follows from the remark that the functor

$$X\mapsto \mathrm{Map}_*\left(S^{\mathbf{l}}(\boldsymbol{\alpha})/\Delta^{\boldsymbol{\alpha}}(\mathbf{l}),S^{\mathbf{l}}_X(\boldsymbol{\alpha})\right)^{\Sigma_{\boldsymbol{\alpha}}}$$

is homogeneous of degree n if $l_j=\infty$ for all j. Therefore, by induction on n,

$$\mathrm{hocolim}_{I^{k+1}}\mathrm{Nat}_{O_kM_n}(S^{\mathbf{l}},S^{\mathbf{l}}_X)$$

is an n-excisive functor of X. It remains to show that it also is a model for the n-th Taylor polynomial of $Q^{k+1}X$. This is the content of the next theorem, which is the main theorem of this section.

Theorem 2.21 *Let* $\mathbf{l}=\underbrace{(\infty,\dots,\infty)}_{k+1}$. *Let X be a c-connected space. The map*

$$p_n: Q^{k+1}X\to Nat_{O_kM_n}(S^{\mathbf{l}},S^{\mathbf{l}}_X)$$

is $(n+1)c+n$ connected. Together with the preceding remarks this implies that

$$P_nQ^{k+1}X\simeq Nat_{O_kM_n}(S^{\mathbf{l}},S^{\mathbf{l}}_X)$$

and p_n is a model for the approximation map.

Proof In the course of the proof we will use shorter notation

$$P'_nQ^{k+1}X=\mathrm{Nat}_{O_kM_n}(S^{\mathbf{l}},S^{\mathbf{l}}_X).$$

It follows from the classical theory of little cubes models that the map

$$QC_n^k X \to Q^{k+1}X$$

is $(n+1)c+n+1$ connected. It is enough to prove that the composed map

$$QC_n^k X \to Q^{k+1}X \to P_n' Q^{k+1}X$$

is an equivalence. The proof is by induction on n. In case $n=1$, it is easy to see that $QC_1^k X \simeq P_1'Q^{k+1}X \simeq QX$ and our map is a homotopy equivalence. For the induction step consider the following diagram where the right vertical map is the one from Remark 2.20.

$$\begin{array}{ccccc}
QC_{n-1}^k X & \to & QC_n^k X & \to & \prod_{\boldsymbol{\alpha}\in[^n I_{k-1}M_n]} Q\left(C(\mathbb{R}^{\mathbf{l}},\boldsymbol{\alpha})_+ \wedge_{\Sigma_{\boldsymbol{\alpha}}} X^{\wedge n}\right) \\
\downarrow & & \downarrow & & \downarrow N \\
P_{n-1}'Q^{k+1}X & \leftarrow & P_n'Q^{k+1}X & \leftarrow & \prod_{\boldsymbol{\alpha}\in[^n O_{k-1}M_n]} \mathrm{Map}_*\left(S^{\mathbf{l}}(\boldsymbol{\alpha})/\Delta^{\boldsymbol{\alpha}}(\mathbf{1}), S_X^{\mathbf{l}}(\boldsymbol{\alpha})\right)^{\Sigma_{\boldsymbol{\alpha}}}
\end{array}$$

Notice that the horizontal arrows go in opposite directions in the two rows and that the right-hand side square does not commute. However, it is easy to see that the right-hand square commutes on elements of highest filtration in $C_n^k X$. Our goal is to prove that the middle vertical map is an equivalence. It is a map between two functors of degree n. It is easy to see that the left square commutes. By our induction assumption, the left vertical map is a weak equivalence. It follows that the middle map induces an equivalence upon passage to $(n-1)$-th Taylor polynomials. It is easy to check, using the fact that the right-hand square commutes on elements of highest filtration, that the right-hand square commutes up to homotopy after passing to n-th cross-effects, and therefore it commutes after passing to n-th differentials. Since the right vertical map is an equivalence, it follows that the central vertical map induces an equivalence after passing to n-th differentials. Thus the central vertical map is a natural transformation between two n-excisive functors which induces an equivalence on the $(n-1)$-th Taylor polynomials and on the n-th differentials and thus is an equivalence. □

3 The Taylor tower of the Bousfield-Kan cosimplicial space

In the previous section we constructed models for P_nQ^kX, the "Taylor polynomials" of Q^kX, and in this section we identify P_nQ^kX with the model given in Theorem 2.21. Our next goal is to show that the spaces P_nQ^kX, $1\le k$, fit into a semi-cosimplicial space

$$[k] \mapsto P_nQ^{k+1}X.$$

Let $\mathbf{l}=(l_k,\dots,l_0)\in \hat{I}^{k+1}$. For $0\le j\le k$, let $d_j\mathbf{l}$ be the sequence $\mathbf{l}$ with l_j omitted. For $1\le j\le k$ define the map

$$d^j : \mathrm{Nat}_{O_{k-1}M_n}(S^{d_j\mathbf{l}}, S_X^{d_j\mathbf{l}}) \to \mathrm{Nat}_{O_kM_n}(S^{\mathbf{l}}, S_X^{\mathbf{l}})$$

as a composition

$$\mathrm{Nat}_{O_{k-1}M_n}(S^{d_j\mathbf{l}}, S_X^{d_j\mathbf{l}}) \to \mathrm{Nat}_{O_kM_n}(S^{d_j\mathbf{l}}\circ d_j, S_X^{d_j\mathbf{l}}\circ d_j) \to \mathrm{Nat}_{O_kM_n}(S^{\mathbf{l}}, S_X^{\mathbf{l}}),$$

where the first map is induced by the face map $d_j : O_kM_n \to O_{k-1}M_n$ and the second map is induced by smashing with the identity map on $S^{l_ji_j}$. Taking the homotopy colimit we obtain maps $d^j : P_nQ^kX \to P_nQ^{k+1}X$ for $j=1,\dots,k$.

The coface map d^0 is special in several ways (in particular, it is the only coface map that is not homotopic to an infinite loop map) and it should not be surprising that d^0 has to be constructed separately. We have to make yet one more definition:

Definition 3.1 Let

$$\boldsymbol{\alpha} = (\underline{i_k} \overset{\alpha_k}{\twoheadrightarrow} \underline{i_{k-1}} \overset{\alpha_{k-1}}{\twoheadrightarrow} \cdots \overset{\alpha_2}{\twoheadrightarrow} \underline{i_1} \twoheadrightarrow \underline{i_0} = \underline{1})$$

be a chain of surjections. Let $0 \leq j < k$. Let $s \in \underline{i_j}$. We let $\boldsymbol{\alpha}^{-1}(s)$ to denote the chain of surjections consisting of the inverse images of s

$$\boldsymbol{\alpha}^{-1}(s) = (\alpha_k^{-1} \cdots (\alpha_{j+2}^{-1}(\alpha_{j+1}^{-1}(s))) \twoheadrightarrow \cdots \twoheadrightarrow \alpha_{j+2}^{-1}(\alpha_{j+1}^{-1}(s)) \twoheadrightarrow \alpha_{j+1}^{-1}(s) \twoheadrightarrow \{s\}).$$

To construct the map $d^0 : P_n Q^k X \to P_n Q^{k+1} X$, we begin with defining, for every $\boldsymbol{\alpha}' = (\underline{i'_k} \twoheadrightarrow \cdots \twoheadrightarrow \underline{1}) \in O_k M_n$, a map

$$g_{\boldsymbol{\alpha}'} \colon \mathrm{Nat}_{O_{k-1}M_n}(S^{d_0 \mathbf{l}}, S_X^{d_0 \mathbf{l}}) \to \mathrm{Map}_*(S^{\mathbf{l}}(\boldsymbol{\alpha}'), S_X^{\mathbf{l}}(\boldsymbol{\alpha}')).$$

Let $\boldsymbol{\beta}, \boldsymbol{\beta}'$ etc. be objects of $O_{k-1}M_n$. Let

$$(\ldots, f_{\boldsymbol{\beta}}, \ldots, f_{\boldsymbol{\beta}'}, \ldots) \in \mathrm{Nat}_{O_{k-1}M_n}(S^{d_0 \mathbf{l}}, S_X^{d_0 \mathbf{l}}).$$

Let $j \in \underline{i'_1}$. Then "morally speaking" ${\boldsymbol{\alpha}'}^{-1}(j)$ can be considered an object of $O_{k-1}M_n$, and the idea is to associate with $(\ldots, f_{\boldsymbol{\beta}}, \ldots, f_{\boldsymbol{\beta}'}, \ldots)$ the map

$$\left(\bigwedge_{j \in \underline{i'_1}} f_{{\boldsymbol{\alpha}'}^{-1}(j)} \right) \wedge \mathrm{Id}|_{S^{l_0}}.$$

However, since we use a skeletal category of chosen finite sets rather than the category of all finite sets, ${\boldsymbol{\alpha}'}^{-1}(j)$ is just isomorphic to some $\boldsymbol{\beta}_j \in O_{k-1}M_n$. Let $h_j \colon \boldsymbol{\beta}_j \to {\boldsymbol{\alpha}'}^{-1}(j)$ be such an isomorphism. Now define

$$g_{\boldsymbol{\alpha}'}(\ldots, f_{\boldsymbol{\beta}}, \ldots) \colon S^{\mathbf{l}}(\boldsymbol{\alpha}') \to S_X^{\mathbf{l}}(\boldsymbol{\alpha}')$$

to be the composite:

$$S^{\mathbf{l}}(\boldsymbol{\alpha}') \cong \bigwedge_{j \in \underline{i'_1}} S^{\mathbf{d_0 l}}({\boldsymbol{\alpha}'}^{-1}(j)) \wedge S^{l_0} \to \bigwedge_{j \in \underline{i'_1}} S^{\mathbf{d_0 l}}(\boldsymbol{\beta}_j) \wedge S^{l_0}$$

$$\to \bigwedge_{j \in \underline{i'_1}} S_X^{\mathbf{d_0 l}}(\boldsymbol{\beta}_j) \wedge S^{l_0} \to \bigwedge_{j \in \underline{i'_1}} S_X^{\mathbf{d_0 l}}({\boldsymbol{\alpha}'}^{-1}(j)) \wedge S^{l_0} \cong S_X^{\mathbf{l}}(\boldsymbol{\alpha}')$$

where the second and the second last maps are induced by the isomorphisms h_j and their inverses respectively and the middle map is $\left(\bigwedge_{j \in \underline{i'_1}} f_{\boldsymbol{\beta}_j}\right) \wedge \mathrm{Id}|_{S^{l_0}}$.

We have to show that the map is independent of the choice of isomorphisms h_j. Let $h'_j \colon \boldsymbol{\beta}'_j \to {\boldsymbol{\alpha}'}^{-1}(j)$ be another family of isomorphisms. Let

$$(\ldots, f_{\boldsymbol{\beta}}, \ldots, f_{\boldsymbol{\beta}'}, \ldots) \in \mathrm{Nat}_{O_{k-1}M_n}(S^{d_0 \mathbf{l}}, S_X^{d_0 \mathbf{l}}).$$

Then for each j the diagram

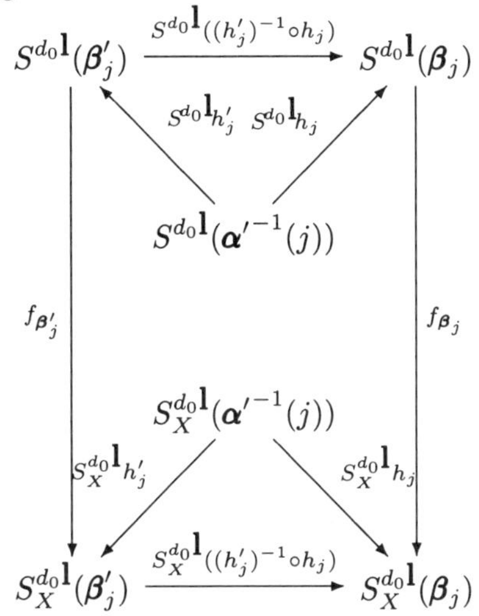

commutes. All the arrows, except for the vertical ones, are isomorphisms. It is easy to check that this implies

$$(S_X^{d_0\mathbf{l}}(h_j))^{-1} \circ f_{\boldsymbol{\beta}_j} \circ S^{d_0\mathbf{l}}(h_j) = (S_X^{d_0\mathbf{l}}(h_j'))^{-1} \circ f_{\boldsymbol{\beta}_j'} \circ S^{d_0\mathbf{l}}(h_j').$$

This means precisely that $g_{\boldsymbol{\alpha}'}$ is independent of the choice of h_j.

The maps $g_{\boldsymbol{\alpha}'}$ assemble to a map

$$\widetilde{d^0}\colon \mathrm{Nat}_{O_{k-1}M_n}(S^{d_0\mathbf{l}}, S_X^{d_0\mathbf{l}}) \to \prod_{\boldsymbol{\alpha}' \in O_k M_n} \mathrm{Map}_*(S^{\mathbf{l}}(\boldsymbol{\alpha}'), S_X^{\mathbf{l}}(\boldsymbol{\alpha}')).$$

Proposition 3.2 *The map $\widetilde{d^0}$ induces a map*

$$d^0\colon \mathit{Nat}_{O_{k-1}M_n}(S^{d_0\mathbf{l}}, S_X^{d_0\mathbf{l}}) \to \mathit{Nat}_{O_k M_n}(S^{\mathbf{l}}, S_X^{\mathbf{l}}),$$

which is natural both in X and $\mathbf{l}$.

Proof It is enough to show that for every morphism $p\colon \boldsymbol{\alpha} \twoheadrightarrow \boldsymbol{\alpha}'$ in $O_k M_n$ the diagram

$$\begin{array}{ccc} S^{\mathbf{l}}(\boldsymbol{\alpha}') & \xrightarrow{S^{\mathbf{l}}(p)} & S^{\mathbf{l}}(\boldsymbol{\alpha}) \\ \downarrow{\scriptstyle g_{\boldsymbol{\alpha}'}} & & \downarrow{\scriptstyle g_{\boldsymbol{\alpha}}} \\ S_X^{\mathbf{l}}(\boldsymbol{\alpha}') & \xrightarrow{S_X^{\mathbf{l}}(p)} & S_X^{\mathbf{l}}(\boldsymbol{\alpha}) \end{array}$$

commutes. In view of the preceding remarks, we may think of $g_{\boldsymbol{\alpha}'}$ as $\left(\bigwedge_{j=1}^{i_1'} f_{\boldsymbol{\alpha}'^{-1}(j)}\right) \wedge \mathrm{Id}|_{S^{l_0}}$ and similarly for $g_{\boldsymbol{\alpha}}$. For every $j \in \underline{i_1'}$, p induces a morphism $p_j\colon \boldsymbol{\alpha}^{-1}(j) \twoheadrightarrow \boldsymbol{\alpha}'^{-1}(p(j))$. Notice that this would not be true if we were working with the category $N_k M_n$ rather than $O_k M_n$. Now

$$S^{\mathbf{l}}(p) = \left(\bigwedge_{j=1}^{i_1'} S^{d_0\mathbf{l}}(p_j)\right) \wedge \mathrm{Id}|_{S^{l_0}} \quad \text{and} \quad S_X^{\mathbf{l}}(p) = \left(\bigwedge_{j=1}^{i_1'} S_X^{d_0\mathbf{l}}(p_j)\right) \wedge \mathrm{Id}|_{S^{l_0}}$$

holds for all j. Since for every j the square

$$\begin{array}{ccc} S^{d_0\mathbf{l}}(\boldsymbol{\alpha}'^{-1}(p(j))) & \xrightarrow{S^{d_0\mathbf{l}}(p_j)} & S^{d_0\mathbf{l}}(\boldsymbol{\alpha}^{-1}(j)) \\ \downarrow{\scriptstyle f_{\boldsymbol{\alpha}'^{-1}(p(j))}} & & \downarrow{\scriptstyle f_{\boldsymbol{\alpha}^{-1}(j)}} \\ S_X^{d_0\mathbf{l}}(\boldsymbol{\alpha}'^{-1}(p(j))) & \xrightarrow{S_X^{d_0\mathbf{l}}(p_j)} & S_X^{d_0\mathbf{l}}(\boldsymbol{\alpha}^{-1}(j)) \end{array}$$

commutes by assumption, the claim follows by "smashing the squares together". □

Proposition 3.3 *The maps $d^j : P_nQ^kX \to P_nQ^{k+1}X$ satisfy the identities $d^jd^i = d^id^{j-1}$ for $i < j$. In other words, they satisfy the relations for the coface maps in a (semi-)cosimplicial object.*

Proof The identities clearly hold for $i > 0$. Let then $i = 0$. Also this case follows rather straightforwardly from the definitions of d^j and d^0. In what follows, let $\boldsymbol{\alpha}, \boldsymbol{\alpha}'$ and $\boldsymbol{\alpha}''$ denote sequences of surjections $\underline{i_{k-2}} \twoheadrightarrow \underline{i_{k-3}} \twoheadrightarrow \cdots \twoheadrightarrow \underline{i_0} = \underline{1}$, $\underline{i'_{k-1}} \twoheadrightarrow \underline{i'_{k-2}} \twoheadrightarrow \cdots \twoheadrightarrow \underline{i'_0} = \underline{1}$ and $\underline{i''_k} \twoheadrightarrow \underline{i''_{k-1}} \twoheadrightarrow \cdots \twoheadrightarrow \underline{i''_0} = \underline{1}$, respectively. Let $\mathbf{l} = (l_k, \ldots, l_0)$. Then $d_0d_j\mathbf{l} = d_{j-1}d_0\mathbf{l}$. Assume $(\ldots, f_{\boldsymbol{\alpha}}, \ldots) \in \mathrm{Nat}_{O_{k-2}M_n}(S^{d_0d_j\mathbf{l}}, S_X^{d_0d_j\mathbf{l}})$. Then, on one hand

$$\begin{aligned} d^jd^0(\ldots, f_{\boldsymbol{\alpha}}, \ldots) = \quad & d^j\left(\left(\left(\bigwedge_{p=1}^{i'_1} f_{\boldsymbol{\alpha}'^{-1}(p)}\right) \wedge \mathrm{Id}|_{S^{l_0}}\right)_{\boldsymbol{\alpha}' \in O_{k-1}M_n}\right) \\ = \quad & \left(\left(\bigwedge_{p=1}^{i''_1} f_{(d_j(\boldsymbol{\alpha}''))^{-1}(p)}\right) \wedge \mathrm{Id}|_{S^{l_0}} \wedge \mathrm{Id}|_{S^{l_ji''_j}}\right)_{\boldsymbol{\alpha}'' \in O_kM_n}. \end{aligned}$$

On the other hand,

$$\begin{aligned} d^0d^{j-1}(\ldots, f_{\boldsymbol{\alpha}}, \ldots) = \quad & d^0\left(\left(f_{d_{j-1}(\boldsymbol{\alpha}')} \wedge \mathrm{Id}|S^{l_ji'_{j-1}}\right)_{\boldsymbol{\alpha}' \in O_{k-1}M_n}\right) \\ = \quad & \left(\left(\bigwedge_{p=1}^{i''_1} f_{d_{j-1}(\boldsymbol{\alpha}''^{-1}(p))}\right) \wedge \mathrm{Id}|_{S^{l_ji''_j}} \wedge \mathrm{Id}|_{S^{l_0}}\right)_{\boldsymbol{\alpha}'' \in O_kM_n}. \end{aligned}$$

Since $(d_j(\boldsymbol{\alpha}''))^{-1}(p) = d_{j-1}(\boldsymbol{\alpha}''^{-1}(p))$ for every p, the claim follows. □

Thus the spaces P_nQ^kX, $1 \le k$, fit into a semi-cosimplicial space. Let P_nX denote the total space of this semi-cosimplicial space. $P_n(-)$ is a homotopy inverse limit of n-excisive functors and therefore is n-excisive.

Lemma 3.4 *The diagrams*

$$\begin{array}{ccc} Q^kX & \xrightarrow{d^j} & Q^{k+1}X \\ \downarrow{\scriptstyle p_n} & & \downarrow{\scriptstyle p_n} \\ P_nQ^kX & \xrightarrow{d^j} & P_nQ^{k+1}X \end{array}$$

commute for $0 \le j \le k$.

Proof Straightforward chasing of the definitions. □

In other words, we have a map of semi-cosimplicial spaces, which induces a map on total spaces $p_n \colon L_{\mathbf{S}}X \to P_nX$. (Recall that $L_{\mathbf{S}}X$ denotes the total space of the Bousfield-Kan cosimplicial space.) Clearly, P_n is n-excisive and the maps are $(n+1)c+1$ connected (see Section 5). Since this is the main result of this section, we formulate it as a theorem.

Theorem 3.5 *Let $P_nQ^{k+1}X = \mathrm{hocolim}_{I^{k+1}} Nat_{O_kM_n}(S^{\mathbf{l}}, S^{\mathbf{l}}_X)$. The maps d^j defined above give rise to a semi-cosimplicial space $[k] \mapsto P_nQ^{k+1}X$. Let P_nX denote the total space of this semi-cosimplicial space. P_n is n-excisive and there is a natural map $X \to P_nX$ that is $(n+1)c+1$ connected, where c is the connectivity of X.*

Thus what we denote P_nX is worthy of its name - it is a model for the n-th Taylor polynomial of the identity.

Remark 3.6 It may seem puzzling that we use O_kM_n rather than the nerve N_kM_n for our definition of $P_n\Omega S(\mathbf{l})(-)$. Indeed, if we defined

$$P_n\Omega S(\mathbf{l})(X) = \mathrm{Nat}_{N_kM_n}(S^{\mathbf{l}}, S^{\mathbf{l}}_X)$$

this would have the right homotopy type. In fact, it would have the right homotopy type as soon as $l_0 = \infty$, for all values $l_1, \ldots, l_k$. However, this construction is not functorial enough. It is not possible to define d^0 in terms of this construction (it is possible to define d^j for $j > 0$).

4 A description of the layers of the identity - proof of Johnson's theorem

The goal of this section is to prove the following theorem.

Theorem 4.1

$$D_nId(X) \simeq \Omega^\infty \left(Map_* \left(K_n, \Sigma^\infty X^{\wedge n}\right)_{h\Sigma_n}\right).$$

Here K_n is a finite polyhedron with an action of Σ_n, whose definition and basic properties we will recall shortly. Theorem 4.1 is equivalent to the main theorem of [**12**] (see [**2**, Section 2] for proof of this equivalence).

Our K_n is going to be the same as the space denoted SK_n in [**2**]. It is essentially the geometric realization of the poset of unordered partitions of $\underline{n}$. More precisely, it is the geometric realization of the simplicial set $k \mapsto T^k_n$ defined in [**2**, Section 1], whose definition we will now recall, in a slightly different form.

Definition 4.2 Let $\mathrm{Sur}(U, V)$ stand for the set of surjections from U to V.

$$T^n_0 = \emptyset_+ = *.$$

$$T^n_k = \left(\coprod_{(n \geq i_{k-1} \geq \cdots \geq i_1 \geq 1)} \mathrm{Sur}(\underline{n}, \underline{i_{k-1}}) \times_{\Sigma_{i_{k-1}}} \cdots \times_{\Sigma_{i_1}} \mathrm{Sur}(\underline{i_1}, \underline{1})\right)_+ \quad \text{if } k > 0.$$

Thus T^n_k is obtained from the set of objects of ${}^nO_{k-1}M_n$ by dividing each connected component by the action of an appropriate group $\Sigma_{i_{k-1}} \times \cdots \times \Sigma_{i_1}$. Clearly, the face and degeneracy maps in ${}^nO_{\bullet-1}M_n$ give rise to face and degeneracy maps in $T^n_\bullet$. There is a map of simplicial sets ${}^nO_{\bullet-1}M_n \to T^n_\bullet$. This map induces a map between the geometric realizations. Because the action of the groups of the form $\Sigma_{i_{k-1}} \times \cdots \times \Sigma_{i_1}$ is free, this map is a homotopy equivalence. In fact, the inverse image of every point is of the form $E(\Sigma_{i_{k-1}} \times \cdots \times \Sigma_{i_1})$ and hence contractible. It is easy to see that this definition of T^n_k coincides with the one in [**2**]. T^n_k has an action of Σ_n, and we may consider the groupoid $\Sigma_n \ltimes T^n_k$.

K_n is defined to be the geometric realization of $T^n_\bullet$. Clearly, K_n has dimension $n-1$ and there is an action of Σ_n on K_n induced by the simplicial action on $T^n_\bullet$. It is well-known that non-equivariantly, $K_n \simeq \bigvee_{(n-1)!} S^{n-1}$.

Corollary 2.9 provides us with a description of the homotopy fiber of the restriction map $\mathrm{Nat}_{O_kM_n}(S^{\mathbf{l}}, S^{\mathbf{l}}_X) \to \mathrm{Nat}_{O_kM_{n-1}}(S^{\mathbf{l}}, S^{\mathbf{l}}_X)$, namely,

$$\prod_{\boldsymbol{\alpha}\in[{}^nO_{k-1}M_n]} \mathrm{Map}_*\left(S^{\mathbf{l}}(\boldsymbol{\alpha})/\Delta^{\boldsymbol{\alpha}}(\mathbf{l}), S^{\mathbf{l}}_X(\boldsymbol{\alpha})\right)^{\Sigma_{\boldsymbol{\alpha}}}. \tag{4.1}$$

Since $\Delta^{\boldsymbol{\alpha}}(\mathbf{l})$ is precisely the subspace of $S^{\mathbf{l}}(\boldsymbol{\alpha})$ on which the action of $\Sigma_{\boldsymbol{\alpha}}$ is not free, it follows that there is a weak homotopy equivalence

$$\mathrm{Map}_*\left(S^{\mathbf{l}}(\boldsymbol{\alpha})/\Delta^{\boldsymbol{\alpha}}(\mathbf{l}), S^{\mathbf{l}}_X(\boldsymbol{\alpha})\right)^{\Sigma_{\boldsymbol{\alpha}}} \simeq \mathrm{Map}_*\left(S^{\mathbf{l}}(\boldsymbol{\alpha}), S^{\mathbf{l}}_X(\boldsymbol{\alpha}) \wedge E\Sigma_{n+}\right)^{\Sigma_{\boldsymbol{\alpha}}}.$$

To recast (4.1) in a more functorial way, recall that by Propositions 1.9 and 1.10

$$\prod_{\boldsymbol{\alpha}\in[{}^nO_{k-1}M_n]} \mathrm{Map}_*\left(S^{\mathbf{l}}(\boldsymbol{\alpha}), S^{\mathbf{l}}_X(\boldsymbol{\alpha}) \wedge E\Sigma_{n+}\right)^{\Sigma_{\boldsymbol{\alpha}}} \cong \mathrm{Nat}_{{}^nO_{k-1}M_n}(S^{\mathbf{l}}, S^{\mathbf{l}}_X \wedge E\Sigma_{n+}).$$

Since the action of $\Sigma_{\boldsymbol{\alpha}}$ on $E\Sigma_n$ is free, it follows that every element of

$$\mathrm{Nat}_{{}^nO_{k-1}M_n}(S^{\mathbf{l}}, S^{\mathbf{l}}_X \wedge E\Sigma_{n+})$$

takes $\Delta^{\boldsymbol{\alpha}}(\mathbf{l})$ to the basepoint and therefore there is a map

$$\mathrm{Nat}_{{}^nO_{k-1}M_n}(S^{\mathbf{l}}, S^{\mathbf{l}}_X \wedge E\Sigma_{n+}) \to \mathrm{Nat}_{O_kM_n}(S^{\mathbf{l}}, S^{\mathbf{l}}_X)$$

which is given by collapsing $E\Sigma_n$ to a point and extending the natural transformation to be the trivial map for all $\boldsymbol{\alpha}\notin{}^nO^{k-1}M_n$. It follows from Theorem 2.5 that

$$\begin{array}{ccc} \mathrm{Nat}_{{}^nO_{k-1}M_n}(S^{\mathbf{l}}, S^{\mathbf{l}}_X \wedge E\Sigma_{n+}) & \to & \mathrm{Nat}_{O_kM_n}(S^{\mathbf{l}}, S^{\mathbf{l}}_X) \\ & & \downarrow \\ & & \mathrm{Nat}_{O_kM_{n-1}}(S^{\mathbf{l}}, S^{\mathbf{l}}_X) \end{array}$$

is a homotopy fibration sequence. The coface maps

$$d^j : \mathrm{Nat}_{O_{k-1}M_n}(S^{d_j\mathbf{l}}, S^{d_j\mathbf{l}}_X) \to \mathrm{Nat}_{O_kM_n}(S^{\mathbf{l}}, S^{\mathbf{l}}_X)$$

induce coface maps

$$d^j : \mathrm{Nat}_{{}^nO_{k-2}M_n}(S^{d_j\mathbf{l}}, S^{d_j\mathbf{l}}_X \wedge E\Sigma_{n+}) \to \mathrm{Nat}_{{}^nO_{k-1}M_n}(S^{\mathbf{l}}, S^{\mathbf{l}}_X \wedge E\Sigma_{n+})$$

between the fibers. These maps are given by the composition

$$\begin{array}{ccc} \mathrm{Nat}_{{}^nO_{k-2}M_n}(S^{d_j\mathbf{l}}, S^{d_j\mathbf{l}}_X \wedge E\Sigma_{n+}) & \to & \mathrm{Nat}_{{}^nO_{k-1}M_n}(S^{d_j\mathbf{l}} \circ d_j, (S^{d_j\mathbf{l}}_X \circ d_j) \wedge E\Sigma_{n+}) \\ & & \downarrow \\ & & \mathrm{Nat}_{{}^nO_{k-1}M_n}(S^{\mathbf{l}}, S^{\mathbf{l}}_X \wedge E\Sigma_{n+}). \end{array}$$

Note that on this level all the coface maps, including d^0, are "almost" induced by the face maps d_j on ${}^nO_{\bullet-1}M_n$. If $l_j = \infty$ for all $0 \le j \le k$ then the spaces $\mathrm{Nat}_{{}^nO_{k-1}M_n}(S^{\mathbf{l}}, S^{\mathbf{l}}_X \wedge E\Sigma_{n+})$ fit into a semi-cosimplicial space $D^\bullet_n(X)$, whose total space has the homotopy type of D_nX. Next we want to replace $D^\bullet_n(X)$ with an equivalent semi-cosimplicial space, whose coface maps are genuinely, rather than almost, dual to the face maps in ${}^nO_\bullet M_n$. Recall that $\mathrm{Nat}_{{}^nO_{k-1}M_n}(S^{\mathbf{l}}, S^{\mathbf{l}}_X \wedge E\Sigma_{n+})$ really stands for the homotopy direct limit

$$\mathrm{hocolim}_{I^{k+1}}\mathrm{Nat}_{{}^nO_{k-1}M_n}(S^{\mathbf{l}}, S^{\mathbf{l}}_X \wedge E\Sigma_{n+})$$

where I is the category of standard finite sets with injections.

For $\mathbf{m} \in I$, let $S^{\mathbf{m}n}$, $S^{\mathbf{m}n}_X$ denote the functors from ${}^nO_{k-1}M_n$ to based spaces given on objects by $\boldsymbol{\alpha} \mapsto S^{mn}$ and $\boldsymbol{\alpha} \mapsto S^{mn} \wedge X^{\wedge n}$ respectively. Clearly, there are maps

$$d^j : \mathrm{Nat}_{{}^nO_{k-2}M_n}(S^{\mathbf{m}n}, S^{\mathbf{m}n}_X \wedge E\Sigma_{n+}) \to \mathrm{Nat}_{{}^nO_{k-1}M_n}(S^{\mathbf{m}n}, S^{\mathbf{m}n}_X \wedge E\Sigma_{n+})$$

induced by the face maps in ${}^nO_{\bullet-1}M_n$. Thus the spaces

$$\mathrm{Nat}_{{}^nO_{k-1}M_n}(S^{\mathbf{m}n}, S^{\mathbf{m}n}_X \wedge E\Sigma_{n+}),$$

$k \geq 0$, fit into a semi-cosimplicial space. We want to compare it with $D^\bullet_n(X)$. We define a map

$$\mathrm{Nat}_{{}^nO_{k-1}M_n}(S^{\mathbf{l}}, S^{\mathbf{l}}_X \wedge E\Sigma_{n+}) \to \mathrm{Nat}_{{}^nO_{k-1}M_n}(S^{(\mathbf{l}_0 \amalg \cdots \amalg \mathbf{l}_k)n}, S_X^{(\mathbf{l}_0 \amalg \cdots \amalg \mathbf{l}_k)n} \wedge E\Sigma_{n+})$$

where $\mathbf{l}_0 \coprod \cdots \coprod \mathbf{l}_k$ is considered an element of I. The map is defined by smashing with the identity map on the orthogonal complement of the image of the map $\mathbb{R}^{nl_k + i_{k-1}l_{k-1} + \cdots + i_0 l_0} \to \mathbb{R}^{n(l_k + \cdots + l_0)}$ induced by $\boldsymbol{\alpha} = (\underline{n} \twoheadrightarrow \underline{i_{k-1}} \twoheadrightarrow \cdots \twoheadrightarrow \underline{1})$. This map induces a map on homotopy direct limits

$$\begin{array}{c}\mathrm{hocolim}_{I^{k+1}}\mathrm{Nat}_{{}^nO_{k-1}M_n}(S^{\mathbf{l}}, S^{\mathbf{l}}_X \wedge E\Sigma_{n+}) \\ \downarrow s \\ \mathrm{hocolim}_{I}\mathrm{Nat}_{{}^nO_{k-1}M_n}(S^{\mathbf{m}n}, S^{\mathbf{m}n}_X \wedge E\Sigma_{n+})\end{array}$$

which is a homotopy equivalence. The passage from direct limit over I^{k+1} to direct limit over I is possible because we use I rather than $\mathbb{N}$. This is one of the key ideas in Bökstedt's way to make the definition of THH rigorous, i.e., to get a simplicial object on the nose rather than only up to homotopy. It is only for the sake of this step that we used I all along. The map s commutes with coface maps. Let us denote $\mathrm{hocolim}_I\mathrm{Nat}_{{}^nO_{k-1}M_n}(S^{\mathbf{m}n}, S^{\mathbf{m}n}_X \wedge E\Sigma_{n+})$ by $\mathrm{Nat}_{{}^nO_{k-1}M_n}(S^{\infty n}, S^{\infty n}_X \wedge E\Sigma_{n+})$. Thus D_nX is equivalent to the total space of the semi-cosimplicial space

$$k \mapsto \mathrm{Nat}_{{}^nO_{k-1}M_n}(S^{\infty n}, S^{\infty n}_X \wedge E\Sigma_{n+}). \tag{4.2}$$

To analyze this cosimplicial space, observe that by Propositions 1.10 and 1.11, Remark 2.4 and the definition of T^k_n it follows that

$$\mathrm{Nat}_{{}^nO_{k-1}M_n}(S^{\infty n}, S^{\infty n}_X \wedge E\Sigma_{n+}) \cong \mathrm{Nat}_{\Sigma_n \ltimes T^k_n}(S^{\infty n}, S^{\infty n}_X \wedge E\Sigma_{n+})$$

which in turn is naturally homeomorphic to

$$\mathrm{Map}_*\left(T^k_n, \mathrm{Map}_*(S^{\infty n}, S^{\infty n} \wedge X^{\wedge n} \wedge E\Sigma_{n+})\right)^{\Sigma_n}. \tag{4.3}$$

We rewrite this as

$$\mathrm{Map}_*\left(T^k_n, Q_{\Sigma_n}(X^{\wedge n} \wedge E\Sigma_{n+})\right)^{\Sigma_n}.$$

The notation $Q_{\Sigma_n}(-)$ is more or less standard in equivariant stable homotopy. It signifies that the stabilization is performed with respect to a Σ_n-universe. Notice, however, that our universe is not complete in the sense of [**15**] - it contains multiples of the standard representation of Σ_n, rather than the regular representation. However, Σ_n embeds into this universe as a Σ_n-set, and therefore, either by [**14**] or more simply by duality on manifolds, the Adams isomorphism, which we need for the final step of the proof, still holds. It is easy to see that the coface maps in terms of (4.3) are dual to the face maps in T_n^k and therefore the total space of (4.2) is given by

$$\mathrm{Map}_*\left(K_n, Q_{\Sigma_n}(X^{\wedge n} \wedge E\Sigma_{n+})\right)^{\Sigma_n}.$$

By the Adams isomorphism (for incomplete universes) [**14**, Theorem 9.2], there is an equivalence:

$$\mathrm{Map}_*\left(K_n, Q_{\Sigma_n}(X^{\wedge n} \wedge E\Sigma_{n+})\right)^{\Sigma_n} \simeq \Omega^\infty\left(\mathrm{Map}_*(K_n, \Sigma^\infty X^{\wedge n})_{h\Sigma_n}\right).$$

5 Convergence

In this section P_nQ^k and D_nQ^k can stand for any model which strictly commutes with the coface maps in the BK cosimplicial space. Let $P_n\mathrm{BK}^\bullet$ denote the semi-cosimplicial space obtained by applying P_n to the BK cosimplicial space degree-wise. Let $P_n\mathrm{BK}(X)$ be the total space of $P_n\mathrm{BK}^\bullet(X)$. We want to show that $P_n\mathrm{BK}(X)$ is a model for $P_n\mathrm{Id}(X)$. As a functor of X, $P_n\mathrm{BK}(-)$ is n-excisive, being a homotopy inverse limit of n-excisive functors. There is a natural map $X \to P_n\mathrm{BK}(X)$, and what remains to show is that Id and $P_n\mathrm{BK}$ *agree to n-th order* in the terminology of [**11**] (this would imply, by results of [**10, 11**], that this map is in fact $(n+1)c+1$ connected if X is c-connected). To this end, consider the Bousfield-Kan tower

$$\begin{array}{lcl}
L_S X & & \\
\downarrow & & \\
\vdots & & \\
\downarrow & & \\
E_i(X) & \longleftarrow & \Omega^{i-1}F_i(X) \\
{\scriptstyle f_i}\downarrow & & \\
E_{i-1}(X) & \longleftarrow & \Omega^{i-2}F_{i-1}(X) \\
{\scriptstyle f_{i-1}}\downarrow & & \\
\vdots & & \\
\downarrow & & \\
E_1(X) & \longleftarrow & F_1(X) \\
{\scriptstyle f_1}\downarrow & & \\
* & &
\end{array}$$

Let us examine the fibers $\Omega^{i-1}F_i(X)$ more closely. It can be easily seen that
$F_1(X) = QX$
$F_2(X) =$ homotopy fiber $(QQX \to QX)$

$$F_3(X) = \text{iterated homotopy fiber} \left\{ \begin{array}{ccc} QQQX & \to & QQX \\ \downarrow & & \downarrow \\ QQX & \to & QX \end{array} \right\}$$

$\vdots$

etc.

A tedious, but straightforward combinatorial analysis of higher iterates of QX leads to the following proposition:

Proposition 5.1 *If X is c-connected then $\Omega^{i-1}F_i(X)$ is ic-connected.*

In general, any homotopy functor F which takes c-connected spaces to ic-connected spaces passes this property to its cross-effects and hence has vanishing n-th differentials for all $n < i$. In particular, this implies that P_nF is contractible for all $n < i$ and thus $P_n\Omega^{i-1}F_i(X)$ is contractible for $i > n$. From this, and the fact that P_n commutes up to homotopy with finite homotopy inverse limits of functors, it follows that the tower of fibrations associated with $P_n\mathrm{BK}^\bullet$ becomes homotopy constant after the n-th stage and that its homotopy inverse limit is equivalent to P_nE_n. On the other hand, it follows from Proposition 5.1 that the map $L_{\mathbf{S}}X \to E_iX$ is $(i+1)c+1$ connected. Thus, the functors $L_{\mathbf{S}}$ and E_i agree to i-th order and therefore the induced map $P_nL_{\mathbf{S}} \to P_nE_i$ is a weak homotopy equivalence of functors for $n \leq i$. In particular, $P_nL_{\mathbf{S}} \to P_nE_n$ is a weak equivalence. Thus $P_n\mathrm{BK}$ and $P_nL_{\mathbf{S}}$ are both equivalent to $P_n\mathrm{E}_n$ and thus are equivalent to each other. Since $P_nL_{\mathbf{S}}$ is equivalent to $P_n\mathrm{Id}$, the claim follows.

We make a remark on convergence of the Goodwillie tower of the identity. It is known, essentially by the theory of configuration space models and their stable splittings, that the Goodwillie tower of Q^iX converges for all connected X. In the terminology of [**10**], Q^i is zero analytic. It follows that E_i, being homotopy equivalent to a finite homotopy inverse limit of zero analytic functors, is zero analytic. We claim that $L_{\mathbf{S}}$ is zero analytic, i.e. that the map $L_{\mathbf{S}}X \to \operatorname{ho}\lim_{n\to\infty} P_nL_{\mathbf{S}}X$ is a weak equivalence for all connected X. Indeed, consider the double inverse limit system $\{P_nE_i\}_{n,i\to\infty}$, the statement then follows from the following chain of equivalences

$$\operatorname{ho}\lim_{n\to\infty} P_nL_{\mathbf{S}} \simeq \operatorname{ho}\lim_{n\to\infty} P_n\operatorname{ho}\lim_{i\to\infty} E_i \simeq \operatorname{ho}\lim_{n\to\infty} \operatorname{ho}\lim_{i\to\infty} P_nE_i \simeq \operatorname{ho}\lim_{i\to\infty} \operatorname{ho}\lim_{n\to\infty} P_nE_i \simeq$$

$$\simeq \operatorname{ho}\lim_{i\to\infty} E_i \simeq L_{\mathbf{S}}.$$

Here the second equivalence follows from the fact that

$$P_n\operatorname{ho}\lim_{i\to\infty} E_i \simeq P_nE_n \simeq \operatorname{ho}\lim_{i\to\infty} P_nE_i$$

and the second to the last equivalence follows from the fact that E_i is 0-analytic (strictly speaking, it is only an equivalence when evaluated on connected spaces). The other equivalences are formal. Thus, if we identify $P_n\mathrm{Id}$ with $P_nL_{\mathbf{S}}$, the Goodwillie tower of the identity always converges to the $\mathbf{S}$-localization of X. In particular, it converges to the identity for nilpotent, and not only simply connected X.

Acknowledgements

It was M. Hopkins who first suggested, in a conversation with the first author, that one should be able to learn something useful about the Goodwillie tower of the identity by considering the Bousfield-Kan tower. Later the same suggestion was made by G. Carlsson.

We would like to thank Randy McCarthy for reading the paper carefully and providing many constructive criticisms.

References

[1] Arone, G. Z., *A generalization of Snaith-type splittings*, preprint.

[2] Arone, G. Z. and Mahowald, M., *The Goodwillie tower of the identity functor and the unstable periodic homotopy of spheres,* preprint, available on the Hopf archive.

[3] Bökstedt, M., *Topological Hochshild homology,* preprint, Bielefeld.

[4] Bousfield, A. K. and Kan, D. M. [1972], *Homotopy limits, completions and localizations,* Lecture Notes in Math. **304**.

[5] Cohen, F. R., Lada, T. J. and May, J. P. [1976], *The homology of iterated loop spaces,* Lecture Notes in Math. 533.

[6] Cohen, F. R. and Taylor, L. R. [1993], *On the representation theory associated to the cohomology of configuration spaces,* Contemp. Math. **146**, 91-109.

[7] Cohen, F. R. and Taylor, L. R. [1978], *Computations of Gelfand-Fuks cohomology, the cohomology of function spaces and the cohomology of configuration spaces,* in "Geometric applications of homotopy theory I", Lecture Notes in Math. **657**, 106-143.

[8] Dwyer, W. G. and Kan, D. M. [1983], *Function complexes for diagrams of simplicial sets,* Indagationes Mathematicae **45**, 139-147.

[9] Goodwillie, T. G. [1990], *Calculus I: the first derivative of pseudoisotopy theory,* K-Theory **4**, 1-27.

[10] Goodwillie, T. G. [1992], *Calculus II: analytic functors,* K-Theory **5**, 295-332.

[11] Goodwillie, T. G., *Calculus III: the Taylor series of a homotopy functor,* preprint.

[12] Johnson, B. [1995], *The derivatives of homotopy theory,* Trans. Amer. Math. Soc. **347**, no. 4, 1295-1321.

[13] Kuhn, N. J. [1987], *The transfer and James-Hopf invariants,* Math. Z. **196**, 391-405.

[14] Lewis, G., *Splitting theorems for certain equivariant spectra,* preprint, available on the Hopf archive.

[15] Lewis, G., May, J. P. and Steinberger, M. [1986], *Equivariant stable homotopy theory,* Lecture Notes in Math. 1213.

[16] Mac Lane, S. [1971], *Categories for the working mathematician,* Springer-Verlag.

[17] May, J. P. [1972], *The geometry of iterated loop spaces,* Lecture Notes in Math. **271**.

Fields Institute Communications
Volume 19, 1998

Some Remarks on the Root Invariant

Robert R. Bruner
Mathematics Department
Wayne State University
Detroit, Michigan 48202 USA
rrb@math.wayne.edu

Abstract. We show how the root invariant of a product depends upon the product of the root invariants, give some examples of the equivariant definition of the root invariant, and verify a weakened form of the algebraic Bredon-Löffler conjecture .

1 The Definition and Some Examples

One of the most pleasing aspects of the equivariant point of view is the fact that concepts which are obscure non-equivariantly sometimes have quite clear equivariant meaning. Greenlees' observation [**3**] that Bredon's filtration of stable homotopy by restrictions of equivariant maps is the same as Mahowald's root invariant filtration is a good example, as it leads to an elementary definition of the root invariant as follows.

Let $G = Z/2$, let $\mathbb{R}^{n+k\xi}$ be the G-representation which is trivial on n coordinates and negation on k coordinates, and let $S^{n+k\xi}$ be the one point compactification of $\mathbb{R}^{n+k\xi}$. Let

$$\phi_k : [S^{k\xi}, S^0]_n^G \longrightarrow [S^0, S^0]_n = \pi_n S^0$$

and

$$U_k : [S^{k\xi}, S^0]_n^G \longrightarrow [S^k, S^0]_n = \pi_{n+k} S^0$$

be the fixed point and underlying map homomorphisms, respectively. Then [**3**, Proposition 2.5] shows that if k is maximal such that $x \in \pi_n S^0$ is in $\mathrm{Im}(\phi_k)$, then

$$R(x) = U_k(\phi_k^{-1}(x)).$$

In other words, to compute $R(x)$, extend $x : S^n \longrightarrow S^0$ to an equivariant map $\widetilde{x} : S^{n+k\xi} \longrightarrow S^0$ with k maximal. Then $R(x)$ contains the underlying map $U_k(\widetilde{x}) : S^{n+k} \longrightarrow S^0$. Note that Lin's theorem, that $S^{-1} \simeq \lim_{\longleftarrow k} P_{-k}$, is not required for this definition.

1991 *Mathematics Subject Classification.* Primary 55Q45, 55Q35, 55Q91; Secondary 55P42, 55P91, 55T15.

The simplicity of this definition suggests that we should easily be able to see some examples. The Hopf maps bear this out. Let D be one of the division algebras $\mathbb{R}$, $\mathbb{C}$, $\mathbb{H}$ or the Cayley numbers $\mathcal{C}$, and let $d_D = 1, 2, 4$, or 8, respectively. The associated Hopf map $h_D : S^{2d-1} \longrightarrow DP^1 \cong S^d$ given by $h(z_1, z_2) = z_1 z_2^{-1}$ is 2, η, ν, or σ, respectively. Each of these division algebras is two-dimensional over the preceding one in the sequence. If we write such a pair as $D_1 \subset D_2$, then the elements of D_2 may be written $a + be$, with $a, b \in D_1$, $e \in D_2 - D_1$, and $e^2 = -1$. We may then give D_2 the G-action $a + be \mapsto a - be$, with respect to which h_{D_2} is then an equivariant map $S^{2d-1+2d\xi} \longrightarrow S^{d+d\xi}$ whose restriction to the fixed points is h_{D_1}. This shows that $R(2) = \eta$, $R(\eta) = \nu$, and $R(\nu) = \sigma$, once we verify that these extensions are maximal, which we will do in Corollary 2.3.

2 The Cartan Formula

The equivariant definition also allows an elementary proof of the Cartan formula, independent of the theory worked out in [5], which we present now.

Theorem 2.1 *Let $\alpha_i \in \pi_{n_i} S^0$ and $R(\alpha_i) \in \pi_{n_i+k_i} S^0$, for $i = 1, 2$. Let $k = k_1 + k_2$ and let $i : S^{-k-1} \longrightarrow P_{-k-1}$ be the inclusion of the bottom cell of the stunted projective space P_{-k-1}.*

1. *If $i_*(R(\alpha_1)R(\alpha_2)) \neq 0$ then $R(\alpha_1)R(\alpha_2) \subset R(\alpha_1\alpha_2)$.*
2. *If $i_*(R(\alpha_1)R(\alpha_2)) = 0$ then $R(\alpha_1\alpha_2)$ lies in a higher stem than does $R(\alpha_1)R(\alpha_2)$.*

Proof Let $\widetilde{\alpha_i} : S^{n_i+k_i\xi} \longrightarrow S^0$ be a maximal extension of α_i for each i. Then $\widetilde{\alpha} = \widetilde{\alpha_1} \wedge \widetilde{\alpha_2}$ is an extension of $\alpha = \alpha_1 \wedge \alpha_2$ and, if it is maximal then $R(\alpha_1\alpha_2)$ contains $U_k(\widetilde{\alpha}) = U_k(\widetilde{\alpha_1})U_k(\widetilde{\alpha_2}) = R(\alpha_1)R(\alpha_2)$. To determine whether or not it is maximal, we must analyze the relation between ϕ_k and ϕ_{k+1}.

Recall that the fixed point homomorphism $[S^{k\xi}, S^{\infty\xi}]_n^G \longrightarrow [S^0, S^0]_n$ is an isomorphism by elementary equivariant obstruction theory, so the inclusion of fixed points $i : S^0 \longrightarrow S^{\infty\xi}$ induces the fixed point homomorphism ϕ_k.

$$\begin{array}{ccc} [S^{k\xi}, S^0]_n^G & \xrightarrow{\ i_*\ } & [S^{k\xi}, S^{\infty\xi}]_n^G \\ & \searrow^{\phi_k} & \downarrow \cong \\ & & [S^0, S^0]_n \end{array}$$

The G-equivariant cofiber sequence $EG_+ \longrightarrow S^0 \longrightarrow S^{\infty\xi}$ allows us to embed ϕ_k in a long exact sequence. Then, the cofiber sequence

$$S^k \wedge G_+ \longrightarrow S^{k\xi} \longrightarrow S^{(k+1)\xi},$$

allows us to compare ϕ_k and ϕ_{k+1}. Using the adjunction isomorphism $[S^k \wedge G_+, EG_+]_n^G \cong [S^k, EG_+]_n$, and the isomorphism $[S^k, EG_+]_n \cong [S^0, S^0]_n$ induced by the nonequivariant equivalence from EG_+ to S^0, we obtain the following braid of long exact sequences relating ϕ_k and ϕ_{k+1}. (This is a piece of the diagram used in [3] to show the equivalence of the two definitions of the root invariant.)

$$
\begin{array}{ccccccc}
 & & [S^{(k+1)\xi}, S^0]^G_n & \xrightarrow{\phi_{k+1}} & [S^0, S^0]_n & & \\
 & \nearrow & & \searrow \quad \nearrow{\scriptstyle \phi_k} & & \searrow & \\
[S^{(k+1)\xi}, EG_+]^G_n & & & [S^{k\xi}, S^0]^G_n & & & [S^{(k+1)\xi}, EG_+]^G_{n-1} \\
 & \searrow & & \nearrow \quad \searrow{\scriptstyle U_k} & & \nearrow{\scriptstyle \partial_k} & \\
 & & [S^{k\xi}, EG_+]^G_n & \longrightarrow & [S^k, EG_+]_n & &
\end{array}
$$

Commutativity of the right diamond shows that the obstruction to $\phi_k(\tilde{\alpha})$ being in the image of ϕ_{k+1} is exactly $\partial_k U_k(\tilde{\alpha})$. Thus, cases (i) and (ii) of the theorem are distinguished by the nontriviality or triviality, respectively, of the composite

$$S^{(k+1)\xi} \longrightarrow S^k \wedge G_+ \longrightarrow EG_+$$

of the boundary map of the cofiber sequence and the free G-map induced by $U_k(\tilde{\alpha})$. To express this in more familiar terms, observe that naturality of Adams' isomorphism gives a commutative square

$$
\begin{array}{ccccc}
[S^k, EG_+]_n & \xrightarrow{\cong} & [S^k, S^0]_n & \xrightarrow{\cong} & [S^0, S^{-k}]_n \\
\downarrow{\scriptstyle \partial_k} & & & & \downarrow{\scriptstyle i_*} \\
[S^{(k+1)\xi}, EG_+]^G_{n-1} & \xrightarrow{\cong} & [S^0, EG_+ \wedge S^{-(k+1)\xi}]^G_{n-1} & \xrightarrow{\cong} & [S^0, P_{-k-1}]_{n-1}
\end{array}
$$

where $i : S^{-k} = \Sigma S^{-k-1} \longrightarrow \Sigma P_{-k-1}$ is inclusion of the bottom cell. This implies the form of the theorem we have stated. In fact, we have proved the following.

Proposition 2.2 *Suppose $\phi_k(\tilde{\alpha}) = \alpha$. Then $\tilde{\alpha}$ is maximal, and hence $R(\alpha) = U_k(\tilde{\alpha})$, if and only if $0 \neq i_* U_k(\tilde{\alpha}) \in \pi_{n-1} P_{-k-1}$.*

Corollary 2.3 *$R(2) = \eta$, $R(\eta) = \nu$, and $R(\nu) = \sigma$.*

Proof Given the extensions of 2, η and ν produced in Section 1, we need only verify that η, ν, and σ are nontrivial on the bottom cells of P_{-2}, P_{-3}, and P_{-5}, a task which is easily accomplished.

Previous versions of the Cartan formula have appeared in [**9**, **12**], with the condition for $R(\alpha_1)R(\alpha_2)$ to be contained in $R(\alpha_1\alpha_2)$ stated as $R(\alpha_1)R(\alpha_2) \neq 0$ rather than $i_*(R(\alpha_1)R(\alpha_2)) \neq 0$.

Here is an example which can be found in [**10**]. If $\mu \in \pi_8 S$ is detected by Ph_1, and $\bar{\kappa} \in \pi_{20} S$ detected by $g \in \mathrm{Ext}^{4,24}$ is chosen correctly then we have $R(\mu) = \nu\bar{\kappa}$. To compute $R(\eta\mu)$ we thus consider the product $R(\eta)R(\mu) = \nu^2\bar{\kappa}$. This is nonzero in $\pi_{26} S = \pi_9 S^{-17}$. However $i_*(\nu^2\bar{\kappa}) = 0$ in $\pi_9 P_{-17}$, so we conclude that $R(\eta\mu)$ lies in a stem higher than π_{26}.

The algebraic root invariant, discussed in the next section, exhibits the same behavior. There, we have $R(h_1) = h_2$, $R(Ph_1) = h_2 g$, and $R(h_1)R(Ph_1) = h_2^2 g \neq 0$ in $\mathrm{Ext}(F_2, F_2)$. However,

$$i_* : \mathrm{Ext}(F_2, F_2) \longrightarrow \mathrm{Ext}(H^* P_{-17}, F_2)$$

sends $h_2^2 g$ to zero, so the algebraic root invariant $R(h_1 Ph_1)$ lies in a higher stem as well. In fact, we can calculate that $R(h_1 Ph_1) = r$, which lies in the 30-stem and "lives on" the bottom cell of P_{-21}. These calculations were done by computing the induced maps of Adams spectral sequences for the spectra involved, using the programs described in [**1**] and [**2**]. It was the anomolous behavior of $R(h_1Ph_1)$ in those calculations which alerted me to the correct formulation of the Cartan formula.

In terms of the root invariant spectral sequence of [**5**], the 'exceptional' behavior in case (ii) of the theorem is the usual behavior of products in an associated graded when there are filtration shifts.

The condition for maximality allows us to make systematic conclusions, along the lines of [**11**], generalizing the example above. Let us write $|x| = n$ if $x \in \pi_n S^0$.

Corollary 2.4 *For any* $x \in \pi_* S^0$,

1. $|R(2x)| > 1 + |R(x)|$ *if* $|R(x)| - |x| \equiv -1$ *(mod 4)*,
2. $|R(\eta x)| > 3 + |R(x)|$ *if* $|R(x)| - |x| \equiv -2$ *(mod 8)*,
3. $|R(\nu x)| > 7 + |R(x)|$ *if* $|R(x)| - |x| \equiv -4$ *(mod 16)*.

Proof The map η is trivial on the bottom cell of P_{-j-2} if $-j-1 \equiv -1$ (mod 4). Similarly, ν is trivial on the bottom cell of P_{-j-3} if $-j-3 \equiv -1$ (mod 8), and σ is trivial on the bottom cell of P_{-j-5} if $-j-5 \equiv -1$ (mod 16).

3 The Algebraic Bredon-Löffler Conjecture

As noted in [**3**], the equivariant definition of the root invariant also allows a very simple proof of Jones' [**6**] lower bound

$$|R(x)| \geq 2|x|.$$

Namely, any $x : S^n \longrightarrow S^0$ occurs as the fixed points of $x \wedge x : S^{n+n\xi} = S^n \wedge S^n \longrightarrow S^0 \wedge S^0 = S^0$, where the smash products are given the action which interchanges factors. The *Bredon-Löffler Conjecture* is the upper bound

$$|R(x)| \leq 3|x|$$

when $|x| > 0$. By [**3**], this is equivalent to the assertion that the map $\eta_k : S^0 \longrightarrow \Sigma P_{-k}$ induces a monomorphism of π_j for $0 < j < k/2$. The Adams spectral sequence gives an algebraic analog, which we call the *Algebraic Bredon-Löffler Conjecture*, namely, that

$$\eta_k^* : \mathrm{Ext}_A^{s,t}(F_2, F_2) \longrightarrow \mathrm{Ext}_A^{s,t}(H^*\Sigma P_{-k}, F_2)$$

is a monomorphism for $0 < t - s < k/2$. We are able to prove the following much weaker assertion.

Theorem 3.1 η_k^* *is a monomorphism for* $0 < t - s < -2 + \sqrt{3 + k/2}$.

Proof Let ρ_r be the natural transformation $\mathrm{Ext}_A \longrightarrow \mathrm{Ext}_{A_r}$ induced by the inclusion $A_r \longrightarrow A$ of the subalgebra generated by $Sq^1, \ldots, Sq^{2^r}$. We will show that $\rho_r \eta_k^*$ is a monomorphism of $\mathrm{Ext}_A^{s,t}$ in the stated range if r is minimal such that $n = t - s < 2^r - 1$. We shall use several facts from [**8**]. Let $L = F_2[x, x^{-1}]$ and let $j_k : L_{-k} = H^*P_{-k} \longrightarrow L$ be the obvious inclusion. Let $\lambda : \Sigma L \longrightarrow F_2$ be the coefficient of x^{-1}. It is easy to see that $\lambda j_k = \eta_k^* : \Sigma L_{-k} \longrightarrow F_2$. Further, we have:

1. The induced homomorphism $\lambda^* : \mathrm{Ext}_A(F_2, F_2) \longrightarrow \mathrm{Ext}_A(\Sigma L, F_2)$ is an isomorphism [**8**, Theorem 1.1].
2. There is an isomorphism

$$\mathrm{Ext}_{A_r}(\Sigma L, F_2) \cong \bigoplus_{j \in \mathbb{Z}} \mathrm{Ext}_{A_{r-1}}(\Sigma^{j2^{r+1}} F_2, F_2).$$

See [**8**, Theorem 1.1] and [**4**, Theorem 2.1].

3. The lower left square in the following diagram commutes. That is, the preceding isomorphism composed with λ_* and projected onto the zeroth component, is induced by the inclusion $A_{r-1} \longrightarrow A_r$ [8, Lemma 1.6].

$$
\begin{array}{ccccc}
\operatorname{Ext}_A(F_2,F_2) & \xrightarrow[\lambda^*]{\cong} & \operatorname{Ext}_A(\Sigma L,F_2) & \xrightarrow[j_k^*]{} & \operatorname{Ext}_A(\Sigma L_{-k},F_2) \\
\downarrow \rho_r & & \downarrow \rho_r & & \downarrow \rho_r \\
\operatorname{Ext}_{A_r}(F_2,F_2) & \xrightarrow[\lambda^*]{} & \operatorname{Ext}_{A_r}(\Sigma L,F_2) & \xrightarrow[j_k^*]{} & \operatorname{Ext}_{A_r}(\Sigma L_{-k},F_2) \\
\downarrow & & \downarrow \cong & & \\
\operatorname{Ext}_{A_{r-1}}(F_2,F_2) & \xleftarrow{\pi_0} & \bigoplus_{j\in\mathbb{Z}} \operatorname{Ext}_{A_{r-1}}(\Sigma^{j2^{r+1}}F_2,F_2) & &
\end{array}
$$

(with $\eta_k^*: \operatorname{Ext}_A(F_2,F_2) \to \operatorname{Ext}_A(\Sigma L_{-k},F_2)$ along the top, and $\rho_{r-1}: \operatorname{Ext}_A(F_2,F_2) \to \operatorname{Ext}_{A_{r-1}}(F_2,F_2)$ along the left)

The left hand map ρ_{r-1} is a monomorphism on $\operatorname{Ext}_A^{s,t}(F_2,F_2)$ since $n = t-s < 2^r - 1$. Hence $\rho_r\lambda^*$ is as well. Now Adams' vanishing line implies that $\operatorname{Ext}_A^{s,t}(F_2,F_2) = 0$ unless $s < (n+4)/2$, so we will be done if we can show that the lower map j_k^* is a monomorphism in these filtrations. This will be true if $s < (n+k)/(2^{r+1}-2)$, because then $\operatorname{Ext}_{A_r}^{s,t}(\Sigma(L/L_{-k}),F_2) = 0$. This follows by filtering L/L_{-k} by degrees and using the fact that $\operatorname{Ext}_{A_r}^{s,t}(F_2,F_2) = 0$ if $s < n/(2^{r+1}-2)$ by the May spectral sequence, c.f. [4, proof of Theorem 2.1]. Thus, it suffices to have

$$\frac{n+4}{2} < \frac{n+k}{2^{r+1}-2}.$$

Since r is minimal with respect to $n < 2^r - 1$, we have $2^r - 1 \le 2n$, so it suffices that $2n^2 + 7n < k$, which holds if $n < -2 + \sqrt{3 + k/2}$.

Of course this is far from the actual conjecture, but is a first step toward it. In fact, the calculations in [3] suggest the following sharpening of the algebraic Bredon-Löffler conjecture.

Conjecture 3.2 *(Strong Algebraic Bredon-Löffler Conjecture) The map η_k^* is a monomorphism if*

$$s < (k-n)/2.$$

This conjecture is based on calculations of root invariants [3]. The sparsity of elements in some bidegrees introduces an element of uncertainty in the intercept, but the slope of $-1/2$ and the approximate intercept $n = k$ are clearly evident for $n < 40$.

Adams' vanishing line intersects this line at about $n = k/2$, so this conjecture implies the algebraic Bredon-Löffler conjecture (see Figure 1). Since $k/2$ is the last entire stem which is mapped monomorphically, this is the dimension one sees in homotopy.

This conjecture also correctly predicts that part of the 0-stem which maps monomorphically according to Landweber [7], unlike the algebraic Bredon-Löffler conjecture which says nothing about the 0-stem.

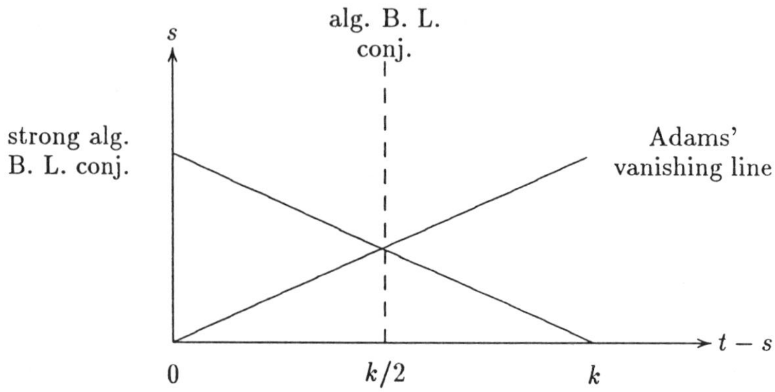

Figure 1 Conjectured limits to the range in which η_k^* is a monomorphism.

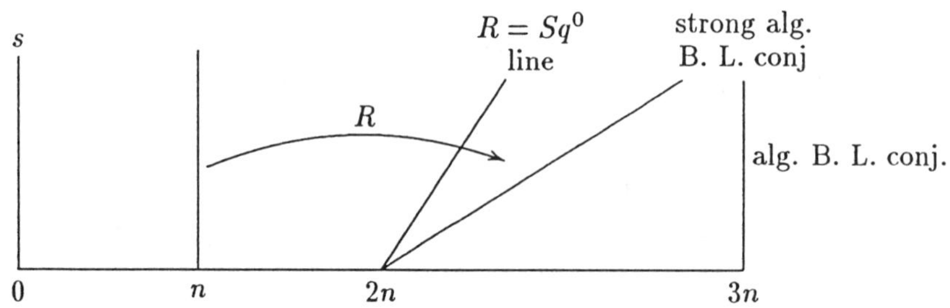

Figure 2 Root invariants of the n-stem

Finally, note that this confines root invariants to a narrow band,

$$2|x| + s \ \leq\ |R(x)| \ \leq\ 2|x| + 2s,$$

as in Figure 2. Here, the lower bound follows from the result of [**9**, 2.5], that $R(x) = Sq^0(x)$ (*if* the latter is nonzero on the appropriate cell of projective space), and the upper bound is equivalent to the strong algebraic Bredon-Löffler conjecture.

Acknowledgements

These remarks were worked out during the Stable Homotopy Theory Workshop at The Fields Institute in Toronto, Canada during January of 1996. The author would like to thank the organizers and The Fields Institute for support and for an environment conducive to doing mathematics.

References

[1] Bruner, R. R. [1989], *Calculation of large Ext modules*, Computers in Geometry and Topology (M. C. Tangora, ed.), Marcel Dekker, New York, 79-104.

[2] Bruner, R. R. [1993], *Ext in the nineties*, in Algebraic Topology, Oaxtepec 1991, pp. 71-90, Contemp. Math. **146**, Amer. Math. Soc., Providence, R.I.

[3] Bruner, R. R. and Greenlees, J. [1995], *The Bredon-Löffler conjecture*, Exper. Math. **4**, 101-109.

[4] Davis, D. M. [1981], *An infinite family in the cohomology of the Steenrod algebra*, J. Pure Appl. Alg. **21**, 145-150.

[5] Greenlees, J. P. C. and May, J. P. [1995], *Generalized Tate cohomology*, Mem. Amer. Math. Soc. **543**.

[6] Jones, J. D. S. [1985], *Root invariants, cup-r products and the Kahn-Priddy theorem*, Bull. London Math. Soc. **17**, 479-483.

[7] Landweber, P. S. [1969], *On equivariant maps between spheres*, Ann. Math. **89**, 125-137.

[8] Lin, W. H., Davis, D., Mahowald, M. E. and Adams, J. F. [1980], *Calculations of Lin's Ext groups*, Math. Proc. Camb. Phil. Soc. **87**, 459-469.

[9] Mahowald, M. E. and Ravenel, D. C. [1993], *The root invariant in homotopy theory*, Topology **32**, 865-898.

[10] Mahowald, M. E. and Shick, P. [1983], *Some root invariants in* $\mathrm{Ext}_A(Z/2, Z/2)$, Proceedings of the Northwestern Homotopy Theory Conference, Contemp. Math. **19**, 227-231.

[11] Sadofsky, H. [1992], *The root invariant and* v_1*-periodic families*, Topology **31**, 65-112.

[12] Shick, P. [1984], *Periodic Phenomena in the Adams Spectral Sequence*, Thesis, Northwestern University.

Fields Institute Communications
Volume **19**, 1998

On the Lusternik-Schnirelmann Category of an Iterated Loop Space

F. R. Cohen
Department of Mathematics
University of Rochester
Rochester, NY 14627 USA
cohf@math.rochester.edu

Let X be a space which has the homotopy type of a CW complex and let $\Omega_0^j X$ denote the component of the constant map in the jth pointed loop space of X. A question posed by Ed Fadell and Viera Benci is whether $\Omega_0^j X$ has infinite Lusternik-Schnirelmann category for simply-connected finite complexes X (which are non-trivial).

Theorem 1 *If X is a simply-connected finite complex which is not contractible, then the Lusternik-Schnirelmann category of $\Omega_0^j X$ is infinite for $j \geq 1$.*

It seems likely that the Lusternik-Schnirelmann category of the pointed function space $\mathrm{map}_*(M, X)$ is infinite for simply-connected finite complexes M and X for which the reduced mod$-p$ homology groups of M and X are both non-trivial. Evidence for this statement is given in the work of Felix and Thomas [**2**]. However, the proof of Theorem 1 breaks down in this generality.

The proof of Theorem 1 gives a slightly stronger conclusion than that stated. For example, if X is a simply-connected space of finite type, and quasi-bounded over the mod p Steenrod algebra such that for any integer q, there is an integer $n(q) > q$ with $\pi_{n(q)}(X_{(p)}) \neq 0$, then the Lusternik-Schnirelmann category of $\Omega_0^j(X)$ is never finite for $j \geq 2$. Some examples of such spaces X are of course simply-connected closed manifolds.

Assume that X is a simply-connected finite complex which is not contractible, and thus $\bar{H}^*(X; \mathbf{F}_p)$ is non-trivial for some prime p. Fix such a p throughout the remainder of this note. The proof of Theorem follows from the following 3 lemmas.

Lemma 2 *If $j \geq 1$, then $H^*(\Omega_0^j X; \mathbf{F}_p)$ is isomorphic to a tensor product of exterior algebras and/or truncated polynomial algebras.*

Lemma 3 *The algebra $H^*(\Omega X; \mathbf{F}_p)$ is not finitely generated. Hence the cup-length for $H^*(\Omega X; \mathbf{F}_p)$ is not finite.*

1991 *Mathematics Subject Classification.* 55M30, 55P35.

Partially supported by the NSF.

Lemma 4 *If $j \geq 2$, the algebra $H^*(\Omega_0^j X; \mathbf{F}_p)$ is not finitely generated. Hence the cup-length for $H^*(\Omega_0^j X; \mathbf{F}_p)$ is not finite.*

Proof of Theorem 1 If $j \geq 1$, the cup-length for $H^*(\Omega_0^j X; \mathbf{F}_p)$ is never finite by (3) and (4). The theorem follows as the finiteness of the Lusternik-Schnirelmann category of $\Omega_0^j X$ implies finite cup-length. □

Sketch of Lemma 2. The mod$-p$ cohomology ring of $\Omega_0^j X$ is quasi-bounded over the mod$-p$ Steenrod algebra by [**5**]. Thus $H^*(\Omega_0^j X; \mathbf{F}_p)$ does not contain an element of infinite height. The lemma follows from the Borel structure theorem.

Sketch of Lemma 3. If $H^*(\Omega X; \mathbf{F}_p)$ were finitely generated as an algebra, then it is totally finite dimensional as a vector space by Lemma 2. Let f denote a nonzero cohomology class of maximal degree in $H^*(\Omega X; \mathbf{F}_p)$. As X is finite dimensional, let b denote a nonzero cohomology class in $H^*(X; \mathbf{F}_p)$ of maximal degree. Notice that $b \otimes f$ survives to give a non-trivial class in the Serre spectral sequence abutting to $H^*(PX; \mathbf{F}_p)$ as X is simply-connected where PX denotes the path space over X. This is a contradiction and so $H^*(\Omega X; \mathbf{F}_p)$ is not finitely generated as an algebra. Thus the cup-length is not finite by Lemma 2.

Proof of Lemma 4 Two proofs are given below. The first proof applies to the case $p = 2$.

If $j \geq 2$, then $\Omega_0^j X$ is a homotopy abelian H-space. Thus if $H^*(\Omega_0^j X; \mathbf{F}_2)$ is finitely generated, then it is a totally finite dimensional vector space and, by Hubbuck's Theorem ([**3**][**6**]), $\Omega_0^j X$ is a torus (localized at 2). Thus the homotopy groups of $\Omega_0^j X_{(2)}$ quit. This contradicts Serre's theorem [**7**].

The argument in case p is greater than 2 is somewhat different, but also applies to the case $p = 2$. Note that by the above remarks, it suffices to see that $H^*(\Omega_0^j X; \mathbf{F}_p)$ is not a finite dimensional vector space for $j \geq 2$.

Thus suppose that (i) $H^*(\Omega_0^j X; \mathbf{F}_p)$ is a finite dimensional vector space and (ii) $j \geq 2$. By the definitions in [**1**], $H^*(\Omega_0^j X; \mathbf{F}_p)$ is a p-compact group. Thus there is an essential map

$$s : B\mathbf{Z}/p \to \Omega^{j-1} X\langle j \rangle$$

by [**1**]. Furthermore, this last map composed with the natural map $\Omega^{j-1} X\langle j \rangle \to \Omega^{j-1} X\langle j-1 \rangle$ is null by [**4**] and [**5**] because every element in the mod-p cohomology ring of $\Omega^{j-1} X\langle j-1 \rangle$ has finite height.

Next, consider the fibration

$$X\langle j \rangle \to X\langle j-1 \rangle \xrightarrow{q} K(\pi, j)$$

where (iii) $X\langle j \rangle$ denotes the j-connected cover of X, and (iv) q induces an isomorphism on the level of jth homotopy groups with π assumed to be non-trivial (as the case for which this group is trivial is immediate). There is an associated fibration

$$K(\pi, 1) \to \Omega^{j-2} X\langle j \rangle \to \Omega^{j-2} X\langle j-1 \rangle \ .$$

Since $j \geq 2$, passage to the adjoint of s gives

$$\sigma : \Sigma B\mathbf{Z}/p \to \Omega^{j-2} X\langle j \rangle$$

which is essential, but is null in the base. Thus, this map lifts to an essential map

$$\tau : \Sigma B\mathbf{Z}/p \to K(\pi, 1) \ .$$

Since $j \geq 2$, and X is of finite type, π is a finitely generated abelian group. Hence τ is null-homotopic as it is trivial in cohomology by the simple-connectivity of $\Sigma B\mathbf{Z}/p$. This last statement contradicts the fact that s is essential and the lemma follows. □

Acknowledgement

The author would like to thank The Fields Institute for providing a fertile mathematical environment.

References

[1] Dwyer, W. G. and Wilkerson, C. W. [1994], *Homotopy fixed-point methods for Lie groups and finite loop spaces*, Ann. of Math. **139**, 395–442.

[2] Felix, Y. and Thomas, J.-C. [1993], *Mapping spaces at large primes*, Topology **32**, 198–232.

[3] Hubbuck, J. R. [1969], *On homotopy commutative H-spaces*, Topology **8**, 119–126.

[4] Lannes, J. [1992], *Sur les espaces fonctionnels dont la source est le clasifiant d'un pgroupe abélien élémentaire*, Publ. I.H.E.S. **75** , 135–244.

[5] Lannes, J. and Schwartz, L. [1986], *A propos de conjectures de Serre et Sullivan*, Invent. Math. **83**, 593–603.

[6] Lin, J. P. and Williams, F. [1991], *Homotopy commutative H-spaces*, P.A.M.S. **113**, 857–865.

[7] Serre, J. P. [1953], *Cohomologie modulo 2 des complexes d'Eilenberg-MacLane*, Comm. Math. Helv. **27**, 198–232.

Fields Institute Communications
Volume **19**, 1998

Thom Spectra which are Wedges of Eilenberg-MacLane Spectra

F. R. Cohen
Department of Mathematics
University of Rochester
Rochester, NY 14627 USA
`cohf@math.rochester.edu`

V. V. Vershinin
Sobolev Institute of Mathematics
Novosibirsk, 630090, Russia
`versh@math.nsc.ru`

Abstract. Thom spectra generated by a system of monoids or by a map from a space to BO are considered. Mahowald's theorem that $K\mathbb{Z}/2$ is a Thom spectrum has consequences for many examples within homotopy theory as well as outside of the subject. Examples such as: (1) groups of automorphisms of free groups, (2) the Baez-Birman monoid, and (3) loop spaces over the James stages of a sphere are considered here.

1 Introduction

One of the high points of cobordism theory occurred when René Thom [29] proved that the spectrum of the unoriented cobordism ring is equivalent to an infinite sum of suspensions of the Eilenberg-MacLane spectrum $H\mathbb{Z}/2$. Another important result is Mahowald's theorem which gives a natural way to see that $K\mathbb{Z}/2$ is a Thom spectrum which splits off of MO [18, 19]. It was observed in [8, 6] that the Eilenberg-MacLane spectrum $H\mathbb{Z}/2$ itself can be realized as a spectrum of some cobordism theory. The construction of the corresponding spectrum can be done in two ways the first of which is homotopy theoretic while the second is group theoretic in nature.

The infinite orthogonal group O is an infinite loop space, so it is possible to consider the classifying spaces B^iO for all natural numbers i. Let a map

$$g : S^3 \to B^3O$$

1991 *Mathematics Subject Classification.* 55N22, 20F36, 20F38.
FRC was partially supported by the NSF.
VVV was supported in part by RFFR Grants 96-01-016331 and 96-01-00062G.

represent the generator of $\pi_3(B^3O) \cong \pi_1(BO) \cong \mathbb{Z}/2$. The double loop map of g is the map

$$\Omega^2 g : \Omega^2 S^3 \to BO.$$

Mark Mahowald proved that the Thom spectrum of this map is equivalent to the Eilenberg-MacLane spectrum $H\mathbb{Z}/2$ [18, 19].

Subsequently, one considered the spectrum formed by the Thom spaces $M(Br_n)$, $n = 1, 2, \ldots$, of the bundles over the classifying spaces of the braid groups Br_n, obtained by the maps

$$BBr_n \to BO_n,$$

induced by the canonical homomorphisms

$$Br_n \to O_n.$$

The Thom spectrum here is equivalent to the Eilenberg-MacLane spectrum $H\mathbb{Z}/2$ [8, 6]. These facts are based essentially on results of Arnold [1] and Fuks [15] on cohomology of the braid groups. The connection between these two constructions is given by the theorem of Graeme Segal [26] that plus construction of the classifying space of the infinite braid group BBr is homotopy equivalent to $\Omega^2 S^3$.

Later Bullett [6] proved that the Thom spectrum of the symmetric groups induced by the canonical inclusion

$$\Sigma_n \to O_n$$

is also equivalent to a wedge of suspensions of the Eilenberg-MacLane spectrum $H\mathbb{Z}/2$. This fact was also known to Nigel Ray. It also turned out [31] that for the generalized braid groups in the sense of Brieskorn [5] and Deligne [11] the corresponding Thom spectra are equivalent to bouquets of suspensions of $H\mathbb{Z}/2$.

In the present paper we give natural examples of Thom spectra which are wedges of suspensions of $H\mathbb{Z}/2$. The basic principle here is to use Mahowald's theorem together with classical properties of Hopf algebras. The first series of examples is given by various objects connected with ether the braid group or the symmetric group. These are groups of automorphisms of free groups and the series of Baez–Birman monoids. On the other hand it is possible to use loop spaces $\Omega^n S^{n+1}$ or $\Omega^n J_k S^{n+1}$, $n = 2, 3, \ldots$, where J_k are the stages of James construction for a sphere. One is led to wonder whether these methods can be used to give further information about the homology of these groups and monoids as the situation discovered by Thom, that some Thom spectrum is a wedge of suspensions of the Eilenberg-MacLane spectrum $H\mathbb{Z}/2$ occurs frequently in nature.

2 Thom spectra of systems of monoids

We denote by Br_m the braid group on m strings, by i_m the standard inclusion

$$i_m : Br_m \to Br_{m+1}$$

and by i_{n+m}^m the composition $i_{n+m-1} \ldots i_m$. We denote by $\beta_{l,m}$ the standard morphism of pairing of the groups Br_l and Br_m into the group Br_{l+m}

$$\beta_{l,m} : Br_l \times Br_m \to Br_{l+m}.$$

Suppose that we are given a sequence of finitely presented monoids H_n for $n = 0, 1, 2, \ldots$; and homomorphisms

$$g_n : H_n \to H_{n+1}.$$

We denote by g^n_{n+m} the composition $g_{n+m-1} \ldots g_n$

$$g^n_{n+m} : H_n \to H_{n+m}$$

and by H the direct limit of the system $\{H_n, g_n\}$. Suppose also that for all m and n there exist homomorphisms

$$\nu_{m,n} : Br_m \times H_n \to H_{m+n}$$

such that $\nu_{0,n}(e, x) = Id(x), x \in H_n$ and e is the unit of the group $Br_0 = \{e\}$ and that the following diagrams

$$\begin{array}{ccc} Br_l \times Br_m \times H_n & \xrightarrow{Id \times \nu_{m,n}} & Br_l \times H_{m+n} \\ \downarrow \beta_{l,m} \times Id & & \downarrow \nu_{l,m+n} \\ Br_{l+m} \times H_n & \xrightarrow{\nu_{l+m,n}} & H_{l+m+n}, \end{array}$$

$$\begin{array}{ccc} Br_l \times H_m & \xrightarrow{i^l_{l+k} \times g^m_{m+n}} & Br_{l+k} \times H_{m+n} \\ \downarrow \nu_{l,m} & & \downarrow \nu_{l+k,m+n} \\ H_{l+m} & \xrightarrow{g^{l+m}_{l+k+m+n}} & H_{l+k+m+n} \end{array}$$

are commutative up to conjugation by invertible elements of monoids.

Definition 1 *We call such a system of monoids a braid-module system of monoids. If the braid groups are replaced by the symmetric groups in the above construction, then such a system of monoids will be called a symmetric module system of monoids.*

Suppose that a system of monoids $\{H_n, g_n\}$ is such that H_0 is the group consisting of one element $H_0 = \{e\}$. Then $\nu_{m,0}$ is a homomorphism from Br_m to H_m. There exists the canonical homomorphism τ_m from Br_m to symmetric group Σ_m

$$\tau_m : Br_m \to \Sigma_m.$$

Thus every symmetric module system of monoids inherits the structure of a braid-module system of monoids by means of the above homomorphisms. Suppose now that a system of monoids $\{H_n, g_n\}$ is such that for all m and n there exist homomorphisms

$$\mu_{m,n} : H_m \times H_n \to H_{m+n}, \tag{1}$$

such that $\mu_{e,n}(e, x) = Id(x)$ and $\mu_{m,n}(x, e) = Id(x)$, where $x \in H_n$ and e is the unit of the monoid H_0 and that the following diagrams

$$\begin{array}{ccc} H_l \times H_m \times H_n & \xrightarrow{Id \times \mu_{m,n}} & H_l \times H_{m+n} \\ \downarrow \mu_{l,m} \times Id & & \downarrow \mu_{l,m+n} \\ H_{l+m} \times H_n & \xrightarrow{\mu_{l+m,n}} & H_{l+m+n} \ , \end{array} \tag{2}$$

$$\begin{array}{ccc} H_m \times H_n & \xrightarrow{g^m_{m+k} \times g^n_{n+l}} & H_{m+k} \times H_{n+l} \\ \downarrow \mu_{l,m} & & \downarrow \mu_{m+k,n+l} \\ H_{l+m} & \xrightarrow{g^{l+m}_{m+k+n+l}} & H_{m+k+n+l} \end{array}$$

are commutative up to conjugation by invertible elements of monoids. If the diagram (2) is strictly commutative (without conjugation), then a strict monoidal (tensor) category $\mathcal{H}$ is defined in a usual way. Its objects $\{\bar{0}, \bar{1}, ...\}$ correspond to integers from 0 to infinity and morphisms are defined by the formula:

$$hom(\bar{k}, \bar{l}) = \begin{cases} H_k, & \text{if} \quad k = l, \\ \emptyset, & \text{if} \quad k \neq l. \end{cases}$$

The product in $\mathcal{H}$ is defined on objects by the sum of numbers and on morphisms by the pairing μ.

Definition 2 *We call such a system of monoids a monoidal system of monoids. If monoidal and braid-module structures of the system are compatible in the evident way then we say that $\{H_n\}$ is a braid-algebra.*

Morphisms of monoidal systems of monoids are defined in the evident way. The braid groups and the symmetric groups form monoidal systems of groups $\{Br_n, i_n\}$ and $\{\Sigma_n, i_n\}$ in a usual way. If we have a monoidal system of monoids $\{H_n, g_n\}$ and morphism of monoidal systems

$$\phi : \{Br_n, i_n\} \to \{H_n, g_n\},$$

then the system $\{H_n, g_n\}$ becomes a braid-module system of monoids. Let K be an associative ring with the unit 1 and σ_m be the canonical representation from symmetric group Σ_m to $GL_n(K)$. The standard inclusions of $GL_m(K)$ and $GL_n(K)$ as diagonal blocks in $GL_{m+n}(K)$ define a monoidal structure for the system of groups $GL_n(K)$. Morphisms σ_m and $\sigma_m\phi_m$ define symmetric module and braid-module structures on the system of groups $GL_n(K)$. We denote this braid-module structure by ψ. Let us suppose now that we have a braid-module system of monoids $\{H_n, g_n\}$ and a braid-module morphism ρ from $\{H_n, g_n\}$ to $\{GL_n(K)\}$. This means that we have a system of homomorphisms (representations)

$$\rho_n : H_n \to GL_n(K).$$

We suppose that the diagrams

$$\begin{array}{ccc} H_n & \xrightarrow{\rho_n} & GL_n(K) \\ \downarrow g_n & & \downarrow i_n \\ H_{n+1} & \xrightarrow{\rho_{n+1}} & GL_{n+1}(K), \end{array}$$

$$\begin{array}{ccc} Br_m \times H_n & \xrightarrow{Id \times \rho_n} & Br_m \times GL_n(K) \\ \downarrow \nu_{m,n} & & \downarrow \psi_{m,n} \\ H_{m+n} & \xrightarrow{\rho_{m+n}} & GL_{m+n}(K) \end{array}$$

are commutative up to conjugation.

Definition 3 *We say in this situation that the system of representations ρ_n agree with the braid-module structure of the system of monoids H_n. Analogously we require that the system of representations agree with the symmetric module structure of the system of monoids.*

There exists a large variety of examples of the structure described above. Some of them have been discussed already: general linear groups $\{GL_n(K)\}$ form a structure of algebra over symmetric group, symmetric groups $\{\Sigma_n\}$ form a structure of braid-algebra. Any system of finite Coxeter groups $\{W_n\}$ acting on $\mathbb{R}^n$ form a symmetric module and the corresponding system of generalized braid groups $\{Br(W_n)\}$ form a braid-module. Braid groups $\{Br(C_n)\}$ of the type C form a structure of braid-algebra (see [31] for details). The Coxeter group C_n is isomorphic to the wreath product $\Sigma_n \int \mathbb{Z}/2$. Let G be a group and γ a homomorphism from G to $\mathbb{Z}/2$:

$$\gamma : G \to \mathbb{Z}/2.$$

We consider the wreath product $\Sigma_n \int G$. The homomorphism γ induces the homomorphism

$$\gamma_n : \Sigma_n \int G \to \Sigma_n \int \mathbb{Z}/2.$$

The last wreath product $\Sigma_n \int \mathbb{Z}/2$ is isomorphic to the Coxeter group C_n and so, is imbedded in the orthogonal group O_n. Thus we have a composition

$$\Sigma_n \int G \to \Sigma_n \int \mathbb{Z}/2 \to O_n \to GL_n(\mathbb{R}), \tag{3}$$

which gives a necessary representation of $\Sigma_n \int G$. We have the canonical pairings

$$(\Sigma_k \times \Sigma_n) \int G \to \Sigma_{n+k} \int G, \tag{4}$$

$$\Sigma_k \times (\Sigma_n \int G) \to \Sigma_{n+k} \int G, \tag{5}$$

which give $\Sigma_n \int G$ a structure of algebra over symmetric group and the representation (3) agrees with this structure. Analogously the wreath products of a group G with braid groups can be considered.

Let F_n be a free group of the rank n with the set of generators $\{x_1, ..., x_n\}$. Other examples of structures described above give the groups $AutF_n$ $n = 1, 2, ...,$ of automorphisms of F_n and its subgroups. We have the canonical imbeddings

$$AutF_m \times AutF_n \to AutF_{m+n},$$

which generate a structure of monoidal system of groups. There are the standard inclusions of the symmetric group Σ_n and the braid group Br_n into $AutF_n$. They can be characterized as follows. Let $\xi_i \in AutF_n, i = 1, 2, ..., n-1$, be given by the formula which describes their action on generators:

$$\begin{cases} x_i & \mapsto x_{i+1}, \\ x_{i+1} & \mapsto x_i, \\ x_j & \mapsto x_j, j \neq i, i+1. \end{cases}$$

Let $\sigma_i \in AutF_n, i = 1, 2, ..., n-1$, be given by the analogous formula of their action on generators:

$$\begin{cases} x_i & \mapsto x_{i+1}, \\ x_{i+1} & \mapsto x_{i+1}^{-1} x_i x_{i+1}, \\ x_j & \mapsto x_j, j =\neq i, i+1. \end{cases}$$

If we map the standard generators of the symmetric group to ξ_i and the standard generators of the braid group to σ_i, then we get the monomorphisms s and t of monoidal systems of groups:

$$t : \{\Sigma_n\} \to \{AutF_n\},$$

$$s : \{Br_n\} \to \{AutF_n\}.$$

If we take the composition of morphisms t and τ we get one more homomorphism of monoidal systems of groups

$$\tau t : \{Br_n\} \to \{AutF_n\}.$$

The homomorphism t defines a structure of algebra over symmetric group for the system of groups $AutF_n$ and homomorphisms s and τt define two different structures of braid algebra for the system of groups $AutF_n$. If we consider the abelianization homomorphism for the free group F_n, we get a free abelian group of the rank n and also we get the epimorphism of $AutF_n$ onto $GL_n\mathbb{Z}$:

$$\rho_n : AutF_n \to GL_n\mathbb{Z}.$$

It is evident that the system of representations ρ_n agrees with the structure of algebra over symmetric group and two structures of braid algebra on the system of groups $AutF_n$. Some of this structure has been exploited by Hatcher in his study of the automorphism group of the free group and in [10] where F_n is also replaced by free products of cyclic groups.

If we have a system of monoids $H_n < AutF_n$, $H_n < H_{n+1}$, which contain symmetric or braid groups:

$$\Sigma_n < H_n < AutF_n,$$

or

$$Br_n < H_n < AutF_n,$$

such that the compositions

$$\Sigma_n \to H_n \to AutF_n \to GL_n\mathbb{Z},$$

$$Br_n \to H_n \to AutF_n \to GL_n\mathbb{Z},$$

are the canonical representations, then they define symmetric or braid systems of monoids. The more concrete example of this situation is the braid-permutation group BP_n of Fenn, Rimányi and Rourke [13, 14]. This is the subgroup of $AutF_n$, generated by the symmetric and braid subgroups, that is, generated by both sets of elements ξ_i and σ_i. Fenn, Rimányi and Rourke proved in [13, 14] that this group is given by the following set of generators: $\{\xi_i, \sigma_i, i = 1, 2, ..., n-1\}$ and relations:

$$\begin{cases} \xi_i^2 & = 1, \\ \xi_i\xi_j & = \xi_j\xi_i, \ \text{if } |i-j|>1, \\ \xi_i\xi_{i+1}\xi_i & = \xi_{i+1}\xi_i\xi_{i+1}. \end{cases}$$

The symmetric group relations

$$\begin{cases} \sigma_i\sigma_j & = \sigma_j\sigma_i, \ \text{if } |i-j|>1, \\ \sigma_i\sigma_{i+1}\sigma_i & = \sigma_{i+1}\sigma_i\sigma_{i+1}. \end{cases}$$

The braid group relations

$$\begin{cases} \sigma_i\xi_j & = \xi_j\sigma_i, \ \text{if } |i-j|>1, \\ \xi_i\xi_{i+1}\sigma_i & = \sigma_{i+1}\xi_i\xi_{i+1}, \\ \sigma_i\sigma_{i+1}\xi_i & = \xi_{i+1}\sigma_i\sigma_{i+1}. \end{cases}$$

The mixed relations

As in the case of groups $AutF_n$ the homomorphism t induces a structure of algebra over symmetric group for the system of groups BP_n. The homomorphisms s and τt induce two different structures of braid algebra for the system of groups BP_n. The system of representations arising from $AutF_n$ agrees with the structures of the braid algebra and the algebra over symmetric group on BP_n.

Next consider the system of groups $\overline{BP}_m$ with the same generators as the braid-permutation group and which satisfy (i) the above defining relations for BP_n together with (ii) the additional two series of relations given next:

$$\xi_i\sigma_{i+1}\sigma_i = \sigma_{i+1}\sigma_i\xi_{i+1},$$

$$\xi_i\sigma_i = \sigma_i\xi_i.$$

We call this group the *reduced braid-permutation group* [33]. Thus we have the mixed relations for the reduced braid-permutation group in the following form:

$$\begin{cases} \sigma_i\xi_j & = \xi_j\sigma_i, \ \text{for all } i,\ j, \\ \xi_i\xi_{i+1}\sigma_i & = \sigma_{i+1}\xi_i\xi_{i+1}, \\ \sigma_i\sigma_{i+1}\xi_i & = \xi_{i+1}\sigma_i\sigma_{i+1}, \\ \xi_i\sigma_{i+1}\sigma_i & = \sigma_{i+1}\sigma_i\xi_{i+1}. \end{cases}$$

The mixed relations for $\overline{BP}_m$

Although $\overline{BP}_m$ is not a subgroup of $AutF_m$, using the presentation of this group, a structure of algebra over symmetric group and two different structures of braid algebra for the system of groups $\overline{BP}_m$ can be defined. The representations

$$AutF_m \to GL_m\mathbb{Z}, \quad m = 0, 1, ...,$$

map the elements ξ_i and σ_i to the same matrix. So the representations

$$\overline{BP}_m \to GL_m\mathbb{Z}, \quad m = 0, 1, ...,$$

are well defined and are compatible with the structures of the algebra over symmetric group and the braid algebra structures on $\overline{BP}_n$.

Let $Aut(\vee^k S^n)$ be the monoid of pointed homotopy equivalences of the bouquet of n-dimensional spheres $\vee^k S^n$. Let V_k be defined as follows

$$V_k = \lim_{\to n} Aut(\vee^k S^n).$$

Then the pairing (1) is defined for the system of $\pi_0(V_k)$ evidently. The maps

$$Aut(\vee^k S^0) \to Aut(\vee^k S^1) \to V_k$$

are defined by the suspension functor. They define the homomorphisms

$$Aut(\vee^k S^0) \to \pi_0 Aut(\vee^k S^1) \to \pi_0 V_k.$$

The H-space $Aut(\vee^k S^1)$ is homotopy equivalent to $AutF_k$, so we have a homomorphism

$$\chi : AutF_k \to \pi_0 V_k,$$

equal to the composition

$$AutF_k \cong \pi_0 Aut(\vee^k S^1) \to \pi_0 V_k.$$

The map

$$\pi_0 Aut(\vee^k S^n) \to Aut(H_n(\vee^k S^n; \mathbb{Z})) \cong GL_k\mathbb{Z}$$

induces the homomorphism

$$\pi_0 V_k \to GL_k\mathbb{Z},$$

such that the composition

$$AutF_k \xrightarrow{\chi} \pi_0 V_k \to GL_k\mathbb{Z}$$

is equal to ρ_k. With the use of homomorphism χ the structure of algebra over symmetric group and two structures of braid algebras for the system of groups $\pi_0 V_n$ are defined.

Another example of a system of monoids, which do not arise from groups, is given by the system of Baez-Birman monoids. The *Baez-Birman monoid* SB_n (or *generalized braid monoid*, or *singular braid monoid*) [2, 3] is defined as a monoid with generators g_i, g_i^{-1}, a_i, $i = 1, ..., n-1$, and relations

$$\begin{aligned}
&[g_i, g_j] = [a_i, a_j] = [a_i, g_j] = 0, \text{ if } |i-j| > 1,\\
&[a_i, g_i] = 0,\\
&g_i g_{i+1} g_i = g_{i+1} g_i g_{i+1},\\
&g_i g_{i+1} a_i = a_{i+1} g_i g_{i+1},\\
&g_{i+1} g_i a_{i+1} = a_i g_{i+1} g_i,\\
&g_i g_i^{-1} = g_i^{-1} g_i = 1.
\end{aligned}$$

In pictures g_i corresponds to the canonical generator of the braid group (right-handed crossing) and a_i represents an intersection of the ith and $(i+1)$st strand. The homomorphism j_n from the braid group Br_n to SB_r is defined in the evident way:

$$j_n : Br_n \to SB_n.$$

The pairing

$$\mu_{m,n} : SB_m \times SB_n \to SB_{m+n}$$

is defined in a usual way by the formulas

$$\mu_{m,n}(g_i') = g_i,\ \mu_{m,n}(a_i') = a_i;\ g_i', a_i' \in SB_m;\ g_i,\ a_i \in SB_{m+n},$$

$$\mu_{m,n}(g_j'') = g_{j+m},\ \mu_{m,n}(a_j'') = a_{j+m};\ g_i'',\ a_i'' \in SB_n;\ g_{j+m},\ a_{j+m} \in SB_{m+n}.$$

This pairing agrees with the corresponding pairing for the braid groups and defines a braid-algebra structure for the system of monoids SB_n.

The pairing μ is commutative up to conjugation analogously as for the braid groups. More precisely, it was proved in [33] that the category

$\mathcal{SB}$ is a *braided category*, as it is defined by Joyal and Street [16]. That means that (1) there exists a system of isomorphisms of objects

$$c_{\bar{m},\bar{n}} : \overline{m+n} \to \overline{m+n}\ ,$$

natural with respect to morphisms from $\bar{m}$ to itself and from $\bar{n}$ to itself, and (2) the properties of coherence B1 and B2 from [16], which are sometimes called the Drinfeld triangles must be satisfied. The braid-algebra structure for $\{SB_n\}$ is defined. The homomorphism

$$h_n : SB_n \to \overline{BP}_n$$

is defined by the formulas:

$$h_n(g_i) = \sigma_i,$$
$$h_n(a_i) = \xi_i.$$

The composition $h_n \circ j_n$ is equal to the canonical homomorphism

$$\kappa_n : Br_n \to \overline{BP}_n.$$

The system of representations for the reduced braid-permutation group defines representations for the system of monoids SB_n. This system of representations is compatible with the braid-algebra structure.

Suppose now that there exists a homomorphism f from the ring K to the real numbers $\mathbb{R}$ which sends the unit of K to the unit of $\mathbb{R}$. Then f generates the homomorphism

$$GL_n(f) : GL_n(K) \to GL_n(\mathbb{R}).$$

We denote it shortly by f_n. There is the canonical retraction r_n from the classifying space of the general linear group over the real numbers to the classifying space of the orthogonal group

$$r_n : BGL_n(\mathbb{R}) \to BO_n.$$

The following classical lemma concerning the map r_n is useful.

Lemma 1 *The diagram*

$$\begin{array}{ccc} BGL_m(\mathbb{R}) \times BGL_n(\mathbb{R}) & \xrightarrow{r_m \times r_n} & BO_m \times BO_n \\ \downarrow & & \downarrow \\ BGL_{m+n}(\mathbb{R}) & \xrightarrow{r_{m+n}} & BO_{m+n}, \end{array} \qquad (6)$$

where the vertical arrows are the canonical inclusions, is homotopy commutative.

Proof We denote $BGL_n(\mathbb{R})$ by BGL_n in this proof. For all cellular spaces X we have the following bijection:

$$[X; BO_n] \xrightarrow{Bi_*} [X; BGL_n]$$

which is induced by the canonical inclusion $i_n : O_n \to GL_n$ (see [28], Proposition 11.45). So homotopy class of r_n can be considered as $Bi_*^{-1}(Id)$, where Id is the identical map of BGL_n. Consider the following commutative diagram

$$\begin{array}{ccc} [BGL_m \times BGL_n; BO_m \times BO_n] & \longrightarrow & [BGL_m \times BGL_n; BGL_m \times BGL_n] \\ \downarrow & & \downarrow \\ [BGL_m \times BGL_n; BO_{m+n}] & \longrightarrow & [BGL_m \times BGL_n; BGL_{m+n}] \\ \uparrow & & \uparrow \\ [BGL_{m+n}; BO_{m+n}] & \longrightarrow & [BGL_{m+n}; BGL_{m+n}]. \end{array} \tag{7}$$

One composition in the diagram (6) is the image of $r_m \times r_n \in [BGL_m \times BGL_n; BO_m \times BO_n]$ in (7) by the left arrow, going down and another one is the image of r_{m+n} by the left arrow going up. Using the commutativity of the diagram (7), we see that their images by the bijection Bi_{m+n} coincide in $[BGL_m \times BGL_n; BGL_{m+n}]$. Hence these elements coincide in $[BGL_m \times BGL_n; BO_{m+n}]$. Lemma is proved.

Let H_n be a braid-module system of monoids, compatible with the representations ρ_n. For the monoids H_n we consider the functor of classifying space B [22, 20, 21] and we take the following composition

$$r_n B(f_n) B(\rho_n) : BH_n \to BGL_n(K) \to BGL_n(\mathbb{R}) \to BO_n. \tag{8}$$

It induces the vector bundle

$$F_n \to E_n \xrightarrow{p} BH_n.$$

Now we add one more map $B\mu_{m,n}$ to the composition (8) for $m+n$:

$$r_{m+n} B(f_{m+n}) B(\rho_{m+n}) B\mu : BBr_m \times BH_n \to BH_{m+n} \to BGL_{m+n}(\mathbb{R}) \to BO_{m+n}. \tag{9}$$

From the Lemma 1 we get the following commutative diagram

$$\begin{array}{ccc} BBr_m \times BH_n & \longrightarrow & BO_m \times BO_n \\ \downarrow & & \downarrow \\ BG_{m+n} & \longrightarrow & BO_{m+n}, \end{array}$$

where the left-bottom composition is the map (9). The canonical representation of the braid group in the orthogonal group O_n generates the map

$$BBr_n \to BO_n.$$

We apply the functor given by the Thom space construction to this map:

$$MBr_n \to MO_n.$$

It was proved in [18, 19] and [8, 6] that the Thom spectrum $\{MBr_n\}$ is isomorphic to the Eilenberg-MacLane spectrum for the group $\mathbb{Z}/2$:

$$MBr_n \cong H\mathbb{Z}/2.$$

Apply the functor of the Thom space to the compositions (8) and (9). We get:

$$Mr_n M(f_n) M(\rho_n) : MH_n \to MGL_n(K) \to MGL_n(\mathbb{R}) \to MO_n,$$

$$Mr_{m+n} M(f_{m+n}) M(\rho_{m+n}) M\mu : MBr_m \wedge MH_n \to MH_{m+n} \to MGL_{m+n}(\mathbb{R}) \to MO_{m+n}.$$

Our conditions on the system of monoids H_n imply that the Thom spectrum MH of the system of monoids H_n becomes a module spectrum over the Thom spectrum of the braid groups, i.e., over $H\mathbb{Z}/2$. It is well known that such a spectrum is itself a wedge of suspensions of the Eilenberg-MacLane spectrum for the group $\mathbb{Z}/2$. If the system of monoids H_n is monoidal and its monoidal structure agrees with the braid-module structure and with the system of representations, then the spectrum MH becomes a multiplicative spectrum. Suppose in addition that MH is commutative and D is a multiplicative discrete (wedge of spheres) spectrum, such that

$$H_*(D; \mathbb{Z}/2) \cong \pi_*(MH)$$

as rings. Then we construct a multiplicative morphism of spectra

$$H\mathbb{Z}/2 \wedge D \to MH$$

by the formula

$$H\mathbb{Z}/2 \wedge D \to MH \wedge MH \xrightarrow{\mu} MH.$$

The next theorem follows.

Theorem 1 *The Thom spectrum MH of a braid-module system of monoids with a system of representations which agree with the braid-module structure is equivalent to a wedge of suspensions over the Eilenberg-MacLane spectrum for the group $\mathbb{Z}/2$. If the system of monoids H_n has a monoidal structure which agrees with the two previous ones, then the spectrum MH is multiplicative. If in addition MH is commutative, then it is equivalent to a smash product of $H\mathbb{Z}/2$ with some multiplicative discrete spectrum D:*

$$MH \simeq H\mathbb{Z}/2 \wedge D.$$

Corollary 1 *The Thom spectrum MH of a symmetric module system of monoids with a system of representations which agrees with the symmetric module structure is equivalent to a wedge of suspensions over the Eilenberg-MacLane spectrum for the group $\mathbb{Z}/2$. The spectrum MH is a module spectrum over the Thom spectrum $M\Sigma$ for the symmetric group. If the system of monoids H_n has a monoidal structure which agrees with the two previous ones, then the spectrum MH is multiplicative. If in addition MH is commutative, then it is equivalent to a smash product of $H\mathbb{Z}/2$ with some multiplicative discrete spectrum D:*

$$MH \simeq H\mathbb{Z}/2 \wedge D.$$

Consider the homology of the classifying space of a monoid H_n with coefficients in some ring regarded as a trivial H_n-module K: $H_*(BH_n; K)$. Homomorphisms g_n induce the maps between the classifying spaces

$$Bg_n : BH_n \to BH_{n+1}.$$

Let BH be the direct limit

$$BH \cong \lim_{\to n} BH_n.$$

The homomorphisms $\mu_{m,n}$ induce the pairing

$$\mu : H_*(Br; K) \otimes H_*(BH; K) \to H_*(BH; K).$$

The pairing β of the braid groups induce the ring structure in $H_*(Br; K)$, so μ gives a module structure for the homology groups $H_*(BH; K)$. Now, let the ring K be equal to $\mathbb{Z}/2$. Then the Thom class $t_n \in H^*(E_n, E_n - i(BH_n); \mathbb{Z}/2)$ is defined, where i is the zero section. It defines the Thom homomorphism

$$T_n : H_*(MH_n; \mathbb{Z}/2) \to H_*(BH_n; \mathbb{Z}/2) \tag{10}$$

by the usual formula:

$$T_n(x) = p_*(t_n \cap x).$$

The Thom homomorphism is compatible with the pairings of $H_*(MG_n; \mathbb{Z}/2)$ with $H_*(MBr_m; \mathbb{Z}/2)$ and $H_*(BH_n; \mathbb{Z}/2)$ with $H_*(BG_n; \mathbb{Z}/2)$ through the Thom homomorphism for the braid group, i.e. the following diagram

$$\begin{array}{ccc} H_*(MBr_m; \mathbb{Z}/2) \otimes H_*(MG_n; \mathbb{Z}/2) & \xrightarrow{M\mu} & H_*(MG_{m+n}; \mathbb{Z}/2) \\ \downarrow T \otimes T & & \downarrow T \\ H_*(BBr_m; \mathbb{Z}/2) \otimes H_*(BG_n; \mathbb{Z}/2) & \xrightarrow{B\mu} & H_*(BG_{m+n}; \mathbb{Z}/2). \end{array}$$

commutes. The space BH_n, as a geometric realization, is a CW-complex. It is path-connected, and paracompact, so the Thom homomorphism is an isomorphism (see, for example, [28], Theorem 14.6). The system of isomorphisms (10) generates the stable isomorphism

$$T : H_*(MH; \mathbb{Z}/2) \to H_*(BH; \mathbb{Z}/2),$$

which is compatible with the module structures of $H_*(MH; \mathbb{Z}/2)$ over $H_*(MBr; \mathbb{Z}/2)$ and $H_*(BH; \mathbb{Z}/2)$ over $H_*(BBr; \mathbb{Z}/2)$. It follows from Theorem 1, that $H_*(MG; \mathbb{Z}/2)$ is isomorphic to $H_*(H\mathbb{Z}/2; \mathbb{Z}/2) \otimes N$, where N is some $\mathbb{Z}/2$-vector space. So, for $H_*(BH; \mathbb{Z}/2)$ we have the following isomorphism

$$H_*(BH; \mathbb{Z}/2) \cong H_*(Br; \mathbb{Z}/2) \otimes N.$$

The ring $H_*(Br; \mathbb{Z}/2)$ is isomorphic to the polynomial ring on infinitely many generators

$$H_*(Br; \mathbb{Z}/2) \cong \mathbb{Z}/2[a_1, ..., a_i, ...], \quad \deg a_i = 2^i - 1.$$

Hence the following theorem is proved.

Theorem 2 *Let H_n be a system of braid-module monoids with a system of representations, which agrees with its braid-module structure. Then the homology with the coefficients in $\mathbb{Z}/2$ of the direct limit of the classifying spaces of the monoids H_n is a free module over the homology ring of the infinite braid group, which is isomorphic to the polynomial ring $P = \mathbb{Z}/2[a_1, ..., a_i, ...]$, $\deg a_i = 2^i - 1$. So there is a isomorphism of P-modules*

$$H_*(BH; \mathbb{Z}/2) \cong P \otimes Q$$

for some $\mathbb{Z}/2$-module Q. If the system of monoids H_n has monoidal structure which agrees with the two previous ones, then $H_*(BH; \mathbb{Z}/2)$ is isomorphic to a graded algebra over P. If in addition $H_*(BH; \mathbb{Z}/2)$ is commutative, then the above isomorphism can be considered as an isomorphism of algebras for some $\mathbb{Z}/2$-algebra Q.

The composition (3) induces the maps of classifying spaces:

$$B(\Sigma_n \int G) \to B(\Sigma_n \int \mathbb{Z}/2) \to BO(n),$$

and the maps of the corresponding Thom spaces:

$$M(\Sigma_n \int G) \to M(\Sigma_n \int \mathbb{Z}/2) \to MO(n).$$

The Thom spaces $M(\Sigma_n \int G)$ generate the Thom spectrum $M(\Sigma \int G)$. The pairings give $M(\Sigma \int G)$ a multiplicative structure and a module structure over $M(\Sigma)$. The spectrum $M(\Sigma \int G)$ becomes a module spectrum over $H\mathbb{Z}/2 = MBr$ through the canonical maps $Br_n \to \Sigma_n$. So this spectrum is homotpy equivalent to a wedge of suspensions over the Eilenberg-MacLane spectrum $H\mathbb{Z}/2$.

The same is true if we replace the group Σ_n by the braid group on n strings Br_n.

The plus construction of the classifying space $B(\Sigma \int G)$ is equivalent to $\Omega_0^\infty S^\infty((BG)_+)$, where the lower plus denotes the disjoint union of a space with a single point and $\Omega_0^\infty S^\infty((BG)_+)$ is one path-component of $\Omega^\infty S^\infty((BG)_+)$. So in order to calculate generators of the coefficient ring $M(\Sigma \int G)_*$ we must take the generators of $H_*(\Omega_0^\infty S^\infty((BG)_+); \mathbb{Z}/2)$ and delete the generators, which come from the dual of the Steenrod algebra. In terms of Araki-Kudo-Dyer-Lashof operations (which were known originally to Nakaoka [23] in his work on homology of symmetric groups) we delete the generators $Q_1^{k_1} ... Q_n^{k_n} a_0$, where a_0 is the additive generator of $H_0(BG; \mathbb{Z}/2)$. We get the following theorem.

Theorem 3 *The ring $M(\Sigma \int G)_*$ is a polynomial ring over $\mathbb{Z}/2$ on infinitely many generators, which correspond to the following elements in $H_*(\Omega_0^\infty S^\infty((BG)_+); \mathbb{Z}/2)$:*

$$Q_1^{k_1} ... Q_n^{k_n} x,$$

where Q_i are the Araki-Kudo-Dyer-Lashof operations, $x \in H_*(BG; \mathbb{Z}/2)$ is an additive generator and $k_i > 0$ for some $i > 1$, if $x \in H_0(BG; \mathbb{Z}/2)$.

Now let A be an abelian group. We consider the homomorphism $h : \Sigma_n \int A \to A$, defined by the formula

$$h(a_1, ..., a_n; \sigma) = \sum_{i=1}^{n} a_i.$$

We denote by

$$\widetilde{\Sigma_n \int A}$$

the kernel of h. Thus we have the exact sequence

$$\widetilde{\Sigma_n \int A} \to \Sigma_n \int A \to A.$$

In the case when $A = \mathbb{Z}/2$ the group $\widetilde{\Sigma_n \int Z/2}$ is isomorphic to the Coxeter group of the type D_n. Hence we have the composition of homomorphisms

$$\widetilde{\Sigma_n \int A} \to D_n \to O_n,$$

which generates the Thom spectrum $M(\widetilde{\Sigma \int A})$. Canonical pairings analogous to the pairings described by the formulae (4) and (5) give $M(\widetilde{\Sigma \int A})$ a multiplicative structure and a module structure over $M(\Sigma)$. The spectrum $M(\widetilde{\Sigma \int A})$ also becomes a module spectrum over $H\mathbb{Z}/2 = MBr$ through the canonical maps $Br_n \to \Sigma_n$. Thus it is isomorphic to a wedge of suspensions over the Eilenberg-MacLane spectrum $H\mathbb{Z}/2$.

Now let the group A be such, that for some fixed natural number m, $ma = 0$ for any element a of the group A. Then

$$\Sigma_{km+1} \int A = (\widetilde{\Sigma_{km+1} \int A}) \oplus A$$

and this decomposition is correct for the infinite group

$$\Sigma_\infty \int A = (\widetilde{\Sigma_\infty \int A}) \oplus A.$$

So we have

$$\Omega_0^\infty S^\infty((BA)_+) \cong (B\widetilde{\Sigma \int A})^+ \times BA.$$

We get the following theorem.

Theorem 4 *The ring $M(\widetilde{\Sigma \int A})_*$ is a polynomial ring over $\mathbb{Z}/2$ on infinitely many generators, which correspond to the following elements in $H_*(\Omega_0^\infty S^\infty((BA)_+); \mathbb{Z}/2)$:*

$$Q_1^{k_1} \ldots Q_n^{k_n} x,$$

where Q_i are the Araki-Kudo-Dyer-Lashof operations, $x \in H_*(BG; \mathbb{Z}/2)$ is an additive generator, and $k_i > 0$ for some $i > 1$, if $x \in H_0(BG; \mathbb{Z}/2)$, and $k_i > 0$ for some $i \geq 1$, if $x \in H_j(BG; \mathbb{Z}/2)$ for $j \geq 1$.

Example 1 Suppose that the group G considered in our constructions is equal to the group consisting of one element e. Then we obtain the spectrum which corresponds to the infinite symmetric group Σ_∞. This case was studied by Bullett [6]. He proved that the ring $M\Sigma_*$ is polynomial (without calculating generators). The plus construction of the classifying space $B\Sigma_\infty$ is equivalent to $\Omega_0^\infty S^\infty$. Let a_0 be the generator of the reduced zero homology group $\widetilde{H}_0(S^0; \mathbb{Z}/2)$. The polynomial ring $M\Sigma_*$ has generators

$$Q_1^{k_1} \ldots Q_n^{k_n} a_0,$$

with $k_i > 0$ for some $i > 1$.

Example 2 Let us suppose now that the group G is equal to the group $\mathbb{Z}/2$. Then we have the spectrum which corresponds to the Coxeter groups of the type C. The plus construction of the classifying space BC_∞ is equivalent to $\Omega_0^\infty S^\infty(\mathbb{R}P_+^\infty)$. We denote by a_j, $j = 0, 1, \ldots$, the additive generator of $H_j(\mathbb{R}P^\infty; \mathbb{Z}/2) \cong \mathbb{Z}/2$. The polynomial ring MC_* has generators

$$Q_1^{k_1} \ldots Q_n^{k_n} a_0,$$

with $k_i > 0$ for some $i > 1$, and

$$Q_1^{k_1} \ldots Q_n^{k_n} a_j, \quad j > 0,$$

for any sequence of the Araki-Kudo-Dyer-Lashof operations.

Example 3 Let G be equal to the group of integers $\mathbb{Z}$. Plus construction of the classifying space $B(\Sigma \int \mathbb{Z})$ is equivalent to $\Omega_0^\infty S^\infty(S_+^1)$. We denote by a_j, $j = 0, 1$, the additive generator of $H_j(S^1; \mathbb{Z}/2) \cong \mathbb{Z}/2$. Then the polynomial ring $M(\Sigma \int \mathbb{Z})_*$ has the generators

$$Q_1^{k_1} \ldots Q_n^{k_n} a_0,$$

with $k_i > 0$ for some $i > 1$, and

$$Q_1^{k_1} \ldots Q_n^{k_n} a_1,$$

for any sequence of the Araki-Kudo-Dyer-Lashof operations. The canonical epimorphism $\mathbb{Z} \to \mathbb{Z}/2$ induces the map of the ring spectra

$$M(\Sigma \int \mathbb{Z}) \to M(\Sigma \int \mathbb{Z}/2)$$

which on the level of the coefficient rings is expressed by the fact that a_0 goes to a_0 and a_1 goes to a_1.

Example 4 Let us suppose that the abelian group A is equal to the group $\mathbb{Z}/2$. Then the Thom spectrum $M(\widetilde{\Sigma \int A})$ is the spectrum MD which corresponds to the Coxeter groups of the type D. We have

$$\Omega_0^\infty S^\infty(\mathbb{R}P_+^\infty) \cong (BD)^+ \times \mathbb{R}P^\infty.$$

The polynomial ring MD_* has generators

$$Q_1^{k_1} \ldots Q_n^{k_n} a_0,$$

with $k_i > 0$ for some $i > 1$, and

$$Q_1^{k_1} \ldots Q_n^{k_n} a_j, \quad j > 0,$$

with $k_i > 0$ for some $i \geq 1$.

For a system of groups (monoids) to construct a Thom spectrum which is a module spectrum over MBr it is not always necessary to have a system of representations

$$\rho_n : H_n \to GL_n(\mathbb{R}).$$

It is possible to do this using Borel construction. We explain this on the example of the braid group in the handlebody of the genus g and of the braid group of a surface. We denote the first group by Br_n^g. It was studied in [27] and [32]. Its presentation is given by the generators τ_k, $k = 1, 2, ..., g$; σ_i, $i = 1, 2, ..., n-1$; and relations

$$\begin{cases} \sigma_i\sigma_j & = \sigma_j\sigma_i \text{ if } |i-j| > 1, \\ \sigma_i\sigma_{i+1}\sigma_i & = \sigma_{i+1}\sigma_i\sigma_{i+1}, \\ \tau_k\sigma_i & = \sigma_i\tau_k \text{ if } k \geq 1, \ i \geq 2, \\ \tau_k\sigma_1\tau_k\sigma_1 & = \sigma_1\tau_k\sigma_1\tau_k, \ k = 1, 2, ..., g; \\ \tau_k\sigma_1^{-1}\tau_{k+l}\sigma_1 & = \sigma_1^{-1}\tau_{k+l}\sigma_1\tau_k, \ k = 1, 2, ..., g-1; \ l = 1, 2, ..., g-k. \end{cases}$$

The geometrical construction for the classifying space of Br_n^g is the following. Let Q_g be a subset of the complex plain $\mathbb{C}$ consisting of g different points: $z_1^0, ..., z_g^0$, say, $z_i^0 = i$. We denote by $F(\mathbb{C} \setminus Q_g, n)$ the space of n-tuples of pairwise different points in $\mathbb{C} \setminus Q_g$:

$$F(\mathbb{C} \setminus Q_g, n) = \{p_1, ..., p_n) \in (\mathbb{C} \setminus Q_g)^n : p_i \neq p_j \text{ for } i \neq j\}.$$

It is the space of regular points of the canonical action of the symmetric group Σ_n on Cartesian power $(\mathbb{C} \setminus Q_g)^n$ of $\mathbb{C} \setminus Q_g$. The configuration space $B(\mathbb{C} \setminus Q_g, n)$ is the space of regular orbits of this action. It was proved by Fadell and Neuwrith [12] that the spaces $B(\mathbb{C}\setminus Q_g, n)$ and $F(\mathbb{C}\setminus Q_g, n)$ are $K(\pi, 1)$-spaces with fundamental groups isomorphic to Br_n^g and to the corresponding pure braid group. Hence the homology of the braid group Br_n^g is isomorphic to the homology of $B(\mathbb{C}\setminus Q_g, n)$ and the homology of the corresponding pure braid group is isomorphic to the homology of $F(\mathbb{C} \setminus Q_g, n)$. The space $F(\mathbb{C} \setminus Q_g, n)$ can be interpreted as the complement of the arrangement of hyperplanes in $\mathbb{C}^{g+n}$ given by the formulas:

$$H_{j,j} : z_j - z_k = 0 \text{ for all } j, k;$$

$$H_j^i : z_j = z_i^0 \ \text{ for } \ i = 1, ..., g; \ j = 1, ..., n.$$

Using Borel construction we have a vector bundle

$$b_n^g : F(\mathbb{C} \setminus Q_g, n) \times_{\Sigma_n} \mathbb{R}^n \to \ B(\mathbb{C} \setminus Q_g, n),$$

where the action of Σ_n on $\mathbb{R}^n$ is given by the permutation of coordinates. The series of Thom spaces $M(b_n^g)$, $n = 1, 2, ...,$ of these vector bundles b_n^g generates the Thom spectrum MBr^g.

It follows from the presentation of Br_n^g that there exists a well defined pairing

$$\mu_{m,n} : Br_m \times Br_n^g \to Br_{n+m}^g.$$

This pairing can be viewed on the level of configuration spaces the same way as it was done for the ordinary braid groups in [26] and [30]. Before the factorization we have the pairing

$$F(\mathbb{C}, m) \times F(\mathbb{C} \setminus Q_g, n) \to F(\mathbb{C} \setminus Q_g, m+n).$$

After applying Borel construction we obtain the pairing for the spaces of corresponding vector bundles and then the pairing for Thom spaces

$$MBr_m \wedge M(b_n^g) \to M(b_{m+n}^g).$$

Hence the spectrum MBr^g becomes a module spectrum over $MBr \cong H\mathbb{Z}/2$ and so it becomes a wedge of suspensions over $H\mathbb{Z}/2$. The number of summands can be calculated using the theorem of [32]: plus construction of the classifying space of the infinite braid BBr^g group in the handlebody of the genus g is equivalent to the product

$$\Omega^2 S^3 \times \Omega S^2 \times ... \times \Omega S^2,$$

where the number of factors ΩS^2 is equal to the genus g. So the following theorem is proved.

Theorem 5 *The Thom spectrum MBr^g is the wedge of suspensions over the Eilenberg-MacLane spectrum $H\mathbb{Z}/2$, such that its group of coefficients MBr^g_* is isomorphic to $H_*(\Omega S^2 \times ... \times \Omega S^2; \mathbb{Z}/2)$, where the number of factors ΩS^2 is equal to the genus g.*

Now let M_g be a closed surface of the genus g and $M_{g,1}$ be the surface M_g without a disc D^2: $M_{g,1} = M_g \setminus D^2$. The manifold $M_{g,1}$ is a manifold with a boundary $\partial M_{g,1}$. The space $F(M_{g,1}, k)$ is homeomorphic to the space $F(M_{g,1} \setminus \partial M_{g,1}, k)$. We consider the natural imbeddings

$$F(M_{g,1}, k) \to F(M_{g,1}, k+1)$$

which lead to maps

$$B(M_{g,1}, k) \to B(M_{g,1}, k+1).$$

We define

$$B(M_{g,1}, \infty) = \lim_{\to} B(M_{g,1}, k).$$

The fundamental group $\pi_1(B(M_{g,1}, k))$ is called the braid group on k strings of the surface $M_{g,1}$ and is denoted by $Br(M_{g,1}, k)$. The same way as in the previous example we have a vector bundle

$$c^g_n : F(M_{g,1}, n) \times_{\Sigma_n} \mathbb{R}^n \to B(M_{g,1}, n),$$

where the action of Σ_n on $\mathbb{R}^n$ is given by the permutation of coordinates. The series of Thom spaces $M(c^g_n)$, $n = 1, 2, ...$, of these vector bundles c^g_n generates the Thom spectrum $MBr(M_{g,1})$. From the geometrical construction we define a pairing

$$\mu_{m,n} : Br_m \times Br(M_{g,1}, n) \to Br(M_{g,1}, n+m),$$

and the same way we have the pairing of Thom spaces

$$MBr_m \wedge M(c^g_n) \to M(c^g_{m+n}).$$

The spectrum $MBr(M_{g,1})$ becomes a module spectrum over $MBr \cong H\mathbb{Z}/2$. So it is also a wedge of suspensions over $H\mathbb{Z}/2$. The number of summands can be calculated using the results of the work [4]. It was proved there that the plus construction of $B(M_{g,1}, \infty)$ is equivalent to the component of the constant map in the pointed mapping space from the surface M_g to the two dimensional sphere:

$$B(M_{g,1}, \infty)^+ \simeq \mathrm{map}^0_*(M_g, S^2).$$

The homology mod 2 of the space $\mathrm{map}^0_*(M_g, S^2)$ were also calculated in [4].

$$H_*(\mathrm{map}^0_*(M_g, S^2), \mathbb{Z}/2) \cong H_*(\Omega S^2, \mathbb{Z}/2)^{\otimes^{2g}} \otimes H_*(\Omega^2 S^3, \mathbb{Z}/2).$$

So we have the following theorem.

Theorem 6 *The Thom spectrum $MBr(M_{g,1})$ is the wedge of suspensions over the Eilenberg-MacLane spectrum $H\mathbb{Z}/2$, such that its group of coefficients $MBr(M_{g,1})_*$ is isomorphic to $H_*(\Omega S^2 \times \ldots \times \Omega S^2; \mathbb{Z}/2)$, where the number of factors ΩS^2 is equal to $2g$.*

3 Thom spectra in homotopy theory

In this section all spaces and spectra should be considered as localized at the prime 2. First of all we shall give a short proof of Mahowald's theorem.

Let

$$S^3 \to B^3 O$$

be a nontrivial map. We denote by g the double loop map of it

$$g : \Omega^2 S^3 \to BO.$$

We denote the Thom spectrum of the map g by $M(S^3, 2)$.

Theorem of M. Mahowald *The Thom spectrum $M(S^3, 2)$ is equivalent to the Eilenberg-MacLane spectrum $H\mathbb{Z}/2$*

$$M(S^3, 2) \simeq H\mathbb{Z}/2.$$

Proof There is the usual map from $\mathcal{A}$ (the mod 2 Steenrod algebra) to $H^*(M(S^3, 2); \mathbb{Z}/2)$ given by sending Sq^I to $Sq^I(U)$ where U is the Thom class. For the proof of M. Mahowald's theorem it suffices to see that this map is an isomorphism.

Notice:

1) The Euler-Poincaré series for $H^*(\Omega^2 S^3; \mathbb{Z}/2)$ and $\mathcal{A}$ are equal. Thus it suffices to see that the map is a monomorphism as each series is of "finite type".

2) The map is a morphism of coalgebras and so it suffices to see that the map injects on the module of primitives.

3) The module of primitives in $\mathcal{A}$ is given by the Milnor Q_i and there is at most one of these in any dimension. Thus it suffices to check that each Q_i has a nonzero image.

4) We denote by w_i the i-th Stiefel-Whitney class and $x(2^k - 1)$ is the unique primitive element in dimension $2^k - 1$ of $H_*(\Omega^2 S^3; \mathbb{Z}/2)$. From the work of Peterson and Toda [24] we have that

$$Q_k(U) = (w_{2^k-1} + \text{decomposables})U$$

in $H^*(MO; \mathbb{Z}/2)$.

5) Let $<,>$ be the usual Kronecker pairing which evaluates a cohomology class against a homology of the same dimension. We shall check below that

$$< g^*(w_{2^k-1}), x(2^k - 1) >= 1 \tag{11}$$

and

$$< g^*(\text{decomposables}), x(2^k - 1) >= 0. \tag{12}$$

Then we have that $Q_k(U)$ is not zero in $H^*(M(S^3, 2); \mathbb{Z}/2)$ and the theorem follows.

6) Because $x(2^k - 1)$ is primitive, the definition of $< g^*(\text{decomposables}), x(2^k - 1) >$ gives that it is zero. Thus formula (12) follows.

7) To check formula (11), let $Sq^I = Sq^1 Sq^2 Sq^4 ... Sq^{2^{k-2}}$. Thus

$$Sq^I w_{2^{k-1}} = w_{2^k-1} + \text{decomposables}$$

by the Wu formulas.

Thus

$$< g^*(w_{2^k-1}, x(2^k-1) >=< g^* Sq^I w_{2^{k-1}}, x(2^k-1) >=< Sq^I w_{2^{k-1}}, g_* x(2^k-1) >$$

$$=< w_{2^{k-1}}, g_* Sq_*^{2^{k-2}} ... Sq_*^2 Sq_*^1 x(2^k-1) >=< w_{2^{k-1}}, g_* x(1)^{2^{k-1}} >= 1$$

as $w_{2^{k-1}}$ is dual to the primitive class in $H_*(BO; \mathbb{Z}/2)$ given by the 2^{k-1}-st power of the one dimensional class.

Remark 1) This proof uses nothing about the Araki-Kudo-Dyer-Lashof operations. It only requires some *loop* map from $\Omega^2 S^3$ to BO which is nonzero in the first homology group. 2) This proof was sketched in [9].

We now generalize the situation described above. Let us suppose that we have a CW complex X together with maps $f : S^{n+1} \to X$ and $g : X \to B^{n+1}O$ such that the following condition holds. Consider the composition

$$gf : S^{n+1} \to X \to B^{n+1}O$$

and apply the functor of nth loops to it:

$$\Omega^n gf : \Omega^n S^{n+1} \to \Omega^n B^{n+1} O \simeq BO.$$

Let

$$S^1 \to \Omega^n S^{n+1}$$

be the inclusion which is the nth adjoint to the identity

$$S^{n+1} \to S^{n+1}.$$

We suppose that the composition

$$S^1 \to \Omega^n S^{n+1} \to BO$$

is a nontrivial map. The map $\Omega^n g : \Omega^n X \to BO$ generates the Thom spectrum $M(X, n)$. In this case we have

Theorem 7 *The Thom spectrum $M(X,n)$, $n \geq 2$, is a wedge of suspensions over the Eilenberg-MacLane spectrum $H\mathbb{Z}/2$.*

Proof We consider the composition of the map

$$S^3 \to \Omega^{n-2} X$$

which is $n-2$ times adjoint map to $f : S^{n+1} \to X$ with $\Omega^{n-2} g$:

$$S^3 \to \Omega^{n-2} X \to B^3 O$$

and take the double loop map of this composition:

$$\Omega^2 S^3 \to \Omega^n X \to BO.$$

The Thom spectrum of the last composition is $H\mathbb{Z}/2$ and we have the multiplicative map

$$H\mathbb{Z}/2 \to M(X, n).$$

The assertion of the theorem follows from this fact.

The simplest example of such a situation is when X is the sphere S^{n+1} itself $(n \geq 2)$. We get a ring spectrum $M(S^{n+1}, n)$ which is also a module spectrum over $M(S^{k+1}, k)$, $k \leq n$. In this case it is possible to calculate the number of summands in $M(S^{n+1}, n)$. This is the same as to calculate the coefficient ring $M(S^{n+1}, n)_*$. We use the Lemma 3.8 of [7], where the homology of $\Omega^n S^{n+1}$ was calculated in terms of Araki-Kudo-Dyer-Lashof operations. We need the case when the homology is considered with coefficients in $\mathbb{Z}/2$. It is a polynomial algebra over $\mathbb{Z}/2$ with generators

$$Q_1^{k_1} \ldots Q_{n-1}^{k_{n-1}} a_1, \tag{13}$$

where a_1 is the generator of $H_1(S^1; \mathbb{Z}/2)$. To calculate $M(S^{n+1}, n)_*$, as in the Theorem 3, we need to take the generators, which do not come from $\Omega^2 S^3$. So the generators of $M(S^{n+1}, n)_*$ correspond to the elements (13) with $k_i > 0$ for some $i > 1$. The degree of such a generator is given by a little bit of an awkward formula:

$$2^{k_1+\ldots+k_{n-1}} + 2^{k_1+\ldots+k_{n-2}}(2^{k_{n-1}} - 1) \cdot (n-1) + \ldots + 2^{k_1}(2^{k_2} - 1) \cdot 2 + 2^{k_1} - 1.$$

We obtain the simple expression of degree if we use the upper Araki-Kudo-Dyer-Lashof operations. In this case, let $I = (s_1, \ldots, s_k)$ be a finite sequence of non-negative integers. The degree and excess of I are defined by the usual formulas:

$$\deg(I) = \sum_{j=1}^{k} s_j,$$

$$e(I) = s_k - \sum_{j=2}^{k} (2s_j - s_{j-1}) = s_1 - \sum_{j=2}^{k} s_j.$$

The sequence is called admissible, if $2s_j \geq s_{j-1}$ for $2 \leq j \leq k$. Then $H_*(\Omega^n S^{n+1}; \mathbb{Z}/2)$ is the polynomial ring on the generators $Q^I a_1$, where $I = (s_1, \ldots, s_k)$ is an admissible sequence of Araki-Kudo-Dyer-Lashof operations, with excess greater than or equal to 2: $e(I) \geq 2$, and $s_k \leq n$. We denote by $\mathcal{I}_n$ the set of all finite sequences $I = (s_1, \ldots, s_k)$ with $e(I) \geq 2$ and $3 \leq s_k \leq n$ and by $\mathcal{I}_\infty$ the set of all finite sequences $I = (s_1, \ldots, s_k)$ with $e(I) \geq 2$ and $3 \leq s_k$. Then we have the following theorem.

Theorem 8 *The ring $M(S^{n+1}, n)_*$, $n \geq 3$, is isomorphic to the polynomial ring on an infinite number of generators $\mathbb{Z}/2[\omega_I]$, $I \in \mathcal{I}_n$, $\deg \omega_I = \deg(I) + 1$.*

Let $M(\Omega^\infty S^{\infty+1})$ be the Thom spectrum of the limit map

$$\Omega^\infty S^{\infty+1} \to BO.$$

The space $\Omega^\infty S^{\infty+1}$ was identified by Wagoner [34] and Priddy [25] with plus construction of the classifying space of the infinite group of Fredholm permutations.

Corollary 2 *The ring $M(\Omega^\infty S^{\infty+1})_*$, is isomorphic to the polynomial ring on infinite number of generators $\mathbb{Z}/2[\omega_I]$, $I \in \mathcal{I}_\infty$, $\deg \omega_I = \deg(I) + 1$.*

We have the evident map $S^1_+ \to S^1$ which give rise to the map of Thom spectra

$$M(\Omega_0^\infty S^\infty(S^1_+)) \to M(\Omega_0^\infty S^\infty(S^1)) = M(\Omega^\infty S^{\infty+1}),$$

such that for the coefficient rings of these spectra it generates the epimorphism: a_0 goes to 0 and $Q_1^{k_1} \ldots Q_n^{k_n} a_1$ goes to the corresponding generator of $M(\Omega^\infty S^{\infty+1})_*$.

Another example of the situation described at the beginning of this section gives the James stages J_kS^{n+1} of ΩS^{n+2}, $n \geq 2$, i.e., J_kS^{n+1} play the role of the complex X. Starting with the map $S^{n+2} \to B^{n+2}O$ we consider the composition

$$J_kS^{n+1} \to \Omega S^{n+2} \to B^{n+1}O$$

as the map g and we take the canonical inclusion $S^{n+1} \to J_kS^{n+1}$ as the map f. So, by Theorem 7 we get that the Thom spectrum of the map

$$\Omega^n J_kS^{n+1} \to BO$$

is a wedge of suspensions over the Eilenberg-MacLane spectrum $H\mathbb{Z}/2$. Evidently, $M(J_kS^{n+1}, n)$, $n \geq 2$, is a ring spectrum, which is also a module spectrum over $M(S^{n+1}, n)$.

Let $k = 2^i - 1$, then we can apply the results of the paper of Cohen and Peterson [10] on the homology of $J_{2^i-1}S^{n+1}$ to calculate the ring $M(J_{2^i-1}S^{n+1}, n)_*$. The following fibre sequence was considered there:

$$J_{2^i-1}S^{n+1} \to \Omega S^{n+2} \xrightarrow{h_2^i} \Omega S^{2^i(n+1)+1},$$

where h_2 is the second James-Hopf invariant

$$h_2 : \Omega S^{n+2} \longrightarrow \Omega S^{2(n+1)+1}.$$

It is proved in [10] (Lemma 5.5) that the map $\Omega^n h_2^i$ induces an epimorphism in homology and so we have

$$H_*(\Omega^{n+1}S^{n+2}; \mathbb{Z}/2) \cong H_*(\Omega^n J_{2^i-1}S^{n+1}; \mathbb{Z}/2) \oplus H_*(\Omega^{n+1}S^{2^i(n+1)+1}; \mathbb{Z}/2)$$

and $H_*(\Omega^n J_{2^i-1}S^{n+1}; \mathbb{Z}/2)$ as a submodule of $H_*(\Omega^{n+1}S^{n+2}; \mathbb{Z}/2)$ is generated by $Q_1^{k_1}...Q_n^{k_n}a_1$, with $k_n \leq i-1$, where a_1 is the generator of $H_1(S^1; \mathbb{Z}/2)$. Hence generators of the ring $M(J_{2^i-1}S^{n+1}, n)_*$ can be described as follows: they correspond to the generators of $H_*(\Omega^n J_{2^i-1}S^{n+1}; \mathbb{Z}/2)$ with $k_i > 0$ for some $i > 1$. The degree of such a generator is given by the expression

$$2^{k_1+...+k_n} + 2^{k_1+...+k_{n-1}}(2^{k_n} - 1) \cdot n + ... + 2^{k_1}(2^{k_2} - 1) \cdot 2 + 2^{k_1} - 1. \qquad (14)$$

Thus, the following theorem is proved.

Theorem 9 *The ring $M(J_{2^i-1}S^{n+1}, n)_*$ is a polynomial ring over $\mathbb{Z}/2$ on infinitely many generators which correspond to the elements of $Q_1^{k_1}...Q_n^{k_n}a_1$ in $H_*(\Omega^{n+1}S^{n+2}; \mathbb{Z}/2)$ with $k_i > 0$ for some $i > 1$ and $k_n \leq i - 1$. The degree of such a generator is given by the formula (14).*

Let us consider the image of the ring $M(\Omega^\infty S^{\infty+1})_*$ in the ring MO_* by the map f_*. We denote by a_i the generators of $H_*(BO; \mathbb{Z}/2)$ dual to the degrees of the first Stiefel-Whitney class. The following formula was proved by Kochman [17]:

$$Q^r(a_i) \equiv C_i^{r-1}a_{r+i} \qquad \text{modulo decomposables.}$$

Lemma 2 *i) If an odd number m has the following dyadic decomposition $m = 2^{m_r} + ... + 2^{m_1} + 1$, $m_r > ... > m_1 > 0$, then*

$$Q^{2^{m_r}} ... Q^{2^{m_1}}(a_1) \equiv a_m \qquad \textit{modulo decomposables.}$$

ii) It is impossible to obtain by means of the Araki-Kudo-Dyer-Lashof operations a generator (modulo decomposables) with even index a_{2i} from a generator with odd index a_{2k-1}.

Proof The first assertion of the lemma follows from the direct application of Kochman's formula. For the second assertion we have

$$Q^{2l+1}(a_{2k-1}) \equiv C^{2l}_{2k-1} e_{2k+2l} \equiv 0 \qquad \text{modulo decomposables}$$

for any k and l. Thus, the lemma is proved.

Let W be a subring of the unoriented cobordism ring MO_*. We denote by $\operatorname{Ind} W$ the module of all indecomposable with respect to MO_* elements of W:

$$\operatorname{Ind} W = IW / IW \cap (IMO_*)^2,$$

where IW and IMO_* denote the augmentation ideals of the corresponding rings. The following fact is obtained directly from the Lemma 2.

Proposition 1 *The $\mathbb{Z}/2$-basis of the module of indecomposable elements of the image of the ring $M(\Omega^\infty S^{\infty+1})_*$ in the ring MO_* consists of all generators of MO_* in odd dimensions:*

$$\operatorname{Ind} \operatorname{Im} f_* = \{a_5, a_9, ...\}.$$

Acknowledgements

The authors are thankful to The Fields Institute (Toronto) and the organizers of the year on Homotopy Theory in this institute for the invitation to take part in this meeting where part of this work has been done. The second author is especially thankful to Richard Kane and Stan Kochman for their help in getting a travel grant for him to visit The Fields Institute.

References

[1] Arnold, V. I. [1970] *On some topological invariants of algebraic functions* Trudy Moscov. Mat. Obshch. **21**, 27–46 (in russian), English transl. in Trans. Moscow Math. Soc. **21** (1970), 30–52.

[2] Baez, John C. [1992] *Link Invariants of Finite Type and Perturbation Theory* Letters in Math. Physics **26**, 43–51.

[3] Birman, Joan S. [1993] *New points of view in knot theory* Bull. of the Amer. Math. Soc. **28** No 2, 253–387.

[4] Bödigheimer, C.-F., Cohen, F. R. and Milgram, R. J. [1993] *Truncated symmetric products and configuration spaces* Math. Z. **214** No 2, 179–216.

[5] Brieskorn, E. *Sur les groupes de tresses* Sém. Bourbaki, n° 401, novembre 1971 (Lecture Notes in Math., No 317, 1973, 21–44).

[6] Bullett, Shaun [1979] *Permutations and braids in cobordism theory* Proc. of the London Math. Soc. **38,** Part 3, 517–531.

[7] Cohen, F. R. [1976] *The homology of C_{n+1}-spaces* Lecture Notes in Math. **533**, 207–253.

[8] Cohen, F. R. [1978] *Braid orientations and bundles with flat connections* Inventiones Math. **46**, 99–110.

[9] Cohen, F. R. [1985] *Artin's braid groups and classical homotopy theory* Cont. Math. **44**, 207–220.

[10] Cohen, F. and Peterson, F. P. [1989] *On the homology of certain spaces looped beyond their connectivity* Israel J. of Math. **66**, 105–131.

[11] Deligne, P. [1972] *Les immeubles des groupes de tresses généralisés* Inventiones Math. **17**, 273–302.

[12] Fadell, E. and Neuwirth, L. [1962] *Configuration spaces* Math. Scand. **10 Fasc. I**, 111–118.

[13] Fenn,, R., Rimányi, R. and Rourke, C. *Some remarks on the Braid-Permutation Group* Topics in Knot Theory. Kluwer Academic Publishers, 1993, 57–68.

[14] Fenn, R., Rimányi, R. and Rourke, C. *The Braid-Permutation Group* Preprint, 1993; to appear in Topology.

[15] Fuks, D. B. [1970] *Cohomology of the braid group mod 2* Functional. Anal. i Prilozh. **4,** No 2, 62–75 (Russian), English transl. in Functional Anal. Appl. **4** (1970), 143–151.

[16] Joyal, A. and Street, R. [1993] *Braided tensor categories* Advances in Math. **102**, 20–78.

[17] Kochman, Stanley O. [1973] *Homology of the classical groups over the Dyer-Lashof algebra* Trans. of Am. Math. Soc. **185**, 83–136.

[18] Mahowald, M. [1977] *A new family in* π_*^s Topology **16**, 249–254.

[19] Mahowald, M. [1977] *Ring spectra which are Thom complexes* Duke Math. J. **46,** No 3, 549–259.

[20] May, J. P. [1972] *The Geometry of iterated loop spaces*, (Lecture Notes in Math.; No 271), Springer-Verlag, Berlin a. o..

[21] May, J. P. E_∞ *spaces, group completions, and permutative categories* New Developments in Topology (London Math. Soc. Lecture Notes Series; No 11) Cambridge. 1974, 61–93.

[22] Milgram, R. J. [1967] *The bar construction and abelian H-spaces* Ill. J. of Math. **11**, 242-250.

[23] Nakaoka, M. [1961] *Homology of the infinite symmetric group* Ann. of Math. **73**, 229–257.

[24] Peterson, F. P. and Toda, H. [1967] *On the structure of* $H^*(BSF; Z/p)$ J. Math. Kyoto Univ. **7**, 113-121.

[25] Priddy, Stewart [1973] *Fredholm permutations and stable homotopy* Michigan Math. J. **20**, 187–192.

[26] Segal, G. [1973] *Configuration spaces and iterated loop spaces* Inventiones Math. **21**, 213–221.

[27] Sossinsky, A. B. [1992] *Preparation theorems for isotopy invariants of links in 3-manifolds* in Quantum Groups, Proceedings (Lecture Notes in Math.; No 1510), Springer-Verlag, Berlin a. o., 354-362.

[28] Switzer, R. M. [1975] *Algebraic Topology – Homotopy and Homology*, Springer-Verlag, Berlin a. o..

[29] Thom, R. [1954] *Quelques propriétés globales des variétés différentiables* Comm. Math. Helv. **28**, 17–86.

[30] Vassiliev, V. A. [1992] *Complements of discriminants of smooth maps: topology and applications*, (Translations of mathematical monographs; v. 98), AMS, Providence.

[31] Vershinin, V. V. [1995] *Thom spectra of generalized braid groups* Preprint No 95/02-2, Université de Nantes.

[32] Vershinin, V. V. [1996] *Homology of Braid Groups in Handlebodies* Preprint No 96/06-2, Université de Nantes.

[33] Vershinin, V. V. *Homological Properties of Singular Braids* Preprint No 96/06-3, Université de Nantes, 1996, to be published in Trans. of the Amer. Math. Soc..

[34] Waggoner, J. B. [1972] *Delooping classifying spaces in algebraic K-theory* Topology **11**, 349–370.

Fields Institute Communications
Volume **19**, 1998

Some Properties of the Relative Lusternik-Schnirelmann Category

Octavian Cornea
Université de Lille 1
U.F.R. de Mathématiques & U.R.A. 751 au CNRS
59655 Villeneuve D'Ascq, France
cornea@gat.univ-lille1.fr

Abstract. We study some properties of the relative Lusternik-Schnirelmann category of Fadell. In particular, it is shown that it admits a Ganea type description. We also show that, in a compact setting, it bounds the minimal number of critical points of a certain type of functions that have a more complicated boundary behaviour than that usually allowed. This is used to relate some forms of the Arnold conjecture in symplectic geometry to the Ganea conjecture in homotopy theory.

1 Homotopical Remarks

We work in the pointed category of the spaces with the homotopy type of a CW-complex. Let $f : A \longrightarrow X$ be a fixed inclusion. Following Fadell the relative Lusternik-Schnirelmann category of X, $cat(X, A)$, is the minimal n for which there is a covering of X with $n+1$ open subsets U_i of X, $0 \leq i \leq n$, such that there is a homotopy of pairs $(U_0 \times I, A \times I) \longrightarrow (X, A)$ that contracts U_0 to A and, for $i \geq 1$, U_i is contractible in X. The absolute, classical case corresponds to $A = *$.

This invariant has been used in many variational problems (originally, in the absolute case, by Lusternik and Schnirelmann [13]) and has been generalized to the equivariant setting by Clapp and Puppe [1].

The properties of the relative invariant parallel those of the absolute case (compare with [12]). In particular [7] one immediately obtains a Whitehead type definition by adapting the proof from the absolute setting. We first define the fat wedge relative to A: $T^1(X, A) = A$, $T^{n+1}(X, A) = (A \times X^n) \bigcup (X \times T^n(X)) \subset X^{n+1}$. Where $T^n(X)$ is the standard fat wedge. For further use note that $T^n(X, A) = (T^{n-1}(X, A) \times X) \bigcup (X^{n-1} \times *)$. We suppose that X is connected. Then $cat(X, A) \leq n$ iff the diagonal $X \longrightarrow X^{n+1}$ can be deformed in the fat wedge $T^{n+1}(X, A)$ by a homotopy $H : (X, A) \times I \longrightarrow (X^{n+1}, T^{n+1}(X, A))$ which verifies $p_1 \circ H_t \circ f = f$ for all $0 \leq t \leq 1$ where p_1 is the projection on the first factor.

1991 *Mathematics Subject Classification.* Primary 55P50; Secondary 57R70.

Now we want to give another description of the relative category. We consider the following sequence of fibrations: $E_i(X,A) \xrightarrow{j_i} B_i(X,A) \xrightarrow{p_i} X$, $0 \le i$ with $A \simeq B_0(X,A)$; $p_0 \simeq f$; $E_i(X,A)$ is the fibre of p_i; j_i is the inclusion of the fibre; $B_{i+1}(X,A)$ is of the homotopy type of the cofibre of j_i and p_{i+1} is, up to homotopy, the map canonically induced by p_i. Notice that if f is $* \longrightarrow X$ this sequence of fibrations is equivalent to that given by the succesive filtrations in Milnor's classifying construction for ΩX. We also denote by q_i the obvious map $A \longrightarrow B_i(X,A)$ ($q_0 \simeq id$). We have that q_{i+1} factors as $t_i \circ q_i$ where $t_i : B_i(X,A) \longrightarrow B_i(X,A)/E_i(X,A)$ is the collapsing map.

This construction is commonly called the fibre-cofibre construction and is due to Ganea [9] (see also [10]).

Proposition 1.1 *If X is connected* $\mathrm{cat}(X,A)$ *equals the least n such that, with the notations above, p_n admits a section s that verifies $s \circ f \simeq q_n$.*

Proof The key point is to show that:

Lemma 1.2 *Let p'_n be the fibration obtained by transforming the inclusion $T^{n+1}(X,A) \subset X^{n+1}$ into a fibration. The fibration p_n is the pull back of p'_n over the diagonal.*

Indeed, assuming the lemma one easily sees that the existence of a deformation like the one in the Whitehead type definition is equivalent to the existence of a section of p_n that respects the maps out of A.

Proof of the Lemma We proceed by induction. This is clear for $n = 0$. We suppose it holds for $n \le k$. Recall that $T^{k+1}(X,A)$ is the push out of the inclusions: $i : T^k(X,A) \times * \longrightarrow X^k \times *$ and $j : T^k(X,A) \times * \longrightarrow T^k(X,A) \times X$. We transform all inclusions in X^{k+1} into fibrations and then take the pull backs over the diagonal. In this way we obtain spaces U, V, W, K being respectively the pull backs of $T_k(X,A) \times *$, $X^k \times *$, $T^k(X,A) \times X$, $T^{k+1}(X,A)$. There are also maps $i' : U \longrightarrow V$, $j' : U \longrightarrow W$ and $V \longrightarrow K$, $W \longrightarrow K$. Moreover, one sees that by applying the cube theorem (on the pull back of a push out square) we find that the square given by these maps is a homotopy push out square. By induction, the fibration $W \longrightarrow X$ coincides with p_{k-1}. The space V is contractible and j' is identified with the inclusion of the fibre $E_{k-1}(X,A) \longrightarrow B_{k-1}(X,A)$. □

In the following we will assume that the biggest set in the pairs of spaces that we use is connected.

Remarks We recall a few immediate consequences of the three descriptions above.

1. If the pair (X,A) is a pair retract of (Y,B), then $cat(X,A) \le cat(Y,B)$.
2. If f factors as $A \to X' \to X$, then we have $cat(X,A) \le cat(X,X') + cat(X',A)$.
3. We have $cat(X,A) \ge min\{n : H^*(X,A)(\overline{H}^*(X))^n = 0\}$.
4. If $cat(X,A) \ne 0$, then $cat(X,A) = cat(X \vee \Sigma T, A)$ where the map $A \longrightarrow X \vee \Sigma T$ is the obvious composition.

Another invariant that is useful in relation with the category is the cone-length (see [4,14] for more details).

Definition 1.3 The cone-length of X is the least n (or ∞) such that there are cofibration sequences $\Sigma^i Z_i \longrightarrow X_i \longrightarrow X_{i+1}$ with $0 \le i < n$, $X_0 \simeq *$, $X_n \simeq X$.

The relation between category and cone-length is described next [4]:

Proposition 1.4 *Assume* $\mathrm{cat}(X)= n$. *There is a space* Z *such that* $\mathrm{Cl}(X \vee \Sigma^n Z)= n$.

The covering description of the relative category implies $cat(X, A) \leq cat(X)+1$. A special case is of interest:

Corollary 1.5 *If* f *admits a homotopy section, then* $\mathrm{cat}(X, A) \leq \mathrm{cat}(A)$.

Proof We use the interpretation of $Cl-$ in terms of obstruction theory (see [5] for details). Let $g : Y \longrightarrow X$ be a map and suppose that g_1 and g_2 are two lifts of g up to A. We can certainly obtain maps $g_1^i = q_i \circ g_1$, $g_2^i = q_i \circ g_1$. They are liftings of g through p_i. The key point is that if $i \geq cat(Y)$ we have $g_1^i \simeq g_2^i$. This is seen in the following way. First, after possibly wedging Y with a suspension we may assume $Cl(Y)=cat(Y)$. Consider now a minimal cone decomposition of Y, $T_n \longrightarrow Y_n \longrightarrow Y_{n+1}$, $(Y_0 \simeq *)$ and suppose inductively that we have constructed a homotopy between the restrictions to Y_k of g_1^k and g_2^k. The obstruction to extend this homotopy to Y_{k+1} lies in $[\Sigma T_k, E_k(X, A)]$ and is killed when passing to $B_{k+1}(X, A)$. So we get the next step. The process starts for $k = 0$ because all our maps are pointed.

We turn now to our context and apply the above argument to the identity $A \longrightarrow A$ and to the composition of the provided section with f. □

2 Critical Points

Here we show how one can use the relative category to bound the minimal number of critical points in cases when the boundary is present and when the behaviour of the function on the boundary does not allow a direct application of the standard proof of the Lusternik-Schnirelmann inequality. In fact, we obtain the inequality as a consequence of a somewhat stronger result.

We consider a smooth, riemannian, compact manifold (possibly with boundary) N and a smooth function $h : N \longrightarrow \mathbf{R}$. Inside N we consider a topological compact manifold with boundary, M, of the same dimension as that of N. We also assume that M is smooth everywhere except that its boundary might have some corners. Precisely, $\partial M = E \bigcup I$ where E and I are closed, smooth and the union is taken over $\partial E = \partial I$ (which could be void). Moreover, we assume that there is an $\epsilon > 0$ such that $(\nabla h(x)/||\nabla h(x)||) \cdot \nu_E \leq -\epsilon$ for $x \in Int(E)$ and $(\nabla h(x)/||\nabla h(x)||) \cdot \nu_I \geq \epsilon$ for $x \in Int(I)$. Here ν_E, ν_I are respectively normal, unit vectors to E and I, pointing out of M. We will refer to this as condition **T**.

These conditions are not too restrictive: for example if N is a cobordism such that ∂N is the disjoint union of V_1 and V_2 and h is maximal, constant and regular on V_2 and minimal, constant and regular on V_1, then M can be obtained from N by just deleting a small colar of the boundary.

We fix a notation: $\gamma_t(x)$ is the flow induced by $-\nabla h$. It points out on E which will be thus called the exit set of M with respect to γ and in on I which will be called the entrance set.

We are also going to assume that E is connected (which implies that M is connected too).

Assume that S is the maximal invariant set of γ in the interior of M. Let $M^-(S)$ be the set of points x of M that verify $lim_{t \to -\infty} \gamma_t(x) \in S$.

Proposition 2.1 *If* U *is a closed neighbourhood of* $M^-(S)$, *then the pair* (M, E) *is a homotopy retract of pairs of* $(U \bigcup E, E)$.

Proof The idea is the same as that of Proposition 2.1 in [6] applied here with some caution needed to control the boundary behaviour.

Define the function: $t : M \longrightarrow \mathbf{R}\bigcup\{\infty\}$, $t(x) = sup\{\tau \in \mathbf{R} : \gamma_\tau(x) \in M\}$. Call this function the departure-time function relative to E. It is obvious that $t(x) = \infty$ iff $lim_{\tau\to\infty}\gamma_\tau(x) \in S$. Also, if $t(x) \neq \infty$, then $\gamma_{t(x)}(x) \in E$. Moreover, each point x such that $t(x) \neq \infty$ has a neigborhood on which the function t is continuous. Indeed, the continuity of γ implies that there is a neighborhood of x containing only points for which the departure time is finite. As ∇h is never tangent to E it follows that for a given $\epsilon > 0$ there is a possibly smaller neighborhood of x, U, such that for all $y \in U$ we have that $t(x) - \epsilon < t(y) < t(x) + \epsilon$. This implies the continuity we are looking for.

For further use notice also that all these arguments do not actually depend on the smootheness of E. It is enough for E to be a topological manifold admitting a dense set of points where the normal is defined and for condition **T** to be satisfied with respect to these points.

Notice that the pair (K, E) is an index pair in the sense of the Conley index theory [2] for the flow γ and the pair (K, I) is an index pair for the inverse flow that we denote by γ' (it is induced by ∇h). We use Lemma 5.3 of [16] and apply it to γ'. It provides a continuous function $g : M \longrightarrow [0, 1]$ that has the following properties: $g(x) = 1$ iff $x \in M^-(S)$, $g(x) = 0$ iff $x \in I$; for $t < 0$ and x satisfying $g(x) \in (0, 1)$, $\gamma_\tau x \in M$ for $\tau \in [t, 0]$, we have $g(\gamma_t(x)) < g(x)$. Put $L = E\bigcup g^{-1}([\delta, 1])$. We choose δ close enough to 1 such that $L \subset E\bigcup U$.

We end the proof by showing that L is a deformation retract of M relative to E.

For this, define the following function: $j : M \longrightarrow \mathbf{R}$, $j(x) = min(t(x), u(x))$ with $u(x) = sup\{t \in \mathbf{R} : g(\gamma_t(x)) \leq \delta\}$ if $g(x) \leq \delta$ and $u(x) = 0$ in the opposite case. We see that u is continuous around all points where it is finite. This implies that j is continuous and by compacity there is a $T = max\{j(x) : x \in M\}$. In turn, this allows us to define the deformation $d(t, x) = \gamma_t(x)$ if $t \leq j(x)$ and $d(t, x) = \gamma_{j(x)}(x)$ otherwise. □

Corollary 2.2 cat$(M, E)\leq$cat$(U\bigcup E, E)$.

Corollary 2.3 *The number of critical points of h is at least equal to* cat(M, E).

Proof Assume that the number of critical points of h is finite. We may also assume by modifying a bit the function in small neighbourhoods of the critical points that they are situated on different level sets. Consider two regular values of h, $a < b$. Take $h^{-1}(a)\bigcap M$ and push it a little bit such that it will cut transversely E and I as well as their common boundary. We denote the resulting hypersurface by H_a. We clearly may assume that ∇h is never tangent to H_a. We proceed similarly for b thus obtaining H_b. Let K be the portion of M comprised between H_b and H_a. We put $E' = H_a\bigcup(E\bigcap K)$ and $I' = H_b\bigcup(I\bigcap K)$. Notice that the proof of the proposition above applies to K (relative to E' and I'). In particular, if there is no critical point in K we have $cat(K, H_a)= 0$. If there is just one critical point and a is sufficiently close to b then we may take as neighbourhood of $K^-(S)$ a disk that intersects transversely H_a and is disjoint from $\partial(M)$. Hence $cat(K, H_a)\leq 1$. This implies the statement.

Here is the relation with the Arnold conjecture.

Corollary 2.4 *If (M, ω) is a compact symplectic manifold with Euclidean cover, then any smooth 1-periodic Hamiltonian system has at least* $\mathrm{cat}(M, M \times S^n)$ *contractible 1-periodic orbits for any $n \in \mathbf{N}^*$ sufficiently large.*

Proof We use the function g constructed in the proof of Theorem 3.2 in [15]. We see that this function is defined on $M \times D^{2m}$, it points in on $M \times \partial D^m \times D^m$ and out on $M \times D^m \times \partial D^m$ and the estimates at the end of the proof of Theorem 3.2 show that it satisfies condition $\mathbf{T}$. We can multiply $M \times D^{2m}$ by a disk D^q and add to g a quadratic form q defined on D^q and which is negative definite. This sum $g \oplus q$ satisfies again condition $\mathbf{T}$ and has the same number of critical points as g, the exit set in this case is $M \times D^m \times S^{m+q-1}$. We apply our theorem to $g \oplus q$. □

Final comments. A form of Arnold's conjecture claims that even in the absence of the Euclidean cover condition the number of 1-periodic orbits is at least equal to the least number of critical points of a smooth function on M (crit(M)). Results in this direction started with the work of Conley and Zehnder [3] and were then pursued by Floer [8] and Hofer [11]. They show that if $[\omega]|\pi_2(M) = 0$, then the number of 1-periodic orbits is strictly bigger than the cup-length of the cohomology of M (with arbitrary coeficients). Clearly, this cup-length is smaller or equal than $cat(M)$ and $cat(M)+1 \leq crit(M)$. It was proven by McCord and Oprea [15] that the cup-length can be replaced in the Euclidean-cover case by a homotopy invariant: the rational Toomer invariant.

The relative category $cat(X, X \times S^n)$ is bigger or equal than the cup-length of X+1 (with any coefficients). Indeed, take a class of maximal length in $H^*(X)$ and multiply it by $\delta(i^n)$ with i^n the fundamental class of S^n; δ being the connectant $H^n(X \times S^n) \longrightarrow H^{n+1}(X, X \times S^n)$). It can also be shown that for all simply connected spaces X and in all the cases covered by the results of McCord and Oprea $cat(X, X \times S^n)$ is also bigger or equal than the rational Toomer invariant +1 of X (for all $n > 0$). Another study of this problem has been recently done by Rudyak [16].

At the other end of the inequality conjectured by Arnold, the methods in [6] show that in general one might expect $cat(M, M \times S^n)$ to be a close estimate of the minimal number of critical points of a function g like the one in the proof of the corollary above (however, possibly not coming from a Hamiltonian).

This immediately suggests the following statement (that one could call the "homotopical Arnold conjecture"):

Conjecture: $cat(X, X \times S^n)$=$cat(X)$+1 for all connected, finite type CW-complexes X ($n \in \mathbf{N}^*$).

By Corollary 1.4 if this equality is true then $cat(X \times S^n)$=$cat(X)$+1 so the homotopical Arnold conjecture implies the Ganea conjecture.

Clearly, the following question is also open:

Problem: Is the statement of Corollary 2.4 valid in the absence of the Euclidean cover condition?

References

[1] Clapp, M. and Puppe D. [1991], *Critical point theory with symmetries*, J. reine angew. Math. **418**, 1-29.

[2] Conley, Ch. [1976], *Isolated invariant sets and the Morse index*, CBMS Regional Conf.Ser. in Math 38, Amer. Math. Soc. , Providence, R. I.

[3] Conley, Ch. and Zehnder, E. [1983], *The Birkhoff-Lewis Fixed Point Theorem and a conjecture of V. I. Arnold*, Invent. Math. **73**, 33-49.

[4] Cornea, O. [1994], *Cone-length and Lusternik Schnirelmann category*, Topology **33**, 95-111.

[5] Cornea, O. [1995], *Lusternik-Schnirelmann categorical sections*, Ann. Scient. Ec. Norm. Sup. **28**, 689-704.

[6] Cornea, O. [1995],*Cone-decompositions and degenerate critical points*, preprint.

[7] Fadell, E. and Husseini, S. Y. [1994], *Relative category, products and coproducts*, preprint.

[8] Floer, A. [1989], *Cuplength estimates on Lagrangian intersections*, Comm. Pure Appl. Math. **42**, 335-356.

[9] Ganea, T. [1967], *Lusternik-Schnirelmann category and strong category*, Ill. J. of Math. **11**, 417-427.

[10] Gilbert, W. [1968], *Some examples for weak category and conilpotency*, Ill. J. of Math. **12**, 421-432.

[11] Hofer, H. [1988], *Ljusternik-Schnirelmann theory for Lagrangian intersections*, Ann. Inst. H. Poincaré **5**, 465-499.

[12] James, I. M. [1978], *On category in the sense of Lusternik and Schnirelmann*, Topology **17**, 331-348.

[13] Ljusternik, L. and Schnirelmann, L. [1934], *Methodes topologiques dans les problèmes variationels*, Hermann, Paris.

[14] Marcum, H. [1996], *Cone-length of the exterior join*, preprint.

[15] McCord, Ch. and Oprea, J. [1993], *Rational Ljusternik-Schnirelmann category and the Arnold conjecture*, Topology **32**, 701-717.

[16] Rudyak, Y. [1997], *On analytical applications of stable homotopy*, preprint.

[17] Salamon, D. [1985], *Connected simple systems and the Conley index of isolated invariant sets*, Trans of the A.M.S. **291**, 1-41.

Fields Institute Communications
Volume **18**, 1998

The Generating Hypothesis Revisited

Ethan S. Devinatz
Department of Mathematics
University of Washington
Seattle, Washington USA 98195
devinatz@math.washington.edu

Abstract. We show that, at large primes, a localized version of the generating hypothesis can be reduced to the analysis of certain universal examples. We explicitly determine these examples "mod (p)", assuming a certain finiteness result for the stable homotopy groups of the Moore spectrum.

Introduction

Let X and Y be finite spectra, and let $f : X \to Y$ be a stable map. Recall that Freyd's generating hypothesis [**6**] asserts that if $\pi_* f = 0$, then f is stably null homotopic. Since f is trivial if and only if $f_{(p)}$ is trivial for each prime p, the generating hypothesis can be confirmed by proving its p-local analogue one prime at a time. We thus fix a prime p and work in the p-local stable category for the rest of this paper; in particular, by a finite spectrum we mean the p-localization of one.

In much of our work (see, for example, [**2**]), we have alluded to a framework, originally proposed by M. J. Hopkins, for proving the generating hypothesis when the target is S^0. In this framework, one must determine a certain collection of essential stable maps $X \to S^0$ with X a finite spectrum, and then prove that these maps induce nontrivial homomorphisms in stable homotopy. These maps are called the "universal examples", and the search for them motivated the work of [**5**], [**7**], and [**8**]. Hopkins' chromatic splitting conjecture (see [**12**], 4.2) also appears to be relevant. Assuming part of this conjecture, one can explicitly determine some of the universal examples; one can also conjecture as to how these maps might be proved to be nontrivial on π_*. It is the purpose of this paper to give an account of our current state of knowledge regarding the generating hypothesis.

Let E be a spectrum, and let L_E denote E_*-localization. Recall that the "E-approximate generating hypothesis" is the assertion that a map f between finite spectra with $\pi_* f = 0$ satisfies $L_E f = 0$. If the $E(n)$-approximate generating hypothesis is true for all $n \geq 0$, then by the chromatic convergence theorem ([**22**], 7.5.7), so is the generating hypothesis. Here and throughout the paper, $E(n)$ is

1991 *Mathematics Subject Classification.* Primary 55P42, 55Q10; Secondary 55N22, 55T15.
Partially supported by NSF grant DMS–954476.

the spectrum representing the Landweber exact cohomology theory with coefficient ring $\mathbb{Z}_{(p)}[v_1, \ldots, v_n, v_n^{-1}]$. We also write L_n for $L_{E(n)}$.

Unfortunately, the failure of the telescope conjecture for $n \geq 2$ makes the original program for proving the $E(n)$-approximate generating hypothesis impractical; this will be discussed in Section 2. However, the telescope conjecture does hold $E(N)_*$-locally (see Section 1), leading us to modify our program.

Say that an $E(N)_*$-local spectrum X is L_N-finite if it can be constructed from (suspensions of) $L_N S^0$ via a finite number of cofibration sequences and retractions. Our new conjecture is then the following.

Conjecture 1 *Fix $n \geq 2$. If N is sufficiently large, and $f : X \to L_N S^0$ is any map with X L_N-finite and with $\pi_* f = 0$, then*

$$L_n f : L_n X \to L_n L_N S^0 = L_n S^0$$

is trivial.

The requirement that N be sufficiently large is necessary; we will give a counterexample in Appendix 2 to the case $N = n \geq 2$. The $n = 1$ case of Conjecture 1 was proved in [**2**] for p odd.

Of course, one cannot prove the generating hypothesis by proving Conjecture 1 for each n; one needs to show in addition, that, if X is a finite spectrum and $f : X \to S^0$ satisfies $\pi_* f = 0$, then $\pi_* L_N f = 0$ for N sufficiently large (possibly depending on X). This problem may be quite formidable.

Now Hopkins' chromatic splitting conjecture implies (cf. [**12**], 4.3) that if f is a map between finite spectra with $L_{K(n)} f = 0$, then $L_{n-1} f = 0$. If this conjecture were true, we would only need then consider the following conjecture.

Conjecture 2 *Fix $n \geq 2$. If N is sufficiently large, and $f : X \to L_N S^0$ is any map with X L_N-finite and with $\pi_* f = 0$, then*

$$L_{K(n)} f : L_{K(n)} X \to L_{K(n)} L_N S^0 = L_{K(n)} S^0$$

is trivial.

If we assume further the conjecture that $\pi_* L_{K(n)} S^0$ is a module of finite type over the p-adic integers, then the problem of determining the universal examples for Conjecture 2 begins to appear tractable. In fact, we are able to explicitly describe, for large primes p, the universal examples for the following mod(p) version of Conjecture 2. Let $M(p)$ denote the mod(p) Moore spectrum.

Conjecture 3 *Fix $n \geq 2$. If N is sufficiently large, and $f : X \to L_N S^0$ is any map with X L_N-finite and with $\pi_*(f \wedge M(p)) = 0$, then $L_{K(n)} f \wedge M(p)$ is trivial.*

Now let $M(p, v_1^{i_1}, \ldots, v_{n-1}^{i_{n-1}})$ be a finite complex with

$$BP_* M(p, v_1^{i_1}, \ldots, v_{n-1}^{i_{n-1}}) = BP_*/(p, v_1^{i_1}, \ldots, v_{n-1}^{i_{n-1}}) \tag{0.1}$$

constructed as in [**4**] (Section 4). Also as in [**4**] (Section 4), there exists a cofinal sequence T_1 of multi-indices $I = (i_1, \ldots, i_{n-1})$ such that

$$M_n M(p) = \operatorname*{holim}_{\overrightarrow{T_1}} \Sigma^{-n_I+1} L_n M(p, v_1^{i_1}, \ldots, v_{n-1}^{i_{n-1}}), \qquad n_I = \sum_{r=1}^{n-1} 2 i_r (p^r - 1). \tag{0.2}$$

In Section 4, we construct certain elements

$$h_I : L_N S^0 \to L_N M(p, v_1^{i_1}, \ldots, v_{n-1}^{i_{n-1}})$$

detected in $\mathrm{Ext}^{n^2,*}_{BP_*BP}(BP_*S^0, BP_*L_NM(p, v_1^{i_1}, \ldots, v_{n-1}^{i_{n-1}}))$, where suspensions have been deleted from the notation. We then prove the following result. Let

$$h_I^{\#} : L_NDM(p, v_1^{i_1}, \ldots, v_{n-1}^{i_{n-1}}) \to L_NS^0$$

denote the adjoint of h_I, where $DM(p, v_1^{i_1}, \ldots, v_{n-1}^{i_{n-1}})$ denotes the Spanier-Whitehead dual of $M(p, v_1^{i_1}, \ldots, v_{n-1}^{i_{n-1}})$. (In this case it is just a desuspension of $M(p, v_1^{i_1}, \ldots, v_{n-1}^{i_{n-1}})$.)

Theorem 1 *Suppose that $p > \frac{1}{2}(n^2+2)$, $n \geq 2$. If $\pi_*L_{K(n)}M(p)$ is of finite type and, for each multi-index I, $\pi_*(h_I^{\#} \wedge M(p)) \neq 0$, then Conjecture 3 is true.*

We also make some conjectural remarks concerning how these universal examples may be dealt with.

Although we have apparently strayed far from the generating hypothesis—and have even needed to assume an unproven result to make our final reduction—let us emphasize that in the end we have exhibited certain essential maps from an L_N-finite spectrum to L_NS^0. If Conjecture 1 is true, these maps must be nonzero on stable homotopy. Furthermore, if one could prove their nontriviality on π_*, we feel that this would be strong evidence for the generating hypothesis and would indicate that, at least from the chromatic point of view, any failure would have to be rather subtle. Finally, any proof of their nontriviality on π_* would surely involve interesting global methods in stable homotopy theory.

Let us conclude this section by indicating some notation that we will use. Since we will often be dealing with BP_*BP-comodules, we will write $H^{**}(N)$ for $\mathrm{Ext}^{**}_{BP_*BP}(BP_*, N)$ and, more generally, $\mathrm{Ext}^{**}(M, N)$ for $\mathrm{Ext}^{**}_{BP_*BP}(M, N)$. (When M is not projective as a BP_*-module, this is to be computed using an injective resolution of N in the category of BP_*BP-comodules.) If E is an appropriate ring spectrum, and X and Y are any spectra, we will denote the rth stage of the modified Adams spectral sequence for $[X, Y]_*$ [**4**] by $E_r^{**}(E; X, Y)$. If $X = S^0$, we will just write $E_r^{**}(E; Y)$.

1 $E(N)_*$-local Homotopy Theory

In this section we recall the basic theory of nilpotence and periodicity in the $E(N)_*$-local homotopy category and prove that the telescope conjecture holds here.

As in the introduction, we define a spectrum to be L_N-finite if it can be built out of $E(N)_*$-local spheres by a finite number of cofibrations and retractions. More precisely, the class of L_N-finite spectra is the smallest class $\mathcal{G}$ of ($E(N)_*$-local) spectra having the following properties.

(i) If $X \simeq L_NS^k$, $k \in \mathbb{Z}$, then X is in $\mathcal{G}$.
(ii) If $X \to Y \to Z$ is a cofibration sequence with X and Y in $\mathcal{G}$, then Z is in $\mathcal{G}$.
(iii) If X is a retract of Y and Y is in $\mathcal{G}$, then so is X.

These spectra are small in the sense that if X is L_N-finite and $\{Z_\alpha : \alpha \in J\}$ is any collection of $E(N)_*$-local spectra, then the map

$$\bigoplus_\alpha [X, Z_\alpha]_* \to [X, \bigvee_\alpha Z_\alpha]_* \tag{1.1}$$

is an isomorphism. To prove this, observe that $\bigvee_\alpha Z_\alpha$ is $E(N)_*$-local since $E(N)_*$-localization is smashing ([**22**], 7.5.6). Hence the map in (1.1) is an isomorphism for $X = L_NS^k$; this easily implies the desired result.

Furthermore, each L_N-finite spectrum is finite in the sense of [**16**] (III,1.1) in the category of $E(N)_*$-local spectra. This means that the function spectrum $D_N X = F(X, L_N S^0)$ for X L_N-finite has the formal properties of the Spanier-Whitehead dual of a finite spectrum. Note that if $X = L_N Y$ with Y a finite spectrum, then $D_N X = L_N DY$. In general, $D_N X$ is L_N-finite whenever X is.

Now let $\langle Z \rangle$ denote the Bousfield class of Z. Then

$$\langle L_N S^0 \rangle = \langle E(N) \rangle = \langle K(0) \rangle \vee \cdots \vee \langle K(N) \rangle$$

by [**21**] (2.1d) and [**22**] (7.5.6); it therefore follows easily (cf. [**14**], 5.1.3) that if Z is an L_N-finite spectrum and $f : X \to Y$ is a map between L_N-finite spectra with $K(n)_*(Z \wedge f) = 0$ for all $0 \le n \le N$, then

$$Z \wedge f^{(m)} : Z \wedge \underbrace{X \wedge \cdots \wedge X}_{m \text{ times}} \to Z \wedge Y \wedge \cdots \wedge Y$$

is trivial for some m.

Let $\mathcal{C}_m^N$ denote the class of L_N-finite spectra which are $K(m-1)_*$-acyclic, where $0 \le m \le N+1$. We then have the following result.

Proposition 1.2 *$\mathcal{C}_{m+1}^N \subsetneq \mathcal{C}_m^N$ whenever $0 \le m \le N$.*

The proof of Proposition 1.2 requires a lemma. Recall that $P(m)$ is a BP-module spectrum with coefficient ring $P(m)_* = \mathbb{F}_p[v_m, v_{m+1}, \ldots]$.

Lemma 1.3 *Let X be L_N-finite. Then $v_m^{-1}P(m-1)_*X$ is a finitely generated $v_m^{-1}P(m-1)_*$-module whenever $1 \le m \le N$.*

Proof Begin by observing that $v_m^{-1}P(m-1)_*$ is coherent, since it is the direct limit of a sequence of noetherian rings and is flat over each (see [**1**]). Hence it suffices to show that $v_m^{-1}P(m-1)_*L_N S^0$ is a finitely presented $v_m^{-1}P(m-1)_*$-module. But $\langle v_m^{-1}BP \rangle = \langle E(m) \rangle$ ([**15**], 0.5); thus $v_m^{-1}BP_*S^0 \xrightarrow{\approx} v_m^{-1}BP_*L_N S^0$ if $m \le N$. Since $v_m^{-1}P(m-1)_*(?)$ is connected to $v_m^{-1}BP_*(?)$ via a succession of long exact sequences, it follows that

$$v_m^{-1}P(m-1)_*S^0 \xrightarrow{\approx} v_m^{-1}P(m-1)_*L_N S^0.$$

This completes the proof. □

Proof of 1.2 That $\mathcal{C}_{m+1}^N \neq \mathcal{C}_m^N$ follows from the well known global result [**9**]. To complete the proof, it suffices to show that if X is L_N-finite and $K(m)_*X = 0$, then $K(m-1)_*X = 0$. But $K(m)_*X = 0$ implies $v_m^{-1}P(m)_*X = 0$ ([**22**], 2.1a); it therefore follows from the long exact sequence

$$\cdots \longrightarrow v_m^{-1}P(m-1)_*X \xrightarrow{v_{m-1}} v_m^{-1}P(m-1)_*X \longrightarrow v_m^{-1}P(m)_*X \longrightarrow \cdots$$

that

$$v_m^{-1}P(m-1)_*X \xrightarrow{\approx} v_m^{-1}v_{m-1}^{-1}P(m-1)_*X \equiv v_m^{-1}B(m-1)_*X.$$

Now $B(m-1)_*X \approx B(m-1)_* \otimes_{K(m-1)_*} K(m-1)_*X$ as $B(m-1)_*$-modules ([**26**], 3.13), and therefore $v_m^{-1}v_{m-1}^{-1}P(m-1)_*X$ is a direct sum of suspensions of copies of $v_{m-1}^{-1}v_m^{-1}P(m-1)_*$. This cannot be a finitely generated module over $v_m^{-1}P(m-1)_*$ as required by Lemma 1.3 unless it is trivial. Hence $K(m-1)_*X = 0$, completing the proof. □

With this result, one can prove, just as in [**9**], that any class of L_N-finite spectra closed under cofibrations and retracts (a thick subcategory in the language of [**22**]) must be $\mathcal{C}_m^N$ for some m. In particular, all spectra in $\mathcal{C}_m^N - \mathcal{C}_{m+1}^N$ are

Bousfield equivalent and have a v_m self-map. As in [**9**], one can prove the essential uniqueness and centrality ($m > 0$) of these maps.

We now turn to the $E(N)_*$-local telescope conjecture in its various forms. Although this material is well known, we include a sketch for the convenience of the reader.

Proposition 1.4 *Let X be in $\mathcal{C}_m^N - \mathcal{C}_{m+1}^N$, $1 \leq m \leq N$, and let $\widehat{X}$ be the infinite mapping telescope of a v_m self-map on X. Then the canonical map $\widehat{X} \to L_{K(m)}X$ is an equivalence.*

Proof Standard techniques (see [**17**], 2.2) show that it suffices to prove that $\langle \widehat{X} \rangle = \langle K(m) \rangle$. Since $\langle \widehat{X} \rangle$ is independent of the choice of X in $\mathcal{C}_m^N - \mathcal{C}_{m+1}^N$, we may as well assume that $X = L_N Z$, where Z is a finite $K(m-1)_*$-acyclic spectrum with $K(m)_* Z \neq 0$. It is proved in [**12**],(1.9) that $\langle BP \wedge \widehat{Z} \rangle = \langle K(m) \rangle$. Hence $\langle E(N) \wedge \widehat{Z} \rangle = \langle K(m) \rangle$. But $\langle E(N) \rangle = \langle L_N S^0 \rangle$; therefore $\langle L_N \widehat{Z} \rangle = \langle L_N S^0 \wedge \widehat{Z} \rangle = \langle K(m) \rangle$. This completes the proof. □

Proposition 1.5 *Let W be an $E(N)_*$-local spectrum, and suppose $N \geq n$. Then W is $E(n)_*$-local if and only if $[Z, W]_* = 0$ for some (and hence all) L_N-finite spectrum Z in $\mathcal{C}_n^N - \mathcal{C}_{n+1}^N$.*

Proof The "only if" part is obvious. Thus suppose W is an $E(N)_*$-local spectrum with $[Z, W]_* = 0$ for Z as in the hypotheses. In particular, we may take $Z = Z_{n+1}$, where $Z_0, Z_1, \ldots, Z_{n+1}$ are finite spectra with $Z_0 = L_N S^0$, $Z_1 = L_N M(p)$, and Z_{i+1} the cofiber of a v_i self-map g_i on Z_i. Now suppose X is $E(n)_*$-acyclic. It then follows from the proof of Proposition 1.4 that $X \wedge \widehat{Z}_i \simeq *$ for $0 \leq i \leq n$. We will prove by downward induction on i that $[X \wedge Z_i, W]_* = 0$. This proves that W is $E(n)_*$-local as $Z_0 = L_N S^0$. Indeed, $F(Z_{n+1}, W) \simeq *$; hence $[X \wedge Z_{n+1}, W]_* = 0$. Next assume that $[X \wedge Z_{i+1}, W]_* = 0$. Then

$$(X \wedge g_i)^* : [X \wedge Z_i, W]_* \overset{\approx}{\to} [X \wedge Z_i, W]_*$$

is an isomorphism. But $X \wedge \widehat{Z}_i \simeq *$ implies that $\underset{(X \wedge g_i)^*}{\varprojlim} [X \wedge Z_i, W]_* = 0$. Hence $[X \wedge Z_i, W]_* = 0$, completing the induction and the proof. □

2 The $E(n)$-approximate generating hypothesis

In this section we discuss (a slightly modified version of) Hopkins' original framework for proving Conjecture 1. This framework involves an analysis of the "dualizing complex" I_n and will serve as a template for our approach to Conjectures 2 and 3.

Recall that I_n is defined to be the Brown-Comenetz dual of $L_n S^0$; that is, I_n is characterized by

$$\begin{aligned} [X, I_n]_0 &= \operatorname{Hom}(\pi_0(X \wedge L_n S^0), \mathbb{Q}/\mathbb{Z}_{(p)}) \\ &= \operatorname{Hom}(\pi_0 L_n X, \mathbb{Q}/\mathbb{Z}_{(p)}), \end{aligned} \tag{2.1}$$

where the last equality holds because $E(n)_*$-localization is smashing. One now wishes to present I_n explicitly as

$$I_n = \underset{\alpha \in I}{\operatorname{holim}_{\to}} Y_\alpha, \tag{2.2}$$

where I is a directed set and $Y_\alpha = X_\alpha \wedge S\mathbb{Z}_p^\wedge$ for some L_N-finite spectrum X_α. $S\mathbb{Z}_p^\wedge$ denotes the Moore spectrum for the p-adic integers $\mathbb{Z}_p^\wedge$. Of course, any spectrum can be written as the homotopy direct limit of finite spectra; however, one ordinarily cannot say much about these finite spectra other than that they are finite subspectra of the original CW-spectrum. This is not adequate for our purposes; we want each X_α to be an explicitly described spectrum with few cells.

Roughly speaking, a sufficient understanding of I_n will allow us to find appropriate spectra Y_α and an $E(n)_*$-equivalence $\underset{\overrightarrow{\alpha\in I}}{\operatorname{holim}} Y_\alpha \to I_n$. It then suffices to show that $\underset{\overrightarrow{\alpha\in I}}{\operatorname{holim}} Y_\alpha$ is $E(n)_*$-local. But $\underset{\overrightarrow{\alpha\in I}}{\operatorname{holim}} Y_\alpha$ is $E(N)_*$-local—since $E(N)_*$-localization is smashing—therefore Proposition 1.5 gives us a convenient criterion for proving that it is $E(n)_*$-local. Note that the truth of the $E(N)_*$-local telescope conjecture is invoked here; if the Y_α's were merely assumed to be p-completions of finite spectra, it would be extremely difficult to prove that $\underset{\overrightarrow{\alpha\in I}}{\operatorname{holim}} Y_\alpha$ is $E(n)_*$-local. The $E(N)_*$-local telescope conjecture will also be needed to construct the $E(n)_*$-equivalence $\underset{\overrightarrow{\alpha\in I}}{\operatorname{holim}} Y_\alpha \to I_n$. This will be seen when we discuss the $K(n)$-approximate generating hypothesis.

Now suppose a presentation of I_n as in (2.2) has been found. If $p > \max\{n+1, \frac{1}{2}(n^2+n)\}$, then $\pi_0 L_n S^0 = \mathbb{Z}_{(p)}$ generated by the localization map $\iota_n : S^0 \to L_n S^0$ (see A1.6). Let

$$h \in \pi_0 I_n = \operatorname{Hom}(\pi_0 L_n S^0, \mathbb{Q}/\mathbb{Z}_{(p)}) \approx \mathbb{Q}/\mathbb{Z}_{(p)}$$

be the homomorphism with $h(\iota_n) = 1/p$. Fix a sufficiently large $\alpha \in I$, and choose $h_\alpha \in \pi_0(Y_\alpha)$ mapping to h. If $\beta > \alpha$, set h_β to be the image of h_α under the direct system map $Y_\alpha \to Y_\beta$. Let

$$h_\beta^\# : D_N X_\beta \to L_N(S\mathbb{Z}_p^\wedge)$$

be the adjoint of h_β. Note that $L_n h_\beta^\#$ is essential since $L_n h_\beta$ is. The $h_\beta^\#$'s are our universal examples.

Proposition 2.3 *If $\pi_* h_\beta^\# \neq 0$ for all β sufficiently large, then Conjecture 1 is true.*

Proof We will show that if X is an L_N-finite spectrum and $f : X \to L_N S^0$ with $L_n f \neq 0$, then $\pi_* f \neq 0$. Indeed, let $f^\# : L_N S^0 \to D_N X$ be the adjoint of f. Then $L_n f^\#$ is essential, so $(L_n f^\#)_*(\iota_n) \in \pi_0(L_n D_N X)$ is nontrivial. (This follows from the fact, easily proven by induction on the $E(N)_*$-local cells of X, that the evident map $L_n D_N X \to D_n L_n X$ is an equivalence whenever X is L_N-finite.) Hence there exists a homomorphism $g : \pi_0 L_n D_N X \to \mathbb{Q}/\mathbb{Z}_{(p)}$ with

$$g((L_n f^\#)_*(\iota_n)) = 1/p.$$

Let us also write g for the corresponding map $D_N X \to I_n$. Now for β sufficiently large, g factors as the composite

$$D_N X \xrightarrow{g_\beta} Y_\beta \to I_n$$

and the composition

$$L_N S^0 \xrightarrow{f^\#} D_N X \xrightarrow{g_\beta} Y_\beta$$

is h_β. This means that there is a factorization

$$D_N X_\beta \to X \wedge S\mathbb{Z}_p^\wedge \xrightarrow{f \wedge S\mathbb{Z}_p^\wedge} (L_N S^0) \wedge S\mathbb{Z}_p^\wedge$$

of $h^{\#}_{\beta}$. But $\pi_* h^{\#}_{\beta} \neq 0$; therefore $\pi_* f \neq 0$. This completes the proof. □

In [**2**], we showed that, for p odd, $I_1 \simeq (L_1 S^2) \wedge S\mathbb{Z}_p^\wedge$ and provided an explicit description of $L_1 S^0$ as the homotopy direct limit of finite spectra. (We did not need to work with L_N, since the telescope conjecture holds for $n = 1$.) We then determined the maps $h^{\#}_{\beta}$ and proved that they were nonzero on π_*, thus proving the $E(1)$-approximate generating hypothesis.

If $n \geq 2$, then I_n is no longer the $E(n)_*$-localization of a p-completed sphere (see [**8**], (Corollary 7) and 4.8–4.11). However, if $n = 2$, one can use the computations of [**23**] and [**24**] to present $I_2 \wedge M(p)$ as the homotopy direct limit of L_N-finite spectra. One can also determine the universal examples needed to prove the $E(2)$-analogue of Conjecture 3. Presumably an understanding of [**25**] would allow us to present I_2 as in (2.2) and determine the universal examples; however even in the mod(p) case, the analysis of these seems inaccessible. We thus turn to the $K(n)$-analogue of Conjecture 1, where we will see that—at least mod(p)—the universal examples are more tractable.

3 The $K(n)$-approximate generating hypothesis

In this section we consider Conjecture 2. Our approach is analogous to that of Section 2. Since $\mathbb{Q}/\mathbb{Z}_{(p)} = (p^{-1}\mathbb{Z}_p^\wedge)/\mathbb{Z}_p^\wedge$ is a divisible and hence injective $\mathbb{Z}_p^\wedge$-module, we may let $I_{K(n)}$ be the spectrum characterized by

$$[X, I_{K(n)}]_0 = \mathrm{Hom}_{\mathbb{Z}_p^\wedge}(\pi_0(X \wedge L_{K(n)} S^0), \mathbb{Q}/\mathbb{Z}_{(p)}). \tag{3.1}$$

Note that $L_{K(n)} S^0$ is an $S\mathbb{Z}_p^\wedge$-module spectrum, so that $\pi_*(X \wedge L_{K(n)} S^0)$ is indeed a $\mathbb{Z}_p^\wedge$-module. There is also a canonical map $\tau_n : I_{K(n)} \to I_n$ induced by the natural transformation $L_n \to L_{K(n)}$.

As in Section 2, we hope to find an explicit presentation

$$I_{K(n)} = \operatorname*{holim}_{\vec{\alpha}} Z_\alpha, \tag{3.2}$$

where each Z_α is an L_N-finite spectrum. (By [**12**] (5.2), $\pi_* L_{K(n)} S^0$ has no divisible summands; therefore, we need not smash with the p-adic Moore spectrum.) Furthermore, if $p > \max\{n+1, \frac{1}{2}(n^2+2)\}$, $\pi_0 L_{K(n)} S^0 = \mathbb{Z}_p^\wedge$ generated by the localization map $\iota_{K(n)} : S^0 \to L_{K(n)} S^0$ (see A1.7); thus we may define

$$h \in \pi_0 I_{K(n)} = \mathrm{Hom}_{\mathbb{Z}_p^\wedge}(\pi_0 L_{K(n)} S^0, \mathbb{Q}/\mathbb{Z}_{(p)})$$

by $h(\iota_{K(n)}) = 1/p$. As before, choose $h_\alpha \in \pi_0 Z_\alpha$ mapping to h, and, if $\beta > \alpha$, take h_β to be the image of h_α under $Z_\alpha \to Z_\beta$. Let $h^{\#}_{\beta} : D_N Z_\beta \to L_N S^0$ be the adjoint of h_β. We then have the following analogue of Proposition 2.3.

Proposition 3.3 *If $\pi_* h^{\#}_{\beta} \neq 0$ for all β sufficiently large, then Conjecture 2 is true.*

Remark 3.4 Since $I_{K(n)}$ is $E(n)_*$-local by Proposition 3.5, it follows that $L_n h_\beta$ is nontrivial. Moreover, if Z_β is $E(n-1)_*$-acyclic, then $L_{K(n)} Z_\beta = L_n Z_\beta$, and hence $L_{K(n)} h_\beta \neq 0$. But, if Z is L_n-finite, then $L_{K(n)} Z \simeq L_{K(n)} S^0 \wedge Z$, so that $L_{K(n)} h_\beta$ is just

$$L_{K(n)} S^0 \wedge L_n h_\beta : L_{K(n)} S^0 \wedge L_n S^0 \to L_{K(n)} S^0 \wedge L_n Z_\beta.$$

Thus

$$L_{K(n)} h^{\#}_{\beta} : L_{K(n)} D_n L_n Z_\beta = L_{K(n)} S^0 \wedge D_n L_n Z_\beta \to L_{K(n)} S^0 \wedge L_n S^0 = L_{K(n)} S^0$$

is essential, and so the above proposition can be amended in this case to become an "if and only if" statement. We will see in the discussion following Proposition 3.5 that each Z_β can be taken to be $E(n-1)_*$-acyclic provided $\pi_* L_{K(n)} S^0$ is of finite type over $\mathbb{Z}_p^\wedge$. □

We now try to determine these universal examples. Recall from [**21**] (Section 5) that $N_i X$ is defined inductively by $N_0 X = X$ and $N_{i+1}X$ is the cofiber of $N_i X \to L_i N_i X$. We also define $M_i X = L_i N_i X$ and recall that

$$M_i S^0 = \operatorname*{holim}_{\vec{J}} \Sigma^{-n_J} L_i M(p^{j_0}, \ldots, v_{i-1}^{j_{i-1}}), \qquad n_J = \sum_{k=1}^{i-1} 2j_k(p^k - 1),$$

where $M(p^{j_0}, \ldots, v_{i-1}^{j_{i-1}})$ is a certain finite complex with

$$BP_* M(p^{j_0}, v_1^{j_1}, \ldots, v_{i-1}^{j_{i-1}}) = BP_*/(p^{j_0}, \ldots, v_{i-1}^{j_{i-1}}),$$

and $J = (j_0, j_1, \ldots, j_{i-1})$ varies over a cofinal sequence of multi-indices. (See [**4**] (Section 4) for details.) Let ∂_n be the composite of the boundary maps

$$N_n S^0 \to \Sigma N_{n-1} S^0 \to \cdots \to \Sigma^n N_0 S^0 = S^0.$$

Proposition 3.5 *$I_{K(n)}$ is $E(n)_*$-local, and, if $\pi_* L_{K(n)} S^0$ is a $\mathbb{Z}_p^\wedge$-module of finite type, then $I_{K(n)} \wedge \partial_n$ is a homotopy equivalence. Hence $\Sigma^{-n} I_{K(n)} \wedge M_n S^0 \simeq I_{K(n)}$.*

Proof $E(n)$ is Bousfield equivalent to $K(0) \vee K(1) \vee \cdots \vee K(n)$ which is in turn Bousfield equivalent to $L_{K(n)} S^0$ by [**12**] (2.4). It follows that

$$[X, I_{K(n)}]_* = \operatorname{Hom}_{\mathbb{Z}_p^\wedge}(\pi_*(X \wedge L_{K(n)} S^0), \mathbb{Q}/\mathbb{Z}_{(p)}) = 0$$

whenever X is $E(n)_*$-acyclic. This means that $I_{K(n)}$ is $E(n)_*$-local.

We next prove that, if $\pi_* L_{K(n)} S^0$ is of finite type over $\mathbb{Z}_p^\wedge$, $I_{K(n)} \wedge M_i S^0 \simeq *$ for all $i < n$; this implies that $I_{K(n)} \wedge \partial_n$ is an equivalence. First observe that

$$\pi_*(I_{K(n)} \wedge M_0 S^0) = p^{-1} \pi_* I_{K(n)} = p^{-1} \operatorname{Hom}_{\mathbb{Z}_p^\wedge}(\pi_* L_{K(n)} S^0, \mathbb{Q}/\mathbb{Z}_{(p)}) = 0,$$

since $\mathbb{Q}/\mathbb{Z}_{(p)}$ is all p-torsion and $\pi_* L_{K(n)} S^0$ is assumed to be a $\mathbb{Z}_p^\wedge$-module of finite type. Now suppose $0 < i < n$, and let ν_i be a v_i self-map of $M(p^{j_0}, \ldots, v_{i-1}^{j_{i-1}})$. Then by Proposition 1.4,

$$L_i M(p^{j_0}, \ldots, v_{i-1}^{j_{i-1}}) = \nu_i^{-1} L_n M(p^{j_0}, \ldots, v_{i-1}^{j_{i-1}});$$

hence

$$\begin{aligned}
\pi_*(I_{K(n)} \wedge M_i S^0) &= \lim_{\vec{J}} \pi_*(I_{K(n)} \wedge L_i M(p^{j_0}, \ldots, v_{i-1}^{j_{i-1}}) \\
&= \lim_{\vec{J}} \nu_{i*}^{-1} \pi_*(I_{K(n)} \wedge L_n M(p^{j_0}, \ldots, v_{i-1}^{j_{i-1}})) \\
&= \lim_{\vec{J}} \nu_{i*}^{-1} \pi_*(I_{K(n)} \wedge M(p^{j_0}, \ldots, v_{i-1}^{j_{i-1}})),
\end{aligned}$$

where the last equality holds since $I_{K(n)}$ is $E(n)_*$-local, and we have omitted suspensions from the notation. But

$$\nu_{i*}^{-1} \pi_*(I_{K(n)} \wedge M(p^{j_0}, \ldots, v_{i-1}^{j_{i-1}})) = v_i^{-1} \operatorname{Hom}_{\mathbb{Z}_p^\wedge}(\pi_* L_{K(n)} DM(p^{j_0}, \ldots, v_{i-1}^{j_{i-1}}), \mathbb{Q}/\mathbb{Z}_{(p)}),$$

and it follows from our assumption on $\pi_* L_{K(n)} S^0$ that $\pi_* L_{K(n)} DM(p^{j_0}, \ldots, v_{i-1}^{j_{i-1}})$ is finite in each degree. Therefore, $I_{K(n)} \wedge M_i S^0$ is contractible provided that $\pi_* L_{K(n)} DM(p^{j_0}, \ldots, v_{i-1}^{j_{i-1}})$ has no v_i-divisible elements; in other words,

$$\lim_{\overleftarrow{(D\nu_i)_*}} \pi_* L_{K(n)} DM(p^{j_0}, \ldots, v_{i-1}^{j_{i-1}}) = 0.$$

We will prove that $\operatorname{holim}_{\overleftarrow{D\nu_i}} L_{K(n)} DM(p^{j_0}, \ldots, v_{i-1}^{j_{i-1}})$ is contractible. Let Z be a $K(n-1)_*$-acyclic finite spectrum with $K(n)_* Z \neq 0$. Then $\operatorname{holim}_{\overleftarrow{D\nu_i}} L_{K(n)} DM(p^{j_0}, \ldots, v_{i-1}^{j_{i-1}})$ is Z_*-local since it is $K(n)_*$-local. Moreover,

$$\begin{aligned}(\operatorname{holim}_{\overleftarrow{D\nu_i}} L_{K(n)} DM(p^{j_0}, \ldots, v_{i-1}^{j_{i-1}})) \wedge Z &= \operatorname{holim}_{\overleftarrow{D\nu_i \wedge Z}} L_{K(n)}(DM(p^{j_0}, \ldots, v_{i-1}^{j_{i-1}}) \wedge Z) \\ &\simeq *,\end{aligned}$$

since the self-map $D\nu_i \wedge Z$ on $DM(p^{j_0}, \ldots, v_{i-1}^{j_{i-1}}) \wedge Z$ induces the zero homomorphism on all Morava K-theories and is therefore nilpotent ([**9**], [**10**]). But a spectrum which is both Z_*-local and Z_*-acyclic must be contractible. This completes the proof. □

Remark 3.6 If X is any spectrum,

$$L_{K(n)} X = \operatorname{holim}_{\overleftarrow{I}} (L_n X \wedge M(p^{i_0}, \ldots, v_{n-1}^{i_{n-1}})),$$

where $I = (i_0, \ldots, i_{n-1})$ varies over an appropriate cofinal sequence of multi-indices (see [**12**], Section 2). In particular, if X is finite, it follows from Morava's theory –together with the fact that the cohomology of the p-Sylow subgroup of the nth Morava stabilizer group (with coefficients in $\mathbb{F}_p$) is of finite type–that $E_2^{s,t}(BP; L_n X \wedge M(p^{i_0}, \ldots, v_{n-1}^{i_{n-1}}))$ is a finite group. But $E_\infty^{s,t}(BP; L_n X \wedge M(p^{i_0}, \ldots, v_{n-1}^{i_{n-1}}))$ has a horizontal vanishing line (see [**22**], Chapter 8); therefore $\pi_*(L_n X \wedge M(p^{i_0}, \ldots, v_{n-1}^{i_{n-1}}))$ is finite and

$$\pi_* L_{K(n)} X = \lim_{\overleftarrow{I}} \pi_*(L_n X \wedge M(p^{i_0}, \ldots, v_{n-1}^{i_{n-1}})).$$

This implies that $\pi_i L_{K(n)} X$ is either finite or uncountable. But the chromatic splitting conjecture implies that if X is a rationally acyclic finite spectrum, then $\pi_i L_{K(n)} X$ is countable. This would then prove that $\pi_* L_{K(n)} X$ is of finite type.

We do not know whether the chromatic splitting conjecture implies that $\pi_* L_{K(n)} S^0$ is of finite type over $\mathbb{Z}_p^\wedge$. However, $\pi_* L_{K(1)} S^0$ is known to be of finite type over $\mathbb{Z}_p^\wedge$ for all primes p. As for $\pi_* L_{K(2)} S^0$, first observe that the calculations of [**25**] imply that $\pi_*(L_2 S^0)_p$, the homotopy groups of the p-completion of $L_2 S^0$, is of finite type over $\mathbb{Z}_p^\wedge$ for $p \geq 5$. Now there is a cofiber sequence

$$F(L_{n-1} S^0, L_n S^0)_p \to (L_n S^0)_p \to (L_{K(n)} S^0)_p,$$

and the chromatic splitting conjecture (which is known to be true for $n = 2$ and $p \geq 5$) implies that there is a natural equivalence

$$\Sigma^{-2} L_1 X \to F(L_1 S^0, L_2 X)$$

for any rationally acyclic finite spectrum X. Hence

$$\begin{aligned}F(L_1 S^0, L_2 S^0)_p \simeq \operatorname{holim}_{\overleftarrow{i}} F(L_1 S^0, L_2 M(p^i)) &\simeq \operatorname{holim}_{\overleftarrow{i}} \Sigma^{-2} L_1 M(p^i) \\ &= \Sigma^{-2}(L_1 S^0)_p.\end{aligned}$$

In particular, $\pi_* F(L_1 S^0, L_2 S^0)_p$ is of finite type over $\mathbb{Z}_p^\wedge$; therefore so is $\pi_* L_{K(2)} S^0 = \pi_*(L_{K(2)} S^0)_p$. □

Proposition 3.5 thus implies that

$$I_{K(n)} \simeq \underset{\overrightarrow{J}}{\operatorname{holim}}\, \Sigma^{-n-n_J} I_{K(n)} \wedge M(p^{j_0}, \dots, v_{n-1}^{j_{n-1}}).$$

The next result puts us in the situation of [**5**] (Section 5), [**4**], [**8**].

Proposition 3.7 *Let Z be a $K(n-1)_*$-acyclic finite spectrum. Then $\tau_n \wedge Z : I_{K(n)} \wedge Z \to I_n \wedge Z$ is an equivalence.*

Proof Observe that $(L_{K(n)} S^0) \wedge DZ = L_{K(n)} DZ \xleftarrow{\simeq} L_n DZ$. □

If $p > \frac{1}{2}(n^2 + n + 2)$, it is proved in [**8**] that the Morava module of $I_n \wedge M(p^{j_0}, \dots, v_{n-1}^{j_{n-1}})$ is the Pontryagin dual of the Morava module of $M(p^{j_0}, \dots, v_{n-1}^{j_{n-1}})$; this in principle determines its homotopy type (see [**4**]). Furthermore, it follows from the proof of Theorem 4.3 in [**4**] that

$$I_n \wedge M(p^{j_0}, \dots, v_{n-1}^{j_{n-1}}) = Z'_J, \tag{3.8}$$

where Z'_J is an $E(n-1)_*$-acyclic L_n-finite spectrum built up from $L_n M(p, v_1^{j_1}, \dots, v_{n-1}^{j_{n-1}})$ by a finite sequence of cofibrations. But, according to the $E(N)_*$-local telescope conjecture (Proposition 1.4), any map between the $E(n)_*$-localizations of $E(n-1)_*$-acyclic L_N-finite spectra is induced by a map between the spectra themselves followed by the inverse of a v_n self-map. It then follows by an induction on the sequence of cofibrations building Z'_J that

$$Z'_J = L_n Z_J$$

for some $E(n-1)_*$-acyclic L_N-finite spectrum Z_J. Using Proposition 1.4 again, one can identify $I_n \wedge M(p^{i_0}, \dots, v_{n-1}^{j_{n-1}})$ with $L_n Z_J$ in such a way that the maps $L_n Z_I \to L_n Z_J$ in the direct system can be lifted to maps $Z_I \to Z_J$ such that $\underset{\overrightarrow{J}}{\operatorname{holim}}\, Z_J$ satisfies the hypotheses of Proposition 1.5 and is therefore $E(n)_*$-local. Here of course suspensions have been suppressed from the notation. Since $\underset{\overrightarrow{J}}{\operatorname{holim}}\, L_n Z_J = L_n \underset{\overrightarrow{J}}{\operatorname{holim}}\, Z_J$, this proves that

$$I_{K(n)} \simeq \underset{\overrightarrow{J}}{\operatorname{holim}}\, Z_J, \tag{3.9}$$

provided, of course, that $\pi_* L_{K(n)} S^0$ is of finite type over $\mathbb{Z}_p^\wedge$.

Unfortunately, an explicit description of Z_J and the corresponding universal example seems out of reach. However, if $j_0 = 1$, Gross and Hopkins [**8**] have shown that Z_J is a suspension of $L_N M(p, v_1^{j_1}, \dots, v_{n-1}^{j_{n-1}})$. This provides hope for proving Conjecture 3, which we consider in the next section.

4 The mod(p) $K(n)$-approximate generating hypothesis

As before, we try to write

$$I_{K(n)} \wedge M(p) = \underset{\overrightarrow{\alpha}}{\operatorname{holim}}\, W_\alpha, \tag{4.1}$$

where once again the W_α's are explicitly determined L_N-finite spectra. The $K(n)_*$-localization of the inclusion $\eta : S^0 \to M(p)$ of the bottom cell generates $\mathbb{Z}/(p) = \pi_0 L_{K(n)} M(p)$ if $p > \max\{n+1, \frac{1}{2}(n^2+2)\}$ (see A1.7); let

$$h \in \pi_1(I_{K(n)} \wedge M(p)) = \mathrm{Hom}(\pi_0 L_{K(n)} M(p), \mathbb{Q}/\mathbb{Z}_{(p)})$$

be the element with $h(\eta) = 1/p$, where we have made an evident abuse of notation. Choose $h_\alpha \in \pi_1 W_\alpha$ so that h_α maps to h, and, for $\beta > \alpha$, let h_β be the image of h_α in $\pi_1 W_\beta$. As before, $h_\beta^\# : \Sigma D_N W_\beta \to L_N S^0$ denotes the adjoint of h_β. These are the universal examples for Conjecture 3.

Proposition 4.2 *If $\pi_*(h_\beta^\# \wedge M(p)) \neq 0$ for all β sufficiently large, then Conjecture 3 is true.*

Proof Let $f : X \to L_N S^0$ with X L_N-finite, and suppose that $L_{K(n)}(f \wedge M(p))$ is nontrivial. We claim that this implies that the $K(n)_*$-localization of the composite

$$X \wedge M(p) \xrightarrow{f \wedge M(p)} L_N S^0 \wedge M(p) \xrightarrow{L_N S^0 \wedge \partial} \Sigma L_N S^0 \tag{4.3}$$

is also nontrivial, where ∂ projects $M(p)$ onto its top cell.

Assuming this claim, it follows by an argument similar to the proof of Proposition 2.3 (cf. also Remark 3.4) that there exists a map

$$D_N X \wedge M(p) \to I_{K(n)}$$

whose pullback via $D_N f \wedge M(p)$ is h. This implies that there exists a factorization

$$\Sigma D_N W_\beta \longrightarrow X \xrightarrow{f} L_N S^0$$

of $h_\beta^\#$ for β sufficiently large. Thus, if $\pi_*(h_\beta^\# \wedge M(p)) \neq 0$, then $\pi_*(f \wedge M(p))$ is nontrivial as well.

We now verify our claim that the $K(n)_*$-localization of the map in 4.3 is essential. Indeed, since the cofibration sequence

$$S^0 \wedge M(p) \longrightarrow M(p) \wedge M(p) \xrightarrow{\partial \wedge M(p)} S^1 \wedge M(p)$$

splits, it follows that the $K(n)_*$-localization of $f \wedge M(p) \wedge M(p)$ is nontrivial, and furthermore that the $K(n)_*$-localization of

$$X \wedge M(p) \wedge M(p) \xrightarrow{f \wedge M(p) \wedge M(p)} L_N S^0 \wedge M(p) \wedge M(p) \longrightarrow L_N S^0 \wedge S^1 \wedge M(p)$$

is nontrivial. But this map is just $[(L_N S^0 \wedge \partial) \circ (f \wedge M(p))] \wedge M(p)$, completing the proof of the claim. □

Remark 4.4 If each W_β is $E(n-1)_*$-acyclic, it follows as in Remark 3.4 that the above proposition can also be amended to become an "if and only if" statement. If $\pi_* L_{K(n)} M(p)$ is of finite type, we will see in the next proposition that this can be arranged. □

Recall the notation of (0.2). We then have the following result.

Proposition 4.5 *Suppose that $p > \frac{1}{2}(n^2+2)$, $n \geq 2$. If $\pi_* L_{K(n)} M(p)$ is of finite type, then there exists a direct system indexed over T_1 such that*

$$I_{K(n)} \wedge M(p) \simeq \operatorname*{holim}_{\overrightarrow{T_1}} L_N M(p, v_1^{i_1}, \ldots, v_{n-1}^{i_{n-1}}),$$

where suspensions have been deleted from the notation.

Proof It follows as in the proof of Proposition 3.5 that $\Sigma^{-n} I_{K(n)} \wedge M_n M(p) \simeq I_{K(n)} \wedge M(p)$ provided that $\pi_* L_{K(n)} M(p)$ is of finite type. Hence

$$I_{K(n)} \wedge M(p) \simeq \operatorname*{holim}_{\overrightarrow{T_1}} \Sigma^{-n_I - n+1} I_{K(n)} \wedge M(p, v_1^{i_1}, \ldots, v_{n-1}^{i_{n-1}}).$$

But by [**8**] (Corollary 7) and 3.7,

$$I_{K(n)} \wedge M(p, v_1^{i_1}, \ldots, v_{n-1}^{i_{n-1}}) \approx \Sigma^{r_I} L_n M(p, v_1^{i_1}, \ldots, v_{n-1}^{i_{n-1}}), \tag{4.6}$$

for some integer r_I. The discussion following the proof of Proposition 3.5 gives us our desired result. □

We now complete the proof of Theorem 1 in the introduction by describing the universal examples h_I. We begin with some preliminaries.

Recall [**20**] that if $p-1 \nmid n$, then $v_n^{-1} H^*(BP_*/(p, v_1, \ldots, v_{n-1}))$ is a Poincaré algebra of dimension n^2, with internal grading cyclic mod$(2p^n - 2)$. Write

$$c_n \in v_n^{-1} H^{n^2,0}(BP_*/(p, v_1, \ldots, v_{n-1}))$$

for the fundamental class. Now suppose $(i_0, \ldots, i_{n-1})$ is a multi-index with $v_j^{i_j}$ invariant mod$(p^{i_0}, \ldots, v_{j-1}^{i_{j-1}})$ for $1 \leq j \leq n-1$. Then (if $p-1 \nmid n$), the map

$$v_n^{-1} H^{n^2,*}(BP_*/(p^{i_0}, \ldots, v_{n-1}^{i_{n-1}})) \xrightarrow{\rho_*} v_n^{-1} H^{n^2,*}(BP_*/(p, \ldots, v_{n-1})) \tag{4.7}$$

is onto; hence there exists a lift $\widetilde{c}_n$ of c_n. Next observe that $p^{i_0-1} v_1^{i_1-1} \cdots v_{n-1}^{i_{n-1}-1} v_n^k$ is invariant mod$(p^{i_0}, \ldots, v_{n-1}^{i_{n-1}})$ for all k; write $p^{i_0-1} v_1^{i_1-1} \cdots v_{n-1}^{i_{n-1}-1} v_n^k c_n$ for the product of $p^{i_0-1} v_1^{i_1-1} \cdots v_{n-1}^{i_{n-1}-1} v_n^k$ and $\widetilde{c}_n$ in $v_n^{-1} H^*(BP_*/(p^{i_0}, \ldots, v_{n-1}^{i_{n-1}}))$. Note that this product is independent of the lift $\widetilde{c}_n$.

Lemma 4.8 *If $I = (1, i_1, \ldots, i_{n-1})$ is a multi-index as above, then there exists k_I such that $v_1^{i_1-1} \cdots v_{n-1}^{i_{n-1}-1} v_n^{k_I} c_n$ is nontrivial.*

Proof By [**8**], the Morava module $E_{n_*}^{\wedge}/(p, v_1^{i_1}, \ldots, v_{n-1}^{i_{n-1}})$ is Pontryagin self-dual; it therefore follows from the theory of Poincaré pro-p-groups (see [**5**], Section 5, especially the proof of Equation 5.4) that there exists a map

$$g_I : v_n^{-1} H^{n^2, m_I}(BP_*/(p, v_1^{i_1}, \ldots, v_{n-1}^{i_{n-1}})) \to \mathbb{Q}/\mathbb{Z}_{(p)}$$

such that

$$\begin{aligned} v_n^{-1} H^{i,*}(BP_*/(p, v_1^{i_1}, \ldots, v_{n-1}^{i_{n-1}})) \quad \otimes \quad & v_n^{-1} H^{n^2-i,*}(BP_*/(p, v_1^{i_1}, \ldots, v_{n-1}^{i_{n-1}})) \\ & \to v_n^{-1} H^{n^2,*}(BP_*/(p, v_1^{i_1}, \ldots, v_{n-1}^{i_{n-1}})) \\ & \xrightarrow{g_I} \mathbb{Q}/\mathbb{Z}_{(p)} \end{aligned}$$

is a perfect pairing. (The number m_I will be the degree of $v_1^{i_1-1} \cdots v_{n-1}^{i_{n-1}-1} v_n^{k_I}$.) In particular, there exists an element d in $v_n^{-1} H^{n^2}(BP_*/(p, v_1^{i_1}, \ldots, v_{n-1}^{i_{n-1}}))$ whose product with $v_1^{i_1-1} \cdots v_{n-1}^{i_{n-1}-1} \in v_n^{-1} H^0(BP_*/(p, v_1^{i_1}, \ldots, v_{n-1}^{i_{n-1}}))$ is nonzero. Now $\rho_* d \neq 0$; otherwise $d \in v_n^{-1} H^{n^2}(I_n BP_*/(p, v_1^{i_1}, \ldots, v_{n-1}^{i_{n-1}}))$, $I_n = (p, v_1, \ldots, v_{n-1})$, and hence $v_1^{i_1-1} \cdots v_{n-1}^{i_{n-1}-1} d = 0$. Therefore d must be a lift of $v_n^{k_I} c_n$ for some k_I. This completes the proof. □

Remark 4.9 If $v_n^{p^j}$ is invariant mod$(p, v_1^{i_1}, \ldots, v_{n-1}^{i_{n-1}})$, then k_I can be taken to be $(1 - p^{nj})/(p-1)$ ([**8**]). □

and hence $v_1^{i_1-1}\cdots v_{n-1}^{i_{n-1}-1}d = 0$. Therefore d must be a lift of $v_n^{k_I}c_n$ for some k_I. This completes the proof. □

Remark 4.9 If $v_n^{p^j}$ is invariant $\bmod(p, v_1^{i_1}, \ldots, v_{n-1}^{i_{n-1}})$, then k_I can be taken to be $(1-p^{nj})/(p-1)$ ([**8**]). □

Remark 4.10 Let k_I be as in the preceding lemma. Then

$$v_1^{i_1-1}\cdots v_{n-1}^{i_{n-1}-1}v_n^{k'_I}c_n \in v_n^{-1}H^{n^2}(BP_*/(p, v_1^{i_1}, \ldots, v_{n-1}^{i_{n-1}})$$

is nontrivial if and only if there exists an invariant

$$w \in v_n^{-1}BP_*/(p, v_1^{i_1}, \ldots, v_{n-1}^{i_{n-1}})$$

such that $w \equiv v_n^{k_I-k'_I} \bmod(p, v_1, \ldots, v_{n-1})$. □

Remark 4.11 It follows from [**19**] (5.10) that there exist integers m_j with $\lim_{j\to\infty} m_j = \infty$, such that, if w is an invariant element in $v_n^{-1}BP_*/(p, v_1, \ldots, v_{n-2}, v_{n-1}^j)$, then $w \equiv a v_n^{sp^{m_j}} \bmod(p, v_1, \ldots, v_{n-1})$, $a \in \mathbb{F}_p$. Hence by Remarks 4.9 and 4.10, there does not exist k such that $v_{n-1}^{j-1}v_n^k c_n$ is nonzero in $v_n^{-1}H^{n^2}(BP_*/(p, v_1, \ldots, v_{n-2}, v_{n-1}^j))$ for all j. □

Now let $h_I^0 \in \pi_* L_n M(p, v_1^{i_1}, \ldots, v_{n-1}^{i_{n-1}})$ be detected by the nonzero cohomology class $v_1^{i_1-1}\cdots v_{n-1}^{i_{n-1}-1}v_n^{k_I}c_n$, and let ν_n be a v_n self-map of $M(p, v_1^{i_1}, \ldots, v_{n-1}^{i_{n-1}})$. By changing k_I if necessary, we conclude that there exists $h_I \in \pi_* L_N M(p, v_1^{i_1}, \ldots, v_{n-1}^{i_{n-1}})$ whose image in $\nu_n^{-1}\pi_* L_N M(p, v_1^{i_1}, \ldots, v_{n-1}^{i_{n-1}}) = \pi_* L_n M(p, v_1^{i_1}, \ldots, v_{n-1}^{i_{n-1}})$ is h_I^0.

Proof of Theorem 1 By virtue of Proposition 4.2, we need only track down the identifications made in proving Proposition 4.5. Doing so yields the desired result, although once again k_I may need to be changed. □

Let us now indicate a program, relying on the chromatic splitting conjecture, for proving that $\pi_*(h_I^\# \wedge M(p)) \neq 0$ and thus verifying Conjecture 3. This amounts to finding a map

$$g_I : L_N M(p, v_1^{i_1}, \ldots, v_{n-1}^{i_{n-1}}) \to L_N M(p)$$

such that $g_I \circ h_I$ is nontrivial. Here and for the rest of this section, we suppress suspensions from the notation.

Assuming the chromatic splitting conjecture (see [**12**], Conjecture 4.2 part 5), the evident map

$$L_n M(p, v_1^{i_1}, \ldots, v_{n-1}^{i_{n-1}}) \to L_n L_{K(n+r)} M(p, v_1^{i_1}, \ldots, v_{n-1}^{i_{n-1}})$$

is a split monomorphism for $r \geq 1$. Now note that

$$\begin{array}{ccc} L_{n+r} M(p, v_1^{i_1}, \ldots, v_{n-1}^{i_{n-1}}) & \longrightarrow & L_{K(n+r)} M(p, v_1^{i_1}, \ldots, v_{n-1}^{i_{n-1}}) \\ \downarrow & & \downarrow \\ L_n M(p, v_1^{i_1}, \ldots, v_{n-1}^{i_{n-1}}) & \longrightarrow & L_n L_{K(n+r)} M(p, v_1^{i_1}, \ldots, v_{n-1}^{i_{n-1}}) \end{array}$$

commutes, and that

$$\begin{aligned} \pi_* L_{K(n+r)} M(p, v_1^{i_1}, \ldots, v_{n-1}^{i_{n-1}}) &= \pi_* \operatorname{holim}_{\overleftarrow{J}} L_{n+r} M(p, v_1^{i_1}, \ldots, v_{n-1}^{i_{n-1}}, v_n^{j_1}, \ldots, v_{n+r-1}^{j_r}) \\ &= \lim_{\overleftarrow{J}} \pi_* L_{n+r} M(p, v_1^{i_1}, \ldots, v_{n-1}^{i_{n-1}}, v_n^{j_1}, \ldots, v_{n+r-1}^{j_r}), \end{aligned}$$

where $J = (j_1, \ldots, j_r)$ ranges over an appropriate cofinal sequence of multi-indices (see Remark 3.6). Then, if $n + r \le N$, it follows from the chromatic splitting conjecture that the image of h_I in $\pi_* L_{K(n+r)} M(p, v_1^{i_1}, \ldots, v_{n-1}^{i_{n-1}})$ is nonzero and hence that the image of h_I under the map

$$\begin{array}{ccc} \pi_* L_N M(p, v_1^{i_1}, \ldots, v_{n-1}^{i_{n-1}}) & \longrightarrow & \pi_* L_{n+r} M(p, v_1^{i_1}, \ldots, v_{n-1}^{i_{n-1}}, v_n^{j_1}, \ldots, v_{n+r-1}^{j_r}) \\ & & \Vert \\ & & (\nu_{n+r})_*^{-1} \pi_* L_N M(p, v_1^{i_1}, \ldots, v_{n+r-1}^{j_r}) \end{array}$$

is nontrivial for J sufficiently large, where once again ν_{n+r} is a v_{n+r} self-map. This means that the composition

$$\begin{array}{rcl} S^0 \xrightarrow{h_I} L_N M(p, v_1^{i_1}, \ldots, v_{n-1}^{i_{n-1}}) & \to & L_N M(p, v_1^{i_1}, \ldots, v_{n+r-1}^{j_r}) \\ & \xrightarrow{\nu_{n+r}^t} & L_N M(p, v_1^{i_1}, \ldots, v_{n+r-1}^{j_r}) \end{array} \tag{4.12}$$

is nontrivial for any t. One might therefore hope that r, J, t can be chosen so that the composition of the map in (4.12) with the projection

$$L_N M(p, v_1^{i_1}, \ldots, v_{n+r-1}^{j_r}) \to L_N M(p)$$

onto the top Moore spectrum is also nontrivial. Then one could take g_I to be the composition

$$\begin{array}{rcl} L_N M(p, v_1^{i_1}, \ldots, v_{n-1}^{i_{n-1}}) & \to & L_N M(p, v_1^{i_1}, \ldots, v_{n+r-1}^{j_r}) \\ & \xrightarrow{\nu_{n+r}^t} & L_N M(p, v_1^{i_1}, \ldots, v_{n+r-1}^{j_r}) \to L_N M(p). \end{array} \tag{4.13}$$

Such a program seems well suited for chromatic spectral sequence techniques.

Example 4.14 In this example we show that the choice of r, J, and t must allow for a certain amount of "noncentrality". Indeed, using the notation as above, it follows from 4.9–4.11 that the composition

$$\begin{array}{rcl} S^0 \xrightarrow{h_I} L_N M(p, v_1^{i_1}, \ldots, v_{n-1}^{i_{n-1}}) & \to & L_N M(p, v_1^{i_1}, \ldots, v_{n-2}^{i_{n-2}}, v_{n-1}^{\infty}) \\ & \to & L_n M(p, v_1^{i_1}, \ldots, v_{n-2}^{i_{n-2}}, v_{n-1}^{\infty}) \end{array}$$

is trivial. We may therefore assume that

$$S^0 \xrightarrow{h_I} L_N M(p, v_1^{i_1}, \ldots, v_{n-1}^{i_{n-1}}) \to L_N M(p, v_1^{i_1}, \ldots, v_{n-2}^{i_{n-2}}, v_{n-1}^{\infty})$$

is also trivial. (Remember, there was a good deal of indeterminacy in defining h_I.) Now the diagram

$$\begin{array}{ccccc} M(p, v_1^{i_1}, \ldots, v_{n-2}^{i_{n-2}}) & \xrightarrow{v_{n-1}^{i_{n-1}}} & M(p, v_1^{i_1}, \ldots, v_{n-2}^{i_{n-2}}) & \longrightarrow & M(p, v_1^{i_1}, \ldots, v_{n-1}^{i_{n-1}}) \\ \Vert & & \downarrow & & \downarrow \\ M(p, v_1^{i_1}, \ldots, v_{n-2}^{i_{n-2}}) & \longrightarrow & \nu_{n-1}^{-1} M(p, v_1^{i_1}, \ldots, v_{n-2}^{i_{n-2}}) & \longrightarrow & M(p, v_1^{i_1}, \ldots, v_{n-2}^{i_{n-2}}, v_{n-1}^{\infty}) \end{array}$$

of cofibration sequences implies that

$$S^0 \xrightarrow{h_I} L_N M(p, v_1^{i_1}, \ldots, v_{n-1}^{i_{n-1}}) \xrightarrow{\partial} L_N M(p, v_1^{i_1}, \ldots, v_{n-2}^{i_{n-2}})$$

is trivial, so that there exists a factorization

$$S \xrightarrow{h_I'} L_N M(p, v_1^{i_1}, \ldots, v_{n-2}^{i_{n-2}}) \to L_N M(p, v_1^{i_1}, \ldots, v_{n-1}^{i_{n-1}})$$

of h_I. By sparseness, h_I' is detected by

$$\overline{h}_I' \in E_\infty^{n^2,*}(E(N); L_N M(p, v_1^{i_1}, \ldots, v_{n-2}^{i_{n-2}}))$$

(see Appendix 1). But by A1.1

$$E_\infty^{s,*}(E(N); v_{n-1}^{-1} L_N M(p, v_1^{i_1}, \ldots, v_{n-2}^{i_{n-2}})) = 0$$

for $s > (n-1)^2$; therefore, h_I' is v_{n-1}-torsion. We will use this fact to prove that if J is "sufficiently large" and ν_{n+r}^t satisfies an appropriate centrality condition, then $g_I \circ h_I = 0$.

Begin by choosing m such that $(L_N v_{n-1}^m) h_I' = 0$ and such that v_{n-1}^m is in the center of the ring $[M(p, v_1^{i_1}, \ldots, v_{n-2}^{i_{n-2}}), M(p, v_1^{i_1}, \ldots, v_{n-2}^{i_{n-2}})]_*$. That this is possible follows from nilpotence technology of **[9]**, **[10]**. Here v_{n-1}^m denotes a v_{n-1} self-map inducing multiplication by v_{n-1}^m on BP-homology. Now suppose J has been chosen so that there exists an i_{n-1}' with the following properties:

(i) $i_{n-1}' = i_{n-1} + sm$ for some positive integer s.

(ii) There exists a spectrum $M(p, v_1^{i_1}, \ldots, v_{n-2}^{i_{n-2}}, v_{n-1}^{i_{n-1}'}, v_n^{j_1}, \ldots, v_{n+r-1}^{j_r})$ together with a map

$$M(p, v_1^{i_1}, \ldots, v_{n-2}^{i_{n-2}}, v_{n-1}^{i_{n-1}'}, v_n^{j_1}, \ldots, v_{n+r-1}^{j_r}) \to M(p, v_1^{i_1}, \ldots, v_{n-1}^{i_{n-1}}, v_n^{j_1}, \ldots, v_{n+r-1}^{j_r}).$$

This map is to be constructed in the usual way from maps of cofiber sequences and induces the evident map in BP-homology.

There exists a cofinal sequence of multi-indices J having these properties.

Then choose t so that

$$\begin{array}{ccc} M(p, v_1^{i_1}, \ldots, v_{n-1}^{i_{n-1}'}, \ldots, v_{n+r-1}^{j_r}) & \xrightarrow{\nu_{n+r}'} & M(p, v_1^{i_1}, \ldots, v_{n-1}^{i_{n-1}'}, \ldots, v_{n+r-1}^{j_r}) \\ \downarrow & & \downarrow \\ M(p, v_1^{i_1}, \ldots, v_{n-1}^{i_{n-1}}, \ldots, v_{n+r-1}^{j_r}) & \xrightarrow{\nu_{n+r}^t} & M(p, v_1^{i_1}, \ldots, v_{n-1}^{i_{n-1}}, \ldots, v_{n+r-1}^{j_r}) \end{array}$$

commutes, where ν_{n+r}' is an appropriate v_{n+r} self-map.

Observe that the diagrams

$$\begin{array}{ccc} M(p, v_1^{i_1}, \ldots, v_{n-2}^{i_{n-2}}) & \xrightarrow{\eta} & M(p, v_1^{i_1}, \ldots, v_{n-1}^{i_{n-1}'}) \\ & \searrow^{\eta} & \downarrow \\ & & M(p, v_1^{i_1}, \ldots, v_{n-1}^{i_{n-1}}) \end{array}$$

and

$$\begin{array}{ccc} M(p, v_1^{i_1}, \ldots, v_{n-1}^{i_{n-1}'}, \ldots, v_{n+r-1}^{j_r}) & \xrightarrow{\partial} & M(p, v_1^{i_1}, \ldots, v_{n-2}^{i_{n-2}}) \\ \downarrow & & \downarrow{\scriptstyle v_{n-1}^{ms}} \\ M(p, v_1^{i_1}, \ldots, v_{n-1}^{i_{n-1}}, \ldots, v_{n+r-1}^{j_r}) & \xrightarrow{\partial} & M(p, v_1^{i_1}, \ldots, v_{n-2}^{i_{n-2}}) \end{array}$$

commute, where maps denoted by η are inclusions of bottom objects, and maps denoted by ∂ are projections onto top objects. Then $g_I \circ h_I$ factors as

$$\begin{aligned} S^0 \xrightarrow{h_I'} L_N M(p, v_1^{i_1}, \ldots, v_{n-2}^{i_{n-2}}) & \xrightarrow{L_N q} L_N M(p, v_1^{i_1}, \ldots, v_{n-2}^{i_{n-2}}) \\ & \xrightarrow{L_N v_{n-1}^{ms}} L_N M(p, \ldots, v_{n-2}^{i_{n-2}}) \xrightarrow{\partial} L_N M(p), \end{aligned}$$

where q is the composition

$$\begin{aligned} M(p, v_1^{i_1}, \dots, v_{n-2}^{i_{n-2}}) &\xrightarrow{\eta} M(p, v_1^{i_1}, \dots, v_{n-1}^{i'_{n-1}}, \dots, v_{n+r-1}^{j_r}) \\ &\xrightarrow{\nu'_{n+r}} M(p, v_1^{i_1}, \dots, v_{n-1}^{i'_{n-1}}, \dots, v_{n+r-1}^{j_r}) \xrightarrow{\partial} M(p, \dots, v_{n-2}^{i_{n-2}}). \end{aligned}$$

But the centrality of v_{n-1}^{ms} and the fact that $L_N v_{n-1}^{ms} \circ h'_I = 0$ imply that $g_I \circ h_I = 0$. □

Appendix 1. The $E(n)$-based Adams spectral sequence

In this section we study the (modified) $E(n)$-based Adams spectral sequence, and we compute $\pi_0 L_n S^0$, $\pi_0 L_{K(n)} S^0$, and $\pi_0 L_{K(n)} M(p)$. Some of the results of this section will be used in Appendix 2, where we provide a counterexample to Conjecture 1 for $N = n \geq 2$.

Originally, our approach to making computations in $E(n)_*$-local homotopy theory involved the modified BP-based Adams spectral sequence converging to $[X, L_n Y]_*$. This necessitated the use of [**4**] (1.14). However, Hovey and Sadofsky have recently proved a change of rings theorem which allows us to work directly with the $E(n)$-based Adams spectral sequence.

Let N be a BP_*BP-comodule with $v_{k-1}^{-1} N = 0$, where k is some nonnegative integer. Hovey and Sadofsky [**13**] prove that the natural map

$$H^{**}(v_k^{-1} N) \to \operatorname{Ext}^{**}_{E(n)_*E(n)}(E(n)_*, E(n)_* \otimes_{BP_*} v_k^{-1} N) \tag{A1.1}$$

is an isomorphism whenever $n \geq k$. (The case $n = k$ is the "classical" case of [**18**].) It then follows as in the proof of [**4**] (3.3) that if M is any BP_*BP-comodule,

$$\operatorname{Ext}^{**}(M, v_k^{-1} N) \xrightarrow{\approx} \operatorname{Ext}^{**}_{E(n)_*E(n)}(E(n)_* \otimes_{BP_*} M, E(n)_* \otimes_{BP_*} v_k^{-1} N). \tag{A1.2}$$

One can also use the isomorphism in A1.1 to prove as in [**22**] (Chapter 8) that every spectrum is $E(n)_*$-prenilpotent, and therefore the spectral sequence $E_r^{**}(E(n); Y)$ converges conditionally and strongly to $\pi_* L_n Y$. By [**4**] (1.3, 1.21), it follows that there is a modified $E(n)$-based Adams spectral sequence $E_r(E(n); X, Y)$ converging conditionally to $[X, L_n Y]_*$.

Now let M be as above, and consider the short exact sequences

$$\begin{aligned} 0 \to E(n)_*/(p^\infty, \dots, v_{k-1}^\infty) \to v_k^{-1} E(n)_*/(p^\infty, \dots, v_{k-1}^\infty) \\ \to E(n)_*/(p^\infty, \dots, v_k^\infty) \to 0, \end{aligned}$$

where $0 \leq k \leq n$. Applying the functor $\operatorname{Ext}_{E(n)_*E(n)}(E(n)_* \otimes_{BP_*} M, ?)$ yields a spectral sequence converging to $\operatorname{Ext}^{**}_{E(n)_*E(n)}(E(n)_* \otimes_{BP_*} M, E(n)_*)$ with

$$E_1^{s,t,*} = \begin{cases} \operatorname{Ext}^{t,*}(M, v_s^{-1} BP_*/(p^\infty, \dots, v_{s-1}^\infty)) & s \leq n \\ 0 & s > n. \end{cases}$$

This is, of course, just a truncated chromatic spectral sequence. The following result now follows easily.

Proposition A1.3 *Let M be as above. Then*

$$\operatorname{Ext}^{s,*}_{E(n)_*E(n)}(E(n)_* \otimes_{BP_*} M, E(n)_*) = 0$$

if $\operatorname{Ext}^{s-i,*}(M, v_i^{-1} BP_*/(p^\infty, \dots, v_{i-1}^\infty)) = 0$ *for all* $0 \leq i \leq n$.

If $n < p - 1$, then $\operatorname{Ext}^{s,*}(BP_*X, v_i^{-1} BP_*/(p^\infty, \dots, v_{i-1}^\infty)) = 0$ whenever $s > i^2 + i$, $0 \leq i \leq n$, provided that $E(i)_*X$ is a finitely generated $E(i)_*$-module ([**4**], 3.9). This implies that $\operatorname{Ext}^{s,*}_{E(n)_*E(n)}(E(n)_*X, E(n)_*) = 0$ for $s > n^2 + 2n$, and hence that the $E(n)$-based modified Adams spectral sequence converges strongly to $[X, L_nS^0]_*$. If $X = S^0$, we can obtain a slightly better horizontal vanishing line.

Proposition A1.4 *Let* $2 \le n < p-1$. *Then* $\mathrm{Ext}^{s,*}_{E(n)_*E(n)}(E(n)_*, E(n)_*) = 0$ *if* $s \ge n^2 + n$.

Proof First observe that $H^s(v_i^{-1}M) = 0$ for $s > i^2$ whenever $0 \le i \le n$ and M is a BP_*BP-comodule with $v_{i-1}^{-1}M = 0$. It thus suffices to show that

$$H^{n^2}(v_n^{-1}BP_*/(p^\infty, \dots, v_{n-1}^\infty)) = 0. \qquad \text{(A1.5)}$$

Let

$$\begin{aligned} N_j^i &= BP_*/(p, v_1, \dots, v_{j-1}, v_j^\infty, \dots, v_{j+i-1}^\infty) \\ M_j^i &= v_{i+j}^{-1}N_j^i. \end{aligned}$$

Consideration of the exact sequences

$$0 \to M_j^{n-j} \to M_{j-1}^{n-j+1} \xrightarrow{v_{j-1}} M_{j-1}^{n-j+1} \to 0$$

for $1 \le j \le n-1$ shows that it suffices to prove that

$$H^{n^2}(v_n^{-1}BP_*/(p, v_1, \dots, v_{n-2}, v_{n-1}^\infty)) = 0.$$

Next consider the short exact sequence

$$0 \to M_n^0 \to M_{n-1}^1 \xrightarrow{v_{n-1}} M_{n-1}^1 \to 0.$$

If $H^{n^2}(M_{n-1}^1) \ne 0$, then there exists a t such that $v_n^t c_n \in H^{n^2}(M_n^0)$ maps nontrivially to $H^{n^2}(M_{n-1}^1)$. But this implies that $v_n^t v_{n-1}^{k-1} c_n$ is nonzero in the cohomology of $v_n^{-1}BP_*/(p, \dots, v_{n-2}, v_{n-1}^k))$, contradicting Remark 4.11. This proves A1.5. □

We end this section by giving the promised computations of $\pi_0 L_n S^0$, $\pi_0 L_{K(n)}S^0$, and $\pi_0 L_{K(n)}M(p)$.

Proposition A1.6 *Suppose* $n \ge 1$. *If* $p > \max\{n+1, \frac{1}{2}(n^2+n)\}$, *then* $\pi_0 L_n S^0 = \mathbb{Z}_{(p)}$ *generated by the localization map* $S^0 \to L_n S^0$.

Proof If $n = 1$ this is well known, so suppose $n \ge 2$. Clearly the map

$$\mathbb{Z}_{(p)} = \pi_0 S^0 \to \pi_0 L_n S^0$$

is a monomorphism detected in $\mathrm{Ext}^{0,0}_{E(n)_*E(n)}(E(n)_*, E(n)_*) = \mathbb{Z}_{(p)}$. The proof is now completed by observing that sparseness and A1.4 imply that the group $\mathrm{Ext}^{s,s}_{E(n)_*E(n)}(E(n)_*, E(n)_*)$ is trivial for all $s \ne 0$. □

Proposition A1.7 *Let* $n \ge 1$, *and suppose* $p > \max\{n+1, \frac{1}{2}(n^2+2)\}$. *Then*

(i) $\pi_0 L_{K(n)}M(p) = \mathbb{Z}/(p)$, *generated by the map* $S^0 \xrightarrow{\eta} M(p) \to L_{K(n)}M(p)$.
(ii) $\pi_0 L_{K(n)}S^0 = \mathbb{Z}_p^\wedge$, *generated by the localization map* $S^0 \to L_{K(n)}S^0$.

Proof (i) There is a spectral sequence

$$\varprojlim_{(i_1, \dots, i_{n-1})} H^{**}(v_n^{-1}BP_*/(p, v_1^{i_1}, \dots, v_{n-1}^{i_{n-1}})) \Rightarrow \pi_* L_{K(n)}M(p)$$

(see [**11**], 7.4). By sparseness and a vanishing line argument, it suffices to show that

$$\varprojlim_{(i_1, \dots, i_{n-1})} H^{0,0}(v_n^{-1}BP_*/(p, v_1^{i_1}, \dots, v_{n-1}^{i_{n-1}})) = \mathbb{F}_p.$$

But Theorem 5.10 of [**19**] implies that

$$\lim_{\overleftarrow{j}} H^{0,*}(v_n^{-1}BP_*/(p, v_1, \dots, v_{n-2}, v_{n-1}^j)) = \mathbb{F}_p[[v_{n-1}]];$$

in particular, $\lim_{\overleftarrow{j}} H^{0,*}(v_n^{-1}BP_*(p, \dots, v_{n-2}, v_{n-1}^j))$ is concentrated in nonnegative degrees. A straightforward induction then shows that, for any appropriate multi-index $(i_1, \dots, i_{n-2})$, $\lim_{\overleftarrow{j}} H^{0,*}(v_n^{-1}BP_*/(p, v_1^{i_1}, \dots, v_{n-2}^{i_{n-2}}, v_{n-1}^j))$ is concentrated in nonnegative degrees and that

$$\lim_{\overleftarrow{j}} H^{0,0}(v_n^{-1}BP_*/(p, v_1^{i_1}, \dots, v_{n-2}^{i_{n-2}}, v_{n-1}^j)) = \mathbb{F}_p.$$

Hence

$$\begin{aligned}\lim_{\overleftarrow{(i_1,\dots,i_{n-1})}} & H^{0,0}(v_n^{-1}BP_*/(p, v_1^{i_1}, \dots, v_{n-1}^{i_{n-1}})) \\ &= \lim_{\overleftarrow{(i_1,\dots,i_{n-2})}} \lim_{\overleftarrow{i_{n-1}}} H^{0,0}(v_n^{-1}BP_*/(p, v_1^{i_1}, \dots, v_{n-1}^{i_{n-1}})) \\ &= \mathbb{F}_p.\end{aligned}$$

(ii) As before, there is a spectral sequence

$$\lim_{\overleftarrow{(i_0,\dots,i_{n-1})}} H^{**}(v_n^{-1}BP_*/(p^{i_0}, v_1^{i_1}, \dots, v_{n-1}^{i_{n-1}})) \Rightarrow \pi_* L_{K(n)}S^0,$$

and as before, it suffices to show that

$$\lim_{\overleftarrow{(i_0,\dots,i_{n-1})}} H^{0,0}(v_n^{-1}BP_*/(p^{i_0}, v_1^{i_1}, \dots, v_{n-1}^{i_{n-1}})) = \mathbb{Z}_p^\wedge. \tag{A1.8}$$

Continuing the induction of Part (i), one easily obtains

$$\lim_{\overleftarrow{(i_1,\dots,i_{n-1})}} H^{0,0}(v_n^{-1}BP_*/(p^{i_0}, v_1^{i_1}, \dots, v_{n-1}^{i_{n-1}})) = \mathbb{Z}/(p^{i_0})$$

for any $i_0 > 0$. The proof is now completed as before. □

Remark A1.9 Alternatively, one can prove A1.8 by observing that

$$\lim_{\overleftarrow{(i_0,\dots,i_{n-1})}} H^{0,0}(v_n^{-1}BP_*/(p^{i_0}, v_1^{i_1}, \dots, v_{n-1}^{i_{n-1}})) = W\mathbb{F}_{p^n}[[u_1, \dots, u_{n-1}]]^{S_n \rtimes \mathrm{Gal}},$$

where $W\mathbb{F}_{p^n}$ is the ring of Witt vectors with coefficients in the field $\mathbb{F}_{p^n}$ of p^n elements. Gal is the Galois group of $\mathbb{F}_{p^n}/\mathbb{F}_p$, and S_n is the nth Morava stabilizer group (see [**3**], 6.3). But, using the notation of [**5**],

$$W\mathbb{F}_{p^n}[[u_1, \dots, u_{n-1}]]^{S_n \rtimes \mathrm{Gal}} \subset W\mathbb{F}_{p^n}\langle\langle w_1, \dots, w_{n-1}\rangle\rangle^{S_n \rtimes \mathrm{Gal}}.$$

Now the action of S_n on the w_i's is easy to describe [**5**] (3.3); using this description one sees easily that

$$W\mathbb{F}_{p^n}\langle\langle w_1, \dots, w_{n-1}\rangle\rangle^{S_n \rtimes \mathrm{Gal}} = \mathbb{Z}_p^\wedge.$$

□

Appendix 2. A counterexample

Let $p > \max\{n+1, \frac{1}{2}(n^2+1)\}$. We now give a counterexample to Conjecture 1 for $N = n \geq 2$. As the considerations of Section 4 suggest, our counterexample will be the adjoint $h_I^{\#}$ of the map

$$h_I : S^0 \to L_n M(p, v_1^{i_1}, \ldots, v_{n-1}^{i_{n-1}})$$

of Theorem 1. Again, suspensions will be suppressed from the notation. The following result proves, upon taking adjoints, that $\pi_* h_I^{\#} = 0$.

Proposition A2.1 *If* $g : L_n M(p, v_1^{i_1}, \ldots, v_{n-1}^{i_{n-1}}) \to L_n S^0$ *is any map, then* $g \circ h_I = 0$.

Proof We use the Yoneda pairing μ in the $E(n)$-based modified Adams spectral sequence [**4**] and prove that $\mu(g \otimes h_I) = 0$ for any

$$g \in E_2^{**}(E(n); M(p, v_1^{i_1}, \ldots, v_{n-1}^{i_{n-1}}), L_n S^0),$$

where $h_I \in E_2^{n^2,*}(E(n); L_n M(p, v_1^{i_1}, \ldots, v_{n-1}^{i_{n-1}}))$. By sparseness and the vanishing line of A1.4, this will establish the proposition.

Again by A1.4, we may assume that $g \in E_2^{i,*}$, where $i < n$. But $M(p, v_1^{i_1}, \ldots, v_{n-1}^{i_{n-1}})$ is $E(i)_*$-acyclic for $i < n$; hence

$$\mathrm{Ext}^{**}(BP_* M(p, v_1^{i_1}, \ldots, v_{n-1}^{i_{n-1}}), v_i^{-1} BP_*/(p^\infty, v_1^\infty, \ldots, v_{i-1}^\infty)) = 0$$

whenever $i < n$ (cf. A1.2). It then follows from Proposition A1.3 that

$$\mathrm{Ext}^{s,*}_{E(n)_* E(n)}(E(n)_* M(p, v_1^{i_1}, \ldots, v_{n-1}^{i_{n-1}}), E(n)_*) = 0$$

if $s < n$. This completes the proof. □

References

[1] Cohen, J. M. [1969] *Coherent graded rings and the non-existence of spaces of finite stable homotopy type*, Comment. Math. Helv. **44**, 217–228.

[2] Devinatz, E. S. [1990] *K-theory and the generating hypothesis*, Amer. J. Math. **112** 787–804.

[3] Devinatz, E. S. [1995] *Morava's change of rings theorem*, The Cech Centennial, Contemp. Math., vol. 181, Amer. Math. Soc., Providence, R. I., 83–118.

[4] Devinatz, E. S. [1997] *Morava modules and Brown-Comenetz duality*, Amer. J. Math. **119**.

[5] Devinatz, E. S. and Hopkins, M. J. [1995] *The action of the Morava stabilizer group on the Lubin-Tate moduli space of lifts*, Amer. J. Math. **117**, 669–710.

[6] Freyd, P. [1966] *Stable homotopy*, Proc. Conf. Categorical Algebra (La Jolla, 1965), Springer-Verlag, Berlin, 121–172.

[7] Gross, B. H. and Hopkins, M. J. [1994a] *Equivariant vector bundles on the Lubin-Tate moduli space*, Topology and Representation Theory, Contemp. Math., vol. 158, Amer. Math. Soc., Providence, R. I., 23–88.

[8] Gross, B. H. and Hopkins, M. J. [1994b] *The rigid analytic period mapping, Lubin-Tate space, and stable homotopy theory*, Bull. Amer. Math. Soc. **30**, 76–86.

[9] Hopkins, M. J. [1987] *Global methods in homotopy theory*, Homotopy Theory—Proceedings of the Durham Symposium 1985, London Math. Soc. Lecture Note Series 117 (E. Rees and J. D. S. Jones, eds.), Cambridge University Press, Cambridge, UK, 73–96.

[10] Hopkins, M. J. and Smith, J. H. [to appear] *Nilpotence and stable homotopy theory II*, Ann. of Math.

[11] Hopkins, M. J., Mahowald, M., and Sadofsky, H. [1994] *Constructions of elements in Picard groups*, Topology and Representation Theory, Contemp. Math. **158**, (E. Friedlander and M. Mahowald, eds.), Amer. Math. Soc., Providence, R. I., 89–126.

[12] Hovey, M. [1995] *Bousfield localization functors and Hopkins' chromatic splitting conjecture*, The Cech Centennial, Contemp. Math., vol 181, Amer. Math. Soc., Providence, Rhode Island, 225–250.

[13] Hovey, M. and Sadofsky, H. [to appear] *Invertible spectra in the $E(n)$-local stable homotopy category*, J. London Math. Soc.

[14] Hovey, M., Palmieri, J. H., and Strickland, N. P. [to appear] *Axiomatic stable homotopy theory*, Mem. Amer. Math. Soc.

[15] Johnson, D. C. and Yosimura, Z. [1980] *Torsion in Brown-Peterson homology and Hurewicz homomorphisms*, Osaka J. Math. **17**, 117–136.

[16] Lewis, L. G., May, J. P., and Steinberger, M., with contributions by McClure, J. E. [1986] *Equivariant Stable Homotopy Theory*, Lecture Notes in Math., vol. 1213, Springer-Verlag, Berlin.

[17] Mahowald, M. and Sadofsky, H. [1995] *v_n telescopes and the Adams spectral sequence*, Duke Math. J. **78**, 101–129.

[18] Miller, H. R. and Ravenel, D. C. [1977] *Morava stabilizer algebras and the localization of Novikov's E_2-term*, Duke Math. J. **44**, 433–447.

[19] Miller, H. R., Ravenel, D. C., and Wilson, W. S. [1977] *Periodic phenomena in the Adams-Novikov spectral sequence*, Ann. of Math. **106**, 469–516.

[20] Morava, J. [1985] *Noetherian localisations of categories of cobordism comodules*, Ann. of Math. **121**, 1–39.

[21] Ravenel, D. C. [1984] *Localization with respect to certain periodic homology theories*, Amer. J. Math. **106**, 351–414.

[22] Ravenel, D. C. [1992] *Nilpotence and Periodicity in Stable Homotopy Theory*, Ann. of Math. Stud., vol. 128, Princeton University Press, Princeton, New Jersey.

[23] Shimomura, K. [1986] *On the Adams-Novikov spectral sequence and products of β-elements*, Hiroshima Math. J. **16**, 209–224.

[24] Shimomura, K. and Tamura, H. [1986] *Non-triviality of some compositions of β-elements in the stable homotopy of the Moore spaces*, Hiroshima Math. J. **16**, 121–133.

[25] Shimomura, K. and Yabe, A. [1995] *The homotopy groups $\pi_*(L_2S^0)$*, Topology **34**, 261–289.

[26] Würgler, U. [1978] *On the relation of Morava K-theories to Brown-Peterson homology*, Topology and Algebra, Proceedings of a Colloquium in Honor of B. Eckmann, Monographie de L'Enseignement Mathématique, no. 26, Geneva, 269–280.

Fields Institute Communications
Volume **19**, 1998

The Periodic Lambda Algebra

Brayton Gray
Department of Mathematics
University of Illinois at Chicago
850 S. Morgan
Chicago, Illinois USA 60607-7045

Abstract. The lambda algebra has been a useful tool both for making *Ext* calculations in a limited range and for understanding *Ext* on the chain level, both stably and unstably. For odd primes, the lambda algebra is inherently more complex and much less intuitive. This article presents a replacement for the lambda algebra at odd primes which is smaller, simpler, yields unstable analogs, and displays v_m periodicity both stably and unstably.

1 Introduction

The object of this work is to present a differential graded algebra (dga) $\overline{\Lambda}$ at primes $p > 2$ together with unstable approximations $\overline{\Lambda}(n)$. These objects are a significant improvement over the classical Λ algebra [B]. The advantages are:

(1) It is considerably smaller.

(2) The relations, and consequently the description of a basis is simpler.

(3) There is a natural generalization to a dga $\overline{\Lambda}_{(m)}$ for $m \geq -1$ ($\overline{\Lambda} = \overline{\Lambda}_{(-1)}$) whose homology is $\mathrm{Ext}^{**}_{A_*}\left(H_*\left(V(m)\right)\right)$, where $V(m)$ is the Smith-Toda complex.[1] $\overline{\Lambda}_{(m)}$ is actually smaller than $\overline{\Lambda}$, as is its homology. This is contrary to what happens when a standard approach is used to calculate the Ext groups of a finite complex [B].

(4) The classes v_n which figure prominently in the periodic approach to homotopy theory are clearly visible as generators in $\overline{\Lambda}$.

Some important features of the classical lambda algebra are preserved. In particular:

1991 *Mathematics Subject Classification.* Primary 55Q40; Secondary 55T15.

[1]Although $V(m)$ only exists under certain restrictions on m and $p, H_*(V(m)) \simeq E(\tau_0, \dots, \tau_m)$ always exists.

(1) There is a multiplicative epimorphic chain equivalence

$$\overline{C}^*(A_*) \xrightarrow{\iota^*} \overline{\Lambda}$$

where $\overline{C}^*(A_*)$ is the cobar construction on the dual of the Steenrod algebra. Consequently, $\overline{\Lambda}$ can be used to calculate Massey products.

(2) There is a natural filtration by subcomplexes $\overline{\Lambda}(2n+1)$ whose homology is the appropriate unstable Ext group for S^{2n+1}.

(3) $\overline{\Lambda}_{(m)}$ is likewise filtered by subcomplexes $\overline{\Lambda}_{(m)}(2n+1)$ in a way suggestive of unstable analogs to the spectra $V(m)$ with corresponding EHP sequences, periodicity operators, and unstable compositions as in [G 2]. (The complexes described in [G2] are a generalization of results of Mahowald [M] and Harper and Miller [HM]).

We now state our result:

Theorem 3.2. There is a trigraded dga $\overline{\Lambda}^{s_1,s_2,t}$ and a multiplicative epimorphic chain equivalence:

$$\iota^* : \overline{C}^*(A_*) \longrightarrow \overline{\Lambda}$$

where $\overline{C}^*(A_*)$ is the cobar construction on the dual of the Steenrod algebra. Furthermore $\overline{\Lambda}$ is generated by classes $\lambda_k \in \overline{\Lambda}^{0,1,kq}$ for $k \geq 1$ and $v_n \in \overline{\Lambda}^{1,0,2p^n-1}$ for $n \geq 0$ subject to the relations:

$$\text{(3C)}\quad
\begin{array}{ll}
\text{(i)} & \lambda_i\lambda_{pi+k} = \Sigma(-1)^{j+1}\binom{(p-1)(k-j)-1}{j}\lambda_{k+i-j}\lambda_{pi+j} \\
\text{(ii)} & v_n v_k = v_k v_n \\
\text{(iii)} & \lambda_k v_n = v_n\lambda_k + v_{n-1}\lambda_{k+p^{n-1}} \text{ if } n > 0 \\
\text{(iv)} & \lambda_k v_0 = v_0\lambda_k \\
\text{(v)} & \partial v_n = v_{n-1}\lambda_{p^{n-1}} \text{ if } n > 0 \\
\text{(vi)} & \partial v_0 = 0 \\
\text{(vii)} & \partial\lambda_k = \Sigma(-1)^{j+1}\binom{(p-1)(k-j)-1}{j}\lambda_{k-j}\lambda_j.
\end{array}$$

Furthermore, let $\overline{\Lambda}_{(m)} = \overline{\Lambda}/_{(v_0,v_1,\dots,v_m)}$; write $A_{(m)}$ for the subalgebra of A generated by the P^n and Q_k with $k > m \geq -1$. Then there is an epimorphic multiplicative chain equivalence:

$$\overline{C}^*(A_{(m)*}) \xrightarrow{\iota_m^*} \overline{\Lambda}_{(m)}.$$

Finally, if L is a right $A_{(m)}$ module ($m \geq -1$), $\text{Ext}_A^{**}(E(\tau_0,\dots,\tau_m)\otimes L)$ is isomorphic with $H_*(\Lambda_{(m)}\otimes L)$ where the differential is determined by

$$\partial(1\otimes x) = \sum \lambda_j \otimes xP^j + (-1)^{|x|}\sum v_n \otimes xQ_n.$$

Stably the complex $\overline{\Lambda}$ appears in Haynes Miller's thesis [Mi 1] without the algebra structure. In fact the multiplicative structure can be derived by applying the results of [Mi 2]. It does not appear explicitly in the literature, however. We choose an alternative route which is, in some sense, more elementary: we show that the Milnor basis for the Steenrod algebra [Miln] is a Koszul basis and apply the results of Priddy [P 2].

This material was developed after conversations with S. Pemmaraju [P 1] who, without the aid of the multiplicative structure, has been making calculations in Ext for $V(1)$ when $p = 3$ for use in his thesis. Our interest in this is as a theoretical and practical calculational tool, both stably and unstably. We believe that a systematic use of these results, using a downward induction on m, is a considerable improvement over standard methods of obtaining the Ext groups. For example, Tangora [T] calculates the homology of Λ in dimensions less than 94 and the homology of the subalgebra Λ' generated by all λ_i, in dimensions less than 242 when $p = 3$.

2 Koszul Resolutions

We begin by paraphrasing and summarizing the results in [P 2]. Let A be a pre-Koszul algebra; i.e., a connected augmented algebra over $\mathbf{Z}/p$ together with a presentation as an algebra with generators a_i, $i \in I$ modulo relations of the form:

$$\sum f_i a_i + \sum f_{jk} a_j a_k = 0.$$

Thus A has a $\mathbf{Z}/p$ basis consisting of words in the a_i. Filter A by minimal word length. Then $E^0 A$, the associated graded, is generated by $a_i \in E^0_1 A$ subject to the relations:

$$\sum f_{jk} a_j a_k = 0.$$

If $J = (j_1, \dots, j_n)$ with $j_k \in I$, write a_J for the word $a_{j_1} \dots a_{j_n}$. Suppose that we have choosen a $\mathbf{Z}/p$ basis B for $E^0 A$ consisting of all words a_J for $J \in S$, where S is a distinguished set of sequences of indices from I. We will call the sequences $J \in S$ and the corresponding words a_J admissible. We define coefficients $f\begin{pmatrix} k & \ell \\ i & j \end{pmatrix}$ by the rule:

$$a_k a_\ell = \sum_{(i,j) \in S} f\begin{pmatrix} k & \ell \\ i & j \end{pmatrix} a_i a_j \quad \text{in } E^0 A$$

and $f\binom{k,\ell}{m}$ by the consequent formula:

$$a_k a_\ell = \sum_m f\begin{pmatrix} k,\ell \\ m \end{pmatrix} a_m + \sum_{(i,j) \in S} f\begin{pmatrix} k & \ell \\ i & j \end{pmatrix} a_i a_j \quad \text{in } A.$$

We suppose that the indexing set I is countable and totally ordered. We order the set of sequences $J \in S$ lexicographically. If $J_1 = \{j_1 \dots j_s\}$ and $J_2 = \{j_{s+1}, \dots, j_k\}$ we will write $J = J_1 J_2$ for the sequence $\{j_1, \dots, j_k\}$. We will now say that $E^0 A$ is a Poincaré-Birkhoff-Witt (PBW) algebra if the following criteria are met:

(1) Cutting: If $J = J_1 J_2 \in S$, then $J_1 \in S$ and $J_2 \in S$,

(2) Expansion: If J_1, $J_2 \in S$ and $J = J_1 J_2$, then either $J \in S$ or each term of the expansion of a_J in the basis of admissible sequence, involves only a_K with $K > J$.

With this in mind, Priddy [P 2] shows that each PBW algebra is a Koszul algebra (which we needn't define) and concludes with the following result, which we summarize and paraphrase.

Theorem 2.1 [P 2]. Suppose A is a pre-Koszul algebra and E^0A is a PBW algebra. Then there is a dga $\overline{K}^*(A)$ called the co-Koszul complex satisfying the following conditions:

a) There is an epimorphism of dga's

$$\iota^* : \overline{C}(A^*) \longrightarrow \overline{K}^*(A)$$

where $\overline{C}(A^*)$ is the cobar construction on A^*.

b) ι^* induces an isomorphism in homology, and hence $H^*(A) \cong H_*(\overline{K}^*(A))$.

c) $\overline{K}^*(A)$ is generated by classes $\beta_i \in E_1^0A$ for $i \in I$ corresponding to $[a_i^*]$ in the cobar construction. Write $|\beta_i| = 1 + \deg a_i$. Then $\overline{K}^*(A)$ has defining relations:

$$(-1)^{\nu_{i,j}}\beta_i\beta_j + \sum_{(k,\ell)\notin S} (-1)^{\nu_{k,\ell}} f\begin{pmatrix} k & \ell \\ i & j \end{pmatrix} \beta_k\beta_\ell = 0$$

for each $(i,j) \in S$, where $\nu_{u,v} = |\beta_u| + \big(|\beta_u| - 1\big)\big(|\beta_v| - 1\big)$.

d) The differential in $\overline{K}^*(A)$ is defined by:

$$\delta\beta_m = \sum_{(k,\ell)\notin S} (-1)^{\nu_{k,\ell}} f\begin{pmatrix} k,\ell \\ m \end{pmatrix} \beta_k\beta_\ell.$$

e) Suppose L is a left A module. Then $H^*(A;L)$ is the homology of the complex $\overline{K}^*(A) \otimes L^*$, where L^* has the induced right A module structure, and the differential is determined by d) above and

$$\delta(\iota \otimes \lambda) = \sum (-1)^{|\lambda||\beta_j|} \beta_j \otimes \lambda a_j.$$

Proof Everything except the last statement can be found in [P 2]. The last statement follows since if $\mu_L : A \otimes L \longrightarrow L$ is the left action with dual $\mu_L^* : L^* \longrightarrow A^* \otimes L^*$, then

$$\mu_L^*(\lambda) = \sum_{\xi \in B} (-1)^{|\xi|(|\xi|+|\lambda|)} \xi^* \otimes \lambda\xi$$

using the sign conventions in [P 2]. Thus the projection onto E_1^0A, $\overline{\mu}_L^*$, is given by

$$\overline{\mu}_L^*(\lambda) = \sum (-1)^{|\beta_j|(|\beta_j|+|\lambda|)} \beta_j \otimes \lambda a_j.$$

The formula then follows from [P 2; 4.2]. □

3 Application to the Milnor basis

We now apply the above theory to some subalgebras of the Steenrod algebra. Milnor [Miln] has described the Steenrod algebra as having generators P^n, Q_k for $n > 0$ and $k \geq 0$ where $|Q_k| = 2p^k - 1$ and $|P^n| = nq$ subject to the following relations:

(3A)

$$\text{(i) if } a < pb,\ P^aP^b = (-1)^a\binom{(p-1)b-1}{a}P^{a+b} + \sum_{t=1}^{[a/p]}(-1)^{a+t}\binom{(p-1)(b-t)-1}{a-pt}P^{a+b-t}P^t$$

$$\text{(ii) } Q_kQ_\ell = -Q_\ell Q_k$$

$$\text{(iii) } P^nQ_k = \begin{cases} Q_kP^n + Q_{k+1}P^{n-p^k}, & n > p^k \\ Q_kP^n + Q_{k+1}, & n = p^k \\ Q_kP^n, & n < p^k. \end{cases}$$

We define the subalgebra $A_{(m)}$ to be the algebra generated by all P^n and all Q_k with $k > m \geq -1$. Clearly $\mathrm{Ext}_{A_{(m)*}}(\mathbf{Z}/p, L) \cong \mathrm{Ext}_{A_*}(\mathbf{Z}/p, E(\tau_0, \dots, \tau_m) \otimes L)$ for each left $A_{(m)*}$ comodule L.

It is clear from (3A) that $A_{(m)}$ is a pre-Koszul algebra for each $m \geq -1$. Now let I be the ordered set $2\omega = \{1, 2, \dots, \omega, \omega+1, \dots\}$ and define $a_n = P^n$ for $n < \omega$ and $a_{\omega+n} = Q_{m+n+1}$. We now write any monomial in $A_{(m)}$ as a sum of monomials in which:

(3B)

(a) all the Q_k's occur prior to all the P^n's using (3A iii),

(b) the Q_k's are ordered in decreasing order using (3A ii),

(c) the P^n's are in standard admissible ordering using (3A i).

We will call such a monomial an admissible monomial. For a sequence $\{i_1, \dots, i_n\}$ to be admissible, $i_j \in I$ we have $i_k > i_{k+1}$ if $i_{k+1} \geq \omega$ and $i_k \geq pi_{k+1}$ if $i_k < \omega$. Let S be the set of sequences satisfying these conditions and $B_{(m)}$ the basis of admissible monomials. Then we have:

Proposition 3.1. $E^0A_{(m)}$ is a PBW algebra.

Proof Clearly if $J = J_1J_2$ is admissible, J_1 and J_2 are as well. Suppose now that J_1 and J_2 are admissible, $J = J_1J_2$ and a_J is not admissible. Then we can write a_J as a sum of admissible monomials as described in (3B). Each move writes a term as a sum of terms with a higher order in the lexicographic ordering. □

Consequently, we may describe the co-Koszul complex whose homology is the sought after Ext group. As in the usual case, we obtain the opposite algebra structure. Thus we define $\overline{\Lambda}_{(m)} = \overline{K}^*(A_{(m)})^{op}$. This has no consequence since $H_*(A_{(m)})$ is commutative.

Proof of 3.2 (see Section 1). Write $v_n = [Q_n]$ and $\lambda_k = [P^k]$. The relations in (3C) then follow immediately from the corresponding relations in (3A) and 2.1c and d. Since $A_{(m)}$ is a subalgebra of A, $A_{(m)_*}$ is a quotient co-algebra of A_* and hence $\overline{\Lambda}_{(m)}$ is a quotient of $\overline{\Lambda}$. The Koszul generators correspond, excepting that $v_i = 0$ in $\overline{\Lambda}_{(m)}$ if $i \leq m$. The formula for the boundary in $\Lambda_{(m)} \otimes L$ follows from 2.1c. □

Notes

(1) The trigrading is possible since all the formulas in (3C) respect the polynomial degree in the v_i's. This induces a trigrading in the corresponding Ext groups. This trigrading is probably the same as that given by the number of μ's.

(2) By the description in (3B), $\overline{\Lambda}_{(m)}$ has a basis consisting of elements of the form $f(v)\lambda_I$ where $f(v)$ is a monomial in the v_m's, $n > m$ and λ_I is an admissible monomial in Λ'. We could have just as well written as a basis, all $\lambda_I f(v)$, since by (3C)

$$v_n\lambda_k = \lambda_k v_n - \lambda_{k+p^{n-1}}v_{n-1} + \lambda_{k+p^{n-1}+p^{n-2}}v_{n-2} - \dots$$

(3) It is easy to see that there is no multiplicative chain equivalence either from Λ to $\overline{\Lambda}$ or from $\overline{\Lambda}$ to Λ, since in Λ, $\lambda_1\mu_0 \neq 0$ while $\mu_0\lambda_1 = 0$. This suggests an interesting question. Define a partial ordering of dga's by $\Lambda_1 \leq \Lambda_2$ if there is a multiplicative chain equivalence $\gamma : \Lambda_2 \longrightarrow \Lambda_1$. Then $\Lambda < \overline{C}^*(A_*)$ and $\overline{\Lambda} < \overline{C}^*(A_*)$, but Λ and $\overline{\Lambda}$ are not comparable. What are the minimal elements in the set of all M with $M < \overline{C}^*(A_*)$?

Let $v_m : \overline{\Lambda}_{(m-1)} \longrightarrow \overline{\Lambda}_{(m-1)}$ be left multiplication by the central cycle v_m. Then there is an exact sequence:

$$0 \longrightarrow \overline{\Lambda}_{(m-1)} \xrightarrow{v_m} \overline{\Lambda}_{(m-1)} \xrightarrow{\pi} \overline{\Lambda}_{(m)} \longrightarrow 0. \tag{3E}$$

This induces a long exact sequence in homology, and in particular, a connecting homomorphism $\delta : H_*(\Lambda_{(m)}) \longrightarrow H_*(\Lambda_{(m-1)})$ of tridegree $(-1, 1, -(2p^m - 1))$. Using this we can define the rth Bockstein $\beta^{(r)} : H_*(\overline{\Lambda}_{(m)}) \longrightarrow H_*(\overline{\Lambda}_{(m)})$ of tridegree $(-r, 1, -r(2p^m - 1))$ by $\beta^{(r)}(\pi x) = [v_m^{-r}x]$. One can easily see for example that $\beta^{(p^r)}(v_{m+1}^{p^r}) = \lambda_{p^{r+n-1}}$. These Bocksteins are the differentials in the spectral sequence of a filtered dga.

Theorem 3.3. There is a Bockstein spectral sequence which is a spectral sequence of graded rings such that:

$$E^1_{s,t} = Z/p\langle v_m^s\rangle \otimes H_t(\overline{\Lambda}_{(m)})$$
$$d^r_{s,t}(v_m^s \otimes c) = v_m^{s+r} \otimes \beta^{(r)}(c)$$
$$d^r_{s,t} : E^r_{s,t} \longrightarrow E^r_{s+r,t-rq_m-1} \quad \text{where } q_m = 2p^m - 2$$
$E^\infty_{s,t}$ is the associated graded to the filtration
$$F^s = v_m^s H_t(\overline{\Lambda}_{(m-1)}) \subset H_{sq_m+t}(\overline{\Lambda}_{(m-1)}).$$

Proof Filter $\overline{\Lambda}_{(m-1)}$ by $F^s\overline{\Lambda}_{(m-1)} = v_m^s\overline{\Lambda}_{(m-1)}$. This makes $\overline{\Lambda}_{(m-1)}$ into a filtered dga. The spectral sequence is then the standard one in this situation. □

4 An unstable filtration of $\overline{\Lambda}_{(m)}$

In [G 2] we gave a filtration of a dga $\Lambda_{(m)}$ whose homology for each $m \geq -1$ is $\mathrm{Ext}_{A_*}(\mathbf{Z}/p\{\tau_0, \dots, \tau_m\})$. The filtration $F^n\Lambda_{(m)} = \Lambda_{(m)}(n)$ generalized to the context of $V(m)$ the chain complex $\Lambda_{(-1)}(n) = \Lambda(n)$ whose homology is the E^2 term for the unstable Adams spectral sequence converging to $\pi_*(S^n)$. The case $m = 0$ had been considered by Mahowald [Ma] and Harper and Miller [HM]. This filtration provided an algebraic glimpse of unstable periodic homotopy. These constructions used the full lambda algebra Λ, and it is our purpose here to show that $\overline{\Lambda}_{(m)}$ has a similar filtration. In particular, in case $m = -1$, we construct subcomplexes $\overline{\Lambda}(2n+1) \subset \overline{\Lambda}$ whose homology is the appropriate Ext group for $\pi_*(S^{2n+1})$.

Let us write Λ' for the subalgebra of Λ generated by the λ_i for $i > 0$. Write $\Lambda'(2n+1) = \Lambda' \cap \Lambda(2n+1)$; this is spanned by all admissible $\lambda_I = \lambda_{i_1} \ldots \lambda_{i_k}$ with $i_1 \leq n$. Define $H_n : \Lambda'(2n+1) \longrightarrow \Lambda'(2np-1)$ by[2] $H_n(\lambda_I) = 0$ if $\lambda_I \in \Lambda'(2n-1)$ and I is admissible, and $H_n(\lambda_n \lambda_I) = \lambda_I$ if $\lambda_I \in \Lambda'(2np-1)$ and I is admissible. This covers all cases and we have a short exact sequence of chain complexes:

$$0 \longrightarrow \Lambda'(2n-1) \longrightarrow \Lambda'(2n+1) \xrightarrow{H_n} \Lambda'(2np-1) \longrightarrow 0. \tag{5A}$$

By [HM; 1.6], we have

$$\lambda_k \Lambda'(2n+1) \subset \Lambda'(2n-kq+1) \quad \text{if} \quad n \geq pk. \tag{5B}$$

Now let $PV_{(m)} = \mathbf{Z}/p[v_{m+1}, v_{m+2}, \ldots]$ for $m \geq -1$. This is bigraded with $v_i \in PV_{(m)}^{1,2p^i-1}$. For each $n \geq 0$ we define $\overline{\Lambda}_{(m)}(2n+1)$ as a subspace of $\overline{\Lambda}_{(m)}$ by:

$$\overline{\Lambda}_{(m)}(2n+1) = \oplus PV_{(m)}^{s,t} \otimes \Lambda'(2n+s+t+1). \tag{5C}$$

It is easy to check that for each $k > m$:

$$x \in \overline{\Lambda}_{(m)}(2n+1) \quad \text{iff} \quad v_k x \in \overline{\Lambda}_{(m)}(2n-2p^k+1). \tag{5D}$$

Notice that when $m = -1$ and $k = 0$, this corresponds to the geometric fact, due to Cohen, Moore, and Neisendorfer, that $p \cdot \pi_*(S^{2n+1}) \subset \pi_*(S^{2n-1})$.

By induction on the first degree $s(\lambda_I) =$ length of I, and use of (5B) one can easily see that:

$$\lambda_k \overline{\Lambda}_{(m)}(2n+1) \subset \overline{\Lambda}_{(m)}(2n-kq+1) \quad \text{if} \quad n \geq pk. \tag{5E}$$

Combining (5D), (5E), and induction, one can then prove:

$$\overline{\Lambda}_{(m)}^{s_1,s_2,t}(2n+1) \cdot \Lambda_{(m)}(2n+t+s_1+1) \subset \overline{\Lambda}_{(m)}(2n+1). \tag{5F}$$

This is reminiscent of unstable composition and is slightly stronger than [G 2; 6.6].

Consider $x = f(v)\lambda_I \in \overline{\Lambda}_{(m)}(2n+1)$ with $f(v) \in PV_{(m)}^{s,t}$ and $\lambda_I \in \Lambda'(2n+s+t+1)$. Then

$$\begin{aligned}\partial x = \big[\partial f(v)\big]\lambda_I + f(v)\partial\lambda_I &\in \big[\partial f(v)\big]\Lambda'(2n+s+t+1) \\ &+ f(v)\Lambda'(2n+s+t+1) \subset \overline{\Lambda}_{(m)}(2n+1)\end{aligned}$$

when $n \geq 0$ by (5D) and (5E). Hence

$$\overline{\Lambda}_{(m)}(2n+1) \quad \text{is a subcomplex of} \quad \overline{\Lambda}_{(m)} \quad \text{when} \quad n \geq 0. \tag{5G}$$

We now define $\phi_n : \overline{\Lambda}_{(m)}(2n+1) \longrightarrow \overline{\Lambda}_{(m+1)}(2np-1)$ for $n > 0$ by

$$\phi_n(v_{i_1} \ldots v_{i_s}\lambda_I) = v_{v_1+1} \ldots v_{i_s+1} H_{n+p^f}(\lambda_I) \tag{5H}$$

where $p^f = p^{i_1} + \cdots + p^{i_s}$; ϕ_n has tridegree $(0, -1, -nq)$. It is straightforward but somewhat tedious to check that $\phi_n \partial = -\partial\phi_n$, and there is a short exact sequence[3]:

$$0 \longrightarrow \overline{\Lambda}_{(m)}(2n-1) \longrightarrow \overline{\Lambda}_{(m)}(2n+1) \xrightarrow{\phi_n} \overline{\Lambda}_{(m+1)}(2np-1) \longrightarrow 0. \tag{5I}$$

[2] see [G 2; 3.1] where H_n is called H'_n

[3] Compare to [G 2; 3.5]

In case $m=-1$, this corresponds to the fibration sequence:

$$S^{2n-1} \longrightarrow \Omega^2 S^{2n+1} \longrightarrow BW_n$$

of [G1].

Using (5D) we see that there is a short exact sequence of chain complexes:

$$0 \longrightarrow \overline{\Lambda}_{(m-1)}(2n+2p^m+1) \xrightarrow{v_m} \overline{\Lambda}_{(m-1)}(2n+1) \xrightarrow{\pi} \overline{\Lambda}_{(m)}(2n+1) \longrightarrow 0 \quad (5J)$$

which is a desuspension of (3E). In particular, we can desuspend the Bockstein spectral sequence in Theorem 3.3 by setting $F^s\overline{\Lambda}_{(m-1)}(2n+1) = v_m^s\overline{\Lambda}_{(m)}(2n+2sp^m+1)$. This spectral sequence has $E^1_{s,t} = \mathbf{Z}/p\langle v_m^s\rangle \otimes H_t\big(\overline{\Lambda}_{(m)}(2n+2sp^m+1)\big)$ and converges to $H_*\big(\overline{\Lambda}_{(m-1)}(2n+1)\big)$.

In [G 2], we defined $\Lambda_{(m)}(n)$ for all n and it would be desireable to have a suitable definition for $\overline{\Lambda}_{(m)}(2n)$. Such a definition can be obtained in the form of a pull back diagram:

$$\begin{array}{ccccc}
\overline{\Lambda}_{(m)}(2n) & \longrightarrow & \overline{\Lambda}_{(m)}(2n+1) & & \\
\big\downarrow & & \big\downarrow \phi & & \\
\overline{\Lambda}_{(m)}(2np-1) & \longrightarrow & \overline{\Lambda}_{(m+1)}(2np-1) & \longrightarrow & 0 \\
& & \big\downarrow & & \\
& & 0. & &
\end{array} \quad (5K)$$

Although $\overline{\Lambda}_{(m)}(2n)$ contains $\overline{\Lambda}_{(m)}(2n-1)$ as a subcomplex, the natural map $\overline{\Lambda}_{(m)}(2n) \longrightarrow \overline{\Lambda}_{(m)}(2n+1)$ is an epimorphism. Thus $\overline{\Lambda}_{(m)}(2n)$ sits more awkwardly between $\overline{\Lambda}_{(m)}(2n-1)$ and $\overline{\Lambda}_{(m)}(2n+1)$ than in the corresponding case in [G2], where both suspensions are monomorphisms. Nevertheless, there are EHP sequences:

$$\begin{array}{ccccccccc}
0 & \longrightarrow & \overline{\Lambda}_{(m)}(2n-1) & \longrightarrow & \overline{\Lambda}_{(m)}(2n) & \longrightarrow & \overline{\Lambda}_{(m)}(2np-1) & \longrightarrow & 0 \\
0 & \longrightarrow & \overline{\Lambda}_{(m)}(2np+2p^{m+1}-1) & \longrightarrow & \overline{\Lambda}_{(m)}(2n) & \longrightarrow & \overline{\Lambda}_{(m)}(2n+1) & \longrightarrow & 0
\end{array} \quad (5L)$$

corresponding to [G2; 4.2].

References

[B] Bousfield, A. K. and Curtis, E. B. [1972], *The homotopy spectral sequence of a space with coefficients in a ring*, Topology **11**, 79-106.

[CMN] Cohen, F. R., Moore, J. C. and Neisenderfer, J. A. [1979], *Torsion in homotopy groups*, Ann. Math. **109**, 121-168.

[G 1] Gray, B. [1988], *On the Iterated Suspension*, Topology **27**, 301-310.

[G 2] Gray, B. [1993], *EHP Spectra and Periodicity II: Λ-algebra models*, TAMS **340**, 617-640.

[HM] Harper, J. R. and Miller, H. R. [1982], *On the double suspension homomorphism at odd primes*, TAMS **273**, 319-331.

[Ma] Mahowald, M., *On the double suspension homomorphism*, TAMS **214**, 169-178.

[Mi 1] Miller, H. R. [1974], *Some apsects of the Adams-Novikov spectral sequence*, Thesis, Princeton Univ.

[Mi 2] Miller, H. R. [1978], *A localization theorem in homological algebra*, Math. Proc. Phil. Soc. **84**, 73-84.

[Miln] Milnor, J. [1958], *The Steenrod algebra and its dual*, Ann Math. **67**, 150-171.

[P 1] Pemmaraju, S., private conversations, 1994-5.

[P 2] Priddy, S. [1970], *Koszul resolutions*, TAMS **152**, 39-60.

[R] Ravenel, D. [1986], *Complex Cobordism and stable homotopy groups of spheres*, Academic Press.

[T] Tangora, M. C. [1985], *Computing the homology of the lamdba algebra*, Memoirs AMS #337.

Fields Institute Communications
Volume **19**, 1998

Rational $O(2)$-Equivariant Cohomology Theories

J. P. C. Greenlees
School of Mathematics and Statistics
Hicks Building, Sheffield S3 7RH UK
j.greenlees@sheffield.ac.uk

Abstract. We use results from [**2, 3**] to give a complete description of rational $O(2)$-equivariant cohomology theories.

1 Introduction

This paper presents an algebraic model for $O(2)$-equivariant rational cohomology theories. The main mathematical input is the model for $SO(2)$-equivariant theories given in Part I of [**3**], together with special cases of results on rational Mackey functors from [**2**]. Using these ingredients, and the usual formal framework of equivariant homotopy, it is not hard to assemble the model. Nonetheless, several features of the model were unexpected to the author, so it seems desirable to illustrate the issues which arise in this simple case. An exactly analagous model can be given with the continuous quaternion group as the group of equivariance, but it is most unlikely that such an explicit model can be given for groups of larger rank.

First, we should explain that the cohomology theories we refer to are those admitting suspension isomorphisms for arbitrary representations, often known as '$RO(G)$-graded' or 'genuine' cohomology theories. These are represented by G-spectra as in [**4**], and more specifically by those indexed on a complete G-universe. The advantages of considering representing objects rather than the cohomology theories are well understood, and we work with G-spectra without further comment, or more precisely in the homotopy category of G-spectra, in the sense of Quillen. We work exclusively with rational cohomology theories, and rational G-spectra, without always displaying this in the notation. Thus G-**spectra** denotes the homotopy category of rational G-spectra, and $[X, Y]^G$ denotes the rational vector space of morphisms in this category. The G-spectrum representing the equivariant cohomology theory $Y_G^*(\cdot)$ is denoted Y and the correspondence is given by $Y_G^*(X) = [X, Y]_G^*$ in the usual way.

For brevity we let $G = O(2), T = SO(2)$ and $W = G/T$. The other closed subgroups of G are the cyclic subgroups C_n for $n \geq 1$, which are normal, and the dihedral groups. For each $n \geq 1$ there is a single conjugacy class of dihedral groups

1991 *Mathematics Subject Classification.* Primary 55N91; Secondary 55P42, 55P91.
The author is grateful to the Nuffield Foundation for support.

of order $2n$; for definiteness we think of $O(2)$ acting on $\mathbb{R}^2$, and we let D_{2n} denote the representative containing reflection in the x-axis. One calculates $N_G(D_{2n}) = D_{4n}$, so that the conjugacy class of D_{2n} is G-homeomorphic to G/D_{4n}, and hence to the circle $S(1/n)$ of radius $1/n$. It is natural to think of T as a continuous cyclic group, and thus to consider the set

$$\mathcal{C} = \{C_n \mid n \geq 1\} \cup \{T\}$$

of all cyclic subgroups. Similarly, it is natural to think of G as the continuous dihedral group, and thus to consider the set

$$\mathcal{D} = \{D \mid D \text{ is conjugate to } D_{2n} \text{ for some } n \geq 1\} \cup \{G\}$$

of all dihedral subgroups. We consider $\mathcal{C}$ as a discrete topological space, and $\mathcal{D}$ as the union of concentric circles $S(1/n)$ of radius $1/n$ with G at the centre. We consider the set of closed subgroups of G as the topological sum

$$\mathcal{S}_f G = \mathcal{C} \amalg \mathcal{D}.$$

The notation is chosen to correspond to that of [**2**], and the subscript f refers to the fact that we have not used the Hausdorff metric topology. Note that G acts continuously on $\mathcal{S}_f G$ by conjugation.

Roughly speaking, the way to find the model of a rational G-spectrum X is to consider the family of geometric fixed point functors $H \longmapsto \Phi^H X$ as giving an equivariant sheaf over $\mathcal{S}_f G$. With this point of view it is obvious that we should separate behaviour over $\mathcal{C}$ from that over $\mathcal{D}$. It turns out that the model of X over $\mathcal{C}$ is that of a T-spectrum with W-action, which is modelled by considering W-equivariant objects in either of the models of [**3**]. The model over $\mathcal{D}$ splits as a product of Eilenberg-MacLane spectra, and is thus modelled by graded rational Mackey functors. By [**2**], this is equivalent to using graded equivariant sheaves over $\mathcal{D}$.

2 Burnside splitting

Let $\mathcal{F}G$ denote the subspace of $\mathcal{S}_f G$ consisting of subgroups of finite index in their normalizers. Consider the equivalence relation $\sim$ on $\mathcal{S}_f G$ in which any pair of subgroups $H \trianglelefteq \hat{H}$ with $\hat{H}/H$ a torus are $\sim$-related; it turns out that each conjugacy class in this quotient is represented by a unique conjugacy class of elements of $\mathcal{F}G$ so that $\mathcal{F}G/G \xrightarrow{\cong} (\mathcal{S}_f G/\sim)/G$. This is easily checked directly in the present case.

For any stable G-map $f : S^0 \longrightarrow S^0$, the function $\phi(f) : H \longmapsto deg(\Phi^H f)$ is a continuous equivariant function on $\mathcal{S}_f G$. Furthermore, if $H \trianglelefteq \hat{H}$ with $\hat{H}/H$ a torus, we have $\phi(f)(H) = \phi(f)(\hat{H})$ so ϕ passes to the quotient to give tom Dieck's isomorphism

$$\phi : A(G) := [S^0, S^0]^G \xrightarrow{\cong} C^G(\mathcal{F}G, \mathbb{Q}).$$

Thus the idempotents of $A(G)$ correspond to open and closed G-invariant subspaces of $\mathcal{S}_f G$ which are unions of $\sim$-equivalence classes.

In particular, if we let c and d denote the characteristic functions of $\mathcal{C}$ and $\mathcal{D}$, we have the decomposition $1 = c+d$ into orthogonal idempotents. Each idempotent acts on all G-spectra, and thus we have a corresponding decomposition

$$G\text{-}\mathbf{spectra} \simeq (G,c)\text{-}\mathbf{spectra} \times (G,d)\text{-}\mathbf{spectra},$$

where the category (G,c)-**spectra** is the full subcategory of G-spectra X with $X \simeq cX$, and similarly for d.

Checking geometric fixed points, we see that this idempotent splitting is a splitting of the cofibre sequence $EW_+ \longrightarrow S^0 \longrightarrow \tilde{E}W$, so that $cS^0 = EW_+$ and $dS^0 = \tilde{E}W$. Indeed, $d : \tilde{E}W \longrightarrow \tilde{E}W$ is an equivalence since $\tilde{E}W$ is non-equivariantly contractible, so $\tilde{E}W \simeq d\tilde{E}W$. Also, the map $S^0 \longrightarrow \tilde{E}W$ is an equivalence after applying Φ^W, so $dS^0 \simeq d\tilde{E}W$; the other statements follow since idempotents preserve cofibre sequences and c, d are orthogonal idempotents adding to 1. Thus we have $cX = EW_+ \wedge X$ and $dX = \tilde{E}W \wedge X$ for any G-spectrum X.

3 The cyclic part of the model

In this section we describe the category of (G, c)-spectra.

Note that for any two G-spectra X and Y, we have an action of W on $[X, Y]^T$ by conjugation. One way to package this is via the action of W on G/T together with the change of groups isomorphism $[X \wedge G/T_+, Y]^G = [X, Y]^T$. Note that $X \wedge EW_+$ is T-equivalent to X.

Proposition 3.1 *If X and Y are G-spectra then the forgetful map induces a natural isomorphism*

$$[EW_+ \wedge X, Y]^G \xrightarrow{\cong} \{[X, Y]^T\}^W.$$

Proof Since any strictly equivariant map is homotopy equivariant, we have a natural map $[EW_+ \wedge X, Y]^G \longrightarrow \{[X, Y]^T\}^W$. Since we are working over the rationals, both sides are cohomology theories of X, so that it suffices to deal with the case of a finite spectrum X. If X is finite, both sides preserve direct limits in the second variable, so it suffices to deal with the case when Y is finite. But in that case the spectral sequence

$$H^*(W; [X, Y]^T) \Longrightarrow [EW_+ \wedge X, Y]^G$$

obtained from the skeletal filtration of EW_+ is obviously convergent. Over the rationals, the spectral sequence collapses to give the required isomorphism. □

Corollary 3.2 *The category of (G, c)-spectra is equivalent to the category of T-spectra with a homotopy action of W:*

$$(G, c)\textbf{-spectra} \simeq (T\textbf{-spectra})[W].$$

Proof Consider the forgetful functor (G, c)-**spectra** $\longrightarrow$ (T-**spectra**) [W] taking a (G, c)-spectrum to the underlying T-spectrum with G/T giving a homotopy involution. The proof of the proposition shows the functor is full and faithful. It remains to prove that it is essentially surjective.

Suppose Z is a T-spectrum with a homotopy involution z. We may form the induced G-spectrum $G_+ \wedge_T Z$, and note that it is in fact a (G, c)-spectrum. Now, by the proposition we calculate its selfmaps to be $[Z, Z]^T[W]$. We want to select a suitable idempotent summand of $G_+ \wedge_T Z$, motivated by the special case that Z comes from a G spectrum and the induction untwists. Accordingly we choose an element $\hat{w} \in G$ reducing to the non-trivial element of G/T, and note that right multiplication $R_{\hat{w}} : G \longrightarrow G$ is of order 2 up to homotopy; thus $e = (1 \wedge_T 1 + R_{\hat{w}} \wedge_T z)/2$ is idempotent, and we may take $X = eG_+ \wedge_T Z$. □

We can replace the category of T-spectra by either of the algebraic models from **[3]** to see that the category of (G, c)-spectra is the derived category of a suitable abelian category of objects with W-action. We are again using the fact that a W-action in a model category is equivalent to a W-action on the homotopy category, provided 2 is invertible.

4 (G,d)-Mackey functors

First recall that a Mackey functor is a contravariant additive functor $M : \mathbf{Or}_G \longrightarrow \mathbb{Q} - \mathrm{mod}$, where $\mathbf{Or}_G$ is the full subcategory of G-**spectra** on the orbits G/H_+. Before describing the model of (G,d)-spectra, we need to summarize the relevant properties of Mackey functors. This is the special case of the results of [**2**] in which the group is $O(2)$; alternatively, it is easy to deal with this special case directly.

The main result states that the category of rational Mackey functors M is equivalent to the category of equivariant sheaves V on $\mathcal{S}_f G$ equipped with restriction maps $V(T) \longrightarrow V(C_n)$ for all n. More precisely, we should make $\mathcal{S}_f G$ into a topological category with one non-identity morphism $C_n \longrightarrow T$ for each $n \geq 1$, and ask that V be a sheaf on the category. However, since we are only interested in sheaves supported on $\mathcal{D}$, where there are no non-identity morphisms, we do not need this extra precision. To be explicit, the stalks of V may be constructed from M as follows, if H is a finite subgroup $V(H) = e_H M(H)$, where $e_H \in A(H)$ is the idempotent of H supported at H, $V(T) = M(T)$ and $V(G) = \varinjlim_{e(G)=1} eM(G)$, where the direct limit is over all idempotents e with $e(G) = 1$.

Notice that for any Mackey functor M, an idempotent $e \in A(G)$ gives a new Mackey functor eM, defined by $(eM)(H) = (eM)(G/H_+) := M(eG/H_+)$. In particular, the category of rational Mackey functors splits:

$$G\text{-}\mathbf{Mackey} \simeq (G,c)\text{-}\mathbf{Mackey} \times (G,d)\text{-}\mathbf{Mackey}.$$

In terms of sheaves, we find (G,c)-Mackey functors correspond to equivariant sheaves supported on the subcategory $\mathcal{C}$: they are thus specified by W-modules $V(T)$, $V(C_n)$ for all $n \geq 1$, and W-maps $V(T) \longrightarrow V(C_n)$.

Similarly, we find (G,d)-Mackey functors correspond to equivariant sheaves supported on the space $\mathcal{D}$. Since we need to perform calculations, it is useful to be more explicit. The stalks are $V(G)$, and $V(D)$ for finite dihedral groups D. By G-equivariance, the stalks $V(D)$ are specified by $W_G(D_{2n})$-modules $V(D_{2n})$ for each n. Since $W_G(D_{2n})$ is of order 2, we abbreviate it too by W, relying on context to determine which group of order 2 is intended. Now the orbit of D_{2n} is isolated for each n, so the only piece of data required to recover the sheaf from its stalks is the way a germ from $V(G)$ may be represented in some neighbourhood of G. Each germ is defined on D_{2n} for all sufficiently large n, so we consider the end space

$$V(D_\infty) := \varinjlim_i \prod_{n \geq i} V(D_{2n}),$$

of sections of $\mathcal{D} \setminus \{D\}$ near infinity, and the behaviour of sections is described by a W-map $\sigma : V(G) \longrightarrow V(D_\infty)$.

Summary 4.1 *A (G,d)-Mackey functor is given by a vector space $V(G)$ together with W-modules $V(D_{2n})$ for $n \geq 1$ and a W-map $\sigma : V(G) \longrightarrow V(D_\infty)$. This correspondence is natural and gives an equivalence of categories.* □

It is then not hard to give a sufficient condition for a Mackey functor to be injective, and to conclude that all (G,d)-Mackey functors have an injective resolution of length 1.

Lemma 4.2 *A (G,d)-Mackey functor is injective if the map σ is an epimorphism, and hence the category of (G,d)-Mackey functors is of injective dimension 1.*

Proof Consider the category $\mathbb{C}$ of diagrams $\mathbb{V} = (V_0 \xrightarrow{s} V_1)$ of $\mathbb{Q}W$-modules with V_0 fixed by W. It is easy to see that an object is injective if and only if the structure map s is surjective. Indeed, given a monomorphism $\mathbb{A} \longrightarrow \mathbb{B}$ and a map $f : \mathbb{A} \longrightarrow \mathbb{V}$ we first choose an extension $\hat{f}_1$ of f_1 over B_1, then choose a splitting $V_0 = V_1 \oplus V_0'$. This determines the component of $\hat{f}_0$ mapping into V_1, and the other component may be any extension of the V_0' component of f_0.

In the case of a Mackey functor, we first choose the extensions $\hat{f}_D$ for the finite dihedral groups D. This gives the extension $\hat{f}_1$ for the object $V(G) \longrightarrow V(D_\infty)$, and we proceed as for the category $\mathbb{C}$.

It is now clear that an arbitrary sheaf V can be embedded in an associated injective sheaf I_V with $I_V(D) = V(D)$ and $I_V(G) = V(G) \oplus V(D_\infty)$. The cokernel is concentrated at G and thus injective. □

For completeness we record the situation for projective functors, although we do not use it elsewhere.

Remark 4.3 *By contrast, the surjectivity of σ is not a sufficient condition for the corresponding (G,d)-Mackey functor to be projective. However all (G,d)-Mackey functors admit projective resolution of length 1.*

Proof In the category $\mathbb{C}$, an object is projective if and only if s is a monomorphism. The problem is that the proof starts by lifting the component on V_0. If the constant sheaf is denoted $\underline{\mathbb{Q}}$, one may check that the sheaf V with $V(D) = \mathbb{Q}$, $V(G) = \underline{\mathbb{Q}}(D_\infty)$ and whose structural map σ is the identity, is not projective.

However, the constant sheaf is projective, as are the sheaves supported away from G, and hence all (G,d)-Mackey functors admit projective resolution of length 1. □

5 The dihedral part of the model

We may now show that any (G,d)-spectrum splits as a product of Eilenberg-MacLane spectra. Indeed, we may use the same proof as given in Appendix A of [**1**] for finite groups.

In the usual way the nth H-equivariant stable homotopy group of a G-spectrum X is defined by $\pi_n^H(X) = [G/H_+ \wedge S^n, X]^G = [S^n, X]^H$, and these are assembled into the homotopy Mackey functor $\underline{\pi}_n^G(X)$ defined by $H \longmapsto \pi_n^H(X)$. The Eilenberg-MacLane spectrum HM represents ordinary cohomology with coefficients in the Mackey functor M: it is characterized by the dimension axiom, which states that $\underline{\pi}_0^G(HM) = M$ and $\underline{\pi}_n^G(HM) = 0$ for $n \neq 0$.

For the proof we define a related cohomology theory. Given any injective rational Mackey functor I we may define a cohomology theory $hI_G^*(\cdot)$ by

$$hI_G^n(X) = \mathrm{Hom}(\underline{\pi}_n^G(X), I).$$

Evidently, if I is supported over $\mathcal{D}$, the cohomology theory vanishes on (G,c)-spectra.

Lemma 5.1 *If I is an injective (G,d)-Mackey functor then hI is an Eilenberg-MacLane spectrum: $hI = HI$.*

Proof Since $\underline{\pi}_0^G(G/H_+) = [\cdot, G/H_+]^G$ is the free functor, it is clear that hI has the correct homotopy groups in degree 0. We must calculate $hI_G^n(G/H_+)$ for each subgroup H, and show that it is zero if $n \neq 0$. We have already observed that $hI_G^*(\cdot)$ is zero on (G,c)-spectra, so this deals with the case in which H lies in $\mathcal{C}$. This leaves us to deal with the cases $H = G$ and $H = D_{2n}$ for some $n \geq 1$.

For this we need to examine the functor $\underline{\pi}_n^G(G/H_+)$, which is made up from the groups $\pi_n^K(G/H_+)$. Since I is a (G,d)-Mackey functor, the groups with K in $\mathcal{C}$ have no effect, and we need only consider $K = G$ and $K = D_{2m}$ for some $m \geq 1$.

The tom Dieck splitting theorem for the G-space X states

$$\pi_n^G(X) = \pi_n(X^G) \oplus \bigoplus_{(D)} \pi_n(EW_+ \wedge_W X^D)$$
$$\oplus\, \pi_n(EW_+ \wedge_W X^T) \oplus \bigoplus_{C} \pi_n(E(G/C)_+ \wedge_{G/C} \Sigma^\xi X^C),$$

where ξ is the non-trivial representation of W on $\mathbb{R}$. The summands involving X^L with L in $\mathcal{C}$ do not contribute, since I is concentrated over $\mathcal{D}$. Now if $X = G/G_+$ then $X^G = X^D = S^0$, and the relevant part is zero if $n \neq 0$, since W has no positive dimensional homology over $\mathbb{Q}$. If $X = (G/D_{2n})_+$ then $X^G = *$, and X^D is either $*$ if D_{2n} is not subconjugate to D, or W_+ if it is subconjugate. In either case the relevant part is zero if $n \neq 0$.

Similarly, tom Dieck splitting for the G-space X states

$$\pi_n^{D_{2m}}(X) = \bigoplus_{(D)} \pi_n(EW_+ \wedge_W X^D) \oplus \bigoplus_{C} \pi_n(E(D_{2m}/C)_+ \wedge_{D_{2m}/C} X^C),$$

where the sums are now over conjugacy classes of subgroups of D_{2m}. The argument that the relevant part is zero if $n \neq 0$ is a little easier than that above. □

Now if M is any (G,d)-Mackey functor, we may choose an injective resolution $0 \longrightarrow M \longrightarrow I \longrightarrow J \longrightarrow 0$, and by definition of hI there is a map $HM \longrightarrow hI$ lifting the embedding $M \longrightarrow I$. Thus by the lemma we have a cofibre sequence $HM \longrightarrow hI \longrightarrow hJ$, and this gives a Universal Coefficient Theorem.

Corollary 5.2 *For any (G,d)-Mackey functor M there is a short exact sequence*

$$0 \longrightarrow \mathrm{Ext}(\underline{\pi}_{n-1}^G(X), M) \longrightarrow HM_G^n(X) \longrightarrow \mathrm{Hom}(\underline{\pi}_n^G(X), M) \longrightarrow 0.$$

□

Corollary 5.3 *Any (G,d)-spectrum X splits as a product of Eilenberg-MacLane spectra*

$$X \xrightarrow{\simeq} \prod_n \Sigma^n H\underline{\pi}_n^G(X).$$

Proof By 5.2 we may lift the identity of $\underline{\pi}_n^G(X)$ to give a map $X \longrightarrow \Sigma^n H\underline{\pi}_n^G(X)$ inducing the identity of $\underline{\pi}_n^G$. Using these as components we obtain a map into the product. The resulting map is an isomorphism in homotopy groups

π_n^H for all n and H by construction, and thus a weak equivalence. □

Note that since the product in the corollary is equivalent to a sum, this gives a short exact sequence for calculating $[X,Y]^G$ for any two (G,d)-spectra X and Y.

Corollary 5.4 *For any (G,d)-spectra, X and Y there is a short exact sequence*

$$0 \longrightarrow \mathrm{Ext}(\underline{\pi}_*^G(\Sigma X), \underline{\pi}_*^G(Y)) \longrightarrow [X,Y]^G \longrightarrow \mathrm{Hom}(\underline{\pi}_*^G(X), \underline{\pi}_*^G(Y)) \longrightarrow 0.$$

□

We may go a little further.

Corollary 5.5 *The category of (G,d)-spectra is equivalent to the derived category of graded (G,d)-Mackey functors:*

$$(G,d)\text{-}\mathbf{spectra} \simeq D((G,d)\text{-}\mathbf{Mackey}).$$

Proof It remains to comment on the naturality of the splitting. First choose a splitting of each (G,d)-spectrum X as a product of Eilenberg-MacLane spectra. This determines a splitting of the short exact sequence of 5.4, and this allows us to define a functor

$$(G,d)\text{-}\mathbf{spectra} \longrightarrow D((G,d)\text{-}\mathbf{Mackey})$$

to the derived category. The functoriality follows from the fact that the composite of two elements of Ext is zero. It is easy to see the map is an equivalence of categories. □

6 Sheaf theoretic formulation

We may package 5.5 by saying that a rational (G,d)-spectrum is modelled by an equivariant sheaf of graded vector spaces on $\mathcal{D}$, and morphisms are given by morphisms in the derived category of such sheaves. The functor is simply given by $X \longmapsto \underline{\pi}_*^G(X)$ on objects.

To give the best possible picture, we want to package 3.2 as showing (G,c)-spectra may be viewed as sheaves on the *category* $\mathcal{C}$. Indeed, since $\mathcal{C}$ is discrete as a topological space, we simply need to specify the stalks over the finite cyclic subgroups C, the stalk over T, and the restriction maps relating them. For this we can adapt the torsion model of [**3**].

We begin by describing a target category $\mathbf{Sh}_{\mathcal{C}}$ of sheaves on $\mathcal{C}$. First, we have a sheaf $\mathcal{O}$ of rings on $\mathcal{C}$: its stalk over the finite cyclic subgroup C is $\mathcal{O}_C := \mathbb{Q}[c_C]$, where c_C is of degree -2. Thus its space of sections over the space $\mathcal{F}$ of finite subgroups of T is $\mathcal{O}_{\mathcal{F}} = \prod_C \mathbb{Q}[c_C]$, and we may consider the set $\mathcal{E}$ of elements $x = (x_C)_C \in \mathcal{O}_{\mathcal{F}}$ with the components x_C homogeneous, and almost all 1; we refer to these as Euler classes. Alternatively, we may think of $\mathcal{E}$ as the space of sections which are 1 except on a compact set. Now we may define the stalk over T to be $\mathcal{O}_T := \mathcal{E}^{-1}\mathcal{O}_{\mathcal{F}}$. The structure maps for this sheaf are *covariant* and consist of the obvious maps $\mathcal{O}_C \longrightarrow \mathcal{E}^{-1}\mathcal{O}_{\mathcal{F}} = \mathcal{O}_T$ making $\mathcal{O}_T$ into a module over $\mathcal{O}_C = \mathbb{Q}[c_C]$. The category $\mathbf{Sh}_{\mathcal{C}}$ is the full subcategory of the category of sheaves of modules over the sheaf of rings $\mathcal{O}$, consisiting of sheaves $\mathbb{X}$ so that the stalk $\mathbb{X}(C)$ over C is a torsion module over $\mathcal{O}_C = \mathbb{Q}[c_C]$, the stalk $\mathbb{X}(T)$ is a free module over $\mathcal{O}_T$, and so that the map $\mathbb{X}(T) \longrightarrow \mathbb{X}(\mathcal{F}) = \prod_C \mathbb{X}(C)$ of $\mathcal{O}_{\mathcal{F}}$-modules, maps into the subspace $\bigoplus_C \mathbb{X}(C)$ of sections with compact support. The category $\mathbf{Sh}_{\mathcal{C}}$ is clearly equivalent to the torsion model category described in Chapter 6 of [**3**], and thus is an abelian category of injective dimension 2. Accordingly, differential graded sheaves $\tilde{\mathbb{X}}$ are not necessarily determined as objects in the derived category of $\mathbf{Sh}_{\mathcal{C}}$ by their homology in $H_*(\tilde{\mathbb{X}})$ in $\mathbf{Sh}_{\mathcal{C}}$. The main result of [**3**] Part I gives an equivalence T-**spectra** $\simeq D(\mathbf{Sh}_{\mathcal{C}})$, and so we conclude that (G,c)-**spectra** $\simeq D(\mathbf{Sh}_{\mathcal{C}})[W]$ by 3.2. The homology of the object corresponding to the (G,c)-spectrum X is the sheaf $\mathbb{X}$ with stalk $\pi_*^T(E\langle C\rangle \wedge \Sigma X)$ over C. Here $E\langle C\rangle$ is a space with nontrivial isotropy only at C; it may be constructed as the cofibre of the natural map $E[\subset C]_+ \longrightarrow E[\subseteq C]_+$ between the universal spaces for the family of proper subgroups and arbitrary subgroups of C. The presence of the suspension may be explained from one point of view by the fact that $W_T(C)$ is one dimensional. The stalk at T is $\pi_*^T(\tilde{E}\mathcal{F} \wedge DE\mathcal{F}_+ \wedge X) \cong \mathcal{O}_T \otimes \pi_*(\Phi^T X)$, where $DE\mathcal{F}_+$ denotes the functional dual, and the calculation is a direct consequence of [**3**](2.4.1) (see [**3**], (5.2.3)). The structure map is induced by

$$\tilde{E}\mathcal{F} \wedge DE\mathcal{F}_+ \wedge X \longrightarrow \Sigma E\mathcal{F}_+ \wedge DE\mathcal{F}_+ \wedge X \simeq \Sigma E\mathcal{F}_+ \wedge X \simeq \bigvee_C E\langle C\rangle \wedge \Sigma X,$$

where the first equivalence is immediate from the fact $DE\mathcal{F}_+$ is $\mathcal{F}$-equivalent to DS^0, and the second equivalence is the splitting of [**3**] (2.2.3). To see the W-action is the obvious one, note that we may take $E\mathcal{F}_+$ and $\tilde{E}\mathcal{F}$ to be G-spaces. The other important point is that all the subgroups of T are normal in G, so that W preserves the splitting $E\mathcal{F}_+ \simeq \bigvee_C E\langle C\rangle$.

Extending the sheaf $\mathcal{O}$ to a sheaf on $\mathcal{S}_f G$ using the constant sheaf $\underline{\mathbb{Q}}$ on $\mathcal{D}$, and combining the two parts of the model, we obtain a sheaf theoretic model for rational G-spectra.

Summary 6.1 *The category of rational G-spectra is equivalent to the derived category of the category of equivariant sheaves of $\mathcal{O}$-modules on the category $\mathcal{S}_f G$ whose stalks at finite cyclic groups are torsion, whose stalk at T is free, and whose restriction map from $\mathcal{C}$ maps into sections with compact support.* □

The standard model for T-spectra ([**3**], Chapter 5) consists of maps $N \longrightarrow \mathcal{E}^{-1}\mathcal{O}_{\mathcal{F}} \otimes V$ of $\mathcal{O}_{\mathcal{F}}$-modules, which become an isomorphism when $\mathcal{E}$ is inverted, but I do not see at present how to express this in sheaf theoretic terms. It would be more satisfactory to describe a sheaf theoretic model corresponding to this standard model, both because of the fact it is of injective dimension 1, and because of the intuition coming from the localization theorem.

References

[1] Greenlees, J. P. C. and May, J. P. [1995], *Generalized Tate cohomology*, Mem. American Math. Soc. **543**.

[2] Greenlees, J. P. C. [1998], *Rational Mackey functors for compact Lie groups I*, Proc. London Math. Soc., to appear.

[3] Greenlees, J. P. C. [1998], *Rational S^1-equivariant cohomology theories*, Mem. American Math. Soc., to appear.

[4] Lewis, L. G., May, J. P. and Steinberger, M. (with contributions by J. E. McClure) [1986], *Equivariant stable homotopy theory*, Lecture Notes in Mathematics **1213**, Springer-Verlag, Berlin.

Fields Institute Communications
Volume **19**, 1998

The Transcendence Degree of the mod p Cohomology of Finite Postnikov Systems

Jesper Grodal
Department of Mathematics
University of Copenhagen
Universitetsparken 5
DK–2100 Copenhagen Ø, Denmark
jg@math.ku.dk

Abstract. We examine the transcendence degree of the mod p cohomology of a finite Postnikov system E. We prove that, under mild assumptions on E, the transcendence degree of $H^*(E; \mathbf{F}_p)$ is always positive, and give a complete classification of the Postnikov systems where the transcendence degree of $H^*(E; \mathbf{F}_p)$ is finite. More precisely we prove that $H^*(E; \mathbf{F}_p)$ is of finite transcendence degree iff E is $\mathbf{F}_p$-equivalent to the classifying space of a p-toral group. To obtain the results we establish a general formula for determining the transcendence degree of an unstable algebra given in terms of the growth of certain 'unstable Betti numbers'. As an application of these results we derive statements about the n-connected cover $X\langle n\rangle$ of a finite complex X. We show for instance that, under suitable connectivity assumptions on X, the LS category of $X\langle n\rangle$ is always infinite assuming $X\langle n\rangle \neq X$. Finally we discuss generalizations of the obtained results to polyGEMs.

1 Introduction

In 1953 Serre showed his celebrated result that a 1-connected finite Postnikov system E with finitely generated homotopy always has homology in infinitely many dimensions, using his newly invented spectral sequence [**31**]. His methods, however, although revealing the asymptotic size of the Betti numbers (the coefficients in the Poincaré series) of $H^*(E; \mathbf{F}_p)$, did not in general give information about the ring or $\mathcal{A}$-module structure of $H^*(E; \mathbf{F}_p)$ (here $\mathcal{A}$ denotes the Steenrod algebra). Serre's theorem has since then been generalized in several ways by a number of people (Dwyer-Wilkerson [**10**], Lannes-Schwartz [**20, 22**], McGibbon-Neisendorfer [**24**]) all utilizing the theory of unstable modules over the Steenrod algebra as developed by Lannes, Schwartz and others. One of the main advantages of this approach is that, by relating certain properties of the cohomology to questions about mapping spaces, it gives a grip on how these properties behave with respect to fibrations—i.e.,

1991 *Mathematics Subject Classification.* Primary 55S45; Secondary 55S10, 55M30, 55P60.
The author was partially supported by Det internationale kontor, Det naturvidenskablige fakultet, Kund Højgaards fond and Julie Damms fond.

it turns traditional spectral sequence questions into long exact sequence questions. This paper is a contribution along these lines.

We offer the following two main theorems:

Theorem 1.1 *Let E be a connected nilpotent finite Postnikov system with finite $\pi_1(E)$. Assume that $H^*(E;\mathbf{F}_p)$ is of finite type and that $\bar{H}^*(E;\mathbf{F}_p) \neq 0$. Then $\bar{H}^*(E;\mathbf{F}_p)$ contains an element of infinite height.*

Theorem 1.2 *Let E be a connected nilpotent finite Postnikov system with finite $\pi_1(E)$. Assume that $H^*(E;\mathbf{F}_p)$ is of finite type. Then $H^*(E;\mathbf{F}_p)$ has finite transcendence degree iff E is $\mathbf{F}_p$-equivalent to a space E' fitting into a principal fibration sequence of the form*

$$\mathbf{CP}^\infty \times \cdots \times \mathbf{CP}^\infty \to E' \to K(P,1),$$

where P is a finite p-group.

Here we define the transcendence degree, $d(K)$, of a graded algebra K as the maximal number of homogeneous algebraically independent elements in K. When K is connected noetherian, $d(K)$ is equal to the Krull dimension of K. Note that Theorem 1.1 can be reformulated as saying that the transcendence degree of $H^*(E;\mathbf{F}_p)$ is always positive. Theorem 1.1 was previously known in the case $p=2$ for E 1-connected by work of Lannes and Schwartz [**22**], but was actually rediscovered independently by the author and formed the starting point for this work.

From now on $H^*(X)$ denotes the mod p cohomology of X for some fixed but arbitrary prime p. In the rest of this introduction we use some standard notation concerning unstable modules over the Steenrod algebra. In the next section we briefly introduce these concepts, for a general reference see e.g. Schwartz [**30**].

The key topological result needed in proving the results about $H^*(E)$ is the following asymptotic growth formula in $\mathrm{rk}_p V$, where V is an elementary abelian group:

$$\log_p |[BV, K(\pi_n(E),n)]| \lesssim \log_p |[BV,E]| \lesssim C \log_p |[BV, K(\pi_n(E),n)]|,$$

where n is the dimension of the highest homotopy group which is not uniquely p-divisible.

The above growth formula relates to the algebra structure of $H^*(E)$ by the following theorem of Lannes, generalizing earlier work of Miller [**25**]:

Theorem 1.3 [**18, 19**] *Let X be a connected space. Suppose that X is nilpotent with finite $\pi_1(X)$ and $H^*(X)$ of finite type. Then the natural map $f \mapsto f^*$, $[BV,X] \to \mathrm{Hom}_{\mathcal{K}}(H^*(X), H^*(V))$ is a bijection.*

Here $\mathcal{K}$ denotes the category of unstable algebras over the Steenrod algebra.

We have the following theorem of Lannes and Schwartz:

Theorem 1.4 [**29, 30**] *Let M be an unstable module over the Steenrod algebra. Then the following two conditions are equivalent:*

1. *M is nilpotent.*
2. *$\mathrm{Hom}_{\mathcal{U}}(M, H^*(V)) = 0$ for all elementary abelian p-groups V.*

Here $\mathcal{U}$ denotes the category of unstable modules over the Steenrod algebra.

The Lannes linearization principle [**19**]

$$\mathrm{Hom}_{\mathcal{U}}(K, H^*(V))' \simeq \mathbf{F}_p^{\mathrm{Hom}_{\mathcal{K}}(K,H^*(V))},$$

where $'$ denotes the (continuous) vector space dual, now immediately leads to an analogous theorem concerning unstable algebras.

Theorem 1.5 [**21**] *For an augmented unstable algebra K the following conditions are equivalent:*

1. $\mathrm{Hom}_{\mathcal{K}}(K, H^*(V)) = 0$ *for all elementary abelian p-groups V.*
2. *K is F-equivalent to the trivial unstable algebra $\mathbf{F}_p$.*
3. *$\bar{K}$ is nilpotent as an unstable module over the Steenrod algebra.*
4. *$\bar{K}$ is a nil ideal in K, i.e., it consists of nilpotent elements.*

Here the bar denotes the augmentation ideal. The equivalence of 2., 3., and 4. are direct consequences of the definitions. Theorem 1.1 now follows from the growth formula and the preceding general theorems.

It is worth noting that in the preceding theorems one has to consider elementary abelian groups V of an arbitrary size. Restriction to e.g. $V = \mathbf{Z}/p$ would not be enough, which is seen for example, by setting $K = H^*(K(\mathbf{Z}, 3))$. This makes the property 'nilpotent' a bit less well behaved than for example the property 'locally finite'.

To prove Theorem 1.2 we examine the work of Henn, Lannes and Schwartz [**16**] on the structure of unstable algebras and look at it in terms of growth properties. This leads to a characterization of the transcendence degree of an unstable algebra K in terms of the growth of the characteristic numbers $\gamma_v(K) := \log_p |\mathrm{Hom}_{\mathcal{K}}(K, H^*(V))|$ in $v := \mathrm{rk}_p V$, under mild restrictions on K. More precisely we prove:

Theorem 1.6 *Let K be an unstable algebra with $1 \neq 0$. Assume that K has transcendence degree $d(K)$ and that $\mathrm{Hom}_{\mathcal{K}}(K, H^*(V))$ is finite for all V. If $d(K)$ is finite then $\gamma_v(K) \sim d(K)v$. If $d(K)$ is infinite then $\gamma_v(K)$ grows faster than linearly in v.*

Theorem 1.6 is a powerful tool for calculating the transcendence degree of unstable algebras. To demonstrate this we rederive Quillen's theorem in the finite group case, by proving that the Krull dimension of $H^*(G)$ is equal to the p-rank $\mathrm{rk}_p G$ for every finite group G.

We call the numbers $\gamma_v(K)$ (which are not necessarily integers) 'unstable Betti numbers' since they show some similar properties of the traditional Betti numbers of a graded algebra, $\beta_n(K) = \dim_{\mathbf{F}_p} K^n$. One should think of the $\log_p$ as a replacement of $\dim_{\mathbf{F}_p}$ necessary since $\mathrm{Hom}_{\mathcal{K}}(K, H^*(\mathbf{F}_p^{\oplus n}))$ in general is not an $\mathbf{F}_p$-vector space. Theorem 1.6 shows that an analog of the well known formula relating the growth of the Betti numbers and the transcendence degree (Krull dimension) of a *noetherian* connected graded algebra holds for these new numbers, now with much weaker restrictions on the unstable algebra.

Applying Theorem 1.6 to the obtained growth formula for $\log_p |[BV, E]|$, where E is a finite Postnikov system, and doing a bit of work now leads to Theorem 1.2.

We also include a section where we see that the above results imply that the n-connected cover of a finite complex always has infinite Lusternik-Schnirelmann (LS) category, generalizing earlier partial results of McGibbon and Møller [**23**].

Finally we discuss and conjecture generalizations of the above results to polyGEMs, correcting a small mistake put forth in [**9**].

Several times in the proofs of the theorems we need a small but useful fact about nilpotent actions. Since we have been unable to find this fact stated in the literature, and believe that it ought to be better known, we give a proof in a short appendix.

2 Notation

By a space we, for simplicity, mean an object in the pointed homotopy category Ho_* of CW-complexes. The bracket $[-,-]$ means a *free* homotopy class of maps (Theorem 1.3 is the main reason why this is convenient). We will also need to refer to pointed homotopy classes of pointed maps which we denote by $[-,-]_{pt}$.

We define the p-rank, rk_p, of a group G as the maximal rank of an elementary abelian subgroup V contained in G. At all times we employ the convention that $v = \mathrm{rk}_p V$, for the elementary abelian group V in question.

When talking about the asymptotic behavior of some sequence of numbers, we mean by the symbol $\lesssim$ that for all $\epsilon > 0$ there exists an N such that (left-hand side)$_n \leq (1+\epsilon)$(right-hand side)$_n$ for all $n \geq N$. We define $\gtrsim$ and $\sim$ analogously.

In our notation involving unstable modules over the Steenrod algebra, we follow the standard notation used in e.g., Schwartz [**30**]. We quickly review the basic definitions:

Definition 2.1 *An* unstable module M *is a graded module over the Steenrod algebra* $\mathcal{A}$ *satisfying the following instability conditions:*

- *If* $p=2$ *then* $\mathrm{Sq}^i x = 0$ *for* $i > |x|$.
- *If* $p>2$ *then* $\beta^e P^i x = 0$ *for* $e + 2i > |x|$.

Let $\mathcal{U}$ *denote the category whose objects are unstable modules and whose morphisms are (degree 0)* $\mathcal{A}$*-module maps. Note that this is an abelian category.*

Definition 2.2 *An* unstable algebra K *is an unstable module equipped with two maps* $\eta : \mathbf{F}_p \to K$ *and* $\mu : K \otimes K \to K$ *making* K *into a commutative unital* $\mathbf{F}_p$*-algebra such that*

- μ *is* $\mathcal{A}$*-linear (i.e., the Cartan formula holds).*
- $Sq^{|x|}x = x^2$ *for any* $x \in K$ *if* $p=2$; $P^{|x|/2}x = x^p$ *for any* $x \in K$ *of even degree if* $p > 2$.

Let $\mathcal{K}$ *denote the category whose objects are unstable algebras, and whose morphisms are those (degree 0)* $\mathbf{F}_p$*-algebra maps which are* $\mathcal{A}$*-linear.*

Furthermore, throughout this paper, for convenience we make the standing assumption that all unstable algebras have $1 \neq 0$, i.e., we exclude the unstable algebra 0.

Definition 2.3 *We say that an unstable module is* nilpotent *if the following holds:*

- *If* $p=2$ *then for every* $x \in M$ *there exists an integer* N *such that* $\mathrm{Sq}^{2^{N-1}|x|}\,\mathrm{Sq}^{2^{N-2}|x|}\cdots Sq^{|x|}x = 0$.
- *If* $p>2$ *then for every* $x \in M$ *where* $|x|$ *is even there exists an integer* N *such that* $P^{p^{N-1}|x|/2}P^{p^{N-2}|x|/2}\cdots P^{|x|/2}x = 0$.

Let $\mathcal{N}il$ *denote the full subcategory of* $\mathcal{U}$ *whose objects are nilpotent modules.*

Note that this definition extends the usual definition if I is an ideal in an unstable algebra, in the sense that we have that I lies in $\mathcal{N}il$ iff it is nil as an ideal, that is if all its elements are nilpotent. Beware however, for a non-noetherian algebra a nil ideal I need not be nilpotent (i.e., there exists n such that $I^n = 0$), which makes the terminology slightly ambiguous. Note also that an unstable algebra with $1 \neq 0$ of course cannot itself be nilpotent.

We need a last definition:

Definition 2.4 *A morphism of unstable algebras $\varphi : K \to K'$ is said to be an* F-monomorphism *if the kernel of φ in the (abelian) category of unstable modules is nilpotent as an unstable module. We define* F-epimorphism *and* F-isomorphism *analogously.*

This definition coincides with Quillen's original definition of an F-isomorphism, and is also the same as what is sometimes referred to as a (purely) inseparable isogeny.

We note that everything in this paper only depends on the space E up to $\mathbf{F}_p$-equivalence (E here, as everywhere, connected nilpotent with finite π_1), and that Bousfield-Kan $\mathbf{F}_p$-completion [**5**] of E coincides with Bousfield $\mathbf{F}_p$-localization [**2**]. As noted by Miller [**26**] $\mathrm{map}_*(BV, E) \simeq \mathrm{map}_*(BV, \hat{E}_p)$, and of course $H^*(E; \mathbf{F}_p) \simeq H^*(\hat{E}_p; \mathbf{F}_p)$ – Likewise we have that $[BV, E] \simeq [BV, \hat{E}_p]$ [**5, 2**]. Thus we could reformulate everything by writing: "Let E be $\mathbf{F}_p$-equivalent to..." – For the sake of clarity we refrain from doing so and instead leave reformulations like that to the reader.

3 Growth properties

In this section we establish the growth formula for $\log_p |[BV, E]|$ mentioned in the introduction. To do this we first need to prove a couple of lemmas. The first lemma in its present form is due to Lannes and Schwartz [**22**].

Lemma 3.1 *Let G be an abelian group, and assume that $t = \mathrm{rk}_p({}_pG)$ and $s = \mathrm{rk}_p(G \otimes \mathbf{Z}/p)$ are finite. Furthermore let V be an elementary abelian p-group of rank v. Then the Poincaré series of $H^*(V; G)$, $P_G(x) = \sum_{k>0} \dim_{\mathbf{F}_p}(H^k(V; G) \otimes \mathbf{Z}/p)x^k$ is given by*

$$P_G(x) = \frac{t + sx}{1 + x}((1 - x)^{-v} - 1).$$

Especially

$$\mathrm{rk}_p(H^1(V; G)) \sim tv$$

for $v \to \infty$ and

$$\mathrm{rk}_p(H^k(V; G)) \sim \begin{cases} \frac{tv^k}{k!} & \text{if } t > 0 \\ \frac{sv^{k-1}}{(k-1)!} & \text{if } t = 0 \end{cases}$$

for $v \to \infty$, $k \geq 2$.

Proof Note that induction on the rank v of V, using the cohomology of $H^*(\mathbf{Z}/p; \mathbf{Z})$ together with the Künneth formula, gives us that $pH^k(V; \mathbf{Z}) = 0$ for all $k > 0$ and all V.

The universal coefficient theorem now gives us the following exact sequence

$$0 \to H^k(V; \mathbf{Z}) \otimes G \to H^k(V; G) \to \mathrm{Tor}(H^{k+1}(V; \mathbf{Z}), G) \to 0,$$

i.e., for $k \geq 1$

$$0 \to H^k(V; \mathbf{Z}) \otimes (G \otimes \mathbf{Z}/p) \to H^k(V; G) \to H^{k+1}(V; \mathbf{Z}) \otimes ({}_pG) \to 0.$$

This shows that $xP_G(x) = (t + sx)P_{\mathbf{Z}}(x)$. Setting $G = \mathbf{Z}/p$ we obtain

$$P_{\mathbf{Z}}(x) = \frac{x}{1 + x} P_{\mathbf{Z}/p}(x) = \frac{x}{1 + x}((1 - x)^{-v} - 1)$$

and hence

$$P_G(x) = \frac{t+sx}{1+x}((1-x)^{-v}-1).$$

From this formula the growth formulas are immediate. □

Lemma 3.2 *Let X be an arbitrary space and let E be a connected finite Postnikov system. Then*

$$|[X,E]_{pt}| \le \prod_{i>0} |H^i(X,\pi_i(E))|$$

when E is simple.

If more generally E is nilpotent, then there exists, for each i, a filtration $0 = F_{i,0} \trianglelefteq \cdots \trianglelefteq F_{i,t_i} = \pi_i(E)$ such that $\pi_1(E)$ acts trivially on $F_{i,j}/F_{i,j-1}$ and

$$|[X,E]_{pt}| \le \prod_{i>0} \prod_{j=1}^{t_i} |H^i(X;F_{i,j}/F_{i,j-1})| .$$

Proof The proof is by induction on the number of non-trivial homotopy groups. Assume that the top homotopy group sits in dimension n. Consider first the case where E is simple. We have a fibration sequence, which is principal since E is simple:

$$K(\pi_n(E),n) \to E \to P_{n-1}E,$$

where P_kE denotes the kth Postnikov stage of E. Since the fibration is principal we have an action $E \times K(\pi_n(E),n) \to E$, and thus an induced action $* : [X,E]_{pt} \times [X,K(\pi_n(E),n)]_{pt} \to [X,E]_{pt}$. This action has the property that in the exact sequence

$$[X,K(\pi_n(E),n)]_{pt} \xrightarrow{\iota} [X,E]_{pt} \xrightarrow{\kappa} [X,P_{n-1}E]_{pt}$$

we have that $\kappa(f)=\kappa(g)$ iff there exists $h \in [X,K(\pi_n(E),n)]$ such that $f*h=g$. This implies that $|[X,E]_{pt}| \le |[X,P_{n-1}E]_{pt}||H^n(X,\pi_n(E))|$, so by induction we get that $|[X,E]_{pt}| \le \prod_{i>0}|H^i(X,\pi_i(E))|$ as wanted.

Now assume that E is only nilpotent. The fibration $K(\pi_n(E),n) \to E \to P_{n-1}E$ might no longer be principal, but it does have a principal refinement corresponding to a filtration $0 = F_{i,0} \trianglelefteq \cdots \trianglelefteq F_{i,t_i} = \pi_i(E)$ of $\pi_i(E)$ (cf. [**17**]). Using induction as before finishes the proof in this case too. □

We are now ready to prove the key growth theorem:

Theorem 3.3 *Let E be a connected nilpotent finite Postnikov system with finite $\pi_1(E)$. Assume that $H^*(E)$ is of finite type and that $\bar{H}^*(E) \neq 0$. Let n denote the dimension of the highest homotopy group $\pi_n(E)$ which is not uniquely p-divisible and set $k=n$ if $\pi_n(E)$ has p-torsion, $k=n-1$ if not. Then*

$$cv^k \lesssim \log_p |[BV,E]| \lesssim Cv^k ,$$

for $v \to \infty$, where c, C are positive constants.

Remark 3.4 The condition that $H^*(X)$ is of finite type is equivalent to the condition that $H_*(X;\mathbf{F}_p)$ is of finite type. By the universal coefficient theorem this is again equivalent to requiring ${}_pH_n(X;\mathbf{Z})$ and $H_n(X;\mathbf{Z})\otimes\mathbf{Z}/p$ to be finite for all n. If X is connected nilpotent with finite π_1, this is equivalent to asking that ${}_p\pi_n(X)$ and $\pi_n(X)\otimes\mathbf{Z}/p$ are finite for all $n \ge 2$, e.g. by the Whitehead theorem modulo Serre classes. Note also that for a connected nilpotent space, X, the condition $\bar{H}^*(X) \neq 0$ is equivalent to $\pi_*(X)$ not being uniquely p-divisible (by the above or [**5**]). Also note that the k which appears in the theorem also could be defined as the dimension of the highest non-trivial mod p homotopy group minus 1.

Proof of Theorem 3.3 By replacing E by the $\mathbf{F}_p$-equivalent space P_nE, we can assume that n is the dimension of the top non-trivial homotopy group. Since E is connected and $\pi_1(E)$ is finite we might as well show the theorem for $[BV, E]_{pt}$, which is what we do. We have a principal fibration sequence $\Omega P_{n-1}E \to K(\pi_n(E), n) \to E$ so we get an exact sequence of pointed sets

$$[BV, \Omega P_{n-1}E]_{pt} \to [BV, K(\pi_n(E), n)]_{pt} \to [BV, E]_{pt}.$$

We can also write this as

$$[BV, \Omega_0 P_{n-1}E]_{pt} \to H^n(V; \pi_n(E)) \to [BV, E]_{pt}\,, \tag{3.1}$$

where $[BV, \Omega_0 P_{n-1}E]_{pt}$ acts on $H^n(V; \pi_n(E))$ as described in the previous proof and $\Omega_0 P_{n-1}E$ denotes the zero component of $\Omega P_{n-1}E$.

Now by Lemma 3.2 and 3.1 we get:

$$\begin{aligned}\log_p |[BV, \Omega_0 P_{n-1}E]_{pt}| &\leq \sum_{i=1}^{n-2} \log_p |H^i(V; \pi_{i+1}(E))| \\ &= \sum_{i=1}^{n-2} \mathrm{rk}_p H^i(V; \pi_{i+1}(E)) \\ &\lesssim Kv^{n-2}\end{aligned}$$

for $v \to \infty$, since $\Omega_0 P_{n-1}E$ is an H-space and hence simple. Since $\log_p |H^n(V; \pi_n(E))| \sim cv^k$, where k is as defined in the theorem, (3.1) shows us:

$$\begin{aligned}\log_p |H^n(V; \pi_n(E))| &\sim \log_p |H^n(V; \pi_n(E))| - \log_p |[BV, \Omega_0 P_{n-1}E]|_{pt} \\ &\lesssim \log_p |[BV, E]|_{pt}\,.\end{aligned}$$

We want to get the other inequality by appealing to Lemma 3.2. In the simple case it is immediate that

$$\log_p |[BV, E]_{pt}| \leq \sum_{i=1}^{n} \log_p |H^i(V; \pi_i(E))| \lesssim Cv^k\,.$$

If $\pi_n(E)$ does contain p-torsion we can take $C = c$, whereas C in the case where $\pi_n(E)$ does not contain p-torsion is given in terms of $\pi_n(E)$ and $\pi_{n-1}(E)$.

In the non-simple case we need to worry a little bit about actions. We now have that

$$\log_p |[BV, E]_{pt}| \lesssim \sum_{i=1}^{n} \sum_{j=1}^{t_i} \log_p |H^i(V; F_{i,j}/F_{i,j-1})|\,.$$

It is still clear that $\log_p |[BV, E]_{pt}| \leq Cv^n$ for some C. This takes care of the case where $\pi_n(E)$ does contain p-torsion. In the case where $\pi_n(E)$ does not contain p-torsion however we want the better estimate Cv^{n-1}. This is not obvious since the filtration quotients could have p-torsion, even though $\pi_n(E)$ did not. To resolve this problem we have to appeal to Proposition 7.1.

By replacing E by an $\mathbf{F}_p$-equivalent space, we can assume that $\pi_1(E)$ is a p-group, since every finite nilpotent group is a product of p-groups. By Proposition 7.1 $\pi_1(E)$ acts trivially on $\pi_n(E)$, so we can take the filtration of $\pi_n(E)$ to be the trivial filtration, and thus there is no p-torsion introduced in dimension n. This shows that, in this case, we can conclude that $\log_p |[BV, E]_{pt}| \lesssim Cv^{n-1}$. □

Remark 3.5 It follows from the proof of the theorem above that one can actually get concrete estimates for c and C. This does have some interest, especially in the case $k = 1$ where c and C actually turn out to give lower and upper bounds on the transcendence degree of $H^*(E)$.

The first main theorem now follows from the preceding growth theorem.

Theorem 3.6 *Let E be a connected nilpotent finite Postnikov system with finite $\pi_1(E)$. Assume that $H^*(E)$ is of finite type and that $\bar{H}^*(E) \neq 0$. Then $\bar{H}^*(E)$ contains an element of infinite height.*

Proof Theorem 3.3 shows the asymptotic growth of $\log_p |[BV, E]|$—in particular it shows that $[BV, E]$ is nontrivial for some V. But by Theorem 1.3 and 1.5 this now implies that $\bar{H}^*(E) \notin \mathcal{N}il$, so $\bar{H}^*(E)$ has to contain an element of infinite height. □

4 Applications to n-connected covers

We now turn to investigating the consequences for n-connected covers of finite complexes:

Corollary 4.1 *Let X be a finite complex, 1-connected with finite $\pi_2(X)$. Assume furthermore that $\bar{H}^*(P_n X) \neq 0$. Then $\bar{H}^*(X\langle n\rangle)$ contains an element of infinite height.*

Remark 4.2 The assumption $\bar{H}^*(P_n X) \neq 0$ could be formulated in a number of alternative ways. Since X is 1-connected the assumption is equivalent to the natural map $H^*(X) \to H^*(X\langle n\rangle)$ not being an isomorphism by a Serre spectral sequence argument. Also, since X is of finite type, $\bar{H}^*(P_n X) \neq 0$ iff $\bar{H}^*(P_n X; \mathbf{Z})$ is not entirely q-torsion for primes $q \neq p$, which, for example, by the Whitehead theorem modulo Serre classes, is equivalent to $\pi_*(P_n X)_{(p)} \neq 0$.

Proof of Corollary 4.1 We have an exact sequence

$$[BV, \Omega X] \to [BV, \Omega P_n X] \to [BV, X\langle n\rangle] \to [BV, X].$$

By Miller's theorem $[BV, \Omega X] = [BV, X] = 0$, so $[BV, \Omega P_n X] = [BV, X\langle n\rangle]$. Since $\bar{H}^*(P_n X) \neq 0$ we have that $\bar{H}^*(\Omega P_n X) \neq 0$. Now observe that $\Omega P_n X$ satisfies the assumptions of Theorem 3.3, so in particular we get that $[BV, X\langle n\rangle] = [BV, \Omega P_n X] \neq 0$ for some V. By Theorem 1.3 and 1.5 this is equivalent to $\bar{H}^*(X\langle n\rangle)$ containing an element of infinite height. □

Remark 4.3 The relatively strong assumption on the connectivity of X cannot be weakened, as is shown by the Hopf fibration $S^1 \to S^3 \to S^2$.

Corollary 4.4 *Let X be a finite complex, 1-connected with finite $\pi_2(X)$. Assume that $X \neq X\langle n\rangle$. Then there exists a prime q such that $\bar{H}^*(X\langle n\rangle; \mathbf{F}_q)$ contains an element of infinite height.*

Proof If $X \neq X\langle n\rangle$, we have to have that $\bar{H}^*(P_n X; \mathbf{F}_q) \neq 0$ for some prime q. The statement is now obvious from Corollary 4.1. □

Corollary 4.5 *Let X be a finite complex, 1-connected with finite $\pi_2(X)$. Assume that $X \neq X\langle n\rangle$. Then $X\langle n\rangle$ has infinite LS category.*

Proof By Corollary 4.4 there exists a prime q such that $\bar{H}^*(X\langle n\rangle; \mathbf{F}_q)$ has infinite cup-length, so especially $X\langle n\rangle$ has infinite LS category. □

These corollaries generalize earlier partial results of Møller and McGibbon [**23**]. As they point out, it can be interesting to note the radical difference between these results and the results obtained in the rational case. Here the rational LS category of $X\langle n\rangle$ is always less than or equal to the rational LS category of X by the mapping theorem of Félix and Halperin [**13**]. But, on the other hand, in the rational setting we have no Serre's theorem either—indeed the rational cohomology of $K(\mathbf{Z},3)$ does not contain an element of infinite height.

5 The transcendence degree of $H^*(E)$

In this section we give a complete classification of the Postnikov systems whose cohomology is of finite transcendence degree. We start by using the work of Henn, Lannes and Schwartz [**16**] to relate the growth of $\gamma_v(K) := \log_p |\operatorname{Hom}_{\mathcal{K}}(K, H^*(V))|$ in v to the transcendence degree of K, under mild restrictions on K. Combining these results with our growth formula for $[BV, E] = \operatorname{Hom}_{\mathcal{K}}(H^*(E), H^*(V))$ now leads to a classification theorem for finite Postnikov systems of finite transcendence degree.

In order to state and prove our results we need to review some of the work of Henn, Lannes and Schwartz. For the ease of the reader we follow their notation; we refer the reader to [**16**] for more details.

Definition 5.1 *For any unstable algebra K define its transcendence degree $d(K)$ as the maximal number of algebraically independent homogeneous elements in K.*

Remark 5.2 If K is a connected graded noetherian algebra, the maximal number of algebraically independent elements, $d(K)$, is finite. In this case the number $d(K)$ coincides with the Krull dimension of K, and we can furthermore choose $d(K)$ algebraically independent elements such that K will be finite over the algebra spanned by those elements. If K is an integral domain, $d(K)$ is equal to the classical transcendence degree of the field of fractions of K. A nice and graded proof of these standard facts can be found in [**1**].

In the following we need to refer to elementary abelian groups of different rank. Therefore we sometimes equip the elementary abelian group V with a subscript which in those cases indicate the rank of V.

Definition 5.3 *Let $\mathcal{E}$ denote the category of elementary abelian p-groups, i.e., of finite dimensional $\mathbf{F}_p$-vector spaces. Let $\mathcal{PS}$ denote the category of profinite sets and let $\mathcal{G}$ be the category of functors $\mathcal{E} \to (\mathcal{PS})^{op}$. Note that we can view $\mathcal{G}^{op}$ as the category of contravariant functors $\mathcal{E} \to \mathcal{PS}$.*

One should realize that objects in $\mathcal{G}$ contain a rich structure. This stems from the fact that not only do we associate to each vector space a profinite set, but we do this in a natural way, which, loosely speaking, ties the profinite sets together.

By inverting all F-isomorphisms in $\mathcal{K}$ we obtain a quotient category of $\mathcal{K}$ which will be denoted by $\mathcal{K}/\mathcal{N}il$. Let $g : \mathcal{K} \to \mathcal{G}$ be given by $g(K)(V) = \operatorname{Hom}_{\mathcal{K}}(K, H^*(V))$ for all V. Here we equip $\operatorname{Hom}_{\mathcal{K}}(K, H^*(V))$ with the profinite topology induced by writing

$$\operatorname{Hom}_{\mathcal{K}}(K, H^*(V)) = \operatorname{Hom}_{\mathcal{K}}(\operatorname*{colim}_{\alpha} K_\alpha, H^*(V)) = \lim_{\alpha} \operatorname{Hom}_{\mathcal{K}}(K_\alpha, H^*(V)),$$

where α runs over the finitely generated $\mathcal{A}$-subalgebras K_α of K.

By Theorem 1.4 and the Lannes linearization principle we have that a morphism between unstable algebras $K \to K'$ is an F-monomorphism (resp. F-epi) iff $g(K')(V) \to g(K)(V)$ is a surjective (resp. injective) map of sets for all V. This shows that g induces a a faithful functor $\mathcal{K}/\mathcal{N}il \to \mathcal{G}$ (likewise denoted g). In [**16**]

Henn, Lannes and Schwartz actually identify the image of g—we shall however not need this.

Definition 5.4 *Let $\mathcal{PS} - \operatorname{End} V_d$ denote the category whose objects are profinite sets equipped with a continuous right action of the monoid* $\operatorname{End} V_d$ *and whose morphisms are maps of profinite sets respecting the* $\operatorname{End} V_d$*-action.*

To each $G \in \mathcal{G}^{op}$ and each vector space V_d we can associate a profinite right $\operatorname{End} V_d$-set $G(V_d)$. Namely define the right action of $\operatorname{End} V_d$ on $G(V_d)$ by each $(s, \varphi) \in G(V_d) \times \operatorname{End} V_d$ associating $s\varphi = G(\varphi)s$. This gives us a 'restriction functor'

$$e_d : \mathcal{G}^{op} \to \mathcal{PS} - \operatorname{End} V_d .$$

Likewise we can define an 'induction functor'

$$i_d : \mathcal{PS} - \operatorname{End} V_d \to \mathcal{G}^{op} : S \mapsto (W \mapsto S \times_{\operatorname{End} V_d} \operatorname{Hom}(W, V_d)),$$

where $S \times_{\operatorname{End} V_d} \operatorname{Hom}(W, V_d)$ denotes the coequalizer in the category of profinite sets of the action of $\operatorname{End} V_d$ on S and $\operatorname{Hom}(W, V_d)$ respectively.

Now note that the canonical map $G(V_d) \times \operatorname{Hom}(W, V_d) \to G(W)$ induces a well defined map on the coequalizer $G(V_d) \times_{\operatorname{End} V_d} \operatorname{Hom}(W, V_d) \to G(W)$. Therefore, to each map $S \to G(V_d)$ of profinite $\operatorname{End} V_d$-sets and each W, we can associate a map

$$S \times_{\operatorname{End} V_d} \operatorname{Hom}(W, V_d) \to G(V_d) \times_{\operatorname{End} V_d} \operatorname{Hom}(W, V_d) \to G(W) .$$

Likewise a morphism in $\mathcal{G}^{op}$, $i_d(S) \to G$ of course induces a map $S \to G(V_d)$ of profinite $\operatorname{End} V_d$-sets. Since these operations are inverses of each other we get that $\operatorname{Hom}_{\mathcal{G}^{op}}(i_d(S), G) = \operatorname{Hom}_{\mathcal{PS}-\operatorname{End} V_d}(S, e_d(G))$, naturally in $G \in \mathcal{G}^{op}$, $S \in \mathcal{PS} - \operatorname{End} V_d$, so i_d is left adjoint of e_d. It is immediate that we have an equivalence $1_{\mathcal{PS}-\operatorname{End} V_d} \xrightarrow{\simeq} e_d \circ i_d$, given by the unit of the adjunction, so we get an embedding of $\mathcal{PS} - \operatorname{End} V_d$ as a subcategory of $\mathcal{G}^{op}$. Define $s_d = e_d^{op} \circ g$, where $e_d^{op} : \mathcal{G} \to (\mathcal{PS} - \operatorname{End} V_d)^{op}$ is the opposite functor of e_d.

In [**16**] Henn, Lannes and Schwartz prove the following key results about the structure of $\mathcal{G}^{op}$.

Proposition 5.5 [**16**] *The morphism $(i_d \circ e_d)(G) \to G$ given by the counit of the adjunction is a monomorphism in $\mathcal{G}^{op}$ for all $G \in \mathcal{G}^{op}$, i.e. we have that $(i_d \circ e_d)(G)(V) \to G(V)$ is an injective map of profinite sets for all V.*

This gives us a filtration of $\mathcal{G}$ which, it turns out, coincides with the filtration of $\mathcal{K}/\mathcal{N}il$ by transcendence degree.

Proposition 5.6 [**16**] *Let $G \in \mathcal{G}^{op}$. Define the transcendence degree of G as*

$$d(G) := \min\{d \mid ((i_d \circ e_d)G)(W) \to G(W) \textit{ is bijective for all } W\},$$

where we take $d(G) = \infty$ if no such d exists. For an element $G \in \mathcal{G}$ we define the transcendence degree by viewing it as lying in $\mathcal{G}^{op}$. With these definitions we have that $d(K) = d(g(K))$.

We shall need some alternative ways of expressing the transcendence degree of an object in $\mathcal{G}^{op}$. These can be found implicit in [**16**]. Since they are important in their own right we find it useful to state them explicitly—we include proofs for the convenience of the reader. First a useful definition:

Definition 5.7 *Let S be an* $\operatorname{End} V_d$*-set, and let $s \in S$. Define the rank of $s \in S$ as*

$$\operatorname{rk} s = \min\{\operatorname{rk} \varphi \mid \textit{where } s = t\varphi \textit{ for some } t \in S, \varphi \in \operatorname{End} V_d\}.$$

We say that s is regular if $\operatorname{rk} s = d$, i.e., if $s = t\varphi$ implies that φ is a regular (invertible) matrix.

Proposition 5.8 *We have the following formula for the transcendence degree of $G \in \mathcal{G}^{op}$:*

$$\begin{aligned} d(G) &:= \min\{d|((i_d \circ e_d)G)(W) \to G(W) \text{ is bijective for all } W\} \\ &= \max\{\operatorname{rk} s | s \in G(V_d) \text{ for some } d\} \\ &= \max\{d|G(V_d) \text{ contains a regular element}\}. \end{aligned}$$

Proof We begin by proving that

$$\max\{d|G(V_d) \text{ contains a reg. elt.}\} = \max\{\operatorname{rk} s|s \in G(V_d) \text{ for some } d\}. \tag{5.1}$$

First note that '$\leq$' is obvious. To prove '$\geq$' let $s \in G(W)$ and suppose that $\operatorname{rk} s = d$. We can thus choose $\pi \in \operatorname{End} W$, $t \in G(W)$ such that $s = t\pi$ and $\operatorname{rk} \pi = d$. By changing t we can assume that π is a projection. Let $\rho : W \to \pi W$ and $i : \pi W \to W$ be the canonical projection and inclusion associated to π. Now set $s' = G(i)s \in G(\pi W)$ and note that $G(\rho)s' = G(\rho)G(i)s = G(\rho)G(i)G(\pi)t = G(\pi i \rho)t = s$. We claim that s' is regular. Suppose we have $\varphi \in \operatorname{End}(\pi W), u \in G(\pi W)$ such that $s' = u\varphi$. Then we have

$$s = G(\rho)s' = G(\rho)G(\varphi)u = G(\rho)G(\varphi)G(i)G(\rho)u = G(i\varphi\rho)G(\rho)u = (G(\rho)u)(i\varphi\rho)$$

so φ has to be regular, since $\operatorname{rk} s = d$. This shows the wanted inequality.

We now prove that the number given by (5.1) actually coincides with the transcendence degree. We first see that it is less than or equal to the transcendence degree. Suppose therefore that we have d such that $G(V_d) \times_{\operatorname{End} V_d} \operatorname{Hom}(W, V_d) \to G(W)$ is bijective for all W, and let $s \in G(W)$ be arbitrary. Choose $t \in G(V_d)$ and $\varphi \in \operatorname{Hom}(W, V_d)$ such that $G(\varphi)t = s$. Since the rank of $\varphi : W \to V_d$ must be less than or equal to d, we can choose a projection $\pi \in \operatorname{End} W$ of rank less than or equal to d such that $\varphi\pi = \varphi$. But this gives us $s = G(\varphi)t = G(\pi)G(\varphi)t = (G(\varphi)t)\pi$, which shows that $\operatorname{rk} s \leq d$, as wanted.

We finish the proof by showing $d(G) \leq \max\{d|G(V_d) \text{ contains a regular element}\}$. Let d be the maximal number such that $G(V_d)$ contains a regular element (if there exist regular elements in infinitely many dimension we are done). We want to prove that $G(V_d) \times_{\operatorname{End} V_d} \operatorname{Hom}(W, V_d) \to G(W)$ is surjective for all W. Let $s \in G(W)$ be arbitrary and set $n = \operatorname{rk} s$. Note that $n \leq d$ by (5.1). Choose an element $\pi \in \operatorname{End} W$ of rank n and $t \in G(W)$ such that $s = t\pi$. By changing the choice of t we can assume that π is a projection. Letting $\rho : W \to \pi W$ and $i : \pi W \to W$ denote the canonical projection and inclusion we get that $s = G(\pi)t = G(\pi i \rho)t = G(\rho)(G(\pi i)t)$. Therefore s is in the image of $G(V_n) \times_{\operatorname{End} V_n} \operatorname{Hom}(W, V_n)$ which implies that s is in the image of $G(V_d) \times_{\operatorname{End} V_d} \operatorname{Hom}(W, V_d)$, since $n \leq d$. This completes the proof, since $G(V_d) \times_{\operatorname{End} V_d} \operatorname{Hom}(W, V_d) \to G(W)$ is injective by Proposition 5.5. □

Example 5.9 Let X_n be the $\operatorname{End} V_n$-set consisting of two elements x_1 and x_0, with the $\operatorname{End} V_n$-action given as follows:

$$x_0\alpha = x_0 \text{ for all } \alpha \in \operatorname{End} V_n$$

$$x_1\alpha = \begin{cases} x_1 & \text{for } \alpha \in \operatorname{Aut} V_n \\ x_0 & \text{for } \alpha \notin \operatorname{Aut} V_n. \end{cases}$$

We have that x_1 has rank n and x_0 has rank 0. The $\operatorname{End} V_n$-set X_n is thus the smallest $\operatorname{End} V_n$-set containing a regular element. It has the universal property that any $\operatorname{End} V_n$-set containing a regular element surjects onto X_n as an $\operatorname{End} V_n$-set.

Just knowing the growth properties of the 'unstable Betti numbers' $\gamma_v(K) = \log_p |g(K)(V)|$ in v is actually enough to determine the transcendence degree of K, under mild restrictions on K.

Theorem 5.10 *Let K be an unstable algebra and assume that $\mathrm{Hom}_{\mathcal{K}}(K, H^*(V))$ is finite for all V. If $d(K)$ is finite then $\gamma_v(K) \sim d(K)v$. If $d(K)$ is infinite then $\gamma_v(K)$ grows faster than linearly in v.*

Remark 5.11 The assumption that $\mathrm{Hom}_{\mathcal{K}}(K, H^*(V))$ is finite for all V is a technical assumption which (by Theorem 3.3) is satisfied in all the applications we have in mind. Also it is clear that if for instance K is finitely generated as an $\mathcal{A}$-algebra, then $\mathrm{Hom}_{\mathcal{K}}(K, H^*(V))$ is likewise finite. Furthermore if K is $\mathcal{N}il$-closed (cf. [**6, 16**]) and of finite type then $\mathrm{Hom}_{\mathcal{K}}(K, H^*(V))$ is finite [**16**]. Note however that $H^*(\prod_{i=1}^{\infty} K(\mathbf{Z}/p, i))$ serves as an example of an unstable algebra of finite type where $\mathrm{Hom}_{\mathcal{K}}(K, H^*(V))$ is not finite.

Proof of Theorem 5.10 We want to establish the general growth formulas by establishing them for the 'largest' and the smallest finite $\mathrm{End}\, V_n$-set containing a regular element. Let X_n be the $\mathrm{End}\, V_n$-set defined in Example 5.9. For the elements in $i_n(X_n)(V) = X_n \times_{\mathrm{End}\, V_n} \mathrm{Hom}(V, V_n)$ we have the following relations:

$$\begin{aligned}
(x_0, \varphi) &\sim (x_0, 0) \text{ for all } \varphi \in \mathrm{Hom}(V, V_n) \\
(x_1, \varphi) &\sim (x_0, 0) \text{ if } \mathrm{rk}\, \varphi < n \\
(x_1, \varphi) &\sim (x_1, \psi) \text{ if } \mathrm{rk}\, \varphi = \mathrm{rk}\, \psi = n \text{ and } \ker \varphi = \ker \psi\,.
\end{aligned}$$

This shows that $|i_n(X_n)(V)| = \#(\text{ n dimensional subspaces in V }) + 1$.

For $v > n$, the number of n dimensional subspaces of V is as follows:

$$\begin{aligned}
\#(\text{ n dimensional subspaces in V }) &= \frac{(p^v - 1) \cdots (p^v - p^{n-1})}{(p^n - 1) \cdots (p^n - p^{n-1})} \\
&\sim Cp^{nv}
\end{aligned}$$

for $v \to \infty$.

Let T be a finite set and consider the $\mathrm{End}\, V_n$-set $T \times \mathrm{End}\, V_n$. We have that $i_n(T \times \mathrm{End}\, V_n)(V) = T \times \mathrm{Hom}(V, V_n)$ so

$$|i_n(T \times \mathrm{End}\, V_n)(V)| = |T| p^{nv}\,.$$

From the above we conclude that

$$\log_p |i_n(X_n)(V)| \sim \log_p |i_n(T \times \mathrm{End}\, V_n)(V)| \sim nv\,.$$

Now let K be an unstable algebra and assume that $s_n(K)$ contains a regular element. Since $s_n(K)$ is finite we can find a surjection of $\mathrm{End}\, V_n$-sets $T \times \mathrm{End}\, V_n \to s_n(K)$ for some finite set T. Also, since $s_n(K)$ contains a regular element, we can find a surjection $s_n(K) \to X_n$ of $\mathrm{End}\, V_n$-sets. Now this means that, for all V, we have surjections of sets

$$i_n(T \times \mathrm{End}\, V_n)(V) \to i_n(s_n(K))(V) \to i_n(X_n)(V)\,,$$

so $\log_p |i_n^{op}(s_n(K))(V)|$ behaves asymptotically as nv.

If K has transcendence degree $d < \infty$, then $g(K)$ also has transcendence degree d so $\log_p |g(K)(V)| = \log_p |(i_d^{op} \circ e_d^{op} \circ g)(K)(V)| = \log_p |i_d^{op}(s_d(K))(V)|$ which grows asymptotically as dv, since $s_d(K)$ contains a regular element by Proposition 5.8. If K has transcendence degree $d = \infty$ then by Proposition 5.5

$$\log_p |g(K)(V)| \geq \log_p |(i_n^{op} \circ e_n^{op} \circ g)(K)(V)| = \log_p |i_n^{op}(s_n(K))(V)|\,.$$

Since $s_n(K)$ contains a regular element for infinitely many n by Proposition 5.8, we get that $\log_p |g(K)(V)| \gtrsim nv$ for all n as wanted. $\square$

The unstable Betti numbers $\gamma_v(K)$ can be viewed as an unstable algebra alternative to the classical Betti numbers of a graded algebra. In the classical case we have a formula relating the growth of the Betti numbers and the transcendence degree of a *noetherian* connected graded $\mathbf{F}$-algebra given by

$$d(K) = \min\{k \in \mathbf{N}_0 | \text{ there exists a C such that } \dim_{\mathbf{F}} K^n \leq Cn^{k-1} \text{ for all } n\}.$$

Theorem 5.10 establishes a different but analogous formula for these new numbers, which has among its advantages that it holds with much weaker restrictions on the unstable algebra. Moreover, when applied to the cohomology of spaces, it is easy to see how these numbers behave with respect to fibrations of the underlying spaces. For instance we immediately get that the unstable Betti numbers are subadditive over principal fibrations, and hence that the transcendence degree is likewise so. More precisely:

Proposition 5.12 *Let B, F be connected nilpotent spaces with finite fundamental groups and mod p cohomology rings of finite type. Assume that we have a principal fibration $F \to E \to B$. Then $\gamma_v(H^*(E)) \leq \gamma_v(H^*(F)) + \gamma_v(H^*(B))$, and hence, if $\gamma_v(H^*(F))$ and $\gamma_v(H^*(B))$ are finite for all v, $d(H^*(E)) \leq d(H^*(F)) + d(H^*(B))$.*

Proof Note that E is obviously also connected with finite π_1 and by [**5**, Proposition II.4.4] E is nilpotent as well. Since the fibration is principal $\pi_1(B)$ acts nilpotently on $H^*(F)$ (cf. [**5**, p. 62]), so the Serre spectral sequence guarantees that $H^*(E)$ is of finite type. The subadditivity of the unstable Betti numbers follows from Theorem 1.3 and the fact that, by the principal action, $|[BV, E]| \leq |[BV, F]||[BV, B]|$. The subadditivity of the transcendence degree now follows from Theorem 5.10. $\square$

Remark 5.13 In [**16**] Henn, Lannes and Schwartz use Proposition 5.5 and 5.6 to derive a far reaching generalization of Quillen's theorem about the structure of the mod p cohomology ring of a finite group [**28**]. They prove that for any unstable algebra K of transcendence degree less than or equal to d there is an F-isomorphism

$$K \to \lim_{s_d(K)} H^*(V_d).$$

Here we view $s_d(K)$ as a category by taking as objects the elements in $s_d(K)$, and as morphisms the maps induced on $s_d(K)$ from endomorphisms of V_d. (The theorem is strictly speaking only true as stated here when $s_d(K)$ is finite—when $s_d(K)$ is infinite one has only to consider the elements in $\lim_{s_d(K)} H^*(V_d)$ which are continuous as functions $s_d(K) \to H^*(V_d)$.)

The theorem however has the inherent weakness that it requires a priori knowledge of the transcendence degree of K. Theorem 5.10 can remedy this defect, since it gives an easily applicable way of calculating the transcendence of an unstable algebra K.

To illustrate the usefulness of this approach we rederive Quillen's theorem in the finite group case, by showing that the Krull dimension of $H^*(G)$ is equal to $\mathrm{rk}_p\, G$. By Cartan-Eilenberg [**8**], $H^*(G)$ is isomorphic to an inverse limit of the cohomology of its p-subgroups, so it is enough to show the claim for a finite p-group P. It was shown by Hurewicz that $[BV, BP] = \mathrm{Rep}(V, P) = \mathrm{Hom}(V, P)/(\text{conj. by } p \in P)$. It is furthermore an easy exercise to see that $\log_p |\mathrm{Rep}(V, P)|$ grows asymptotically as $\mathrm{rk}_p P$ v (cf. also Proposition 5.17 and its proof). Theorem 5.10 now implies that $d(H^*(P)) = \mathrm{rk}_p P$, and since we know that $H^*(P)$ is noetherian by the Evens-Venkov theorem (cf. [**12**]) we are done.

Tracing back what elements go into this proof one sees that one of the main ingredients is the use of Proposition 5.8, which in some sense can be seen as being the replacement of Serre's theorem about cohomological detection of elementary abelian groups.

Theorem 5.10 now enables us to prove the second main theorem.

Theorem 5.14 *Let E be a connected nilpotent finite Postnikov system with finite $\pi_1(E)$. Assume that $H^*(E)$ is of finite type. Then $H^*(E)$ has finite transcendence degree iff E is $\mathbf{F}_p$-equivalent to a space E' fitting into a principal fibration sequence of the form:*

$$\mathbf{CP}^\infty \times \cdots \times \mathbf{CP}^\infty \to E' \to K(P,1)\,,$$

where P is a finite p-group.

Proof By Theorem 3.3 and 5.10 all spaces E' as in Theorem 5.14 have cohomology of finite transcendence degree. To prove the converse, assume that $H^*(E)$ has finite transcendence degree. We see from Theorem 3.3 that $\mathrm{Hom}_{\mathcal{K}}(H^*(E), H^*(V))$ is finite for all V. Theorem 3.3 and 5.10 now imply that E has to be $\mathbf{F}_p$-equivalent to P_2E, and furthermore that $\pi_2(E)$ cannot contain p-torsion. Since $\pi_1(P_2E)$ is finite nilpotent, and thus a product of p-groups, we can by passing to a covering replace P_2E with an $\mathbf{F}_p$-equivalent space whose fundamental group is a finite p-group. These observations show that, without restriction, we can assume that E has nontrivial homotopy groups only in dimension 1 and 2, $\pi_2(E)$ being without p-torsion and $\pi_1(E)$ being a finite p-group. But now Proposition 7.1 tells us that the action of $\pi_1(E)$ on $\pi_2(E)$ has to be trivial. This means that the Postnikov fibration $K(\pi_2(E),2) \to E \to P_1E$ is principal (cf. [**17**]). We thus have a fibration sequence

$$E \to P_1E \xrightarrow{k} K(\pi_2(E),3)\,.$$

Upon passing to $\mathbf{F}_p$-completion we obtain a fibration sequence (cf. [**5**])

$$\hat{E}_p \to P_1E \xrightarrow{k} K(\mathrm{Ext}(\mathbf{Z}/p^\infty, \pi_2(E)),3)\,.$$

Now $\mathrm{Ext}((\mathbf{Z}/p^\infty, \pi_2(E))$ is a torsion free Ext $-p$ complete abelian group, and hence classified by the $\mathbf{F}_p$-dimension of $\mathrm{Ext}(\mathbf{Z}/p^\infty, \pi_2(E)) \otimes \mathbf{Z}/p$ (cf. [**5**, p. 181], [**15**]), which is finite since $H^*(E)$ is of finite type. In other words $\mathrm{Ext}(\mathbf{Z}/p^\infty, \pi_2(E)) \simeq \hat{\mathbf{Z}}_p \times \cdots \times \hat{\mathbf{Z}}_p$. Note that $H^3(\pi_1(E);\mathbf{Z}) = H^3(\pi_1(E);\hat{\mathbf{Z}}_p)$ by the universal coefficient theorem. We can thus let $k' : P_1E \to K(\mathbf{Z}\times\cdots\times\mathbf{Z},3)$ be the map corresponding to k under the equivalence $[P_1E, K(\mathrm{Ext}(\mathbf{Z}/p^\infty, \pi_2(E)),3)] \simeq [P_1E, K(\mathbf{Z}\times\cdots\times\mathbf{Z},3)]$. Letting E' denote the homotopy fiber of k' we obtain a diagram

$$\begin{array}{ccccc} E' & \longrightarrow & P_1E & \longrightarrow & K(\mathbf{Z}\times\cdots\times\mathbf{Z},3)\\ \downarrow f & & \| & & \downarrow \mathbf{F}_p \simeq \\ \hat{E}_p & \longrightarrow & P_1E & \xrightarrow{\quad k \quad} & K(\mathrm{Ext}(\mathbf{Z}/p^\infty, \pi_2(E)),3) \end{array}$$

where f is any lifting which makes the diagram commute. From this diagram we see that $f : E' \to \hat{E}_p$ has to be an $\mathbf{F}_p$-equivalence as well. By construction E' is

$\mathbf{F}_p$-equivalent to E and fits into a principal fibration sequence of the form stated in the theorem. □

Remark 5.15 Remember that a p-toral group G is a group which arises as a group extension $1 \to \mathbf{T}^n \to G \to P \to 1$, where $\mathbf{T}^n$ is an n-dimensional torus and P is a finite p-group. The spaces E' which appear in Theorem 5.14 are just those classifying spaces of p-toral groups which arise as central extensions. They are classified by n, P and their one extension class (= Postnikov invariant) $k \in [BP, K(\mathbf{Z}\times\cdots\times\mathbf{Z},3)] = H^3(P;\mathbf{Z}\times\cdots\times\mathbf{Z}) \simeq H^2(P;\mathbf{T}^n)$, where in the cohomology P acts trivially on the coefficients.

Remark 5.16 By Venkov's theorem [**32**], classifying spaces of p-toral groups have noetherian cohomology. Theorem 5.14 thus in particular shows that for the cohomology of a finite Postnikov system, being of finite transcendence degree is *equivalent* to being noetherian. This is indeed a very striking and unusual property which of course does not hold for spaces in general. The cohomology of the loop space of a (1-connected, say) finite complex for instance is *always* non-noetherian and has transcendence degree 0. Just knowing this intriguing property of finite Postnikov systems would actually be enough to rederive the above theorem using the original Betti numbers estimates of Serre—knowing this equivalence would secure that the wild growth of the Betti numbers could only be caused by an infinite number of polynomial generators in the ring.

We end this section by showing a proposition which precisely determines the transcendence degree of the cohomology of the spaces E' of Theorem 5.14, and whose proof illustrates a calculation using Theorem 5.10. The proposition can also be obtained by using Quillen's theorem for compact Lie groups.

Proposition 5.17 *Let E be a space which fits into a fibration sequence of the form*

$$\underbrace{\mathbf{CP}^\infty \times \cdots \times \mathbf{CP}^\infty}_{n} \to E \to K(P,1) \xrightarrow{k} K(\mathbf{Z}\times\cdots\times\mathbf{Z},3)\,,$$

where P is a finite p-group. Then

$$d(H^*(E)) = n + \max\{\mathrm{rk}_p V | V \overset{\iota}{\hookrightarrow} P, \iota^*(k) = 0 \in H^3(V;\mathbf{Z}\times\cdots\times\mathbf{Z})\}\,.$$

Especially $n + \mathrm{rk}_p P \geq d(H^*(E)) \geq n+1$, when P is non-trivial.

Proof Consider the sequence

$$[BV,E]_{pt} \to [BV,K(P,1)]_{pt} \xrightarrow{k_*} [BV,K(\mathbf{Z}\times\cdots\times\mathbf{Z},3)]_{pt}\,.$$

Let $\iota \in [BV_d, BP]_{pt} = \mathrm{Hom}(V_d,P)$ be an injection and assume that $\iota \in \ker(k_*)$. We have that ι naturally gives rise to $\sim Cp^{dv}$ elements in $\ker(k_*)$ as $v \to \infty$ coming from maps arising by precomposing ι with projections onto d dimensional subspaces of V. Since $\mathrm{Hom}(V_{\mathrm{rk}_p P},P)$ is finite we obtain that

$$\log_p|\ker(k_*)| \sim \max\{\mathrm{rk}_p V | V \overset{\iota}{\hookrightarrow} P, \iota^*(k) = 0\}v\,.$$

But since it is easy to see that $[BV,K(\mathbf{Z}\times\cdots\times\mathbf{Z},2)]_{pt}$ acts freely on $[BV,E]_{pt}$ we get that

$$\log_p|[BV,E]_{pt}| \sim nv + \max\{\mathrm{rk}_p V | V \overset{\iota}{\hookrightarrow} P, \iota^*(k) = 0\}v\,,$$

so $d(H^*(E)) = n + \max\{\mathrm{rk}_p V | V \overset{\iota}{\hookrightarrow} P, \iota^*(k) = 0\}$ by Theorem 5.10. To get the last part of the proposition, note that $H^3(\mathbf{Z}/p;\mathbf{Z}) = 0$. □

Example 5.18 Consider the space $E = \mathrm{Fib}(K(\mathbf{Z}/p \times \mathbf{Z}/p, 1) \xrightarrow{k} K(\mathbf{Z}, 3))$, where k is some nonzero element in $H^3(\mathbf{Z}/p \times \mathbf{Z}/p; \mathbf{Z}) \simeq \mathbf{Z}/p$. From the above proposition it follows that $d(H^*(E)) = 2$.

Remark 5.19 Note that the formula for the transcendence degree of $H^*(E)$ involves the behavior of a certain class in the *integral* cohomology of P when restricting to elementary abelian subgroups.

6 Generalizations to PolyGEMs

Recall that a GEM is a (possibly infinite) product of $K(G,n)$'s, where G is an abelian group. A space is a polyGEM if it belongs to the smallest full subcategory (polyGEMs) of spaces containing all GEMs and which is closed under taking extensions by fibrations, i.e., which satisfies that if $F \to E \to B$ is an arbitrary fibration then $B, F \in$ polyGEMs $\Rightarrow E \in$ polyGEMs. We say that a space is an oriented polyGEM if it belongs to the smallest full subcategory $\text{polyGEMs}^{\text{ori}}$ of spaces containing all GEMs and which is closed under taking extensions by principal fibrations, i.e., which satisfies that if $F \to E \to B$ is a principal fibration then $F, B \in \text{polyGEMs}^{\text{ori}} \Rightarrow E \in \text{polyGEMs}^{\text{ori}}$. Note that a nilpotent finite Postnikov system is an oriented polyGEM by [**17**].

In [**9**] Dror Farjoun conjectures that if X is a non-trivial p-complete polyGEM then $[B\mathbf{Z}/p, X]$ is always non-trivial. As stated, the conjecture is false—$\hat{S}^1_p$ and $K(\hat{\mathbf{Z}}_p, 3)$ serve as counterexamples. The problem with $\hat{S}^1_p$ is easy—it's a fundamental group problem. The problem with $K(\hat{\mathbf{Z}}_p, 3)$ is deeper, it demonstrates that it in general is not enough to look at maps from just a single $B\mathbf{Z}/p$ (it is however interesting to note that for any nilpotent, connected space X with finite π_1 and with $H^*(X)$ assumed to be *noetherian*, we have that $[BV, X] \neq 0 \Rightarrow [B\mathbf{Z}/p, X] \neq 0$ (cf. [**16**, Proposition II 7.2])). Since the idea that polyGEMs should behave like finite Postnikov systems still seems very plausible, we dare to state the following more modest conjecture:

Conjecture 6.1 *Let E be a connected nilpotent polyGEM with finite $\pi_1(E)$. Assume that $H^*(E)$ is of finite type and that $\bar{H}^*(E) \neq 0$. Then $\bar{H}^*(E)$ contains an element of infinite height.*

We likewise believe that Theorem 5.14 should generalize:

Conjecture 6.2 *Let E be a connected nilpotent polyGEM with finite $\pi_1(E)$. Assume that $H^*(E)$ is of finite type. Then the transcendence degree of $H^*(E)$ is finite iff E is $\mathbf{F}_p$-equivalent to a space E' fitting into a principal fibration sequence of the form:*

$$\mathbf{CP}^\infty \times \cdots \times \mathbf{CP}^\infty \to E' \to K(P, 1)\,,$$

where P is a finite p-group.

One piece of evidence for the first conjecture is a theorem of Félix, Halperin, Lemaire and Thomas [**14**] which (in the language of Dror Farjoun) states that a 1-connected oriented polyGEM E with $H^*(E)$ of finite type and $\bar{H}^*(E) \neq 0$ has infinite LS category.

We will give another piece of evidence for the conjecture, whose proof makes use of the Neisendorfer localization functor [**27**], which we now define. By [**2, 9**] we can choose a map f such that $L_f = L_{H\mathbf{Z}/p}$, where $L_{H\mathbf{Z}/p}$ denotes localization with respect to mod p homology and L_f denotes localization with respect to the map f. Let g be the map $g : B\mathbf{Z}/p \to *$, and note that, by definition, $L_g = P_{B\mathbf{Z}/p}$, the $B\mathbf{Z}/p$-nullification function (cf. [**9**]). We define the Neisendorfer localization fuctor as $L = L_{f\vee g}$, where $f \vee g$ just denotes the wedge of the two maps. Note that, by t he universal properties of the functors; we have a natural transformation of coaugmented functors $L = L_{H\mathbf{Z}/p}P_{B\mathbf{Z}/p} \to L$. This induces a weak equivalence $L_{H\mathbf{Z}/p}P_{B\mathbf{Z}/p}X \xrightarrow{\simeq} LX$ whenever $P_{B\mathbf{Z}/p}X$ is a nilpotent space, since in that case we know that $L_{H\mathbf{Z}/p}P_{B\mathbf{Z}/p}X$ is $B\mathbf{Z}/p$-null. (Our definition differs slightly from the one given by Neisendorfer, since we want to ensure that L is an idempotent localization functor on all of Ho_*.)

We need the following definition.

Definition 6.3 *Let* polyGEMs^{ft} *denote the smallest full subcategory of spaces which is closed under taking extensions by fibrations and which contains all connected GEMs with finite* π_1.

Remark 6.4 The class polyGEMs^{ft} contains all connected finite Postnikov systems with finite solvable π_1, and is probably very close to being equal to all connected polyGEMs with finite π_1.

Remember that an unstable module M is called locally finite if for all $x \in M$ we have that $\mathcal{A}x$ is finite dimensional.

Proposition 6.5 *Let E be a nilpotent polyGEM which belongs to* polyGEMs^{ft}. *Assume that $H^*(E)$ is of finite type and that $\bar{H}^*(E) \neq 0$. Then $H^*(E)$ is not locally finite.*

Before giving the proof we need a lemma:

Lemma 6.6 *We have that $LE = *$ for all $E \in$* polyGEMs^{ft}, *where L denotes the Neisendorfer localization functor.*

Proof of Lemma 6.6 Our main technical tool is a theorem of Dror Farjoun which states that if L_f is any localization functor with respect to some map f, and if $F \to E \to B$ is any fibration sequence, then $L_fF = *$ implies that $L_fE \xrightarrow{\simeq} L_fB$ (cf. [**9**]). Hence the class of spaces which are acyclic with respect to Neisendorfer localization, i.e., the spaces E which satisfies $LE = *$, is closed under taking extensions by fibrations. It is therefore enough to prove the claim for connected GEMs with finite π_1. Furthermore note that it is enough to prove the claim for 1-connected GEMs, since we can apply the theorem of Dror Farjoun to the fibration sequence $E\langle 1\rangle \to E \to K(\pi_1(E), 1)$, where $LK(\pi_1(E), 1)$ is easily seen to be zero.

Now let E be a 1-connected GEM and write this $E = \Omega\tilde{E}$ where $\tilde{E}$ is 2-connected. By a theorem of Dror Farjoun [**9**, Proposition 7.B.5] $P_{B\mathbf{Z}/p}$ and $P_{M(\mathbf{Z}/p,1)}$ coincide on GEMs (where $M(\mathbf{Z}/p, 1)$ denotes the mod p Moore space). This gives us

$$P_{B\mathbf{Z}/p}E = P_{M(\mathbf{Z}/p,1)}E = \Omega P_{M(\mathbf{Z}/p,2)}\tilde{E}\,,$$

where the last equality is by another theorem of Dror Farjoun [**9**, Proposition 3.A.1]. Bousfield [**3**] has shown that

$$\pi_i(P_{M(\mathbf{Z}/p,2)}\tilde{E}) = \pi_i(\tilde{E}) \otimes \mathbf{Z}\left[\frac{1}{p}\right],$$

where we use that $\tilde{E}$ is a 2-connected GEM. But this shows that $P_{B\mathbf{Z}/p}E$ is a connected nilpotent space with uniquely divisible homotopy groups, and hence $LE = L_{H\mathbf{Z}/p}P_{B\mathbf{Z}/p}E = *$. □

Proof of Proposition 6.5 By Lemma 6.6 we have that $LE = *$, so especially $\mathrm{map}_*(B\mathbf{Z}/p, E) \neq *$, since $\mathrm{map}_*(B\mathbf{Z}/p, E) = *$ would imply that $LE = \hat{E}_p \neq *$.

But saying that $\text{map}_*(B\mathbf{Z}/p, E) \neq *$ is equivalent to saying that $H^*(E)$ is not locally finite (cf. [**26, 21**]). □

Remark 6.7 Using recent results of Bousfield [**4**] it is in fact possible to replace the category polyGEMs^{ft} in Proposition 6.5 with the larger and somewhat more natural category consisting of all connected polyGEMs with finite π_1.

Remark 6.8 Neisendorfer's theorem [**27**], stating that $LX\langle n\rangle = \hat{X}_p$ for a 1-connected finite complex X with finite π_2, follows immediately from Lemma 6.6, by applying L to the fibration sequence $\Omega P_n X \to X\langle n\rangle \to X$ and using Miller's theorem to conclude that $LX = \hat{X}_p$.

Remark 6.9 There is no direct relation between the LS category and the property locally finite. For instance $\Sigma K(\mathbf{Z}/2, 2)$ is an example of a space with non-locally finite cohomology but with LS category 1, whereas $\prod_{n\in\mathbf{N}} S^n$ is an example of a space of infinite LS category whose cohomology is locally finite.

7 Appendix: Nilpotent actions

In this short appendix we prove a proposition about nilpotent actions, which we have used several times in the paper. The author thanks W. Chachólski for pointing out a proof somewhat simpler than the author's original proof.

Proposition 7.1 *Let M be an abelian group which does not contain any p-torsion and let P be a p-group. Assume that P acts nilpotently on M. Then P acts trivially on M.*

Proof Suppose that P acts nilpotently on M and let $0 = F_0 \subset \cdots \subset F_n = M$ be a filtration on M such that P acts trivially on the filtration quotients. We want to do an induction on the length of the filtration, the initial step being trivial. Let $g \in P$ and $x \in F_2$ be arbitrary. We can write $gx = x + x_1$ where $x_1 \in F_1$. By iterating this we get that $g^n x = x + nx_1$. Setting $n = |g|$ gives us $|g|x_1 = 0$. But since M does not contain p-torsion we have that multiplication by $|g|$ is injective on M so $x_1 = 0$. Since x and g were arbitrary we conclude that P acts trivially on F_2. Induction on the length of the filtration now finishes the proof. □

Acknowledgements

I am grateful to The Fields Institute for giving me the opportunity to participate in the 'Emphasis year in Homotopy Theory 95/96', and especially to the organizers and participants for making the stay so enjoyable and profitable. Special thanks to K. Andersen, W. Chachólski, P. Lambrechts, J. Scherer and B. Schuster for so many hours of interesting discussions. Also, thanks to R. Kane for introducing me to the beautiful theory of Lannes and Schwartz which this paper is based on.

I would like to thank W. Chachólski for helpful suggestions and references. Likewise, thanks to my advisor, J. Møller for his support and for suggesting improvements in the presentation. Thanks to H. Minh Le for directing my attention to the conjecture of E. Dror Farjoun. Also thanks to C. Casacuberta for pointing out an error in my original description of the Neisendorfer localization functor. Finally, thank you to C. McGibbon and J. Møller whose questions related to Theorem 1.1 got me interested in the cohomology of finite Postnikov systems in the first place.

References

[1] Benson, D. J. [1993], *Polynomial Invariants of Finite Groups*, London Math. Soc.Lecture Note Series **190**, Cambridge Univ. Press.

[2] Bousfield, A. K. [1975], *The localization of spaces with respect to homology*, Topology **14**, 133-150.

[3] Bousfield, A. K. [1994], *Localization and periodicity in unstable homotopy theory*, J. Amer. Math. Soc. **7**, 831-873.

[4] Bousfield, A. K. [1996], *Homotopical localizations of spaces*, preprint.

[5] Bousfield, A. K. and Kan, D. M. [1972], *Homotopy Limits, Completions and Localization*, Springer, Berlin, Heidelberg, New York.

[6] Broto, C. and Zarati, S. [1988], *Nil localization of unstable algebras over the Steenrod algebra*, Math. Z. **199**, 525-537.

[7] Cartan, H. [1956], *Determination des algébres $H_*(\pi, n, Z_p)$ et $H^*(\pi, n, Z_p)$*, Séminaire Henri Cartan; 7e année: 1954/55. Algébres d'Eilenberg-Mac Lane et Homotopie, Exposé no. 10, Secrétariat mathématique, École Normale Supérieure, Paris.

[8] Cartan, H. and Eilenberg, S. [1956], *Homological algebra*, Princeton University Press.

[9] Dror Farjoun, E. [1996], *Cellular Spaces, Null Spaces and Homotopy Localization*, Springer.

[10] Dwyer, W. G. and Wilkerson, C. W. [1990]. *Spaces of null homotopic maps*, in International Conference on Homotopy Theory (Marseille-Luminy, 1988), Astérisque **191**, 97-108.

[11] Eilenberg, S. and Mac Lane, S. [1954], *On the groups $H(\Pi, n)$, II*, Ann. of Math. **60**, 49–139.

[12] Evens, L. [1991], *The Cohomology of Groups*, Clarendon Press, Oxford, New York, Tokyo.

[13] Félix, Y. and Halperin, S. [1982], *Rational L.-S. Category and its Applications*, Trans. Amer. Math. Soc. **273**, 1–37.

[14] Félix, Y., Halperin, S., Lemaire, J.-M. and Thomas, J.-C. [1989], *Mod p loop space homology*, Invent. Math. **95**, 247–262.

[15] Harrison, D. K. [1959], *Infinite abelian groups and homological methods*, Ann. of Math. **69**, 366-391.

[16] Henn, H.-W., Lannes, J. and Schwartz, L. [1993], *The Categories of Unstable Modules and Unstable Algebras over the Steenrod algebra Modulo Nilpotent Objects*, Amer. J. Math. **115**, 1053–1106.

[17] Hilton, P., Mislin, G. and Roitberg, J. [1975], *Localization of nilpotent groups and spaces*, North-Holland, Amsterdam.

[18] Lannes, J. [1987], *Sur la cohomologie modulo p des p-groupes abéliens élémentaires*, Homotopy theory (Durham, 1985), pp. 97–116, London Math. Soc. Lecture Note Ser. **117**, Cambridge Univ. Press.

[19] Lannes, J. [1992], *Sur les espaces fonctionnels dont la source est le classifiant d'un p-groupe abélien élémentaire*, Inst. Hautes Études Sci. Publ. Math. **75**, 135–244.

[20] Lannes, J. and Schwartz, L. [1986], *A propos de conjectures de Serre et Sullivan*, Invent. Math. **86**, 593–603.

[21] Lannes, J. and Schwartz, L. [1989], *Sur la structure des $\mathcal{A}$-modules instables injectifs*, Topology **28**, 153–169.

[22] Lannes, J. and Schwartz, L. [1989], *Sur les groupes d'homotopie des espaces dont la cohomologie modulo 2 est nilpotente*, Israel J. Math. **66**, 260–273.

[23] McGibbon, C. A. and Møller, J. M. [1997], *Connected Covers and Neisendorfer's Localization Theorem*, Fund. Math. **152**, 211-230.

[24] McGibbon, C. A. and Neisendorfer, J. [1984], *On the homotopy groups of a finite dimensional space*, Comment. Math. Helv. **59**, 253–257.

[25] Miller, H. R. [1983], *Massey-Peterson Towers and Maps from Classifying Spaces*, Algebraic Topology Aarhus 1982, Lecture Notes in Math. 1051, Springer, 401–417.

[26] Miller, H. R. [1984], *The Sullivan conjecture on maps from classifying spaces*, Ann. of Math. **120**, 39–87.

[27] Neisendorfer, J. A. [1995], *Localization and connected covers of finite complexes*, The Čech centennial (Boston, MA, 1993), pp. 385–390, Contemp. Math. **181**, Amer. Math. Soc..

[28] Quillen, D. [1971], *The spectrum of an equivariant cohomology ring, I,II*, Ann. of Math. **94**, 549–572; 573–602.

[29] Schwartz, L. [1988], *La filtration nilpotente de la categorie $\mathcal{U}$ et la cohomologie des espaces de lacets*, Algebraic topology—rational homotopy (Louvain-la-Neuve, 1986), pp. 208–218, Lecture Notes in Math. **1318**, Springer.

[30] Schwartz, L. [1994], *Unstable Modules over the Steenrod Algebra and Sullivan's fixed Point Set Conjecture*, The University of Chicago Press, Chicago and London.

[31] Serre, J.-P. [1953], *Cohomologie modulo 2 des complexes d'Eilenberg-Mac Lane*, Comment. Math. Helv. **27**, 198-232.

[32] Venkov, B. B. [1959], *Cohomology algebras for some classifying spaces*, (in Russian), Dokl. Akad. Nauk SSSR **127**, 943-944.

Fields Institute Communications
Volume **19**, 1998

Cogroups which are not Suspensions II

John Harper
Department of Mathematics
University of Rochester
Rochester, NY 14627 USA

Israel Berstein in Memoriam

Abstract. An example is analyzed to verify a conjecture of Ganea concerning the desuspension of cogroups.

1 Introduction

In [**3**], Ganea proved that an $(n-1)$-connected cogroup of dimension less than or equal to $4n-5$, $n \geq 2$, has the primitive homotopy type of a suspension. The question whether this bound is sharp has been raised by Ganea [**1, 3, 4**]. In this paper, we show Ganea's bound is sharp by constructing a 4-connected cogroup of dimension 16 which does not have the primitive homotopy type of any suspension. Indeed, the example we study was produced in [**2**], where it was denoted $Y(\beta_1)$. In that paper it was proven that $Y(\beta_1)$ has the homotopy type of a certain suspension ΣW, and the existence of a primitive homotopy equivalence of $Y(\beta_1)$ and ΣW was ruled out. However, this information failed to settle the question of the sharpness of Ganea's bound because of the putative possibility of other desuspensions for $Y(\beta_1)$. In effect, the analysis in this paper rules out other ways of desuspending $Y(\beta_1)$, leaving only those for which no primitive homotopy equivalence is possible.

We recall some of the basic definitions. A space X with comultiplication σ is called a *cogroup* provided σ is associative up to homotopy and has both left and right inversion (which are automatic if X is simply connected). A map between co-H-spaces which commutes with the comultiplications, up to homotopy, is called *primitive*. Thus a primitive homotopy equivalence is one which is also a co-H-map.

Next, we introduce the spaces to be studied in this paper. In order to construct spaces with appropriate dimensions, we make our constructions globally and then prove results about their localizations. As cell complexes our spaces have the form $S^5 \cup e^9 \cup e^{16}$, let γ be an element of $\pi_7 S^4$. Let Z_γ denote the pushout in the

1991 *Mathematics Subject Classification.* Primary 55P45; Secondary 55P40.

following diagram

$$\begin{array}{ccc} S^5 & \longrightarrow & S^5 \cup_{\beta_1} e^{16} \\ {\scriptstyle k_1}\big\downarrow & & \big\downarrow \\ \Sigma(S^4 \cup_\gamma e^8) & \xrightarrow[k_2]{} & Z_\gamma \end{array}$$

where β_1 is an element of order 9 in $\pi_{15}S^5$. We write σ_γ for a cogroup structure on Z_γ such that the map k_2 is primitive with respect to the suspension comultiplication on its source and σ_γ on its target. In [**2**] we proved that such σ_γ exist provided $\Sigma\gamma \equiv \pm\alpha_1(5) \bmod 3$. Our aim is to prove that some of the cogroups $(Z_\gamma, \sigma_\gamma)$ are not of the primitive homotopy type of any suspension. To say which, we must choose γ with some care. Write $\pi_7 S^4 = Z + Z/3 + Z/4$ and choose generators for the summands as follows. Let ν generate the infinite cyclic summand and be the element of Hopf invariant one constructed from quaternionic multiplication. Now $[\iota_4, \iota_4] - 2\nu$ is an element of order 12 in the image of the suspension map. We take $\alpha_1(3) \in \pi_6 S^3$ so that the following equation holds:

$$[\iota_4, \iota_4] = 2\nu + \alpha_1(4) + \tau$$

where $\alpha_1(4) = \Sigma\alpha_1(3)$ and τ is a choice for the generator in the 2-primary component (which one does not matter).

The point is the sign in front of $\alpha_1(4)$; and either choice is permissible but we must be specific. In terms of these choices, we have "coordinates" for γ,

$$\gamma = x \cdot \nu + z \cdot \alpha_1(4) + \tau'$$

where x is an integer, z is 0 or ± 1 and τ' is 2-primary. Note that with our choices, we have $\Sigma\nu \equiv \alpha_1(5) \bmod 3$. Thus $\Sigma\gamma \equiv \pm\alpha_1(5) \bmod 3$ whenever $x + z \equiv \pm 1 \bmod 3$. We can now state the main result of this paper.

2 Ganea's upper bound is sharp

Proposition 2.1 *If $x = 0$ and $z = \pm 1$, then $(Z_\gamma, \sigma_\gamma)$ does not have the primitive homotopy type of any suspension.*

The remainder of this paper is devoted to a proof of this proposition. First of all, the cases $x = 0$ and $x \neq 0$ are distinguishable over the rationals. It is straightforward to prove that rationally $(Z_\gamma, \sigma_\gamma)$ has the following structure: $Z_\gamma \simeq S^5 \vee S^9 \vee S^{16}$ with ι_5, ι_{16} primitive and $\iota_9 \to \iota_9^1 + \iota_9^2 + x[\iota_5^1, \iota_5^2]$. Since $\pi_9 S^5 \otimes \mathbb{Q} = 0$, there is no self-equivalence of Z_γ which can exchange a structure with $x \neq 0$ with one for which $x = 0$.

A putative desuspension W for Z_γ can be chosen with the following cell structure,

$$W = S^4 \cup e^8 \cup e^{15},$$

where the only requirement on the attaching map for the eight-cell be that its suspension is $\pm\alpha_1(5)$. Let $A_\delta = S^4 \cup_\delta e^8$ and $B = S^5 \cup_{\alpha_1} e^9$. Then we have a desuspension of the space Z_γ if and only if we have the following diagram:

$$\begin{array}{ccccc} S^{14} & \xrightarrow{(k_1\circ\beta_1)} & \Omega B & \longrightarrow & \Omega Z_\gamma \\ \downarrow & & \uparrow & & \uparrow \\ S^{14} & \xrightarrow{\epsilon} & A_\delta & \longrightarrow & W \end{array}$$

where the vertical maps on the right are adjoint to homotopy equivalences and ϵ is a factorization of the adjoint of $k_1 \circ \beta_1$, the attaching map of the top cell in Z_γ. Our analysis of the rationalization yields that for a primitive desuspension of $(Z_\gamma, \sigma_\gamma)$ in the case $x = 0$, we must take δ to have the form $\delta = \pm\alpha_1(4) + \tau''$, where τ'' is 2-primary. So we now assume that δ has this form and prove

Lemma 2.2 *Any map $E : A_\delta \to \Omega B$ induces 0 on the 3-primary component of π_{14}.*

Since it was shown in [**2**] that $k_1 \circ \beta_1 \neq 0$, the proposition follows from the lemma and our previous discussion.

It will now be convenient to localize at 3. We do so, but suppress this step from the notation. The next step is to gain some information on what homotopy there is in the source and target of the map E. A useful tool for the analysis is knowledge of the fibre of the pinch map from a two-cell complex to its top cell. In general, if $K = S^n \cup_\alpha e^r (r \geq n+1\ , n \geq 2)$ and F is the homotopy fibre of the pinch map $K \to S^r$ then homologically F looks like the half-smash product $S^n \rtimes \Omega S^r$ and has a cell-decomposition of the form

$$S^n \cup e^{n+r-1} \cup e^{n+2(r-1)} \cup \cdots$$

and furthermore, the attaching map of the $(n+r-1)$-cell can be taken to be the Whitehead product $[\iota_n, \alpha]$.

For the case of the space B, this analysis gives a space $F \simeq S^5 \cup e^{13} \cup e^{21} \cup \cdots$. Hence $\pi_{15}B$ is the surjective image from $\pi_{15}(S^5 \cup e^{13})$, since $\pi_{15}S^9 = 0$. It follows that we have a surjection $\pi_{15}S^5 \to \pi_{15}B \to 0$. Since $3\beta_1 = \alpha_1(5)\circ\alpha_2(8)$, the image of this element in $\pi_{15}B$ is zero. On the other hand, we know that the image β_1 in $\pi_{15}B$ is not zero, hence $\pi_{15}B = Z/3$.

For the case of the space A_δ, our analysis gives a space $F' \simeq S^4 \cup e^{11} \cup e^{18} \cup \cdots$ with the 11-cell attached by $[\iota_4, \delta]$. Since $\pi_{14}S^8 = 0$, it follows that $\pi_{14}A_\delta$ is the surjective image from $\pi_{14}X$, where

$$X = S^4 \cup_{[\iota_4,\delta]} e^{11}\ .$$

Furthermore, by naturality, the restriction of E to X factors through ΩF. Taking adjoints and observing that $\Sigma X \simeq S^5 \vee S^{12}$ yields the following factorization of the map E:

$$\begin{array}{ccc} A_\delta & \xrightarrow{E} & \Omega B \\ \uparrow & & \uparrow \\ X & \xrightarrow[E']{} & \Omega S^5 \end{array}$$

by some map E'. Now Lemma 2.2 follows from our discussion and

Lemma 2.3 *If $\delta = \pm\alpha_1(4)$ then 3 times the image of E' is 0 on π_{14}.*

To prove this lemma, we use the fact that δ is a suspension to write $[\iota_4, \delta] = \pm[\iota_4, \iota_4] \circ \alpha_1(7)$. Let $Y = S^7 \cup_{\pm\alpha_1} e^{11}$ with agreement of signs. Then we have a map $g : Y \to X$ extending $[\iota_4, \iota_4]$ which is of degree ± 1 on the 11-cell. We can obtain some information about the homotopy of X by taking the fibre of g, call it F, and studying the fibre sequence

$$\Omega Y \to \Omega X \to F\,.$$

Taking homology with $Z/3$ coefficients, the Bott-Samelson theorem yields that $H_*\Omega Y = T[x_6, x_{10}]$. In the Eilenberg-Moore spectral sequence for $H_*\Omega X$ we have $E_2 = \mathrm{Tor}^{*,*}_{H*X}(Z/3, Z/3) = T(S^{-1}\bar{H}_*X) = T[x_3, x_{10}]$ with the 10-class coming from $H_*\Omega Y$. Therefore, both it and x_3 are permanent cycles and this spectral sequence collapses at E_2. Furthermore, since the map $g : Y \to X$ was constructed to extend the Whitehead product $[\iota_4, \iota_4]$, in homology $\Omega g_*(x_6) = (-1)$ times the commutator $\langle x_3, x_3 \rangle$.

Now let us retreat momentarily from a detailed analysis in order to describe the overall strategy of our argument. First we embed $\pi_{14}X$ in the homotopy sequence for the pair (X, S^4), which looks like

$$\pi_{14}S^4 \longrightarrow \pi_{14}X \xrightarrow{j_*} \pi_{14}(X, S^4) \longrightarrow 0\,,$$

and the surjection on the right follows because $\pi_{13}S^4 = 0$ (the sequence is actually a short exact sequence, but this is immaterial to the argument). If we write

$$C_{11} : (D^{11}, \dot{D}^{11}) \to (X, S^4)$$

for the characteristic map, then from [**5**] we have $\pi_{14}(X, S^4) = Z_{(3)} \oplus Z/3$ with the infinite summand generated by the relative Whitehead product $[\iota_4, C_{11}]$ and the torsion of the form $C_{11} \circ \tau$ for $\tau \in \pi_{14}(D^{11}, \dot{D}^{11})$. Thus $\pi_{14}X$ will be built up from extensions of these elements by elements from $\pi_{14}S^4$. Thus the crucial point is to understand how this homotopy is mapped into $\pi_{14}X$ by the inclusion of S^4.

Lemma 2.4 *Decompose $\pi_{14}S^4$ as $\Sigma\pi_{13}S^3 \oplus [\iota_4, \iota_4] \circ \pi_{14}S^7$ which is possible because we have localized at 3. Then the inclusion of S^4 in X induces a split injection on the summand $\Sigma\pi_{13}S^3$ and maps the complementary summand by composition with the Whitehead product $[\iota_4, \iota_4]$.*

Before proving Lemma 2.4, we show how it implies Lemma 2.3 and hence our proposition. Let $C_{11} \in \pi_{11}(X, S^4)$ be the generator described earlier. There is $K_{11} \in \pi_{11}X$ such that $j_*K_{11} = 3C_{11}$.

Now, suppose $\rho \in \pi_{14}X$ satisfies $j_*\rho = [\iota_4, C_{11}]$. Then, in view of Lemma 2.4, we can write

$$3\rho - [\iota_4, K_{11}] = \lambda[\iota_4, \iota_4] \circ \pi_{14}S^7$$

where λ is a mod 3 integer. Thus 3 times $E'\rho$ is zero because Whitehead products are zero on ΩS^5. Likewise, if $\rho \in \pi_{14}X$ satisfies $j_*\rho = C_{11} \circ \tau$, then

$$3\rho = \lambda[\iota_4, \iota_4] \circ \pi_{14}S^7$$

and 3 times $E'\rho$ is zero. This proves Lemma 2.3.

The proof of Lemma 2.4 is accomplished by means of a splitting theorem for ΩX. Lie theory can be used to predict the maps needed for our splitting. Thus $H_*\Omega X$ is the universal enveloping algebra of a free Lie algebra of the form $L(x_3, x_{10})$. Letting $\langle x_3 \rangle$ denote the free abelian Lie algebra, we have a short exact sequence

$$0 \to L(\langle x_3, x_3 \rangle,\ x_{10}, \langle x_3, x_{10} \rangle) \to L(x_3, x_{10}) \to \langle x_3 \rangle \to 0$$

as is readily checked with Poincaré series. We have the inclusion $S^3 \to \Omega X$ realizing $\langle x_3 \rangle$ and $\Omega g : \Omega Y \to \Omega X$ realizing $\langle x_3, x_3 \rangle$ and x_{10} so only the realization of $\langle x_3, x_{10} \rangle$ is left. Lie theory will not be of any help for constructing the remaining map. Instead we proceed indirectly and study the space F.

In the homology Serre spectral sequence with $Z/3$-coefficients for the fiber sequence $\Omega X \to F \to Y$, the 7-class on the base transgresses to a nonzero multiple of $\langle x_3, x_3 \rangle$ in $H_6 \Omega X$ and thus the column hits $H_* \Omega X \cdot \langle x_3, x_3 \rangle$, i.e., the left ideal generated by $\langle x_3, x_3 \rangle$. Likewise the eleven-class on the base cleans out $H_* \Omega X \cdot x_{10}$. Thus additively the homology of F is spanned by x_3 and $H_* \Omega X \cdot \langle x_3, x_{10} \rangle$. Since $\pi_{12} S^3 = 0$, it follows that $H_{13}(F)$ is spherical. By inspection of the fibre of the pinch map $Y \to S^{11}$, any spherical representative for $H_{13}(F)$ must pull back to $H_{13}(\Omega X)$, where it must also be spherical. Since spherical classes are primitive, we have a map $S^{13} \to \Omega X$ whose image is $\langle x_3, x_{10} \rangle$. We can now construct a splitting

$$\Omega X \simeq S^3 \times \Omega \Sigma (S^6 \cup_{\alpha_1} e^{10} \vee S^{13})$$

and Lemma 2.4 follows. This concludes the proof of our proposition.

The reader will have noticed that in passing we have proved that the fibre F of $g : Y \to X$ is an H-space. For the record, one can show that $F \simeq S^3 \times S^{13} \times F'$ and the Poincaré series for F' is $1 \Big/ 1 - \frac{t^{19} + t^{23} + t^{26}}{1 - t^6 - t^{10}}$ which is the Poincaré series for a tensor algebra. At this writing I have no further information about F'.

References

[1] Arkowitz, M. [1995], *Co-H-spaces*, Handbook of Algebraic Topology (I. M. James, ed.), Elsevier, 1143–1175.

[2] Berstein, I. and Harper, J. R. [1989], *Cogroups which are not suspensions*, Algebraic Topology, Lecture Notes in Mathematics **1370**, Springer-Verlag, Berlin, 63–86.

[3] Ganea, T. [1970], *Cogroups and suspensions*, Invent. Math. **9**, 185–196.

[4] Ganea, T. [1971], Lecture Notes in Mathematics **249**, Springer-Verlag, Berlin, 23–30.

[5] James, I. M. [1954], *On the homotopy groups of certain pairs and triads*, Quart. J. Math **5**, 260–270.

Fields Institute Communications
Volume **19**, 1998

Morava Hopf Algebras and Spaces $K(n)$ Equivalent to Finite Postnikov Systems

Michael J. Hopkins
Massachusetts Institute of Technology
Cambridge, Massachusetts USA 02139
mjh@math.mit.edu

Douglas C. Ravenel
University of Rochester
Rochester, New York USA 14627
drav@troi.cc.rochester.edu

W. Stephen Wilson
Johns Hopkins University
Baltimore, Maryland USA 21218
wsw@math.jhu.edu

Abstract. We have three somewhat independent sets of results. Our first results are a mixed blessing. We show that Morava K-theories don't see k-invariants for homotopy commutative H-spaces which are finite Postnikov systems, i.e., for those with only a finite number of homotopy groups. Since k-invariants are what holds the space together, this suggests that Morava K-theories will not be of much use around such spaces. On the other hand, this gives us the Morava K-theory of a wide class of spaces which is bound to be useful. In particular, this work allows the recent work in [**35**] to be applied to compute the Brown-Peterson cohomology of all such spaces. Their Brown-Peterson cohomology turns out to be all in even degrees (as is their Morava K-theory) and flat as a BP^* module for the category of finitely presented $BP^*(BP)$ modules. Thus these examples have extremely nice Brown-Peterson cohomology which is as good as a Hopf algebra.

Our second set of results produces a large family of spaces which behave as if they were finite Postnikov systems from the point of view of Morava K-theory even though they are not. This allows us to apply the above results to an even wider class of spaces than finite Postnikov systems. These examples come from spaces in omega spectra with certain properties. There are many well known examples with these properties. In particular, we compute the $K(n)$ homology of all the spaces in the Ω-spectra for $P(q)$ and $k(q)$ where $q > n$.

In order to prove our results on finite Postnikov systems we need our third set of results; a beginning of an analysis of bicommutative Hopf algebras over $K(n)_*$.

1991 *Mathematics Subject Classification.* Primary 55N20, 57T05; Secondary 55N15, 55P20.
MJH partially supported by the National Science Foundation.
DCR partially supported by the National Science Foundation.

1 Introduction

The Morava K-theories are a collection of generalized homology theories which are intimately connected to complex cobordism ([**17**], [**42**]). It is known that they play a central role in aspects of homotopy theory ([**14**],[**9**],[**15**]). This and their relative computability (due to a Künneth isomorphism) makes them a powerful tool. For each prime, p, and $n > 0$, the coefficient ring for $K(n)_*(-)$ is $K(n)_* \simeq \mathbf{F}_p[v_n, v_n^{-1}]$ where the degree of v_n is $2(p^n - 1)$.

We assume that all of our spaces are homotopy equivalent to a CW complex.

Theorem 1.1 *Let X be a connected p-local space with $\pi_k(X)$ finitely generated over $\mathbf{Z}_{(p)}$, $k > 1$, and nonzero for only a finite number of k. Then $K(n)_*(\Omega X)$ has a natural filtration by normal sub-Hopf algebras:*

$$K(n)_* \simeq F_{-(n+2)} \subset \cdots \subset F_{-2} \subset F_{-1} \subset F_0 \simeq K(n)_*(\Omega X)$$

with

$$F_{-q}//F_{-(q+1)} \simeq K(n)_*(K(\pi_q(\Omega X), q))$$

as Hopf algebras.

If $K(n)_*(\Omega X)$ is commutative (e.g., if X is an H-space), then $K(n)_*(\Omega X)$ is isomorphic, as a Hopf algebra, to the associated graded object above:

$$K(n)_*(\Omega X) \simeq \bigotimes_{0 \leq m \leq n+1} K(n)_*(K(\pi_m(\Omega X), m)).$$

This is natural if either all $\pi_k(X)$, $k > 1$, are finite, or if they are all free.

We believe that if $K(n)_*(X)$ is not commutative then we still have the last isomorphism as coalgebras but have been unable to prove it. It may require a dual Borel theorem for our Hopf algebras and we have only been able to handle the bicommutative case.

We actually state and prove a theorem (Theorem 2.1) which does not require the homotopy groups to be finitely generated over $\mathbf{Z}_{(p)}$. We can have copies of $\mathbf{Q}/\mathbf{Z}_{(p)}$ but our groups can only have a finite number of summands. This is important both for our proofs and for some of our applications. The last naturality statement is true in this case if all the homotopy groups are torsion.

Note also that the $n = 0$ case of this theorem is a familiar result about rational homology.

In [**26**], McCleary-McLaughlin show that for an Eilenberg–Mac Lane space X with finite homotopy group, $K(n)_*(X)$ has the same rank as $K(n-1)_*(\mathcal{L}X)$, where $\mathcal{L}X$ denotes the free loop space of X. In view of the theorem above, the following generalization of their result is immediate.

Corollary 1.2 *Let X be a simply connected H-space with finitely many nontrivial homotopy groups, each of which is finite. Then $K(n)_*(X)$ has the same rank as $K(n-1)_*(\mathcal{L}X)$.*

The following corollary follows from the fact that the Morava K-theory of Eilenberg–Mac Lane spaces is even degree (except for the circle) and from the main results of [**35**]. The condition on π_1 is needed to avoid having copies of the circle in our space, giving us odd degree elements.

Corollary 1.3 *If X is as in Theorem 1.1 and $\pi_1(\Omega X)$ is torsion, then $K(n)_*(\Omega X)$ is even degree and so is $BP_p^{\hat{}\,*}(\Omega X)$ where $BP_p^{\hat{}}$ is the p-adic completion of BP. If $\pi_k(X)$ is finite for $k > 1$ then $BP^*(\Omega X)$ is even degree. In* *either case, it is a flat BP^*-module for the category of finitely presented $BP^*(BP)$-modules.*

We see that Morava K-theory cannot distinguish between the double loops of such a space and a product of Eilenberg–Mac Lane spaces with the same homotopy groups. One cannot expect to generalize this too much to spaces with an infinite number of nonzero k-invariants; the sphere, S^k, is a counter example to that. Inverse limit problems rear their ugly head. Morava K-theory somehow looks at the whole space rather than how it is put together. On the other hand, our result certainly does cover spaces with infinitely many homotopy groups if the k-invariants are zero for all but a finite number of stages. This is because the Eilenberg–Mac Lane spaces split off as a product if the k-invariant is zero. Another, more substantial direction of generalization is to spaces in Ω-spectra with certain stable properties.

Although the abstract isomorphism is interesting from a theoretical point of view, the practical value comes because the Morava K-theory of Eilenberg–Mac Lane spaces is completely known, [**33**]. In particular, it is always even degree (except for the circle). We also know that

$$K(n)_*(K(\pi_m(\Omega X), m)) \simeq K(n)_*$$

if $m > n+1$, so that Morava K-theory does not see the higher homotopy groups of these spaces, and spaces with only higher homotopy groups are acyclic, e.g., if ΩX is n connected and $\pi_{n+1}(\Omega X)$ is torsion. When we have acyclicity we do not need Hopf algebras much:

Theorem 1.4 *Let X an n-connected p-local space with $\pi_k(X)$ a finitely generated $\mathbf{Z}_{(p)}$ module which is nontrivial for only a finite number of k. If $\pi_{n+1}(X)$ is torsion, then*

$$K(n)_*(X) \simeq K(n)_*.$$

This theorem is rather easy and is proven quite quickly directly from [**33**].

The proof of Theorem 1.1 is given in Section 2 modulo certain general results about graded Hopf algebras over $K(n)_*$ which will be proven in Section 5. It mimics a proof for rational homology. The only place where Section 5 is used is in showing the final statement of Theorem 1.1 about the splitting as Hopf algebras.

In Section 4, which is independent of the rest of the paper, we will prove some results about spaces in the Ω-spectrum of a $K(n)_*$-acyclic spectrum X. Recall that an Ω-spectrum $X = \{\underline{X}_i\}$ has $\Omega \underline{X}_{i+1} = \underline{X}_i$. A motivating example for this study was produced by Richard Kramer's work computing $K(n)_*(\underline{k(q)}_*)$ when $n < q$.

Theorem 1.5 *Let $X = \{\underline{X}_i\}$ be a connective Ω-spectrum of finite type with bottom cell in dimension 0 and $K(n)_*(\underline{X}_m) \simeq K(n)_*$ for some m. Let $X \to F$ be a map to a finite Postnikov system which is an equivalence through dimension $n+1$. Then $\underline{X}_q$ is $K(n)_*$-equivalent to $\underline{F}_q$ for all $q \geq 0$.*

This is Theorem 4.7. We can now apply Theorem 1.1 to such a spectrum to get:

Theorem 1.6 *Let X be as in Theorem 1.5, then, for all $k \in \mathbf{Z}$,*

$$K(n)_*(\underline{X}_k) \simeq \bigotimes_{n+1 \geq i \geq 0} K(n)_*(K(\pi_i(\underline{X}_k), i))$$

as Hopf algebras. In particular, if $k > n+1$, then $K(n)_*(\underline{X}_k) \simeq K(n)_*$. Also, if $\pi_0(X)$ is torsion, then $\underline{X}_{n+1}$ is $K(n)_*$-acyclic. Furthermore, whether $K(n)_*(\underline{X}_k)$

is trivial or not for $k \geq 0$ depends only on whether $K(n)_*(K(\pi_0(X), k))$ is trivial or not.

We call such a spectrum with one space $K(n)_*$-acyclic, *strongly $K(n)_*$-acyclic*, because it implies that almost all other spaces are also $K(n)_*$-acyclic. We can get the spaces $\underline{X}_m$, $m > n+1$ to be $K(n)_*$-acyclic without reference to the first section. Note that being strongly $K(n)_*$-acyclic implies that the spectrum is $K(n)_*$-acyclic since $K(n)_*(X) = \operatorname{dir\,lim} K(n)_*(\underline{X}_k)$ using the suspension maps. Bousfield has a generalization of this which may have useful applications together with the rest of our work. We will discuss his results in Section 4.

This does not lead to a calculation of the Brown-Peterson cohomology as in the case of a real finite Postnikov system because it is easy to have spectra which are strongly $K(n)_*$-acyclic but have nontrivial $K(n+1)$ homology. However, following [**35**], it does lead to the calculation of the $E(n)$ cohomology and a host of others.

What we need now is a condition on X which implies that it is strongly $K(n)_*$-acyclic. Associated with the "telescope" conjecture is a functor L_n^f, see [**32**]. It supplies us with a class of examples. From Corollary 4.13 we have:

Theorem 1.7 *A connective spectrum X for which $L_n^f X$ is contractible is strongly $K(n)_*$-acyclic.*

In particular, a suspension spectrum of a finite complex which is $K(n)_*$-acyclic is strongly $K(n)_*$-acyclic. It is possible that the converse of this Theorem is also true. This result is not phrased in the most familiar or applicable of terms. When X is a BP module spectrum then this functor coincides with a more familiar one. Let $E(n)$ be the homology theory with coefficient ring $\mathbf{Z}_{(p)}[v_1, v_2, \ldots, v_n, v_n^{-1}]$. We have, from Theorem 4.14:

Theorem 1.8 *Let X be a connective BP module spectrum which is $E(n)_*(-)$-acyclic, or, equivalently, $K(q)_*(-)$-acyclic for $0 \leq q \leq n$, then X is strongly $K(n)_*$-acyclic.*

What we need now are some concrete examples of interest. This result can be reduced, Corollary 4.15, to a simpler statement which we can use for this purpose.

Corollary 1.9 *If X is a connective BP-module spectrum in which each element of $\pi_*(X)$ is annihilated by some power of the ideal*

$$I_{n+1} = (p, v_1, v_2, \ldots, v_n) \subset BP_*,$$

then X is strongly $K(n)_*$-acyclic.

Thus we see that any connective BP module spectrum with some power of I_{n+1} mapping to zero is strongly $K(n)_*$-acyclic. There are a lot of familiar examples in this collection. Recall that $BP_* \simeq \mathbf{Z}_{(p)}[v_1, v_2, \ldots]$ where the degree of v_n is $2(p^n-1)$. We have the spectra $BP\langle n\rangle$, with coefficient ring $BP\langle n\rangle_* \simeq \mathbf{Z}_{(p)}[v_1, v_2, \ldots, v_n]$, see [**40**] and [**16**]. We also have $P(k,n)$, the spectrum with $P(k,n)_* \simeq BP\langle n\rangle_*/I_k$ for $0 \leq k \leq n$. These theories are constructed using the usual Baas-Sullivan singularities [**1**], and [**2**]. Many of these theories are already familiar. In particular, $P(0,\infty) = BP$, $P(n,n) = k(n)$, $P(k,\infty) = P(k)$, (see [**17**] and [**41**]), and $P(0,n) = BP\langle n\rangle$. The spaces in the Ω-spectra for $P(k,n)$ are important in studying the spaces in the Ω-spectra for $P(k)$ in [**7**]. In addition, the theories $E(k,n) = v_n^{-1}P(k,n)$, play a prominent role in [**35**].

From the above theorem, we see:

Corollary 1.10 *For $m \geq q > n$, $P(q,m)$ is strongly $K(n)_*$-acyclic, so, for $k \in \mathbf{Z}$,*

$$K(n)_*(\underline{P(q,m)}_k) \simeq \bigotimes_{n+1 \geq i \geq 0} K(n)_*(K(\pi_i(\underline{P(q,m)}_k), i)).$$

We have explicitly computed the $K(n)$ homology of all of the spaces in the Ω-spectrum for all of these theories. Recall that this includes the more familiar $P(q)$ and $k(q)$ for $q > n$. The simplicity of the answer is in stark contrast with the calculation of $K(n)_*(\underline{P(n)}_*)$ in [**34**] and of $K(n)_*(\underline{k(n)}_*)$ in [**20**].

The category of graded Hopf algebras over $K(n)_*$ is equivalent to that of Hopf algebras over $\overline{K(n)}_* \simeq \mathbf{F}_p$, graded over $\mathbf{Z}/(2p^n - 2)$ (where we have set $v_n = 1$ in order to be working over a perfect field). Not everything about connected graded Hopf algebras carries over to these Hopf algebras so one must be somewhat careful. Being careful led us to initiate an investigation of the type of Hopf algebras that arise when you take the Morava K-theory of connected homotopy commutative H-spaces. The results of this investigation may well be of more interest than the applications to finite Postnikov systems and Section 5 is dedicated to this study.

There, we study what we call *commutative Morava homology Hopf algebras* (for a given $n > 0$). This is the category, $C(n)$, of bicommutative, biassociative, Hopf algebras over $\mathbf{F}_p$ which are graded over $\mathbf{Z}/2(p^n - 1)$. Furthermore, the primitive filtration is exhaustive and it is the direct limit of finite dimensional $\mathbf{F}_p$ sub-coalgebras. $\overline{K(n)}_*(\Omega X)$, where X is an H-space, is such an object. We restrict our attention to such objects which are concentrated in even degrees. Denote this category by $EC(n)$. We show it is an abelian category. For the sake of completeness, we show that for odd primes the bigger category, $C(n)$, splits into the product of $EC(n)$ and $OC(n)$, where $OC(n)$ consists of exterior algebras on odd degree primitive generators.

Since we deal only with evenly graded objects, our Morava homology Hopf algebras are really graded over $G = \mathbf{Z}/(p^n - 1)$. Let H be the cyclic group of order n. H acts on G via the pth power map. Writing G additively, the map $H \times G \to G$ is given by $(i, j) \to p^i j$. Let γ denote an H-orbit:

$$\gamma = \{j, pj, p^2 j, \ldots\} \subset G.$$

We have:

Theorem 1.11 *For each evenly graded commutative Morava homology Hopf algebra, i.e., for $A \in EC(n)$, there is a natural splitting*

$$A \simeq \bigotimes_{\gamma} A_\gamma$$

where the tensor product is over all H-orbits γ and the primitives of A_γ all have dimensions in 2γ.

To prove this theorem we construct idempotents in our category.

Theorem 1.12 *For every $A \in EC(n)$ there are canonical idempotents e_γ such that $\sum_\gamma e_\gamma = 1_A$ and $e_\gamma e_\beta$ is trivial if $\gamma \neq \beta$. These idempotents are natural. The idempotent e_γ sends all primitives to zero which do not have dimensions in 2γ and is the identity on all primitives which are in dimensions in 2γ.*

Theorem 1.11 follows immediately from this and the fact that tensor products are the sum in this category. Theorem 1.12, in turn, follows immediately from Theorem 1.13 which does even more for us. Let $EC(n)_\gamma$ be the sub-category of

$EC(n)$ whose objects have primitives only in dimensions in 2γ. What we really prove, in Section 5, is the following:

Theorem 1.13 *There are commuting idempotent functors, e_γ, on $EC(n)$ such that $\sum_\gamma e_\gamma$ is naturally equivalent to the identity functor and $e_\gamma e_\beta$ is the trivial functor if $\gamma \neq \beta$. As categories:*

$$EC(n) \simeq \prod_\gamma EC(n)_\gamma.$$

In particular, there is only the trivial map $e_\gamma(A) = A_\gamma \to e_\beta(B) = B_\beta$ if $\gamma \neq \beta$. Furthermore, $EC(n)_\gamma$ is an abelian category.

We can define a function $\overline{\alpha}_p : \mathbf{Z}/(p^n - 1) \to \mathbf{Z}$ by just taking the usual lift to $\mathbf{Z}$ and then the usual α, the sum of the coefficients in the p-adic expansion of our number. $\overline{\alpha}_p$ is constant on an orbit γ. Usually several orbits will have the same image under the map $\overline{\alpha}_p$. The following fact is a consequence of the computations done in [**33**], and is what we need to help prove Theorem 1.1.

Theorem 1.14 *In the Hopf algebras $\overline{K(n)}_*(K(T,q))$ for any abelian torsion group T and $\overline{K(n)}_*(K(F,q+1))$ for any torsion free abelian group with $q \geq 1$, all orbits γ (as in 1.11) with nontrivial factors satisfy $\overline{\alpha}_p(\gamma) = q$.*

This is proved below as Theorem 5.14.

Our results were discovered while pursuing the *Johnson Question*, see [**33**, Section 13]. This assertion is that if $0 \neq x \in BP_n(X)$ where X is a space, then x is not v_n torsion. This is a very strong unstable condition. At present, two of the authors have a good plausibility argument which they hope to turn into a proof some day. The approach which led to the present paper was just one of many dead ends. A cohomology theory can be defined by:

$$\mathrm{Hom}_{BP_*}(BP_*(X), Q/\mathbf{Z}_{(p)}).$$

The classifying space for the n-th group, $\underline{M}_n$, contains the universal example for the degree n *Johnson Question*. In an attempt to get some insight into its Brown-Peterson homology, the general phenomenon of Theorem 1.1 was discovered. Each one of these spaces has only a finite number of nontrivial homotopy groups. Each nontrivial group is a finite sum of copies of $Q/\mathbf{Z}_{(p)}$. Theorem 2.1 still applies to it to give $K(n)_*(\underline{M}_n)$.

Morava K-theory can sometimes be problematic when $p = 2$. Because we are restricted to even degree objects this is not a problem for us, see Remark 5.4.

The authors thank Bill Dwyer, Takuji Kashiwabara, Richard Kramer, Jim McClure, Jean–Pierre Meyer, Haynes Miller, and Hal Sadofsky, all of whom helped this project along in one way or another.

2 Proof of the main theorem

For a group, G, let $R[G]$ be the group ring for G over R. Let $\tilde{X}$ be the universal cover for X. We have a sequence of fibrations, up to homotopy:

$$\Omega\tilde{X} \to \Omega X \to \pi_1(X) \to \tilde{X} \to X.$$

Since $\Omega\tilde{X}$ and $\tilde{X}$ are connected, we see that

$$K(n)_*(\Omega X) \simeq K(n)_*(\Omega\tilde{X}) \bigotimes K(n)_*[\pi_1(X)].$$

From this we see that it is enough to restrict our attention to simply connected spaces; and they all have Postnikov decompositions.

We recall some basic facts about Postnikov towers. Let X be a simply connected space. Then one has a diagram of the form

$$\begin{array}{ccccccc} X = X_2 & \longleftarrow & X_3 & \longleftarrow & X_4 & \longleftarrow & \cdots \\ \downarrow{\scriptstyle f_2} & & \downarrow{\scriptstyle f_3} & & \downarrow{\scriptstyle f_4} & & \\ K_2 & & K_3 & & K_4 & & \end{array}$$

where $K_s = K(\pi_s(X), s)$, f_s is a map inducing an isomorphism in the bottom homotopy group, and X_{s+1} is the fibre of f_s. X_{s+1} is the s-connected cover of X. Theorem 1.1 is a slightly weaker p-local form of the following.

Theorem 2.1 *Suppose X has only finitely many nontrivial homotopy groups. Assume that $\pi_q(X)$, $q > 1$, has finitely generated torsion free quotient, and finitely generated subgroup of exponent p^k for each p and k. Then for each $s \geq 1$ there is a natural Hopf algebra extension*

$$K(n)_* \longrightarrow K(n)_*(\Omega X_{s+1}) \longrightarrow K(n)_*(\Omega X_s) \longrightarrow K(n)_*(\Omega K_s) \longrightarrow K(n)_*.$$

In particular the map $K(n)_*(\Omega X_s) \to K(n)_*(\Omega X)$ is always one-to-one.

If X is an H-space, then we have a natural isomorphism of Hopf algebras:

$$K(n)_*(\Omega X) \simeq \bigotimes_{i \geq 2} K(n)_*(K(\pi_i(X), i-1)).$$

If X is not an H-space, the above isomorphism is still valid additively.

Although this result implies Theorem 1.1 in the Introduction, it is slightly more general. In particular, it allows for homotopy groups with summands like $\mathbf{Q}/\mathbf{Z}_{(p)}$ which we need. The definition of the filtration in Theorem 1.1 comes from the $s-1$-connected cover, X_s:

$$F_{-s} \equiv \text{im } K(n)_*(\Omega X_s) \longrightarrow K(n)_*(\Omega X).$$

This filtration is natural. We are not really using the Postnikov construction of a space, which is not natural, but a Postnikov decomposition which is. The Postnikov construction starts with a point and builds up the space one homotopy group at a time. We are starting with the space and taking it apart one homotopy group at a time. Naturality is clear for the maps on the Eilenberg–Mac Lane spaces as they are determined by the maps on the homotopy groups. We want a unique map from X_{s+1} to Y_{s+1} if we inductively have a unique map from X_s to Y_s. The obstruction to uniqueness is in

$$[X_{s+1}, K(\pi_s(Y), s-1)]$$

but because X_{s+1} is s-connected, this cohomology group is trivial.

Remark 2.2 *A number of people have asked us questions about how these results can be extended. For example, given a fibration of finite Postnikov systems where the maps all can be delooped several times and the homotopy groups of the fibre map split-injective to the homotopy groups of the total space, what can we say about the Morava K-theory of everything? Since we know the Morava K-theory of all the spaces it seems to us that the results and techniques used here should answer any question about this situation.*

In the arguments that follow, it will be convenient to assume that $\pi_2(X)$ is torsion. In particular, the results of [**33**] imply that $K(n)_*(\Omega X)$ is concentrated in even dimensions in this case. This assumption is harmless for the following reason. In general (subject to the hypotheses of Theorem 2.1) we have a fibre sequence

$$X' \longrightarrow X \xrightarrow{f} L_2$$

where $L_2 = K(\pi_2(X)/\text{Torsion}, 2)$ and $\pi_2(f)$ is surjective. Then $\pi_2(X')$ is the torsion subgroup of $\pi_2(X)$, ΩL_2 is a finite product of circles, and

$$\Omega X \simeq \Omega X' \times \Omega L_2 \tag{2.3}$$

because we can lift the maps of the circles to homotopy generators using the H-space structure. Hence it suffices to compute $K(n)_*(\Omega X')$.

The rational case

The corresponding result for rational homology, $K(0)$, is classical and we will sketch its proof now, since it is a model for the proof of Theorem 2.1. Since X has only finitely many nontrivial homotopy groups, X_s is contractible for large s and we can argue by downward induction on s. Suppose ΩX_{s+1} has the prescribed rational homology, and consider the fibre sequence

$$\Omega^2 K_s \xrightarrow{j_s} \Omega X_{s+1} \longrightarrow \Omega X_s. \tag{2.4}$$

For our inductive step we need to prove that

$$H_*(\Omega X_s) \simeq H_*(\Omega X_{s+1}) \otimes H_*(\Omega K_s), \tag{2.5}$$

where all homology groups have rational coefficients. When X is an H-space, we need the above isomorphism to be one of Hopf algebras. Otherwise it is an isomorphism of coalgebras.

For our calculation we need the bar spectral sequence for a principle fibration:

$$\begin{array}{ccc} F & \longrightarrow & E \\ & & \downarrow \\ & & B \end{array} \tag{2.6}$$

which has

$$E^2_{*,*} \simeq \operatorname{Tor}^{K(n)_*(F)}_{*,*}(K(n)_*(E), K(n)_*) \Rightarrow K(n)_*(B). \tag{2.7}$$

If the fibration we use is the loops on a principle fibration, then the bar spectral sequence is a spectral sequence of Hopf algebras. Unfortunately, despite a fascination with the bar construction, e.g., [**21**], [**22**], [**23**] and [**24**], this fact is not in the literature. It can, of course, be patched up easily from what is there about the standard bar construction. Let our fibration be

$$\begin{array}{ccc} \Omega F & \longrightarrow & \Omega E \\ & & \downarrow \\ & & \Omega B \end{array}$$

where (2.6) is a principle fibration. Then ΩF has two products, one from F and one from the loops. When ΩB is constructed using the bar construction, one product can be used in the construction and the other can be used to get a product on the bar construction giving us this spectral sequence as Hopf algebras (the coalgebra structure is no problem). This works for any homology theory with a Künneth isomorphism

In particular, one has the bar spectral sequence converging to $H_*(\Omega X_s)$ with

$$E^2 = \mathrm{Tor}^{H_*(\Omega^2 K_s)}(H_*(\Omega X_{s+1}), \mathbf{Q}). \tag{2.8}$$

Here the $H_*(\Omega^2 K_s)$-module structure on $H_*(\Omega X_{s+1})$ is induced by the map j_s of (2.4). We have

Lemma 2.9 *The map j_s in (2.4) induces the trivial homomorphism in rational homology.*

Proof The map j_s is an H-map, so it must respect the Pontrjagin ring structure in homology. We know that $H_*(\Omega^2 K_s)$ is generated by elements in dimension $s-2$, while ΩX_{s+1} is $(s-1)$-connected, so $H_*(j_s)$ is trivial. □

It follows that (2.8) can be rewritten as

$$\begin{aligned} E^2 &= \mathrm{Tor}^{H_*(\Omega^2 K_s)}(H_*(\Omega X_{s+1}), \mathbf{Q}) \\ &= \mathrm{Tor}^{H_*(\Omega^2 K_s)}(\mathbf{Q}, \mathbf{Q}) \otimes H_*(\Omega X_{s+1})) \\ &= H_*(\Omega K_s) \otimes H_*(\Omega X_{s+1})) \end{aligned} \tag{2.10}$$

The rational homology bar spectral sequence collapses for Eilenberg–Mac Lane spaces and has no extension problems, which explains the last step above. Now it follows formally that the spectral sequence collapses, because differentials must lower filtration by at least 2, but $H_*(\Omega X_{s+1}))$ is concentrated in filtration 0, and $H_*(\Omega K_s)$ is generated by elements in filtration 1. Thus one gets the desired extension of Hopf algebras

$$\mathbf{Q} \longrightarrow H_*(\Omega X_{s+1}) \longrightarrow H_*(\Omega X_s) \longrightarrow H_*(\Omega K_s) \longrightarrow \mathbf{Q}. \tag{2.11}$$

If X is an H-space, then so is X_s, so $H_*(\Omega X_s)$ is bicommutative. The structure of graded connected bicommutative Hopf algebras over $\mathbf{Q}$ is well known (see [**25**, Section 7]). In particular, we have the following splitting theorem.

Theorem 2.12 *Let A be a graded connected bicommutative Hopf algebra over $\mathbf{Q}$. Then there is a canonical Hopf algebra isomorphism*

$$A \simeq \bigotimes_{k>0} A_k$$

where A_k is generated by primitive elements in dimension k. Moreover, A_k is a polynomial algebra for k even and an exterior algebra for k odd.

It follows that the extension (2.11) is split when X is an H-space and the rational case of Theorem 2.1 is proved.

3 The Morava K-theory case for torsion spaces

Now we will give the proof of Theorem 2.1 under the additional assumption that $\pi_*(X)$ is all torsion. The general setup is the same as in the rational case. The Morava K-theory of Eilenberg–Mac Lane spaces was computed in [**33**]. We have the bar spectral sequence as in the rational case, and we have the following analog of Lemma 2.9.

Lemma 3.1 *The map j_s in (2.4) induces the trivial homomorphism in Morava K-theory. (Here we do not require that $\pi_*(X)$ be all torsion.)*

Proof (See the introduction for $\overline{K(n)}_*(-)$.) We will make use of Theorems 1.11 and 1.14. We are studying the map

$$\overline{K(n)}_*(\Omega^2 K_s) \xrightarrow{\overline{K(n)}_*(j_s)} \overline{K(n)}_*(\Omega X_{s+1}).$$

Recall that we are using downward induction on s so we can assume Theorem 2.1 for all $t > s$. Assume our map is nontrivial. Choose the largest t so that its image is contained in $K(n)_*(\Omega X_t)$. Then the composition

$$\overline{K(n)}_*(\Omega^2 K_s) \longrightarrow \overline{K(n)}_*(\Omega X_t) \longrightarrow \overline{K(n)}_*(\Omega X_t)//\overline{K(n)}_*(\Omega X_{t+1}) = \overline{K(n)}_*(\Omega K_t)$$

must be nontrivial. This is a Hopf algebra map, and both the source and target are subject to the splitting theorem 1.11, with the constraints imposed by 1.14. The factors of $\overline{K(n)}_*(\Omega^2 K_s)$ correspond to orbits γ with $\overline{\alpha}_p(\gamma) = s-2$ or $s-3$, while those of $\overline{K(n)}_*(\Omega K_t)$ have $\overline{\alpha}_p(\gamma) = t-1$ or $t-2$. These orbits are distinct since $s < t \leq n+1$. Since there are no nontrivial Hopf algebra homomorphisms between factors corresponding to distinct orbits, the result follows. □

It follows that the analog of (2.10) holds, namely, in the bar spectral sequence,

$$E^2 = \mathrm{Tor}^{K(n)_*(\Omega^2 K_s)}(K(n)_*, K(n)_*) \otimes K(n)_*(\Omega X_{s+1}).$$

Now, $\mathrm{Tor}^{K(n)_*(\Omega^2 K_s)}(K(n)_*, K(n)_*)$ is the E^2-term of the bar spectral sequence converging to $K(n)_*(\Omega K_s)$, and this was completely determined in [**33**]. There is a map to this spectral sequence from the one we are studying, given by the following commutative diagram, in which each row is a fibre sequence.

$$\begin{array}{ccccc}
\Omega^2 K_s & \xrightarrow{j_s} & \Omega X_{s+1} & \longrightarrow & \Omega X_s \\
\big\downarrow{\scriptstyle =} & & \big\downarrow & & \big\downarrow \\
\Omega^2 K_s & \longrightarrow & \text{pt.} & \longrightarrow & \Omega K_s
\end{array}$$

In our spectral sequence (the one for the top row) the extra factor of $K(n)_*(\Omega X_{s+1})$ is concentrated in the even degrees (by induction) of filtration 0. In a spectral sequence of Hopf algebras the basic differentials must go from generators to primitives, see [**37**, page 78]. In the bar spectral sequence, differentials must start in filtration greater than one. All generators here are in even degrees and so must go to odd degree primitives, all of which are in filtration one [**33**, Theorem 11.5]. So, the start and finish of the generating differentials are all in the part which maps isomorphically to the bottom row. It follows that in our spectral sequence,

$$E^\infty = E^0 K(n)_*(\Omega K_s) \otimes K(n)_*(\Omega X_{s+1}),$$

where $E^0 K(n)_*(\Omega K_s)$ denotes the E^∞-term of the lower spectral sequence, which was determined in [**33**]. In particular, our E^∞-term is concentrated in even dimensions.

Again the edge homomorphism gives us the desired Hopf algebra extension, namely

$$K(n)_* \longrightarrow K(n)_*(\Omega X_{s+1}) \longrightarrow K(n)_*(\Omega X_s) \longrightarrow K(n)_*(\Omega K_s) \longrightarrow K(n)_*. \tag{3.2}$$

The following result is where we need the assumption that $\pi_*(X)$ is torsion.

Lemma 3.3 *When X is an H-space with torsion homotopy, the extension (3.2) is split naturally.*

Proof We use Theorems 1.11 and 1.14, and assume inductively that $\overline{K(n)}_*(\Omega X_{s+1})$ is as advertised. This means that its factors under Theorem 1.11 all correspond to orbits γ with $\overline{\alpha}_p(\gamma) \geq s$, while the orbits of $\overline{K(n)}_*(\Omega K_s)$ have $\overline{\alpha}_p(\gamma) = s-1$. Thus Theorem 1.13 assures us that the extension is split naturally. □

Notice that the proof above would fail if the torsion subgroup of $\pi_s(X)$ and the torsion free quotient of $\pi_{s+1}(X)$ were both nontrivial, because in that case, both $\overline{K(n)}_*(\Omega X_{s+1})$ and $\overline{K(n)}_*(\Omega K_s)$ could have factors corresponding to the same orbit γ. In particular, there is a short exact sequence of Hopf algebras,

$$\mathbf{F}_p \to \overline{K(n)}_*(K(\mathbf{Z}/(p^i),j)) \to \overline{K(n)}_*(K(\mathbf{Z}_{(p)},j+1)) \xrightarrow{p^i} \overline{K(n)}_*(K(\mathbf{Z}_{(p)},j+1)) \to \mathbf{F}_p,$$

which shows how maps can drop filtration (from $-j$ to $-(j+1)$). This example prevents us from having a natural Hopf algebra splitting in general.

We do have an unnatural splitting for the general case however, and we produce that now.

Removing the torsion condition

We have given the proof of Theorem 2.1 in the case when $\pi_*(X)$ is all torsion. We needed the torsion condition to get the desired Hopf algebra structure when X is an H-space. We did *not* need it to show that $K(n)_*(\Omega X_s)$ is a subalgebra of $K(n)_*(\Omega X)$. Now we will describe an alternate approach to the Hopf algebra question which does not require $\pi_*(X)$ to be torsion.

We need to use the *rationalization* $X\mathbf{Q}$ of X. For an H-space (more generally, a nilpotent space), Y, one has a fibre sequence

$$TY \longrightarrow Y \longrightarrow Y\mathbf{Q}$$

where $\pi_*(Y\mathbf{Q}) = \pi_*(Y) \otimes \mathbf{Q}$ and $\pi_*(TY)$ is all torsion. For a p-local H-space Y (such as $\Omega X_{(p)}$), $Y\mathbf{Q}$ is the homotopy direct limit of

$$Y \xrightarrow{[p]} Y \xrightarrow{[p]} Y \xrightarrow{[p]} \cdots$$

where $[p]$ is the H-space p^{th} power map. In this case $Y\mathbf{Q}$ and TY are both H-spaces.

Now for X as in Theorem 2.1, we have a fibre sequence

$$\Omega^2 X\mathbf{Q} \longrightarrow \Omega TX \xrightarrow{\ i\ } \Omega X. \tag{3.4}$$

Before we can proceed, we need to bring in the generalized Atiyah–Hirzebruch spectral sequence, see [**10**] and [**11**]. For a fibration like (2.6), we have the Atiyah–Hirzebruch–Serre spectral sequence:

$$E^2 \simeq H_*(B; K(n)_*(F)) \Longrightarrow K(n)_*(E). \tag{3.5}$$

The map of E to B maps this spectral sequence to the usual Atiyah–Hirzebruch spectral sequence:

$$E^2 \simeq H_*(B; K(n)_*) \Longrightarrow K(n)_*(B). \tag{3.6}$$

Lemma 3.7 *If $\pi_2(X)\otimes\mathbf{Q} = 0$, then the map i of (3.4) induces an isomorphism in $K(n)_*$ for all $n > 0$.*

As remarked above (2.3), this assumption on $\pi_2(X)$ can be made without loss of generality.

Proof of Lemma 3.7 We will use the Atiyah–Hirzebruch spectral sequence for Morava K-theory to compute $K(n)_*(\Omega TX)$. We have the following commutative diagram in which each row is a fibre sequence.

$$\begin{array}{ccccc} \Omega^2 X\mathbf{Q} & \longrightarrow & \Omega TX & \xrightarrow{\ i\ } & \Omega X \\ \Big\downarrow & & \Big\downarrow{\scriptstyle i} & & \Big\downarrow{\scriptstyle =} \\ \text{pt.} & \longrightarrow & \Omega X & \xrightarrow{\ =\ } & \Omega X \end{array} \tag{3.8}$$

The hypothesis on $\pi_2(X)$ implies that $\Omega^2 X\mathbf{Q}$ is path connected. We know that any rational loop space is a product of rational Eilenberg–Mac Lane spaces. From [**33**, Corollary 12.2] we have

$$\operatorname{dir\,lim} K(n)_*(K(\mathbf{Z}/(p^i), q)) \simeq K(n)_*(K(\mathbf{Z}, q+1)). \tag{3.9}$$

We see that iterating $[p]$ on the left kills everything, so the product of rational Eilenberg–Mac Lane spaces has the Morava K-theory of a point. This means that the left vertical arrow in (3.8) is a $K(n)_*$-equivalence. The right vertical map is a homology equivalence (since it is the identity map) so we get an isomorphism of Atiyah–Hirzebruch–Serre spectral sequences, so i is a $K(n)_*$-equivalence. □

Now the torsion case of Theorem 2.1 tells us that

$$K(n)_*(\Omega TX) \simeq \bigotimes_{i\geq 2} K(n)_*(K(\pi_i(TX), i-1)),$$

and this is $K(n)_*(\Omega X)$ by Lemma 3.7. Thus Theorem 2.1 will follow from

Lemma 3.10 *For a simply connected space X with $\pi_2(X)$ torsion,*

$$\bigotimes_{i\geq 2} K(n)_*(K(\pi_i(X), i-1)) \simeq \bigotimes_{i\geq 2} K(n)_*(K(\pi_i(TX), i-1)).$$

as Hopf algebras.

Proof There is a split short exact sequence

$$0 \longrightarrow \pi_{i+1}(X) \otimes \mathbf{Q}/\mathbf{Z} \longrightarrow \pi_i(TX) \longrightarrow \mathrm{Tor}_1(\pi_i(X), \mathbf{Q}/\mathbf{Z}) \longrightarrow 0.$$

Note that $\mathrm{Tor}_1(\pi_i(X), \mathbf{Q}/\mathbf{Z})$ is the torsion subgroup of $\pi_i(X)$, while $\pi_{i+1}(X)\otimes\mathbf{Q}/\mathbf{Z}$ is the tensor product of $\mathbf{Q}/\mathbf{Z}$ with the torsion free quotient of $\pi_{i+1}(X)$. With this in mind, we also have a split short exact sequence

$$0 \longrightarrow \mathrm{Tor}_1(\pi_i(X), \mathbf{Q}/\mathbf{Z}) \longrightarrow \pi_i(X) \longrightarrow \pi_i(X)/\mathrm{Torsion} \longrightarrow 0.$$

According to (3.9)

$$K(n)_*(K(\mathbf{Z}, i+1)) \simeq K(n)_*(K(\mathbf{Q}/\mathbf{Z}, i)) \quad \text{for } i \geq 1,$$

so we have

$$K(n)_*(K(\pi_{i+1}(X) \otimes \mathbf{Q}/\mathbf{Z}, i-1)) \simeq K(n)_*(K(\pi_{i+1}(X)/\mathrm{Torsion}, i)).$$

It follows that

$$\begin{aligned}
&\bigotimes_{i\geq 2} K(n)_*(K(\pi_i(TX), i-1)) \\
&\qquad = \bigotimes_{i\geq 2} K(n)_*(K(\pi_{i+1}(X) \otimes \mathbf{Q}/\mathbf{Z}, i-1)) \\
&\qquad\qquad \oplus \\
&\qquad\quad \bigotimes_{i\geq 2} K(n)_*(K(\mathrm{Tor}_1(\pi_i(X), \mathbf{Q}/\mathbf{Z}), i-1)) \\
&\qquad = \bigotimes_{i\geq 2} K(n)_*(K(\pi_{i+1}(X)/\mathrm{Torsion}, i)) \\
&\qquad\qquad \oplus \\
&\qquad\quad \bigotimes_{i\geq 2} K(n)_*(K(\mathrm{Tor}_1(\pi_i(X), \mathbf{Q}/\mathbf{Z}), i-1)) \\
&\qquad = \bigotimes_{i\geq 2} K(n)_*(K(\pi_i(X)/\mathrm{Torsion}, i-1)) \\
&\qquad\qquad \oplus \\
&\qquad\quad \bigotimes_{i\geq 2} K(n)_*(K(\mathrm{Tor}_1(\pi_i(X), \mathbf{Q}/\mathbf{Z}), i-1)) \\
&\qquad = \bigotimes_{i\geq 2} K(n)_*(K(\pi_i(X), i-1)).
\end{aligned}$$

□

Remark 3.11 *The above process lost us our naturality in the splitting, but not if all of the homotopy groups are free or torsion. The problems only come up if we try to mix them.*

Proof of Theorem 1.4 First let us assume that $\pi_{n+1}(X) = \pi_{n+2}(X) = 0$. Theorem 1.1 tells us that $K(n)_*(\Omega X)$ is trivial and our result follows from the bar spectral sequence.

Remark 3.12 *We do not have to use Theorem 1.1 here at all. We can do our downward induction on the Postnikov system directly. Everything will be trivial so there is no difficulty showing the maps are also trivial and the spectral sequence is trivial. One can just use Lemma 4.3 over and over again. We use this result* *several times in our study of spectra; the point is that those results are independent of Theorem 1.1.*

Next we consider the case where $\pi_{n+1}(X)=0$. If $\pi_{n+2}(X)=0$ or is torsion, the same proof works. It is only if $\pi_{n+2}(X)$ has copies of $\mathbf{Z}_{(p)}$ in it that we could have a problem. We take our usual fibration with map:

$$\begin{array}{ccccc} \Omega K_{n+2} & \longrightarrow & X_{n+3} & \longrightarrow & X \\ \| & & \downarrow & & \downarrow \\ \Omega K_{n+2} & \longrightarrow & \text{pt.} & \longrightarrow & K_{n+2}. \end{array}$$

Since $\pi_{n+1}(X_{n+3})=\pi_{n+2}(X_{n+3})=0$, we have $K(n)_*(X_{n+3})$ is trivial. Thus we get an isomorphism on the E^2 terms of the two bar spectral sequences. Thus the E^∞ terms must be isomorphic as well. However, since we know, from [**33**, Theorem 12.3], that

$$\mathrm{Tor}^{K(n)_*(K(Z_{(p)},n+1))}(K(n)_*,K(n)_*)=K(n)_*,$$

we get our result.

All we have left to deal with is the case where $\pi_{n+1}(X)$ is finite. The argument is exactly the same as that just given except that the Tor is trivial by [**33**, Theorem 11.5]. □

4 Strongly $K(n)_*$-acyclic connective spectra

Throughout this section n will be assumed to be positive. For a connective spectrum X, $\underline{X}_m$ for $m\geq 0$ will denote the mth space in the associated Ω-spectrum. Be alert to our (unusual) convention that $\underline{X}_m$ have its bottom cell in dimension m.

Definition 4.1 *A connective spectrum X is* **strongly $K(n)_*$-acyclic** *if $\underline{X}_m$ is $K(n)_*$-acyclic for some $m\in\mathbf{Z}$.*

One might guess that any $K(n)_*$-acyclic spectrum is strongly $K(n)_*$-acyclic, but we will see below, (4.9), that this is not the case. We will show, Theorem 4.7, that for such an X each space $\underline{X}_m$ is $K(n)_*$-equivalent to a suitable finite Postnikov system.

Remark 4.2 *We take a moment to show that we need not restrict ourselves to positive spaces in the Ω-spectrum, such as in Theorem 1.6. If $m\geq 0$, then we see from the bar spectral sequence that $\underline{X}_m$ $K(n)_*$-acyclic implies $\underline{X}_{m+1}$ is $K(n)_*$-acyclic. If we have a negative number, write $\underline{X}_{-m}$. Let $X\langle m+1\rangle$ be the stable m-connected cover of X. Then we have a stable cofibration:*

$$\Sigma^m X\langle m+1\rangle\longrightarrow X\longrightarrow F$$

where F is a finite Postnikov system and we have $\underline{X\langle m+1\rangle}_0=\underline{X}_{-m}$. If this is $K(n)_*$-acyclic, then $\underline{X\langle m+1\rangle}_k$ is $K(n)_*$-acyclic for all $k\geq 0$. We have an unstable fibration

$$\underline{F}_{k-1}\longrightarrow\underline{X\langle m+1\rangle}_{m+k}\longrightarrow\underline{X}_k.$$

For big k, $\underline{F}_{k-1}$ is also $K(n)_*$-acyclic by Theorem 1.4, so $\underline{X}_k$ is $K(n)_*$-acyclic by the next lemma.

Lemma 4.3 *Let*

$$F \xrightarrow{i} E \xrightarrow{j} B$$

be a fibration with F $K(n)_*$-acyclic. Then the map j is a $K(n)_*$-equivalence.

Proof There is an Atiyah–Hirzebruch–Serre spectral sequence, (3.5), converging to $K(n)_*(E)$ with

$$E_2 = H_*(B; K(n)_*(F)).$$

It maps to the usual Atiyah–Hirzebruch spectral sequence converging to $K(n)_*(B)$. Since F is $K(n)_*$-acyclic, this map is an isomorphism, giving the desired result. □

Proposition 4.4 *Let $m \geq 0$, X be a connective spectrum with $\underline{X}_m$ $K(n)_*$-acyclic, and let Y be any connective spectrum. Then $\underline{(X \wedge Y)}_m$ is also $K(n)_*$-acyclic.*

Recall here that our convention is that $\underline{X}_m$ and $\underline{(X \wedge Y)}_m$ have the same connectivity.

Proof We will argue by skeletal induction on Y. Assume for simplicity that the bottom cell of Y is in dimension 0, so the 0-skeleton Y^0 is a wedge of spheres. Thus $\underline{(X \wedge Y^0)}_m$ is a product of $\underline{X}_m$s and is therefore $K(n)_*$-acyclic. For $k > 0$ consider the cofibre sequence

$$\Sigma^{-1} X \wedge Y^k / Y^{k-1} \longrightarrow X \wedge Y^{k-1} \longrightarrow X \wedge Y^k,$$

which we abbreviate by $A \to B \to C$. Now $\underline{A}_{m+k-1}$ (which may be contractible) is $K(n)_*$-acyclic by a similar argument, and we can assume inductively that $\underline{B}_m$ is $K(n)_*$-acyclic. We have a fibration

$$\underline{A}_{m+k-1} \xrightarrow{i} \underline{B}_m \xrightarrow{j} \underline{C}_m$$

and C_m is $K(n)_*$-acyclic by 4.3. This is true for all $k > 0$, so $\underline{(X \wedge Y)}_m$ is $K(n)_*$-acyclic as claimed because it is the direct limit of $K(n)_*$-acyclic spaces. □

We need a standard result which we include here for completeness.

Lemma 4.5 *Let X be a connective spectrum and H be the integral Eilenberg–Mac Lane spectrum. Then $X \wedge H$ is a product of Eilenberg–Mac Lane spectra.*

Proof We argue by skeletal induction. We have

$$\Sigma^{-1} X^k / X^{k-1} \wedge H \longrightarrow X^{k-1} \wedge H \longrightarrow X^k \wedge H.$$

By induction we have $X^{k-1} \wedge H$ is the product of Eilenberg–Mac Lane spectra with homotopy $H_*(X^{k-1}; \mathbf{Z})$. The first term is just a bunch of $\Sigma^{k-1} H$s and so factors through $K(H_{k-1}(X^{k-1}), k-1)$ in the second term so the third term is as claimed. □

Now we can get some control over the connectivity m.

Lemma 4.6 *Let X be a connective strongly $K(n)_*$-acyclic spectrum with bottom cell in dimension 0. Then the space $\underline{X}_{n+3}$ is $K(n)_*$-acyclic.*

Proof We assume that all spectra and spaces in sight are localized at the prime p. Let H denote the integer Eilenberg–Mac Lane spectrum, and let $\overline{H}$ denote the fibre of the map $S^0 \to H$.

$$\overline{H} \longrightarrow S^0 \longrightarrow H$$

The bottom cell of $\overline{H}$ is in dimension $q-1$, where $q = 2(p-1)$. Thus the cofibre sequence

$$\Sigma^{-1}X \wedge H \longrightarrow X \wedge \overline{H} \longrightarrow X$$

gives a fibre sequence

$$\underline{(X \wedge H)}_{m-1} \xrightarrow{i} \underline{(X \wedge \overline{H})}_{m+q-1} \xrightarrow{j} \underline{X}_m.$$

(Recall our convention!) Now $\underline{(X \wedge H)}_{m-1}$ is an $(m-2)$-connected product of Eilenberg–Mac Lane spaces, so it is $K(n)_*$-acyclic for $m-1 > n+1$. Thus j is a $K(n)_*$-equivalence for $m \geq n+3$. Iterating this argument we see that $\underline{X}_{n+3}$ is $K(n)_*$-equivalent to $\underline{(X \wedge \overline{H}^{(s)})}_{n+3+s(q-1)}$ for each $s > 0$. The latter is $K(n)_*$-acyclic for some s by Proposition 4.4, so $\underline{X}_{n+3}$ is also $K(n)_*$-acyclic. □

We can use this lemma to prove the following.

Theorem 4.7 *Let X be a connective strongly $K(n)_*$-acyclic spectrum with bottom cell in dimension 0. Let $X \to F$ be a map to a finite Postnikov system which is an equivalence through dimension $n+1$. Then $\underline{X}_m$ is $K(n)_*$-equivalent to $\underline{F}_m$ for all $m \geq 0$. In particular, $\underline{X}_{n+2}$ is $K(n)_*$-acyclic, and if $\pi_0(X)$ is torsion, $\underline{X}_{n+1}$ is $K(n)_*$-acyclic.*

Proof Stably we have a fibre sequence

$$X' \longrightarrow X \longrightarrow F$$

where X' is a connected cover of X having bottom cell above dimension $n+1$. Consider the fibre sequence

$$\underline{F}_{m-1} \longrightarrow \underline{X'}_{m'} \longrightarrow \underline{X}_m$$

(where $m' > m+n+1$ depends on the connectivity of X'). From Theorem 1.4 we know that F is strongly $K(n)_*$-acyclic. Since X is also strongly $K(n)_*$-acyclic we can use Lemma 4.3 to see that $\underline{X'}_{m'}$ is also strongly $K(n)_*$-acyclic. Hence $\underline{X'}_{n+3}$ is $K(n)_*$-acyclic by Lemma 4.6.

Now consider the fibre sequence

$$\underline{X'}_{m'} \xrightarrow{i} \underline{X}_m \xrightarrow{j} \underline{F}_m.$$

For positive m, $m' > m+n+1 \geq n+2$, so $\underline{X'}_{m'}$ is $K(n)_*$-acyclic. By Lemma 4.3, j is a $K(n)_*$-equivalence as claimed. In particular, $\underline{X}_{n+2}$ is $K(n)_*$ equivalent to $\underline{F}_{n+2}$, which by Theorem 1.4 is $K(n)_*$-acyclic. Since X was arbitrary strongly $K(n)_*$-acyclic, we also know that $\underline{X'}_{n+2}$ is $K(n)_*$-acyclic, so j is also a $K(n)_*$-equivalence for $m = 0$. □

Pete Bousfield has generalized this in more than one way. His results are not restricted to Morava K-theories and he does not need to work with Ω-spectra. In a short note to us he derived these results from [**4**]. More recently, these have been made explicit in [**6**, Section 7]. Restricting our attention to Morava K-theories, his result of interest to us is:

Theorem 4.8 (Bousfield) *An $(n+2)$-connected H-space Y is $K(n)_*$-acyclic if and only if ΩY is $K(n)_*$-acyclic.*

The following example shows that not all connective $K(n)_*$-acyclic spectra are strongly $K(n)_*$-acyclic.

Example 4.9 *Let X be the spectrum $BP\langle n-1\rangle$ from* [**40**] *and* [**16**] *for $n > 1$. The third author showed in* [**40**] *that $\underline{X}_m$ has torsion free homology and therefore is not $K(n)_*$-acyclic for $m = 2(p^n - 1)/(p-1)$. This would contradict Lemma 4.6 if X were strongly $K(n)_*$-acyclic, so it is not. It is, however, $K(n)_*$-acyclic. This can be seen most easily by considering the exterior Hopf algebra on the Milnor Bockstein, Q_n, as a sub-Hopf algebra of the cohomology*

$$H^*(BP\langle n-1\rangle; \mathbf{Z}/(p)) \simeq \mathcal{A}/\mathcal{A}(Q_0, Q_1, \dots, Q_{n-1}).$$

Then, by Milnor–Moore, 4.4, [**25**], *$H^*(BP\langle n-1\rangle; \mathbf{Z}/(p))$ is free over $E(Q_n)$ and the first differential in the Atiyah–Hirzebruch spectral sequence for $K(n)_*(BP\langle n-1\rangle)$ kills everything and we find $BP\langle n-1\rangle$ is $K(n)_*$-acyclic.*

Theorem 4.10

(i) *Let $X \to Y \to Z$ be a cofibre sequence of connective spectra. If any two of X, Y and Z are strongly $K(n)_*$-acyclic, then so is the third.*
(ii) *Any retract X of a connective strongly $K(n)_*$-acyclic spectrum Y is also strongly $K(n)_*$-acyclic.*
(iii) *Any connective direct limit X of connective strongly $K(n)_*$-acyclic spectra is strongly $K(n)_*$-acyclic.*

Proof (i) is an easy consequence of Lemma 4.3. For (ii) note that $\underline{X}_m$ is a retract of $\underline{Y}_{m'}$ for suitable m'. For (iii) we can assume that each spectrum in the direct system has bottom cell in dimension ≥ 0, so $\underline{Y}_{n+2}$ is a direct limit of $K(n)_*$-acyclic spaces. □

We have so far produced no examples of strongly $K(n)_*$-acyclic spectra other than finite Postnikov systems.

Proposition 4.11 *Any finite $K(n)_*$-acyclic spectrum X is strongly $K(n)_*$-acyclic.*

Proof X is a suspension spectrum of a (connected) finite complex Y, so for some m, $\underline{X}_m = QY$, where

$$QY = \varinjlim_k \Omega^k \Sigma^k Y.$$

Since $K(n)_*(Y) \simeq K(n)_*(X)$, Y is $K(n)_*$-acyclic. All we have to do is show that Y $K(n)_*$-acyclic implies that QY is $K(n)_*$-acyclic. There is a stable splitting, [**38**], for QY. Each piece of this stable splitting is identified explicitly as something called $e[\mathcal{C}(j), \Sigma_j, Y]$ from [**21**, Proposition 2.6(ii), page 14]. Since $\mathcal{C}(j)$ is contractible, [**21**, page 5], this is the same as $D_{\Sigma_j} Y$; for which there is a spectral sequence, [**8**, Corollary 2.4, page 7]:

$$E^2 \simeq H_*(\Sigma_j; K(n)_*(Y^{(j)})) \Rightarrow K(n)_*(D_{\Sigma_j} Y)$$

where $Y^{(j)}$ is the jth smash product. For reduced Morava K-theory, $K(n)_*(Y^{(j)}) = 0$ because we have a Künneth isomorphism. Thus each piece of the stable splitting is $K(n)_*$-acyclic and we have that QY is $K(n)_*$-acyclic. This is well known to those familiar with this but, to a novice, perhaps a bit difficult to dig out of [**38**]. □

Corollary 4.12 *Any connective spectrum which is a direct limit of finite $K(n)_*$-acyclic spectra is strongly $K(n)_*$-acyclic.*

Proof This follows immediately from Theorem 4.10 and Proposition 4.11. □

Now recall the localization functors L_n of [**29**] and L_n^f of [**32**]. The former is Bousfield localization with respect to $E(n)$, while the latter is constructed in such a way that the fibre of the map $X \to L_n^f$ is a direct limit of finite $K(n)_*$-acyclic spectra. Thus we get

Corollary 4.13 *A connective spectrum X for which $L_n^f X$ is contractible is strongly $K(n)_*$-acyclic.*

Proof Since $L_n^f X$ is contractible, the fibre of the map $X \to L_n^f$ is equivalent to X. Since X is a connective spectrum which is now the direct limit of finite $K(n)_*$-acyclic spectra, X is strongly $K(n)_*$-acyclic by Corollary 4.12. □

There is a natural transformation $\lambda_n : L_n^f \to L_n$. The telescope conjecture, which is known to be false for $n = 2$, see [**27**] and [**31**], is equivalent to the assertion that λ_n is an equivalence. It is shown that for $n = 2$ there is a spectrum X for which $L_n X$ is contractible but $L_n^f X$ is not. It is necessarily a torsion spectrum (its rational homology must vanish) and its connective cover has the same property. However we do not know if such a spectrum is strongly $K(n)_*$-acyclic or not.

We also know of no counterexample to the converse of Corollary 4.13, so perhaps that is an equivalence. Alternatively, one can ask if a connective spectrum X is strongly $K(n)_*$-acyclic if and only if $E(n)_*(X) = 0$. In [**29**, Theorem 2.1(d)] it was shown that $E(n)_*(X) = 0$ if and only if $K(i)_*(X) = 0$ for $0 \leq i \leq n$. It is also known, [**19**], that $E(n)_*(X) = 0$ if and only if $v_n^{-1}BP_*(X) = 0$.

The following is a consequence of Corollary 4.13.

Theorem 4.14 *If X is a connective spectrum with $E(n)_*(X) = 0$, then $BP \wedge X$ is strongly $K(n)_*$-acyclic. In particular if X is also a BP-module spectrum then it is strongly $K(n)_*$-acyclic. The same holds with BP replaced by any connective spectrum E with Bousfield class dominated by that of BP.*

Proof By the smash product theorem [**30**, 7.5.6], $E(n)_*(X) = 0$ if and only if $X \wedge L_n S^0$ is contractible. We also know [**32**, Theorem 2.7(iii)] that $L_n S^0$ and $L_n^f S^0$ are BP_*-equivalent and therefore E_*-equivalent. Thus we have

$$\text{pt.} \simeq E \wedge X \wedge L_n S^0 \simeq E \wedge X \wedge L_n^f S^0 \simeq L_n^f(E \wedge X)$$

(the last equivalence is Theorem 2.7(ii) of [**32**]) and $E \wedge X$ is strongly $K(n)_*$-acyclic by Corollary 4.13. If E is a ring spectrum and X is an E-module spectrum, then it is a retract of $E \wedge X$,

$$X \simeq S^0 \wedge X \longrightarrow E \wedge X \longrightarrow X$$

so it is strongly $K(n)_*$-acyclic by Theorem 4.10(ii). □

Corollary 4.15 *If X is a BP-module spectrum in which each element of $\pi_*(X)$ is annihilated by some power of the ideal*

$$I_{n+1} = (p, v_1, v_2, \cdots v_n) \subset BP_*,$$

then X is strongly $K(n)_*$-acyclic.

Proof The hypothesis implies that $v_n^{-1}\pi_*(X) = 0$, so $v_n^{-1}BP_*(X) = 0$, which is equivalent to $E(n)_*(X) = 0$ as noted above. □

Examples of spectra satisfying these hypotheses include the $P(k)$ of [**17**] (with $\pi_*(P(k)) = BP_*/I_k$) for $k > n$ and spectra obtained from $BP\langle k\rangle$ by killing an ideal containing some power of I_{n+1}.

5 Morava homology Hopf algebras

We want to study a category of Hopf algebras which includes the objects of our interest: the Morava K-theory (homology) of homotopy commutative, connected H-spaces. We need this study to solve Hopf algebra extension problems in the bar spectral sequence during the inductive step of the proof of our main theorem. To do this we have a general Hopf algebra splitting theorem which is of interest in its own right. We want to give particular thanks to Hal Sadofsky for help with this section.

Although our Hopf algebras will be bicommutative, and so give an abelian category, we can run into serious problems because of the cyclical grading we use. In particular, we can have an element with the property $x = x^p$. (In $\overline{K(n)_*}(K(\mathbf{Z}/(p), n))$ for example.) This wreaks havoc with all of the algebra portion of Milnor–Moore [**25**]. Such an algebra has no generator (i.e., indecomposable), and the first proposition of Milnor–Moore is false for our situation. The coalgebra portion of Milnor–Moore fares much better. Before we move on to Hopf algebras we want to indicate why our coalgebras are still nice by making some definitions and reproving a basic result which still holds in our setting.

Let A be a cocommutative, coassociative, coaugmented coalgebra with counit over a ring, R. At present we are not concerned with gradings so this could be ungraded, graded or cyclically graded. *To avoid unnecessary complications, we assume that A is flat over R.* Let $J(A)$ be the cokernel of the coaugmentation map:

$$0 \to R \to A \to J(A) \to 0.$$

Using the iterated coproduct we define an increasing filtration, F_qA, by

$$F_qA = \ker(A \to J(A)^{\otimes q+1})$$

for all $q \geq 0$. Note that $F_0A \simeq R$ and $F_1A/F_0A \simeq P(A)$, the primitives of A. We call this the *primitive* filtration of A. Dualizing Milnor–Moore, [**25**, page 252], we could call this the coaugmentation filtration. (Milnor and Moore reserve the name *primitive filtration* for their filtration on primitively generated Hopf algebras.)

Following [**3**] we say a filtration, F_*A, is *exhaustive* (or exhausts A) if every element of A is in some F_qA. We say A is a *(connected) homology coalgebra* if A is a cocommutative, coassociative, coaugmented coalgebra over R with counit, its primitive filtration exhausts A, and it is the direct limit of finite dimensional R sub-coalgebras. The exhaustive condition on A replaces connectivity in the graded case quite nicely. To justify the name we have the following observation:

Lemma 5.1 *Let $E_*(-)$ be a multiplicative homology theory and X a connected CW complex of finite type such that $E_*(X)$ is flat over E_*, then $E_*(X)$ is a homology coalgebra over E_*.*

Proof Because $E_*(X)$ is flat over E_*, the Künneth isomorphism holds for finite products of X. The diagonal, $X \to X \times X$, induces a cocommutative, coassociative coalgebra over E_*. The map of a point into X (because X is connected) and the map of X to a point give the coaugmentation and counit respectively. Again, we need connectivity to show the primitive filtration is exhaustive. An element of $E_*(X)$ which lives on the 0-cell maps trivially to $J(E_*(X))$. Since $E_*(X)$ is the direct limit of $E_*(X^q)$ where X^q is the q-skeleton, any element x in $E_*(X)$ comes from $E_*(X^q)$ for some q (this shows the coalgebra is the direct limit of finite dimensional E_* sub-coalgebras). Getting a cellular approximation to the iterated diagonal map,

$X \to \prod^{q+1} X$ we see that on some coordinate X^q maps to the 0-cell and thus our element must be in $F_q E_*(X)$ and the primitive filtration is exhaustive. □

We can also prove a standard result which, as we see, does not depend on a grading, but just on having an exhaustive primitive filtration. This is well known and is even somewhere in the algebraic topology literature but we cannot remember where we have seen it. In the *other* literature, it could well be that Lemma 11.0.1 on page 217 of [**39**] could prove it; but it is a lot easier to reprove it than to be sure of that.

Lemma 5.2 *Let $A \to B$ be a map of coalgebras where A is a homology coalgebra, then the map injects if and only if the induced map on primitives injects.*

Proof Since $P(A) \subset A$, an injection on A is automatically an injection on $P(A)$. In the other direction our proof is by induction on the degree of the primitive filtration of an element. We are given that $P(A) \simeq F_1 A$ injects to ground our induction. If we have an element, $x \in F_q A$ but not in $F_{q-1}A$, then the coproduct takes x to $\sum x' \otimes x''$ in $J(A) \otimes J(A)$ where each nonzero x' and x'' must be in a lower filtration and so they inject to $J(B) \otimes J(B)$. □

We say that A is a *(connected) homology Hopf algebra* if it is an associative Hopf algebra with unit such that the coalgebra structure is that of a homology coalgebra. Clearly, if our X above is an H-space, then $E_*(X)$ is a homology Hopf algebra. If the algebra structure is commutative, we say A is a *commutative homology Hopf algebra* . If X above is a homotopy commutative H-space (e.g., any double loop space) then $E_*(X)$ is a commutative homology Hopf algebra.

Since the exhaustive condition replaces connectivity so nicely, we can define a Hopf algebra conjugation on homology Hopf algebras. This is done inductively on the primitive filtration. Let $x \to x \otimes 1 + 1 \otimes x + \sum x' \otimes x''$, then $C(x) = -x - x'C(x'')$ inductively. Following [**25**, Proposition 8.8, page 260], we have $C \circ C = I_A$ because our coalgebra is cocommutative. The existence of C is essential to show that commutative homology Hopf algebras are an abelian category, see [**12**]. We are in a slightly generalized situation over the usual connected bicommutative Hopf algebras of finite type so it is worth a short discussion of our case. If A is a coalgebra and B is an algebra, then two maps from A to B can be combined to get a third, still following [**25**, Section 8], by

$$A \to A \otimes A \xrightarrow{f \otimes g} B \otimes B \to B.$$

To get an abelian category we need $\mathrm{Hom}(A, B)$ to be an abelian group. In particular, the above map must be a Hopf algebra map. For $A \to A \otimes A$ to be a Hopf algebra map, A must be cocommutative. For $B \otimes B \to B$ to be a Hopf algebra map, B must be commutative. So, we need the bicommutativity of commutative homology Hopf algebras to get our composition back in our category. The Hopf algebra conjugation above gives us our inverse and bicommutativity shows $\mathrm{Hom}(A, B)$ is an abelian group. Following [**25**, Sections 3 and 4], bicommutativity allows us to define, for all $f : A \to B$, a kernel

$$A \backslash\backslash f \simeq R \square_B A \simeq A \square_B R$$

and a cokernel

$$B//f \simeq R \otimes_A B \simeq B \otimes_A R.$$

Tensor product is the product in our category and given maps $f : A \to B$ and $g : A \to C$, we get $(f,g) : A \to A \otimes A \xrightarrow{f\otimes g} B \otimes C$. We have, see [**5**, Appendix, especially Theorem 10.12]:

Theorem 5.3 *The category of commutative homology Hopf algebras is abelian.*

In the graded connected case over a perfect field of characteristic p, commutative homology Hopf algebras of finite type have been classified nicely in [**36**]. This applies to the case where E is standard mod p homology theory. Unaware of Schoeller's work, the second author had also embarked on such a project. His project was aborted when he learned about Schoeller's paper and the only physical remains of this project are in [**28**]. His approach was quite different from hers and quite easily extends to the case at hand, so we are now reaping the benefits of his study in this paper.

What concerns us here are what we will call *commutative Morava homology Hopf algebras.* These are just commutative homology Hopf algebras over $\mathbf{F}_p$ which are cyclically graded over $\mathbf{Z}/(2(p^n - 1))$ for some n. These arise naturally when you take the Morava K-theory, $K(n)_*(X)$, where X is a connected homotopy commutative H-space, and set v_n equal to one as we do throughout the rest of this paper. We do only what we need with these Hopf algebras here, but we hope to return to the problem of classifying them in a future paper. Because of the cyclic grading there is a much richer, more interesting collection of Hopf algebras than in the standard case. The Morava K-theory of Eilenberg–Mac Lane spaces in [**33**] supplies lots of examples unlike anything seen in the normal graded case.

Remark 5.4 *At this stage we must point out a problem and its solution for the prime $p = 2$. $K(n)$ is not a homotopy commutative ring spectrum for $p = 2$, so $K(n)_*(X)$ is not necessarily in our category. However, if $K(n)_*(X)$ is even degree it is. For a discussion of this problem see the Appendix of* [**18**] *where, following* [**41**], [**33**] *is shown to hold for $p = 2$.*

Because of the cyclic grading, it is not unusual to find ourselves dealing with infinite dimensional vector spaces; for example, a polynomial algebra with one primitive generator already gets us into that situation. However, we can use finiteness when we need it because our coalgebras are the direct limit of finite dimensional sub-coalgebras. Many Hopf algebra categories are self dual, an extremely nice property. Unfortunately, ours is not self dual. We can define the dual category, however.

Proposition 5.5 *Let A be a commutative homology Hopf algebra over a field R. It is the direct limit of its finite dimensional subcoalgebras A_α. Let its dual be defined by*

$$A^* = \lim_{\leftarrow} \mathrm{Hom}_R(A_\alpha, R).$$

Then A^* is a compact topological bicommutative Hopf algebra under the inverse limit topology. Moreover A is the continuous linear dual of A^*.

Proof The product and coproduct in A^* are induced respectively by the coproduct and product in A. A^* is compact because it is the inverse limit of finite dimensional vector spaces. The continuous linear dual of an inverse limit is the direct limit of the continuous linear dual, and

$$\mathrm{Hom}_R(\mathrm{Hom}_R(A_\alpha, R), R) = A_\alpha$$

so the continuous linear dual of A^* is A. □

This result enables us to make the following definition.

Definition 5.6 *Let A be a commutative homology Hopf algebra over $\mathbf{F}_p$. The* **Frobenius map** *$F : A \to A$ is the Hopf algebra homomorphism that sends each element to it p^{th} power. The* **Verschiebung map** *$V : A \to A$ is the dual of the Frobenius map on A^**

We are ready to look closely at the category of commutative Morava homology Hopf algebras and say what we need to say about it. We propose to split up every Hopf algebra into canonical components. Because it is all we need for this paper, we restrict our attention to evenly graded Hopf algebras. Thus our Morava Hopf algebras are really graded over $G = \mathbf{Z}/(p^n - 1)$. Denote by $EC(n)$ the category of evenly graded commutative Morava homology Hopf algebras (for a given n). Let H be the cyclic group of order n. H acts on G via the pth power map. Writing G additively, the map $H \times G \to G$ is given by $(i, j) \to p^i j$. Let γ denote an H-orbit:

$$\gamma = \{j, pj, p^2 j, \dots\} \subset G.$$

Recall Theorem 1.11 from the introduction, which uses these definitions.

Remark 5.7 *We illustrate the result with the case $p = 2$ and $n = 3$. Then H and G have orders 3 and 7 respectively and there are three orbits, namely*

$$\begin{aligned} \gamma_1 &= \{1, 2, 4\}, \\ \gamma_2 &= \{3, 6, 5\} \qquad \textit{and} \\ \gamma_3 &= \{0\}. \end{aligned}$$

We also know, from [**33**], *that for $i = 1, 2$ or 3, $K(3)_*(K(T, i))$ has its primitives in dimensions in $2\gamma_i$ for any torsion abelian group T.*

We discuss this more later, but in general, $K(n)_*(K(T, i))$ can have factors with generators in more than one orbit. For example when $p = 2$, $n = 4$ and $i = 2$ there are two such orbits, namely $\{3, 6, 9, 12\}$ and $\{5, 10\}$. However each orbit is associated with a unique value of i, namely $i = \overline{\alpha}_p(j)$, the sum of the digits of the p-adic expansion of any $j \in \gamma$.

Theorem 1.12 follows from Theorem 1.13 by:

Proof of Theorem 1.12 All we need to prove here, after we see how our functors are constructed, is that there are no nontrivial maps $A_\gamma \to B_\beta$ if $\gamma \neq \beta$. Such a map, f, must commute with e_β which is the identity on B_β but is the trivial map on A_γ and so it must be the trivial map. □

Proof of Theorem 1.13 Our concern for the rest of this section is to construct the idempotents e_γ and prove Theorem 1.13. A Hopf algebra, A, in $EC(n)$ is equipped with the usual Frobenius and Verschiebung endomorphisms F and V (see 5.6). The relation, $VF = FV = p$, is a simple calculation. Do not confuse this p with multiplication by p. This is p times the identity in the abelian group of endomorphisms of A. Thus the ring of endomorphisms of A which ignores the grading is a module over $\mathbf{Z}_{(p)}$. Because the primitive filtration is exhaustive every element of A is annihilated by some power of V and hence by some power of p. This means that the endomorphism ring is also a module over the p-adic integers $\mathbf{Z}_p$.

Given an element $a \in \mathbf{F}_p$ we can define an endomorphism $[a]$ of A to be multiplication by a^i in dimension $2i$. If we replace A by $A \otimes \mathbf{F}_{p^n}$, we have an endomorphism $[a]$ for each $a \in \mathbf{F}_{p^n}$. Hence the endomorphism ring of $A \otimes \mathbf{F}_{p^n}$ is a module over

$\mathbf{Z}_p[G]$, the p-adic group ring over $G = \mathbf{Z}/(p^n - 1)$, which is isomorphic to the multiplicative group of the field $\mathbf{F}_{p^n}$.

Remark 5.8 *Although our applications in this paper only require us to study evenly graded Hopf algebras, our study of these Hopf algebras would be incomplete without splitting the category into evenly graded Hopf algebras and exterior Hopf algebras on odd degree generators for odd primes. Classical sign arguments force a Hopf algebra with odd degree primitives to be an exterior algebra. The idempotents necessary for this splitting are just* $([1]\pm[-1])/2$. *To check this we need only evaluate on primitives.* $([1]-[-1])/2$ *is the identity on odd primitives and trivial on even primitives. The reverse is true for* $([1]+[-1])/2$.

We want to use $\mathbf{Z}_p[G]$ to construct idempotents of A. We need some basic facts to do this. Let M be an R module and let H act on R and M. Then we have a map $j : R \to \mathrm{End}(M)$ and H acts on $f \in \mathrm{End}(M)$ by $(hf)(x) = h(f(h^{-1}(x)))$. Then

Lemma 5.9 *If the map j is H-equivariant then the fixed points, R^H act on the fixed points M^H.*

Proof If $m \in M^H$ and $r \in R^H$ then

$$(jr)(m) = (jh(r))(m) = [h(jr)](m) = h(jr(h^{-1}m)) = h(jr(m)).$$

□

We have our group $H = \mathbf{Z}/(n)$. H is isomorphic to the Galois group of $\mathbf{F}_{p^n}$ over $\mathbf{F}_p$, and its action on the ring $\mathbf{Z}_p[G]$ corresponds to the action of the Galois group on the units of $\mathbf{F}_{p^n}$. We will write G multiplicatively from now on. We want to show that if an element $x \in \mathbf{Z}_p[G]$ is fixed by this action then the corresponding endomorphism leaves $A = (A \otimes \mathbf{F}_{p^n})^H \subset A \otimes \mathbf{F}_{p^n}$ invariant. Hence the endomorphism ring of A is a module over the fixed point subring $\mathbf{Z}_p[G]^H$. By our lemma, we must just show that the map $\mathbf{Z}_p[G] \to \mathrm{End}(A \otimes \mathbf{F}_{p^n})$ is H-equivariant. It is enough to show this on $[g]$. Let $a \in A$ and $z \in \mathbf{F}_{p^n}$. Then:

$$\begin{aligned} j(h([g]))(a \otimes z) &= j([g^{p^h}])(a \otimes z) \\ &= a \otimes g^{p^h|a|}z \\ &= h(a \otimes g^{|a|}z^{1/p^h}) \\ &= h(j[g])(a \otimes z^{1/p^h}) \\ &= h(j[g])h^{-1}(a \otimes z). \end{aligned}$$

We construct some idempotents in an even bigger ring, $W(\mathbf{F}_{p^n})[G]$, where $W(\mathbf{F}_{p^n})$ denotes the Witt ring of $\mathbf{F}_{p^n}$, i.e., the extension of $\mathbf{Z}_p$ obtained by adjoining a primitive $(p^n-1)^{\text{th}}$ root of unity ω. See [**13**, page 132, Remark 17.4.18]. We will use the notation ω for both the root of unity in $W(\mathbf{F}_{p^n})$ and its mod p reduction in $\mathbf{F}_{p^n}$. For each $j \in G$ let

$$e_j = \sum_{1 \le i \le p^n-1} \frac{\omega^{-ij}[\omega^i]}{p^n - 1}. \tag{5.10}$$

Note that e_j is independent of the choice of the primitive root ω. One sees easily that the e_j are orthogonal idempotents, i.e.,

$$\begin{aligned} e_j e_{j'} &= \begin{cases} e_j & \text{if } j = j' \\ 0 & \text{otherwise} \end{cases} \\ \text{and} \quad \sum_j e_j &= 1. \end{aligned} \tag{5.11}$$

Now let

$$c_m = \sum_k \omega^{mp^k} \in \mathbf{Z}_p \subset W(\mathbf{F}_{p^n}) \tag{5.12}$$

where the sum is over all distinct roots of the indicated form. Thus the number of summands is some divisor of n depending on m. This element is in $\mathbf{Z}_p$ because it is invariant under the action of the Galois group H, whose generator sends each root of unity to its pth power.

Now let γ be the H-orbit containing j and consider the element

$$\begin{aligned} e_\gamma &= \sum_k e_{jp^k} \\ &= \sum_{i,k} \frac{\omega^{-ijp^k}[\omega^i]}{p^n - 1} \\ &= \sum_i \frac{c_{-ij}[\omega^i]}{p^n - 1}, \end{aligned}$$

where the first sum is as in (5.12). This element is idempotent by (5.11). It is in $\mathbf{Z}_p[G] \subset W(\mathbf{F}_{p^n})[G]$ since the coefficients c_{-ij} all lie in $\mathbf{Z}_p$.

Furthermore e_γ is fixed by the Galois action since $c_{-ijp^k} = c_{-ij}$, so it lies in the invariant subring $\mathbf{Z}_p[G]^H$. This is the idempotent giving the splitting of Theorem 1.11.

In order to verify this, it suffices to show that it behaves appropriately on primitive elements. Let x be a primitive in dimension m. Recall that the endomorphism ring of a Hopf algebra acts additively on its primitives, i.e., for endomorphisms α and β and coefficients $a, b \in \mathbf{Z}_p$, we have

$$(a\alpha + b\beta)(x) = a\alpha(x) + b\beta(x)$$

where the coefficients on the right are reduced mod p. With this in mind we have

$$\begin{aligned} e_\gamma(x) &= \sum_i \frac{c_{-ij}[\omega^i](x)}{p^n - 1} \\ &= \sum_i \frac{c_{-ij}\omega^{im} x}{p^n - 1} \\ &= \sum_{i,k} \frac{\omega^{im-ijp^k} x}{p^n - 1} \\ &= \sum_k \left(\sum_i \frac{\omega^{i(m-jp^k)}}{p^n - 1} \right) x. \end{aligned}$$

The inner sum is 1 when $m = jp^k$ and 0 otherwise, so we have

$$e_\gamma(x) = \begin{cases} x & \text{if } m \in \gamma \\ 0 & \text{otherwise} \end{cases}$$

as desired. □

In the introduction we defined a function $\overline{\alpha}_p : \mathbf{Z}/(p^n - 1) \to \mathbf{Z}$ by taking the sum of the coefficients of the p-adic expansion. We observed that $\overline{\alpha}_p$ is constant on orbits γ. From Theorem 1.13, we have:

Corollary 5.13 *Let* $EC(n)_q = \prod_{\overline{\alpha}_p(\gamma)=q} EC(n)_\gamma$ *for all* $0 \leq q \leq n(p-1)$. *Then:*

$$EC(n) = \prod_{0 \leq q \leq n(p-1)} EC(n)_q.$$

This is now what we need for our Hopf algebras. We actually need very little knowledge about $K(n)_*(K(\mathbf{Z}/(p^i), q))$. We remind the reader that

$$\overline{K(n)}_*(X) \equiv K(n)_*(X) \otimes_{K(n)_*} \mathbf{F}_p$$

where the $K(n)_*$ module structure of $\mathbf{F}_p$ is given by sending v_n to 1. The following is a restatement of Theorem 1.14.

Theorem 5.14 ([**33**]) *$\overline{K(n)}_*(K(\mathbf{Z}/(p^i), q))$ is in $EC(n)_q$ for $0 < q \leq n$, and is $\overline{K(n)}_* \simeq \mathbf{F}_p$ for $q > n$. $\overline{K(n)}_*(K(\mathbf{Z}_{(p)}, q+1))$ is the direct limit of $\overline{K(n)}_*(K(\mathbf{Z}/(p^i), q))$ and is also in $EC(n)_q$.*

Proof The theorem is read off from [**33**, Theorem 11.1] where we see the primitives are all in degrees $2(p^{i_1} + p^{i_2} + \ldots + p^{i_q})$, $0 = i_1 < i_2 < \cdots < i_q < n$, for $\mathbf{Z}/(p^i)$ and [**33**, Corollary 12.2] for $\mathbf{Z}_{(p)}$. In fact, all primitives come from $\overline{K(n)}_*(K(\mathbf{Z}/(p), q))$. □

References

[1] Baas, N. A. [1973], *On bordism theory of manifolds with singularities*, Math. Scand. **33**, 279–302.

[2] Baas, N. A. and Madsen, I. [1971], *On the realization of certain modules over the Steenrod algebra*, Math. Scand. **31**, 220–224.

[3] Boardman, J. M. [1981], *Conditionally convergent spectral sequences*, Johns Hopkins University, preprint.

[4] Bousfield, A. K. [1994],. *Localization and periodicity in unstable homotopy theory*, Journal of the American Mathematical Society **7**, 831–873.

[5] Bousfield, A. K. [1996], *On λ-rings and the K- theory of infinite loop spaces*, K-Theory **10**, 1–30.

[6] Bousfield, A. K. [1996], *Unstable localization and periodicity*, in New trends in localization and periodicity, vol. 136 of Progress in Mathematics, pp. 33–50. Birkhauser Verlag, Basel.

[7] Boardman, J. M. and Wilson, W. S. *$k(n)$-torsion-free H-spaces and $P(n)$-cohomology*, in preparation.

[8] Cohen, F. R., Lada, T. and May, J. P. [1976], *The Homology of Iterated Loop Spaces*, vol. 533 of Lecture Notes in Mathematics, Springer-Verlag, New York.

[9] Devinatz, E., Hopkins, M. J. and Smith, J. H. [1988], *Nilpotence and stable homotopy theory*, Annals of Mathematics **128**, 207–242.

[10] Dold, A. [1962], *Relation between ordinary and extraordinary homology*, in Colloquium on Algebraic Topology, Aarhus 1962, pp. 2–9.

[11] Dyer, E. [1969], *Cohomology Theories*. Benjamin, New York.

[12] Gugenheim, V. K. A. M. [1966], *Cohomology theory in the category of Hopf algebras*, Colloque de Topologie, Centre Belge de Recherches Mathématiques, pp. 137–148.

[13] Hazewinkel, M. [1978], *Formal Groups and Applications*. Academic Press, New York.

[14] Hopkins, M. J. [1987], *Global methods in homotopy theory*, in Homotopy Theory: proceedings of the Durham Symposium 1985 (J. D. S. Jones and E. Rees, eds.), vol. 117 of London Mathematical Society Lecture Note Series, pp. 73–96, Cambridge University Press, Cambridge.

[15] Hopkins, M. J. and Smith, J. H., *Nilpotence and stable homotopy theory II*, Annals of Mathematics, to appear.

[16] Johnson, D. C. and Wilson, W. S. [1973], *Projective dimension and Brown-Peterson homology*, Topology **12**, 327–353.

[17] Johnson, D. C. and Wilson, W. S. [1975], *BP-operations and Morava's extraordinary K-theories*, Mathematische Zeitschrift **144**, 55–75.

[18] Johnson, D. C. and Wilson, W. S. [1985], *The Brown-Peterson homology of elementary p-groups*, American Journal of Mathematics **107**, 427–454.

[19] Johnson, D. C. and Yosimura, Z. [1980], *Torsion in Brown-Peterson homology and Hurewicz homomorphisms*, Osaka Journal of Mathematics **17**, 117–136.

[20] Kramer, R. [1990], *The periodic Hopf ring of connective Morava K-theory*, PhD thesis, Johns Hopkins University.

[21] May, J. P. [1972], *The geometry of iterated loop spaces*, vol. 271 of Lecture Notes in Mathematics, Springer-Verlag, Berlin.

[22] May, J. P. [1975], *Classifying spaces and fibrations*, vol. 155 of Memoirs of the American Mathematical Society, American Mathematical Society, Providence, R. I.

[23] Meyer, J.-P. [1984], *Bar and cobar constructions, I*, Journal of Pure and Applied Algebra **33**, 163–207.

[24] Meyer, J.-P. [1986], *Bar and cobar constructions, II*, Journal of Pure and Applied Algebra **43**, 179–210.

[25] Milnor, J. W. and Moore, J. C. [1965], *On the structure of Hopf algebras*, Annals of Mathematics **81(2)**, 211–264.

[26] McCleary, J. and McLaughlin, D. Z. [1992], *Morava K-theory and the free loop space*, Proceedings of the American Mathematical Society **114**, 243–250.

[27] Ravenel, D. C., *A counterexample to the telescope conjecture*, to appear.

[28] Ravenel, D. C. [1975], *Dieudonné modules for Abelian Hopf algebras*, in Reunion Sobre Teoria de homotopia, (D. Davis, ed.) Universidad de Northwestern, Agosto 1974, number 1 in Serie notas de matemática y simposia, pp. 177–194, Mexico, D.F., 1975. Sociedad Matematica Mexicana.

[29] Ravenel, D. C. [1984], *Localization with respect to certain periodic homology theories*, American Journal of Mathematics **106**, 351–414.

[30] Ravenel, D. C. [1992], *Nilpotence and periodicity in stable homotopy theory*, vol. 128 of Annals of Mathematics Studies, Princeton University Press, Princeton.

[31] Ravenel, D. C. [1992], *Progress report on the telescope conjecture*, in Adams Memorial Symposium on Algebraic Topology Volume 2 (N. Ray and G. Walker, eds.), vol. 176 of London Mathematical Society Lecture Note Series, pp. 1–21, Cambridge University Press, Cambridge.

[32] Ravenel, D. C. [1993], *Life after the telescope conjecture*, in Algebraic K-Theory and Algebraic Topology (P. G. Goerss and J. F. Jardine, eds.), vol. 407 of Series C: Mathematical and Physical Sciences, pp. 205–222. Kluwer Academic Publishers.

[33] Ravenel, D. C. and Wilson, W. S. [1980], *The Morava K-theories of Eilenberg-MacLane spaces and the Conner-Floyd conjecture*, American Journal of Mathematics **102**, 691–748.

[34] Ravenel, D. C. and Wilson, W. S. [1996], *The Hopf ring for $P(n)$*, Canadian Journal of Mathematics **48(5)**, 1044–1063.

[35] Ravenel, D. C., Wilson, W. S. and Yagita, N., *Brown-Peterson cohomology from Morava K-theory*, K-Theory, to appear.

[36] Schoeller, C. [1970], *Etude de la categorie des algebres de Hopf commutatives connexes sur un corps*, Manuscripta Math. **3**, 133–155.

[37] Smith, L. [1970], *Lectures on the Eilenberg-Moore spectral sequence*, vol. 134 of Lecture Notes in Mathematics, Springer-Verlag.

[38] Snaith, V. [1974], *Stable decomposition of $\Omega^n\Sigma^n X$*, Journal of the London Mathematical Society **7**, 577–583.

[39] Sweedler, M. E. [1969], *Hopf algebras*. Benjamin, New York.

[40] Wilson, W. S. [1975], *The Ω-spectrum for Brown-Peterson cohomology, Part II*, American Journal of Mathematics **97**, 101–123.

[41] Würgler, U. [1977], *On products in a family of cohomology theories associated to the invariant prime ideals of* $\pi_*(BP)$, Commentarii Mathematici Helvetici **52**, 457–481.

[42] Würgler, U. [1991], *Morava K-theories: A survey*, in Algebraic topology (S. Jackowski, B. Oliver and K. Pawalowski, eds.), Poznan 1989: proceedings of a conference held in Poznan, Poland, June 22-27, 1989, vol. 1474 of Lecture Notes in Mathematics, pp. 111–138, Springer-Verlag, Berlin.

Fields Institute Communications
Volume **19**, 1998

The Transfer Between Rings of Modular Invariants of Subgroups of $GL(n,p)$

Nondas E. Kechagias
Department of Mathematics
University of Ioannina
Ioannina, 45110 Greece
nkechag@cc.uoi.gr

Abstract. An algorithm for composing Dickson invariants is discussed along with a free basis of the symmetric algebra over the Dickson algebra and its mod-p analogue. Using a theorem of Campbell and Hughes, the transfer between rings of invariants is discussed and its relation with the mod-p cohomology of subgroups of the symmetric group.

1 Introduction

Let $V^n = \langle e_1, \cdots, e_n \rangle$ be an n-dimensional vector space over the field F_p and $\sum_{p^n}$ the symmetric group on the set V^n which is also an elementary abelian subgroup: $V^n \hookrightarrow \sum_{p^n}$. The following p-Sylow subgroup $\overbrace{(\mathbf{Z/pZ}) \int \cdots \int (\mathbf{Z/pZ})}^{n}$ is denoted by $\sum_{p^n,p}$.

Let $U_n \leq B_n \leq GL_n(F_p)$ be the subgroups of the upper triangular matrices with one and nonzeros along the main diagonal respectively. Let $B_n \leq P_n(N) \leq GL_n(F_p)$ be the parabolic subgroup associated with the sequence $N = (n_1, \cdots n_k)$ where $\sum\limits_{l} n_l = n$.

The groups involved in the first and second paragraphs are related to each other by $W_{\sum_{p^n,p}}(V^n) = U_n$ and $W_{\sum_{p^n}}(V^n) = GL_n(F_p)$. Here $W_G(L)$ denotes the Weyl subgroup of L in G, [**11**]. We recall that $W_G(L)$ acts on $H^*(L)$ and a map $i^* : H^*(G) \to H^*(L)^{W_G(L)}$ induces a map $i^* : H^*(L)^{W_G(L)} \to H^*(L)^{W_K(L)}$. Here $L \leq K \leq G$. Similarly, the transfer $tr^* : H^*(K) \to H^*(G)$ induces

$$\tau^* : H^*(L)^{W_K(L)} \to H^*(L)^{W_G(L)} \tag{1.1}$$

(see [**11**]).

1991 *Mathematics Subject Classification.* Primary 13F20, 13A50, 55S10; Secondary 20J05, 18G10.

Let S_n be the graded symmetric algebra of V^n; $S_n = F_p[y_1, \cdots, y_n]$ and degree $|y_i| = 2$ (if $p = 2$, then $|y_i| = 1$). Also let $E_n = F_p(x_1, \cdots, x_n)$ denote the graded exterior algebra for $p > 2$, $|x_i| = 1$. The tensor product $E_n \otimes S_n$ represents the mod$p$ cohomology of n copies of the classifying space $B(\mathbf{Z}/\mathbf{pZ})$.

The following theorems are well known:

Theorem 1.1 [**6**] *$S_n^{GL_n} := D_n = F_p[d_{n,0}, \cdots, d_{n,n-1}]$ is a polynomial algebra where the degrees are $|d_{n,i}| = p^n - p^i$.*

Theorem 1.2 [**13**] *i) $S_n^{U_n} := H_n = F_p[h_1, \cdots, h_n]$ is a polynomial algebra where the degrees are $|h_i| = p^{i-1}$.*

ii) $S_n^{B_n} := F_p\left[h_1^{(p-1)}, \cdots, h_n^{(p-1)}\right]$ is a polynomial algebra.

Theorem 1.3 (Kuhn and Mitchell)

$$\begin{aligned} S_n^{P_n(N)} &:= PD_n(N) \\ &= F_p\left[d_{\nu_j,i} - \left(d_{\nu_{j-1},i-n_j}\right)^{p^{n_j}} \mid 1 \leq j \leq k,\ \nu_{j-1} \leq i \leq \nu_j - 1\right] \end{aligned}$$

is a polynomial algebra.

The generators above are related by the following relations:

Let $f_{k-1}(x) = \prod_{u \in V^{k-1}} (x - u)$, then $f_{k-1}(x) = \sum_{i=0}^{k-1} (-1)^{n-i} x^{p^i} d_{k-1,i}$ and $h_k = \prod_{u \in V^{k-1}} (y_k - u)$. Moreover,

$$d_{n,n-i} = \sum_{1 \leq j_1 < \cdots < j_i \leq n} \prod_{s=1}^{i} \left(h_{j_s}^{p-1}\right)^{p^{n-i+s+j_s}}. \tag{1.2}$$

[**9**].

Note If $p = 2$ then Arnon [**2**] provides a simple description of the generators above in terms of y_i's.

Proposition 1.4 [**2**] $d_{n,i} = \sum_{\substack{s_1 + \cdots s_n = 2^n - 2^i \\ s_j = 0 \text{ or } 2^{r_j}}} y_1^{s_1} \cdots y_n^{s_n}$.

It is known that the rank of $S_n^{U_n}$ as a free module over $S_n^{GL_n}$ is $[GL_n : U_n]$. Campbell and Hughes in [**3**] provide a set of free basis generators. We recover the same result using an elementary method. Namely, decomposing (1.2) successively, we are able to find the top degree for each generator and express it with respect to Dickson and upper triangular generators of lower degree, Theorems 3.6 and 3.7. Our method provides an algorithm for the decomposition above. We shall note that our method works also for the case including the exterior part of the invariants. This approach is technical and provides information about the relation between U_n and GL_n-invariants.

The decomposition above is of interest mainly because it is related to the transfer between the mod-p homology of $\sum_{p^n,p}$ and $\sum_{p^n}$.

Since the transfer has important geometric applications (see [**8**]), we have been led to investigate the mod-p analogue of the map (1.1). It is known that for $p = 2$, the mod-p homology of $\sum_{p^n}$ can be expressed in terms of coinvariants of certain Dickson algebras. For the $p > 2$ case, subrings of certain extended Dickson algebras are needed.

The transfer in $mod - p$ (co-)homology has long been investigated by various people. Kahn and Priddy provided formulas in [**7**] and Kuhn has studied the $mod-p$ invariant analogue in cohomology in [**11**]. Using Kuhn's theorem, we evaluate the transfer on the base of $(E_n \otimes S_n)^{U_n}$ over $(E_n \otimes S_n)^{GL_n}$, Theorem 35, and discuss the case needed for the direct summand of the mod $- p$ cohomology of $\sum_{p^n,p}$ and $\sum_{p^n}$, namely the summand provided by the Steenrod decomposition.

2 Composing and decomposing Dickson invariants

We extend Arnon-Peterson's formulae above for the upper triangular invariants.

We reserve the symbol $\langle e_i, i \in I\rangle$ for the vector space generated by the e_i's over F_p.

Lemma 2.1 [4] *i)* $h_k = \sum_{\substack{s_1+\cdots s_{k-1}=2^{k-1}-2^{r_k} \\ s_j=0 \ or \ 2^{r_j}}} y_1^{s_1}\cdots y_{k-1}^{s_{k-1}} y_k^{2^{r_k}}.$

ii) Let $k < m$, then

$$h_m h_k = \sum_{\substack{s_1(k)+\cdots+s_{k-1}(k)=2^{k-1}-2^{r_k(k)} \\ t_1(m)+\cdots+t_{m-1}(m)=2^{m-1}-2^{r_m(m)} \\ s_j(i)=0 \ or \ 2^{r_j}}} y_1^{s_1+t_1}\cdots y_{k-1}^{s_{k-1}+t_{k-1}} y_k^{2^{r_k}(k)+t_k} y_{k+1}^{t_{k+1}}\cdots y_m^{2^{r_m(m)}}.$$

iii) Let $m = 2^{k_1} + \cdots + 2^{k_r}$, then

$$h_k^m = \sum_{\substack{s_1(i)+\cdots+s_{k-1}(i)=2^{k-1}-u_i \\ u_i\in\langle 1,2,\cdots,2^{k-1}\rangle \\ s_j=0 \ or \ 2^{r_j}}} \prod_{j=1}^{k-1} y_j^{\left(\sum_{i=1}^{r} 2^{k_i} s_j(i)\right)} y_k^{\left(\sum_{i=1}^{r} 2^{k_i u_i}\right)}$$

Proof i) We expand $h_k = f_{k-1}(y_k)$ and use Arnon-Peterson's formulae.
ii) and iii) are obvious. □

Note

i) $y_k^m | h_k^m$.
ii) If $d_{n,i}$ is decomposed in terms of U_n-invariants, then $h_n \cdots h_{n-i+1}$ is one of its linear terms, the leading term.

For $p > 2$, the degree of a Dickson generator is

$$p^n - p^i = p^i(p-1)\left(p^{n-i-1} + \cdots + 1\right)$$

which is not a power of p. We extend the Arnon-Peterson's formulae above for the odd prime case as follows. Let $\Sigma_n(i)$ be the symmetric group acting on the n elements $(0,1,\cdots,i-1,i+1,\cdots,n)$, for $0 \le i \le n$.

Definition 2.2 *Let* $_n\omega_i = \sum_{\sigma\in\Sigma_n(i)} (-1)^{sgn(\sigma)} y_1^{p^{\sigma(0)}}\cdots y_i^{p^{\sigma(i-1)}} y_{i+1}^{p^{\sigma(i+1)}}\cdots y_n^{p^{\sigma(n)}},$
for $0 \le i \le n$.

Lemma 2.3 i) $_n\omega_i = d_{n,i}(\,_n\omega_n)$.
ii) $h_n = \frac{_n\omega_n}{_{n-1}\omega_{n-1}}$.
iii) $_n\omega_i = h_n^p\left(_{n-1}\omega_i\right) + \left(_n\omega_n\right) d_{n-1,i-1}^p$.

Let $\mathbf{N}$ be the groupoid of the non-negative integers. We give $\mathbf{N}^n$ the lexicographic order. Let the natural addition on $\mathbf{N}^n$ be described by $\mathbf{N}^n \oplus \mathbf{N}^n \longrightarrow \mathbf{N}^n$ and the symbol $+$ be reserved for the groupoid ring structure of $\mathbf{F}_p \otimes \mathbf{N}^n$. We embed $\mathbf{N}^{n-1}$ in $\mathbf{N}^n = \mathbf{N} \times \mathbf{N}^{n-1}$ and under this embedding $I \in \mathbf{N}^{n-1}$ denotes the sequence $(0, I) \in \mathbf{N}^n$.

Let $f : S_n \longrightarrow \mathbf{F}_p \otimes \mathbf{N}^n$ given by $f\left(y^I\right) := f\left(y_n^{i_n} \cdots y_1^{i_1}\right) = 1 \otimes I$. Then $f\left(y^I + y^{I'}\right) = I + I'$ and $f\left(y^I \cdot y^{I'}\right) = I \oplus I'$. Then f is an isomorphism which preserves the order, where S_n is given the induced order.

Let $F : S_n \longrightarrow H_n$ given by $F\left(y^I\right) = h^I := \prod h_n^{i_n} \cdots h_1^{i_1}$. F is an isomorphism and H_n is given the same order. Let $\bar{F}$: $H_n \longrightarrow \mathbf{F}_p \otimes \mathbf{N}^n$ be the induced map and $i_n : H_n \longrightarrow S_n$ the usual inclusion. We call $e : i_n(H_n) \longrightarrow \mathbf{F}_p \otimes \mathbf{N}^n$ the induced map. The image of $u \in i_n(H_n)$ under e is given as follows: There exists a maximal sequence $I_{M_1} = (i_n(M_1), \cdots, i_1(M_1))$ which corresponds to the leading term of u. Since u is a H_n-invariant, $p^{j-1} | i_j(M_1)$ and let $\bar{I}_{M_1} = \left(i_n(M_1)/p^{n-1}, \cdots, i_1(M_1)/p^0\right)$. Since the group action preserves the maximality, $u - h^{\bar{I}_{M_1}}$ is an invariant element and has order smaller than u. After a finite number of steps depending on u, we obtain $u - \sum_t h^{\bar{I}_{M_t}} = 0$. We define $e(u) = \sum_t \bar{I}_{M_t}$.

Proposition 2.4 *i) Suppose f is a homogeneous U_n-invariant, $f = \sum_{I \leq I_0} y^I$, $I_0 = (i_n, \cdots, i_1)$, then f contains the term h^{J_0}, where $J_0 = \left(i_n/p^{n-1}, \cdots, i_1/p^0\right)$.*

ii) Suppose f is a homogeneous GL_n-invariant, $f = \sum_{I \leq I_0} y^I$, $I_0 = (i_n, \cdots, i_1)$, then f contains the term $d^{J_0} = \prod_{0 \leq t \leq n-1} d_{n,t}^{j_{n-t}}$, i.e., $J_0 = (j_n, \cdots j_1)$ where $j_t = \frac{i_{n-t+1} - p i_{n-t}}{p^{n-t}(p-1)}$ and $j_n = \frac{i_1}{(p-1)}$.

Proof i) Let $f = y_n^{(p^{n-1})\beta} g_n(\beta) + a_n\left(\left(p^{n-1}\right)\beta - 1\right) y_n^{(p^{n-1})\beta - 1} g_n\left(\left(p^{n-1}\right)\beta - 1\right) + \cdots + a_1 y_n g_1 + a_0 g_0$. We use induction on n. For $n = 2$, $f = y_2^{j_2} y_1^{j_1} + a_1 y_2^{j_2 - 1} y_1^{j_1 - 1} + \cdots + a_{j_2 - 1} y_2 y_1^{j_1 + j_2 - 1} + a_{j_2} y_1^{j_1 + j_2} \Rightarrow f - a_{j_2} y_1^{j_1 + j_2} = y_2(f_1) \Rightarrow y_2(f_1)$ is a U_2-invariant. This implies that $h_2 | f - a_{j_2} y_1^{j_1 + j_2}$ and $f - a_{j_2} y_1^{j_1 + j_2} = h_2(f_1') \Rightarrow f_1'$ is a U_2-invariant and its leading term is $y_2^{j_2 - p} y_1^{j_1}$. As before, $h_2 | f_1' - a_{j_2 - 1} y_1^{j_1 + j_2 - p}$ and finally, $f = h_2^{j_2/p} h_1^{j_1} +$ other terms.

All g's are U_{n-1}-invariants and the assertion holds. Let $I_M(n-1)$ be the maximal sequence in $g_n(\beta)$. Then $g_n(\beta)$ contains the term $h^{I_M(n-1)}$. Since $f - \alpha_0 g_0$ is a U_n-invariant, $h_n | f - \alpha_0 g_0$. Hence, $h_n^\beta h^I$ is a term of f and comparing coefficients the assertion follows.

ii) Again we use induction on n. Since f is a B_n and GL_{n-1}-invariant, f can be written in the form $f = h_n^{(p-1)\beta} d_{n-1}^I + a_{\beta-1} h_n^{(p-1)(\beta-1)} f_{\beta-1} + \cdots + a_1 h_n^{(p-1)} f_1 + a_0 f_0 = \sum_t \alpha_t d_n^{J_t}$. Let $n = 2$, $f = h_2^{(p-1)\frac{i_2}{(p-1)p}} d_{1,0}^{\frac{i_1}{p-1}} + a_1 h_2^{(p-1)\left(\frac{i_2}{(p-1)p} - 1\right)} f_1 + \cdots + d_{1,0}^{\frac{i_2 + i_1}{p-1}}$. Since $\frac{i_2}{(p-1)p} \geq \frac{i_1}{(p-1)}$, $h_2^{(p-1)\frac{i_2}{(p-1)p}} d_{1,0}^{\frac{i_1}{p-1}}$ can be written in the form $d_{2,1}^{\frac{i_2 - p i_1}{(p-1)p}} d_{2,0}^{\frac{i_1}{p-1}} + \mathit{others}$. Since $f - d_{2,1}^{\frac{i_2 - p i_1}{(p-1)p}} d_{2,0}^{\frac{i_1}{p-1}}$ is a GL_2-invariant, we apply the same method until f is expressed as GL_2-invariants.

$f = h_n^{(p-1)\beta} d_{n-1}^{J_0(n-1)} + a_{\beta-1} h_n^{(p-1)(\beta-1)} f_{\beta-1} + \cdots + a_1 h_n^{(p-1)} f_1 + a_0 f_0$. $I_0(n-1) = (i_{n-1}, \cdots, i_1)$, $J_0(n-1) = (j_{n-1}, \cdots j_1)$ where $j_t = \frac{i_{n-t} - p i_{n-1-t}}{p^{t-1}(p-1)}$ and $j_{n-1} = \frac{i_1}{p^{n-2}(p-1)}$. $d_{n-1}^{J_0(n-1)} = \prod d_{n-1}^{j_{n-1-t}}$ and $f_t \in D_{n-1}$. Now we consider the h_n^{p-1}

monomial in the first term of f. $\beta - \sum_{t=1}^{n-1} j_t = j_1$. Since $f = \sum_t \alpha_t d_n^{J_t}$ and its leading term in terms of B_n-invariants is $h_n^{(p-1)\beta} d_{n-1}^{J_0(n-1)}$ using the formula $d_{n,t} = h_n^{p-1} d_{n-1,t} + d_{n-1,t-1}^p$ we conclude that f contains the term $d_n^{J_0}$. Comparing coefficients the assertion follows. □

For the last sentence in the proof we used the following obvious lemma.

Lemma 2.5 *Suppose the monomials d_n^I and $d_n^{I'}$ have the same leading term in S_n, then $I = I'$.*

We note here that the process above has to be applied step by step as the following example indicates.

Example 1 *Let $p = 2$ and $n = 3$, then the invariant*

$$_3\omega_{42} = \sum_{\substack{\alpha(2^3-1)+\beta(2^3-2)+\gamma(2^3-2^2)=42 \\ \alpha,\beta,\gamma\in\mathbf{N}}} d_{3,0}^{\alpha} d_{3,1}^{\beta} d_{3,2}^{\gamma}$$

has $d_{3,1}d_{3,2}^9$ as its leading term in terms of D_3-invariants. The same term contains the maximal and minimal sequence of $d_{3,1}^3 d_{3,2}^6$ in terms of y_i's.

3 A free basis for S_n over H_n and D_n

In this section, a free bases for S_n over H_n and D_n shall be provided. We start with S_n over H_n. Using the formula which decomposes h_i, we see by inspection that $y_i^{p^{i-1}}$ can be expressed as a combination of h_i and smaller powers of y_i. We see that the last assertion is actually true in the next theorem.

Theorem 3.1 *The set $B_{H_n}(S_n) = \left\{ \prod_{0\le r_i\le p^{i-1}-1} y_2^{r_2}\cdots y_n^{r_n} \right\}$ is a free module basis for S_n over H_n.*

Proof Since $|B_{H_n}(S_n)| = [U_n : \langle identity\rangle] = |U_n| = p^{n(n-1)/2}$, we only have to show that the set above generates the polynomial algebra S_n as a module over the polynomial algebra H_n. Let us note that the freeness can easily be proved by a degree argument.

By induction on n. Let $n = 2$ and $y^{I_0} = y_2^{i_2} y_1^{i_1}$ where $i_2 \ge p$. Again we induct on the exponent i_2. For $i_2 = p$, $y_2^{i_2} = h_2 - y_2^{i_2-1}h_1$. Let $i_2 = ap + r$ where $r = i_2 \bmod p$. Then $y_2^{i_2} = y_2^r h_2^a - \sum_{l=0}^{a-1}\left(\binom{a}{l} \bmod p\right) y_2^{p(a-l)+r+l} h_1^{l(p-1)}$.

Hence, $y^{I_0} = y_2^r h_2^a h_1^{i_1} - \sum_{l=0}^{a-1}\left(\binom{a}{l} \bmod p\right) y_2^{p(a-l)+r+l} h_1^{l(p-1)+i_1}$. The sum in the last expression involves smaller exponents of y_2 and by induction the assertion follows.

Let $y^{I_0} = \prod_{2\le t\le n} y_t^{i_t}$ with $i_n \ge p^{n-1}$. Let $i_n = qp^{n-1} + r$ and $y_n^{i_n} = y_n^r h_k^q - \left(\sum_{l=0}^{n-2} (-1)^l y_n^{p^l} d_{n-1,l}\right)^q$. Each $d_{n-1,l}$ is a polynomial in h_i's and by induction on i_n,

$y_n^{i_n} = \sum_{l=0}^{p^{n-1}-1} a_l y_n^l h^{I(l)}$, where $h^{I(l)}$ is a monomial on the h_i's.

$y^{I_0} = \sum_{l=0}^{p^{n-1}-1} a_l y_n^l y_{n-1}^{i_{n-1}} \cdots y_2^{i_2} h^{I(l)}$ and the assertion follows. □

Here we provide some formulas which explains the motivation for the set of generators for H_n over D_n.

Proposition 3.2

i) $d_{n,n-i} = \sum_{j=0}^{i} \sum_{n-i+1 \le t_1 < \cdots < t_j \le n} \prod_{s=1}^{j} \left(h_{t_s}^{p-1}\right)^{p^{n-j+s-t_s}} d^{p^{i-j}}_{n-i,n-i-(i-j)}$.

ii) $(h_n \cdots h_{n-i+1})^{p-1}$

$$= d_{n,n-i} - \sum_{j=0}^{i-1} \sum_{n-i+1 \le t_1 < \cdots < t_j \le n} \prod_{s=1}^{j} \left(h_{t_s}^{p-1}\right)^{p^{n-j+s-t_s}} d^{p^{i-j}}_{n-i,n-i-(i-j)};$$

and in the expression above the polynomial $(h_{n-1} \cdots h_{n-i+1})^{p(p-1)} d^p_{n-i,n-i-1}$ *is involved as a linear term.*

iii) $(h_n \cdots h_{n-i+1})^{p-1} = d_{n,n-i} -$

$(d_{n,n-i+1} - (d_{n,n-i+2} - (\cdots (d_{n,n-1} - d^{p^i}_{n-i,n-i+1}) \cdots) d^{p^3}_{n-i,n-i-1}) d^{p^2}_{n-i,n-i-1})$

$d^p_{n-i,n-i-1} - \sum_{j=0}^{i-1} \sum_{n-i+1 \le t_1 < \cdots < t_j \le n} \prod_{s=1}^{j} \left(h_{t_s}^{p-1}\right)^{p^{n-j+s-t_s}} d^{p^{i-j}}_{n-i,n-i-(i-j)};$

iv) Inductively: $d_{n,n-i}$ *contains the linear term* $(d_{n-i+1,n-i})^{(p^i-1)/(p-1)}$.

v) $d_{n,n-i} = d_{n,n-1} d_{n-1,n-i} - d^p_{n-1,n-2} d_{n-1,n-i} + d^p_{n-1,n-i-1}$.

Proof The first formula is the one which decomposes $d_{n,n-i}$ in terms of h_i's where sums of products of h_i's involved have rewritten in terms of $d_{n-i,n-j}$'s 1.2. It is proved by induction. Formulas ii) and iii) are applications of i).

v) $d_{n,n-i} = \left(d_{n,n-1} - d^p_{n-1,n-2}\right) d_{n-1,n-i} - d^p_{n-1,n-i-1}$. □

Example i) $d_{n,n-1} = h_n^{p-1} + d^p_{n-1,n-2} \Rightarrow h_n^{p-1} = d_{n,n-1} - d^p_{n-1,n-2}$.

ii) $d_{n,n-2} = d_{n,n-1} d_{n-1,n-2} - d^{p+1}_{n-1,n-2} + d^p_{n-1,n-3} \Rightarrow$

$$h_{n-1}^{(p-1)(p+1)} = -d_{n,n-2} + d_{n,n-1} d_{n-1,n-2} - \sum_{j=0}^{p} \binom{p+1}{j} h_{n-1}^{(p-1)j} d^{p(p+1-j)}_{n-2,n-3} + d^p_{n-1,n-3}.$$

iii) $d_{n,n-3} = d_{n,n-2} d_{n-2,n-3} - d^{p^2}_{n-2,n-4} d_{n-2,n-3} + d^{p^2+p+1}_{n-2,n-3} + d^{p^2}_{n-2,n-5} -$

$d_{n,n-1} \left(d^{p+1}_{n-2,n-3} - d^p_{n-2,n-4}\right) - d^p_{n-2,n-4} d^{p^2}_{n-2,n-3}$ and $d^{p^2+p+1}_{n-2,n-3}$

$$= h_{n-2}^{(p-1)(p^2+p+1)} + \sum_{j=0}^{p^2+p} \binom{p^2+p+1}{j} h_{n-2}^{(p-1)j} d^{p(p^2+p+1-j)}_{n-3,n-4}.$$

Hence, $h_{n-2}^{(p-1)(p^2+p+1)} = d_{n,n-3} - d_{n,n-2} d_{n-2,n-3} + d^{p^2}_{n-2,n-4} d_{n-2,n-3} -$

$\sum_{j=0}^{p^2+p} \binom{p^2+p+1}{j} h_{n-2}^{(p-1)j} d^{p(p^2+p+1-j)}_{n-3,n-4}$

$-d^{p^2}_{n-2,n-5} + d_{n,n-1} \left(d^{p+1}_{n-2,n-3} - d^p_{n-2,n-4}\right) + d^p_{n-2,n-4} d^{p^2}_{n-2,n-3}$.

Note. $h_r^{p^{n+1-i}-1+t}$ decomposes in terms of GL_n-invariants and h_j's for $j \le i$. Here $t \ge 0$.

There is a restriction on the top degree for h_i, namely $p^{n-i+1} - 1$, given by the formula decomposing $d_{n,i-1}$. By a good bookkeeping argument, one can get the decomposition of $h_i^{p^{n-i+1}-1}$ with respect to the basis elements. Such a formulae will be important for studying the natural map ξ_n (see proposition below) between H_n and D_n. This is mainly important for its dual map where Dyer-Lashof operations (being mod-p homology classes of symmetric groups) are mapped on operations characterized mainly by the fact that Adem relations are not allowed and they are mod-p homology classes of particular p-Sylow subgroups of symmetric groups.

The proof of the last theorem can be extended to prove the next one which has been proved by Campbell and Hughes using a different method.

Theorem 3.3 [3] $B_{D_n}(H_n) = \left\{h_1^{r_1}\cdots h_n^{r_n} \mid 0 \le r_i < p^{n-i+1}-1\right\}$ *is a free module basis for* H_n *over* D_n.

The following lemma is a consequence of the proposition above.

Lemma 3.4 $h_i^{p^{n-i+1}-1} = (-1)^{n-i} d_{n,i-1} + \sum_{t=1}^{n-i} c_t d_{n,i-1+t} h^{I_t} + \sum c'_s h^{I_s}$, *where* h^{I_t} *and* h^{I_s} *are basis elements.*

The following proposition is obvious.

Proposition 3.5 *Let* $\xi_n : H_n \longrightarrow D_n$ *be the natural* F_p*-epimorphism. Then* $\xi_n\left(h_i^{p^{n-i+1}-1}\right) = (-1)^{n-i} d_{n,i-1}$.

Campbell and Hughes have shown that the map ξ_n is actually the induced transfer map between U_n and $GL_n(F_p)$-invariants.

The following theorem is easily deduced from the two theorems above.

Theorem 3.6 *The set* $B_{D_n}(S_n) = \left\{ \prod_{0 \le r_i < p^n - p^{i-1}} y_1^{r_1} y_2^{r_2} \cdots y_n^{r_n} \right\}$ *is a free module basis of* S_n *over* D_n.

The following theorem follows easily from the methods developed by the author and others.

Theorem 3.7 *The set* $B_{PD_n(N)}(S_n) = \left\{ \prod_{\substack{0 \le r_i < p^{\nu_j} - p^{i-1} \\ \nu_{j-1} \le i \le \nu_j - 1}} y_1^{r_1} y_2^{r_2} \cdots y_n^{r_n} \right\}$ *is a free module basis of* S_n *over* $PD_n(N)$.

4 Bases related to invariants of $E_n \otimes S_n$

In this section, we extend the sets of free bases for the cases $E_n \otimes S_n$ over subalgebras of $(E_n \otimes S_n)^{U_n}$ and $(E_n \otimes S_n)^{GL_n}$. This case is considered independently, because of the different structure of its invariants. Since homology classes of symmetric groups are closely related with homology operations, we study here the following cases:

i) $(R[n])^*$, where $R[n]$ is the coalgebra of Dyer-Lashof operations of fixed length n, [5]. These elements provide classes for the cohomology of the group Σ_{p^n}, [**12**].
ii); homology operations related with $\Sigma_{p^n,p}$ a Sylow group of the symmetric group containing V^n, [**9**]. These elements provide classes for the cohomology of $\Sigma_{p^n,p}$.

Let us repeat the definitions of the invariant generators. Let

$$M_{n;i_1,\cdots,i_k} = \frac{1}{k!}\begin{vmatrix} x_1 & x_2 & \cdots & x_n \\ \vdots & \vdots & \cdots & \vdots \\ x_1 & x_2 & \cdots & x_n \\ y_1 & y_2 & \cdots & y_n \\ \vdots & & & \vdots \\ y_1^{p^{n-1}} & y_2^{p^{n-1}} & \cdots & y_n^{p^{n-1}} \end{vmatrix} \tag{4.1}$$

Here, $1 \le k \le n$, $0 \le i_1 < \cdots < i_k \le n-1$. There are k rows of x_i's and the following k powers of y_i's: $i_1, \cdots, i_k$. are missing.

Relations between the $M's$ are expressed in the following lemma.

Lemma 4.1 [**13**] *Let* $0 \le s_1 < \cdots < s_m \le n-1$.

a) $M_{n;s_1,\cdots s_m} L_n^{m-1} = (-1)^{m(m-1)/2} M_{n;s_1} \dots M_{n;s_m}$.

b) $(M_{n;s})^2 = 0$.

c) $M_{n;0,1,\dots,n-1} L_n^{n-1} = (-1)^{n(n-1)/2} M_{n;0} \dots M_{n;n-1} = x_1 x_2 \dots x_n$.

d) $M_{n-1;s_1,\cdots s_m} h_n = M_{n;s_1,\cdots s_m} - \sum_1^m (-1)^{m+i} M_{n;s_1,\dots,\widehat{s}_i,\dots,n-1} d_{n-1,s_i}$.

e) $M_{k;s_1,\cdots s_m} h_{k+1} \dots h_n = M_{n;s_1,\cdots s_m} + \sum_{(t_1,\dots,t_m)>(s_m-m+1,\dots,s_m)} M_{n;t_1,\cdots t_m} f_{t_1,\dots,t_m}$,

where $f_{t_1,\dots,t_m} \in (S_n)^{U_n}$.

f) $M_{n;s_1,\cdots s_m} L_n^{m-1} = (-1)^{(m(m-1)/2} \prod_1^m (\sum_{r=s_j+1}^{n} M_{r;r-1} h_{r+1} \dots h_n d_{r-1,r-1-s_j})$.

g) $M_{n-1;s_1,\cdots s_{m-1}} x_n = M_{n;s_1,\cdots s_{m-1},n-1}$

$$- \sum_{|I|=m,\ n\notin I} sgn\sigma_I x_I [0,\dots,\widehat{s}_i,\dots,\widehat{s}_{m-1},\dots,n-2]'_I.$$

h) $M_{n-1;s_1,\cdots s_{m-1}} x_n = -M_{n;s_1,\cdots s_{m-1},n-1}$

$$- \sum_{1\le t\le n-m+1} (-1)^{t+1} M_{n-1;s_1,\cdots,j_t,\cdots s_{m-1}} y_n^{p^{j_t}}.$$ *Here* $(j_1,\cdots,j_{n-m+1})$

is the ordered complement of $(s_1,\cdots,s_{m-1})$ *in* $(1,\cdots,n)$.

Theorem 4.2 [**13**] *i) The algebra* $(E_n \otimes S_n)^{U_n}$ *is a tensor product between the polynomial algebra* H_n *and the* $\mathbf{Z/pZ}$*-module* $UM[n]$, *where* $UM[n]$ *is spanned by the set of elements* $\{M_{s;s_1,\cdots,s_k} \mid 1 \le k \le s \le n,\ 0 \le s_1 < \cdots < s_k = s-1\}$.

ii) The algebra $(E_n \otimes S_n)^{GL_n}$ *is a tensor product between the polynomial algebra* D_n *and the* $\mathbf{Z/pZ}$*-module* $GLM[n]$, *where* $GLM[n]$ *is spanned by the set of elements* $\{M_{n;s_1,\cdots,s_k}\ (L_n)^{p-2} \mid 1 \le k \le n,\ 0 \le s_1 < \cdots < s_k \le n-1\}$.

Definition 4.3 [**9**] *i) Let* $SA(U_n)$ *be the subalgebra of* $(E_n \otimes S_n)^{U_n}$ *defined by the tensor product between* H_n *and the* $\mathbf{Z/pZ}$*-module* $SUM[n]$, *where* $SUM[n]$ *is spanned by the set of elements* $\left\{M_{s;s-1}\,(L_{s-1})^{(p-3)/2} \mid 1 \le s \le n\right\}$.

ii) Let $SA(GL_n)$ *be the subalgebra of* $(E_n \otimes S_n)^{GL_n}$ *defined by the tensor product between* D_n *and the* $\mathbf{Z/pZ}$*-module* $SGLM[n]$, *where* $SGLM[n]$ *is spanned by the set of elements*

$$\left\{M_{n;i}\,(L_n)^{(p-2)}, M_{n;i_1,i_2}\,(L_n)^{(p-2)} \mid 0 \le i \le n-1, 0 \le i_1 < i_2 \le n-1\right\}.$$

Remark 1 *We recall the following facts:*

- The algebra generated by $\left\{M_{k+1;k}L_k^{p(3)} \mid 0 \le k \le n-1\right\}$ is isomorphic to an exterior algebra.

- There is a direct sum decomposition $\bigoplus_{k=1}^{n} \bigoplus_{0\le s_1<\cdots s_k\le n-1} M_{n;s_1,\cdots,s_k} \otimes D_n$.

Before we proceed to the main result of this section, we discuss an example where we demonstrate the method which provides a base.

Example. Let $n=3$. First, we consider the submodule of $(E_3 \otimes S_3)^{U_3}$ where only one x_i is involved. This submodule has a free base $\{M_{3;2},\ M_{2;1},\ M_{1;0}\}$ over H_3.

a) $M_{3;2}h_1^{p-2}h_2^{p-2}h_3^{p-2} = M_{3;2}L_3^{p-2}$. Let

$$B_{3,2} = \{M_{3;2}L_3^{p-2}h_1^{i_1}h_2^{i_2}h_3^{i_3} \mid i_1 < p^3-p+1,\ i_2 < p^2-p+1,\ i_3 = 0\}$$

$$\sqcup \{M_{3;2}h_1^{i_1}h_2^{i_2}h_3^{i_3} \mid i_1 < p^3-1,\ i_2 < p^2-1,\ i_3 < p-2\}$$

$$\sqcup \{M_{3;2}h_1^{i_1}h_2^{i_2}h_3^{p-2} \mid i_j < p-2 \text{ and } i_t < p^{4-t}-1, \text{ for } t \ne j \in \{1,2\}\}.$$

b)

$$\begin{aligned} M_{3;1}L_3^{p-2} &= M_{2;1}h_3L_3^{p-2} + M_{3;2}L_3^{p-2}d_{2,1} \\ &= M_{2;1}L_2^{p-2}h_3^{p-1} + M_{3;2}L_3^{p-2}d_{2,1} \\ &= M_{2;1}L_2^{p-2}(d_{3,2} - h_2^{p^2-p} - h_1^{p^3-p^2}) + M_{3;2}L_3^{p-2}d_{2,1}. \end{aligned}$$

i)

$$\begin{aligned} h_2^{p^2-p} &= d_{3,2} - h_3^{p-1} - h_1^{p^3-p^2} \Rightarrow M_{2;1}L_2^{p-2}h_2^{p^2-p} \\ &= M_{2;1}L_2^{p-2}\left(d_{3,2} - h_3^{p-1} - h_1^{p^3-p^2}\right) \\ &= M_{2;1}L_2^{p-2}\left(d_{3,2} - h_1^{p^3-p^2}\right) - M_{3;1}L_2^{p-2}h_3^{p-2} \\ &\qquad + M_{3;2}L_2^{p-2}\left(h_1^{p^2-p} + h_2^{p-1}\right)h_3^{p-2}. \end{aligned}$$

Let

$$A = \left\{ \begin{array}{c} M_{2;1}h_1^{i_1}h_2^{i_2}h_3^{i_3} \mid i_2 < p^2-2 \text{ and } i_t < p^{4-t}-1 \text{ for } t \ne 2; \\ i_2 = p^2-2,\ i_j < p-2 \text{ and } i_t < p^{4-t}-1,\ \text{for } t \ne j \in \{1,3\} \end{array} \right\} \text{ and}$$

$$B = \left\{M_{3;1}L_3^{p-2}h_1^{i_1}h_3^{i_3} \mid i_3 < p-1,\ i_1 < p^3-p+1\right\}.$$

Let $B_{2,1} = A \sqcup B$.

c)

$$\begin{aligned} M_{3;0}L_3^{p-2} &= M_{2;0}h_3L_3^{p-2} + M_{3;2}L_3^{p-2}d_{2,0} \\ &= M_{1;0}h_2h_3L_3^{p-2} + M_{2;1}h_3L_3^{p-2}d_{1,0} + M_{3;2}L_3^{p-2}d_{2,0} \\ &= M_{1;0}h_1^{p-2}(h_2h_3)^{p-1} + M_{2;1}h_3L_3^{p-2}d_{1,0} + M_{3;2}L_3^{p-2}d_{2,0}. \end{aligned}$$

From (Example ii) after Proposition 3.2, we have:

$$\begin{aligned}h_1^{p^3-p} =& -h_2^{p^2-1} - d_{3,1} + d_{3,2}(h_2^{p-1} + h_1^{p(p-1)}) - \sum_{j=1}^{p-1}\binom{p+1}{j} h_2^{j(p-1)} h_1^{p(p-1)(p+1-j)} \\ =& (h_2h_3)^{p-1} - d_{3,1} + d_{3,2}h_1^{p(p-1)} \Rightarrow M_{1;0}h_1^{p^3-2} \\ =& M_{1;0}h_1^{p-2}\left(-d_{3,1} + d_{3,2}h_1^{p(p-1)}\right) + M_{3;0}L_3^{p-2} \\ & -M_{3;1}L_3^{p-2}h_1^{p-1} + M_{3;2}L_3^{p-2}h_1^{p^2-1}.\end{aligned}$$

Let $A = \left\{ \begin{array}{c} M_{1;0}h_1^{i_1}h_2^{i_2}h_3^{i_3} \,|i_1 < p^3 - 2 \text{ and } i_t < p^{4-t} - 1 \text{ for } t \neq 2; \\ i_1 = p^3 - 2,\ i_j < p - 2 \text{ and } i_t < p^{4-t} - 1, \text{ for } t \neq j \in \{2,3\} \end{array} \right\}$.

Let $B = \left\{M_{3;0}L_3^{p-2}h_2^{i_2}h_3^{i_3} \mid i_3 < p-1,\ i_2 < p^2 - 1\right\}$. Let $B_{1,0} = A \sqcup B$.

It is evident that the set $B_{3,2} \cup B_{2,1} \cup B_{1,0}$ consists of a free module base for the submodule of $(E_3 \otimes S_3)^{U_3}$ which contains only elements involving only one x_i over D_3.

The case where more x_i's are involved is analogous to the one considered above.

Theorem 4.4 *a) The set*
$B_{(H_n)}(E_n \otimes S_n) = \left\{x_1^{i_1}x_2^{i_2}\cdots x_n^{i_n}y_2^{r_2}\cdots y_n^{r_n} \mid \leq 0\ i_j \leq 1,\ 0 \leq r_i \leq p^{i-1} - 1\right\}$ *is a free module basis for* $E_n \otimes S_n$ *over* H_n.

b) Let $B_{(D_n)}(E_n \otimes S_n)^{U_n}$ *be the discrete union of the following sets:*

i) $\{M_{s;\ s_1,\cdots,s_k}h_1^{j_1}\cdots h_n^{j_n} \mid 1 \leq k \leq s, 0 \leq s_1 < \cdots < s_k = s-1;\ j_s < p^{n+1-s} - 2,\ j_t < p^{n+1-t} - 1\}$,

ii) $\{M_{s;\ s_1,\cdots,s_k}h_1^{j_1}\cdots h_n^{j_n} \mid 1 \leq k \leq s, 0 \leq s_1 < \cdots < s_k = s-1;\ j_s = p^{n+1-s} - 2,\ \exists j_{s-m} < p-2, j_t < p^{n+1-t} - 1\}$,

iii) $\{M_{n;\ s_1,\cdots,s_k}L_n^{p-2}h_1^{j_1}\cdots h_n^{j_n} \mid 1 \leq k \leq n, 0 \leq s_1 < \cdots < s_k = s \leq n - 1;\ j_{s+1} = 0,\ j_{s-t} < p^{n+t-s} - p + 1,\ j_{s+1+t} < p^{n-s-t} - 1\}$.

Then $B_{(D_n)}(E_n \otimes S_n)^{U_n}$ *is a free module basis for* $(E_n \otimes S_n)^{U_n}$ *over* D_n.

Proof a) This is a consequence of the decomposition of $(E_n \otimes S_n)$ over S_n and S_n over H_n.

b) We only need to prove that $M_{s;\ s_1,\cdots,s_k}h_1^{j_1}\cdots h_n^{j_n}$ decomposes in terms of basis elements for $s_k = s$ and $j_{s+1} = 0$. And this is due to Proposition 3.2 iii):

$$h_s^{p^{n+1-s}-p} = d_{n,s} - (h_{s+1}\cdots h_n)^{p-1} - (\textit{other terms})$$

We note that no element of the form $(h_{s+1}\cdots h_m)^{p-1}d_{m+1,m}^{(p^{n-m-1}+\cdots+1)}$ appears in the last parenthesis. That element would decompose and also contribute an $M_{n;\ s_1,\cdots,s_k}(h_{s+1}\cdots h_n)^{p-2}$.

Proposition 4.5 *i) The algebra* $SA(U_n)$ *is a free* H_n*-module with basis*

$$\left\{\prod_1^k M_{s_i;\ s_i-1}L_{s_i-1}^{p(3)} \mid 1 \leq k \leq n\right\}$$

ii) The algebra $SA(GL_n)$ is a free D_n-module with basis the discrete union between

$$\left\{ \prod_1^m \left(M_{n;k_t,s_t} L_n^{p-2}\right) M_{n;s_{m+1}} L_n^{p-2} \mid 1 \le m \le [\frac{n-1}{2}],\right.$$
$$\left. 0 \le k_1 < s_1 < \cdots < s_m < s_{m+1} \le n-1 \right\}$$

and

$$\left\{ \prod_1^m \left(M_{n;k_t,s_t} L_n^{p-2}\right) \mid 1 \le m \le [\tfrac{n}{2}], 0 \le k_1 < s_1 < \cdots < s_m \le n-1 \right\}.$$

We shall use instead of i) above an isomorphic image which is similar to the one in ii).

Lemma 4.6 *i)* $M_{k,k-1}(\frac{L_{k+s}}{L_k}) = \sum_{t=0}^{s} M_{k+s,k-1+t}$

$$\sum_{\ell=0}^{t} (-1)^\ell \prod_{0 \le i_1 < \cdots < i_{\ell+1} = t} d_{k-1+t-i_j,k-1+t-i_{j+1}}.$$

ii) $M_{k,k-1} L_{k-1}^{p(3)} M_{k+s,k-1+s} L_{k+s-1}^{p(3)} = -\sum_{t=0}^{s-1} M_{k+s;k-1+t,k+s-1} L_{k-1}^{p(1)} h_k L_{k+s-1}^{p(3)}$

$$\left(\sum_{\ell=0}^{t} (-1)^\ell \prod_{0 \le i_1 < \cdots < i_{\ell+1} = t} d_{k-1+t-i_j,k-1+t-i_{j+1}} \right).$$

Proof

i) We apply Lemma 4.1 d).
ii) We apply i) and Lemma 4.1 b).

□

We extend i) above:

Proposition 4.7 $M_{k;s_1,\cdots,s_m=k-1}(L_n/L_k) =$

$$\sum_{\substack{\{t_1,\cdots,t_m\} \subset \{s_1,\cdots,s_m,k,\cdots,n\} \\ (t_1,\cdots,t_m)}} M_{n;t_1,\cdots,t_m} \sum_{\ell \le q \le n-k+1} (-1)^{(q,t_1,\cdots,t_m)} \prod_{\alpha_j,\beta_j} d_{\alpha_j,\beta_j}.$$

Here ℓ is the number of s_i;s not in $\{t_1,\cdots,t_m\}$. The exponent $(q,t_1,\cdots,t_m)$ represents the expression qm + the sum of positions of the missing s_i;s. And the α_j,β_j satisfy the following condition: $\alpha_q = t_m, \cdots, \alpha_{q-\ell+1} = t_{m-\ell+1} > \alpha_{q-\ell} > \cdots \alpha_1 \ge k$; $\alpha_j > \beta_j$; $\beta_1, \cdots \beta_\ell \in \{s_1,\cdots,s_m\} - \{t_1,\cdots,t_m\}$; $k \le \beta_{\ell+1} < \cdots < \beta_q \le t_m - 1$.

The last technical proposition shall be used in the next proposition and in Theorem 5.7.

Lemma 4.8 *i)* $(h_{k+1}\cdots h_n)^{p-1} = \sum_{t=k}^{n-1} d_{n,t} f_t + f_n$. *Here f_t on $\{h_1,\cdots h_k\}$ both with coefficients in D_n, $k \le t \le n$ and f_n contains the term $(d_{k,k-1})^{(p^{n-k}-p)}$.*

ii) $M_{k,k-1} h_k^{p^{n+1-k}-p} =$

$M_{n,k-1} \prod_{s=k+1}^{n} h_s^{p-2} + \sum_{t=1}^{n-k} M_{n,k-1+t} \prod_{s=k+1}^{n} h_s^{p-2} f_{k-1+t} + M_{k,k-1} f_{k-1}$. *Here* f_{k-1} *and* f_{k-1+t} *are polynomials on* $\left\{h_1^{p-1}, \cdots h_k^{p-1}\right\}$ *both with coefficients in* D_n.

Proof

i) By decomposing $d_{n,k}$ and induction on k.

ii) We use i) above: $h_k^{p^{n+1-k}-p} = (h_{k+1} \cdots h_n)^{p-1} + \sum_{t=k}^{n-1} d_{n,t} f_t + f_n'$. The degree of f_n' with respect to h_k is less than $p^{n+1-k} - p$.

□

Proposition 4.9 *i) The algebra* $SA(U_n)$ *is a free* H_n*-module with basis*

$$\left\{ \left(\prod_1^k M_{s_i;t_i,s_i-1} L_{t_i}^{p(3)} L_{t_i+1} L_{s_i-1}^{p(3)}\right) \left(\prod_1^k M_{s_{k+1};s_{k+1}-1} L_{s_{k+1}}^{p(3)}\right)^{\varepsilon} \mid \atop \varepsilon=0\, or\, 1,\ 1 \leq 2k+\varepsilon \leq n, 0 \leq t_1 < s_1 - 1 < t_2 < \cdots < t_k < s_k - 1 < s_{k+1} - 1 \leq n \right\}$$

ii) The algebra $SA(U_n)$ *is a free* H_n*-module with basis*

$$\{M_{s;s_1,\cdots,s_k} L_{s-1}^{p(3)} \left(\prod_1^{k-1} L_{s_j}^{p(3)}\right) \left(\prod_1^{k-1} L_{s_j+1}\right) \|$$
$$1 \leq s \leq n, 1 \leq k \leq s, 0 \leq s_1 < \cdots < s_k = s-1\}$$

iii) The algebra $SA(GL_n)$ *is a free* D_n*-module with basis*

$$\{M_{n;s_1,\cdots,\ s_k} L_n^{p-2} d_{n,0}^{([\frac{k+1}{2}]-1)} \mid 1 \leq k \leq n, 0 \leq s_1 < \cdots < s_k \leq s-1\}.$$

Proof i) First, we consider the case $k = 2$. Using the previous lemma and double induction on k and s, we can replace the old elements by the new ones. ii) follows from Lemma 4.1 b).

We shall note that the base in iii) above is flexible considering computations instead of Proposition 4.5 ii) which is the dual of the length n Dyer-Lashof algebra $R[n]$. □

Theorem 4.10 *The algebra* $SA(U_n)$ *is a free* D_n*-module with basis*

$$\{M_{s;s_1,\cdots,s_k} L_{s-1}^{p(3)} \left(\prod_1^{k-1} L_{s_j}^{p(3)}\right) \left(\prod_1^{k-1} L_{s_j+1}\right) h^I \mid 1 \leq s \leq n, 1 \leq k \leq s,$$
$$0 \leq s_1 < \cdots < s_k = s-1 \text{ and } i_j < p^{n+1-j} - 1, 1 \leq j \leq n\}$$
$$\sqcup \{h^I \mid i_j < p^{n+1-j} - 1, 1 \leq j \leq n\}.$$

5 The transfer map between rings of invariants

Let H be a subgroup of a finite group G, then the inclusion $i : H \hookrightarrow G$ induces the transfer map $tr^* : H^*(H, F_p) \longrightarrow H^*(G, F_p)$ given by $tr^*(u) = \sum_{g \in G/H} gu$. If $W_G(H)$ is the Weyl subgroup, then the inclusion above induces $i^* : H^*(G, F_p) \longrightarrow H^*(H, F_p)^{W_G(H)}$.

The Weyl subgroups of V in $\sum_{p^n,p}$ and $\sum_{p^n}$ respectively are the upper triangular group U_n and the general linear group GL_n. The induced inclusion $W_{\sum_{p^n,p}}(V)$

$\rightarrow W_{\Sigma_{p^n}}(V)$ induces $\tau_n^* : H^*(V,F_p)^{U_n} \rightarrow H^*(V,F_p)^{GL(n,p)}$ given by $\tau_n^*(f) = \sum_{g\in G/H} gf$. Here f is a U_n -invariant polynomial in S_n. The following diagram is commutative [**11**].

$$\begin{array}{ccc} H^*\left(\Sigma_{p^n,p}\right) & \xrightarrow{\tau_n^*} & H^*\left(\Sigma_{p^n}\right) \\ \uparrow & & \uparrow \\ H^*(V,F_p)^{U_n} & \xrightarrow{\tau_n^*} & H^*(V,F_p)^{GL(n,p)} \end{array}$$

It is well known that the following composite is equal to multiplication by $[G:H]$:

$$F_p[V]^G \hookrightarrow F_p[V]^H \stackrel{\tau^*}{\rightarrow} F_p[V]^G.$$

Here V is a F_pG-module. In our case the transfer is surjective and $(E_n \otimes S_n)^{GL_n}$ is a direct summand of $(E_n \otimes S_n)^{U_n}$.

Campbell and Hughes have shown that the map $\xi_n : H_n \longrightarrow D_n$ is equal to the transfer τ_n^* multiplied by $(-1)^n$.

We recall from [**3**] a set of coset representatives for GL_n over U_n. Let $W^{n-m+1} = \langle y_{n-m+1}, \cdots, y_n\rangle$, then $\left(W^{n-m+1}\right)^* = \langle \sigma_m \rangle$ as a multiplicative group and we denote the corresponding linear transformation again by the same symbol.

Theorem 5.1 [**3**] *The set* $\{\sigma_n^{i_n}\cdots\sigma_1^{i_1} \mid 0 \le i_m < p^m - 1\}$ *is a set of left coset representatives for* GL_n *over* U_n.

Next some technical lemmas are discussed.

Lemma 5.2 $\sum\limits_{w\in\langle y_1,\cdots,y_n\rangle^*} \frac{1}{w^r} = 0$ *for* $0 < r < p-1$.

Proof Let $W = \langle y_1,\cdots,y_n\rangle$, then $|W^*| = (p-1)(p^{n-1}+\cdots+1)$. We can write $W = \sqcup_1^n\{a_t(y_t+\beta(t-1)) \mid a_t \in \mathbf{F}_p^*,\ \beta(t-1) \in \langle y_1,\cdots,y_{t-1}\rangle\}$. The argument follows from $\sum\limits_{a_t} \frac{1}{a_t^r(y_t+\beta(t-1))^r} = 0$. □

Lemma 5.3 $\sum\limits_{t=0}^{p^{n-m}-2} \frac{1}{g^t h_m^r} = 0$ *for* $0 < r < p-1$. *Here* g *is a multiplicative generator of* $\langle y_m,\cdots,y_n\rangle^*$.

Proof This is an application of the previous lemma and the fact that $h_m = \prod\limits_{u\in\langle y_1,\cdots,y_{m-1}\rangle} (y_m - u)$ is linear in y_m. □

Lemma 5.4 *Let* $M_{n;s_1,\cdots s_m} L_n^{p-2} \frac{h^I}{L_n^{p-2}}$ *such that there exists a* t *with* $i_t < p-2$, *then* $\tau_n^*\left(M_{n;s_1,\cdots s_m} L_n^{p-2}\frac{h^I}{L_n^{p-2}}\right) = 0$.

Proof The statement reduces to $\tau_n^*\left(\frac{h^I}{L_n^{p-2}}\right) = 0$ which follows from theorem 11.2 in [**3**] and the last lemma. □

Lemma 5.5 *Let* $f(r) := \sum\limits_{a_i}(a_1x_1+\cdots+a_nx_n)(a_1y_1+\cdots+a_ny_n)^r$ *and* $r(j) = (p-1)(1+\cdots+p^{n-1}) - p^j$, $j = 0,\cdots n-1$. *Then* $f(r(j)) = M_{n;j}L_n^{p-2}$ *and* $f(r) = 0$ *for* $r \neq r(j)$ *and* $r < r(0)$.

Proof It is obvious that $f(r)$ is a GL_n-invariant. Hence the second assertion follows for degree reasons. It suffices to prove that $f(r(j)) \neq 0$.

$f(r(j))\,(x_1 = \cdots = x_n = 0, x_j \neq 0) = x_j \sum\limits_{a_i} a_j \left(a_1 y_1 + \cdots + a_n y_n\right)^r$. The sum above contains the following monomial:

$$\sum_{a_i} a_j a_j^{p-2} \frac{r(j)!}{\prod\limits_{i\neq j}[(p-1)\,p^i]![(p-2)\,p^j]} \prod_{i\neq j} y_i^{(p-1)p^i} y_j^{(p-2)p^j}$$
$$= (p-1)^n \prod_{i\neq j} y_i^{(p-1)p^i} y_j^{(p-2)p^j}.$$

□

Combining the last two lemmas, the next proposition is obtained.

Proposition 5.6 *Let $M_{k;\ i_1,\cdots,i_{m-1},k-1}(w)$ be the polynomial $M_{k;i_1,\cdots,i_{m-1},k-1}$ where x_k has been replaced by $w_x \in< x_k,\cdots,x_n>$ and y_k by $w_y \in< y_k,\cdots,y_n>$. Then*
i) $\sum\limits_{w} M_{k;\ i_1,\cdots,i_{m-1},k-1}(w) h_k^r(w_y) \neq 0$ for $r = p^{n-k+1} - 1 - p^s$ and $s = 0,\cdots n-k$.
ii) $\tau_n^* M_{k;\ i_1,\cdots,i_{m-1},k-1} L_{k-1}^{p-2} h_k^r = 0$ for $r \neq p^{n-k+1} - 1 - p^s$ and $s = 0,\cdots n-k$.

Proof Let us decompose:

$$M_{k;\ i_1,\cdots,i_{m-1},k-1}$$
$$= (-1)^{k-1} x_k M_{k-1;\ i_1,\cdots,i_{m-1}} + \sum_{t=1}^{k-m} (-1)^{k+m+t} M_{k-1;\ i_1,\cdots,j_t,\cdots,i_{m-1}} y_k^{p^{j_t}}.$$

Here $1 \leq j_t \leq k-2$ and $j_t \notin \{i_1,\cdots,i_{m-1}\}$. By the last two lemmas, $\sum\limits_{w} w_x M_{k-1;\ i_1,\cdots,i_{m-1}}(w) h_k^r(w_y) \neq 0$ for $r = p^{n-k+1} - 1 - p^s$ and $s = 0,\cdots n-k$. The first claim follows.

ii) Assume $\tau_n^* M_{k;\ i_1,\cdots,i_{m-1},k-1} L_{k-1}^{p-2} h_k^r \neq 0$. Then $\tau_n^* M_{k;\ i_1,\cdots,i_{m-1},k-1} L_{k-1}^{p-2} h_k^r = \pm M_{n;J} L_n^{p-2}$ for some strictly increasing sequence of m elements. (This is because of the direct sum decomposition between M;s and degree reasons.) We consider degrees on both sides.

$$p^{k-1} r + 1 + \cdots + p^{k-2} - p^{i_1} - \cdots - p^{i_{m-1}}$$
$$= (p-2)(p^{k-1} + \cdots + p^{n-1}) + 1 + \cdots + p^{n-1} - p^{j_1} - \cdots - p^{j_m}.$$

Or $p^{k-1} r - p^{i_1} - \cdots - p^{i_{m-1}} = p^{k-1}(p^{n-k+1} - 1) - p^{j_1} - \cdots - p^{j_m}$. We note that $0 \leq i_1 < \cdots < i_{m-1} < k-1$, $0 \leq j_1 < \cdots < j_m$, and $j_m > k-1$. Hence, $r = p^{n-k+1} - 1 - p^{j_m - k+1}$. □

Before we discuss the relation between τ_n^* and ξ_n, let us consider an example.

Example 2 *Let $p = 3$. A set of coset representatives for GL_2 over U_2 is*

$$\left\{\begin{pmatrix} a & 0 \\ 0 & b \end{pmatrix} |ab \neq 0\right\} \cup \left\{\begin{pmatrix} 0 & b \\ a & 0 \end{pmatrix} |ab \neq 0\right\} \cup \left\{\begin{pmatrix} a & 0 \\ c & b \end{pmatrix} |abc \neq 0\right\}.$$

By direct computation, $\tau_n^*(M_{1,0} h_1^{p^2-1-p}) = M_{2,1} L_2^{p-2}$. Let us note that $M_{1,0} h_1^{p^2-1-p}$ is a basis element.

The same element can be computed by a different method:

$$\begin{aligned}(M_{1,0}h_1^{p^2-1-p})d_{2,0} &= (M_{1,0}h_1^{p^2-1-p})h_1^{p-1}h_2^{p-1}\\ &= M_{2,0}L_2^{p-2}h_1^{p^2-p} - M_{2,1}L_2^{p-2}d_{1,0}h_1^{p^2-p}\\ &= M_{2,0}L_2^{p-2}h_1^{p^2-p} - M_{2,1}L_2^{p-2}h_1^{p^2-1}\\ &= M_{2,0}L_2^{p-2}h_1^{p^2-p} - M_{2,1}L_2^{p-2}(d_{2,1}h_1^{p-1} - d_{2,0}).\end{aligned}$$

Hence,

$$\tau_n^*(M_{1,0}h_1^{p^2-1-p})d_{2,0} = M_{2,1}L_2^{p-2}d_{2,0}$$

The example implies that $\tau_n^* \neq \xi_n$ and $\tau_n^*(\_d_{n,0}) = \xi_n(\_d_{n,0})$.

Theorem 5.7 *Let $\xi_n : (E_n \otimes S_n)^{U_n} \longrightarrow (E_n \otimes S_n)^{GL_n}$ be the natural F_p-epimorphism and $\tau_n^* : (E_n \otimes S_n)^{U_n} \longrightarrow (E_n \otimes S_n)^{GL_n}$ the induced transfer. Then $\xi_n(xd_{n,0}) = \tau_n^*(xd_{n,0})$ for any $x \in (E_n \otimes S_n)^{U_n}$.*

Proof Let $M \in (E_n \otimes S_n)$. Then $\tau_n^*(Md_{n,0}) = \tau_n^*(M)d_{n,0}$. But $\tau_n^*(Md_{n,0}) = \tau_n^*(\sum_J M_{n;J}L_n^{p-2}h^{I(J)}) = \sum_J M_{n;J}L_n^{p-2}\tau_n^*(h^{I(J)}) = \sum_J M_{n;J}L_n^{p-2}\xi_n(h^{I(J)})$. It follows that if $\xi_n(h^{I(J)})$ is not divisible by $d_{n,0}$, then it is zero. It follows that τ_n^* and ξ_n agree on the ideal of $(E_n \otimes S_n)^{U_n}$ generated by $d_{n,0}$. □

We close this work with some remarks.

Remark 2 *1) Let $M \in SA(U_n)$, hence $M \in (E_n \otimes S_n)^{U_n}$ and $\hat{\tau}_n^*(M)$ is given by $\hat{\xi}_n(M)$.*

2) Kuhn and Priddy have considered subgroups where the transfer is multiplicative. We will present more results to that direction in [**10**].

Acknowledgements

First, we thank S. Priddy for his remarkable suggestions and encouragement. Also, we would like to thank H. E. A. Campbell and I. Hughes for access to their paper [**3**]. And special thanks to H.-V Henn and the Mathematics Institute of Heidelberg University for their gracious hospitality during our visit where this work started.

References

[1] Adem, A. and Milgram, R. J. [1994], *Cohomology of finite groups*, Springer-Verlag **309**.

[2] Arnon, D. [1994], *Generalized Dickson invariants*, Ph.D. Thesis, Massachusetts Institute of Technology.

[3] Campbell, H. E. A. and Hughes, I. P. [1996], *The ring of upper triangular invariants as a module over the Dickson algebra*, Mathematisches Annalen **306**, 429-443.

[4] Campbell, H. E. A. and Kechagias, Nondas, *Arnon completion of the Dyer-Lashof algebra*, preprint.

[5] Cohen, F., Lada, T. and May, J. P. [1975], *The homology of iterated loop spaces*, LNM 533.

[6] Dickson, L. E. [1911], *A fundamendal system of invariants of the general modular linear group with a solution of the form problem*, Trans. A. M. S. **12**, 75-98.

[7] Kahn, D. S. and Priddy, S. B. [1978], *On the transfer in the homology of symmetric groups*, Math. Proc. Camb. Phil Soc. **83**, 91-102.

[8] Kahn, D. S. and Priddy, S. B. [1978], *The transfer and stable homotopy theory*, Math. Proc. Camb. Phil Soc. **83**, 103-111.

[9] Kechagias, Nondas E. [1994], *Extended Dyer-Lashof algebras and modular coinvariants*, Manuscripta Math. **84**, 261-290.

[10] Kechagias, Nondas E., mod $-\,p$ *multiplicative transfer*, in preparation.

[11] Kuhn, N. J. [1985], *Chevalley group theory and the transfer in the homology of symmetric groups*, Topology Vol 24, No. 3. pp 247-264.

[12] Madsen, I. and Milgram, R. J. [1979], *The classifying spaces for surgery and cobordism of manifolds*, Princeton University Press **92**.

[13] Mui, Huyhn [1975], *Modular invariant theory and the cohomology algebras of the symmetric groups*, J. Fac. Sci. Univ. Tokyo, **IA**, 319-369.

[14] Priddy, S. B. [1973], *Transfer, symmetric groups, and stable homotopy theory*, Springer Lecture Notes in Mathematics **341**, 244-255.

[15] Wilkerson, C. [1983], *A primer on Dickson invariants*, Contemporary Math. **19**, 421-434.

Fields Institute Communications
Volume **19**, 1998

Projectivity of Im J, Cospherical Classes and Geometric Dimension

Kee Yuen Lam
Mathematics Department
University of British Columbia
Vancouver, B. C., Canada V6T 1Z2
lam@math.ubc.ca

Duane Randall
Department of Mathematics and Computer Science
Loyola University New Orleans
New Orleans, LA USA 70118
randall@beta.loyno.edu

Abstract. Cospherical classes in $\widetilde{KO}(P^n)$ are determined via projectivity properties of $Im\,J$ generators. This leads to new results on the geometric dimensions of such classes.

1 Introduction

Let $m\xi_n$ denote the m-fold Whitney sum of the non-trivial line bundle ξ_n based on a cell complex P^n having the homotopy type of n-dimensional real projective space. We recall that the geometric dimension of a stable vector bundle γ based on a finite, connected complex X, denoted by $gd(\gamma)$, is the minimum fiber dimension of all vector bundles stably equivalent to γ. We recall that a class y in the real K-theory of X is called *cospherical* if there exists a map $g : X \to S^t$ such that $y = g^*\gamma$ for some class γ in $\widetilde{KO}(S^t)$. The purpose of this short note is to stress the fact, hitherto unobserved, that some old results in [**5**], concerning projectivity properties of $\operatorname{Im} J$ generators, are all one needs to determine the subgroup $CS(P^n)$ of cospherical classes in $\widetilde{KO}(P^n)$. As a payoff, one can now obtain new estimates on the geometric dimension of these classes, using knowledge of $gd(\gamma)$for spheres in [**3**]. The process is similar to ([**9**], Theorem 3.2), which can be regarded as a forerunner of the present note.

2 Results

We recall that $\widetilde{KO}(P^n)$ is a cyclic group of order $2^{\phi(n)}$ where $\phi(n)$ counts the number of positive $i \leq n$ that are congruent to $0, 1, 2$, or $4 \bmod 8$. The canonical

1991 *Mathematics Subject Classification.* Primary 55S50; Secondary 55Q50, 55S25.
KYL is partially supported by an NSERC grant of Canada.

generator is given by $x_n = \xi_n - 1$. Let P_t^n denote the truncated projective space P^n/P^{t-1}. The principal result of this paper calculates the subgroup $CS(P^n)$ of cospherical classes.

Theorem 2.1 *The order of $CS(P^n)$ is given below for any integer $n > 31$ where c is defined by $2^c \le n < 2^{c+1}$.*

$$|CS(P^n)| = \begin{cases} 2^{c+1} & \text{if } \phi(n) \ge 2^{c-1}+c \\ 2^c & \text{if } \phi(n) = 2^{c-1}+c-1, \text{ or } \phi(n) < 2^{c-1}+c-1 \\ & \text{and } n \not\equiv 3 \bmod 4 \\ 2^c \text{or } 2^{c-1} & \text{if } \phi(n) < 2^{c-1}+c-1 \text{ and } n \equiv 3 \bmod 4. \end{cases}$$

The proof of Theorem 2.1 in Section 3 shows that the sphere-of-origin for a generator of $CS(P^n)$ is S^{2^c} whenever $\phi(n) \ge 2^{c-1}+c-1$, and is $S^{2^{c-1}}$ whenever $\phi(n) < 2^{c-1}+c-1$. Consequently, new geometric dimension estimates for cospherical classes follow from the geometric dimensions of bundles over spheres S^{2^i} by [**3**]. We define $\varepsilon(n)$ to be 1, 0, or -1 whenever $|CS(P^n)|$ in Theorem 2.1 is 2^{c+1}, 2^c, or the undecided case, respectively. A consequence of Theorem 2.1 and [**3**] is the following result.

Theorem 2.2 *Suppose $2^c \le n < 2^{c+1}$ with $c \ge 5$. If θ in $\widetilde{KO}(P^n)$ satisfies $2^{c+\varepsilon(n)}\theta = 0$, then*

$$gd(\theta) \le \begin{cases} 13 & \text{if } \varepsilon(n) \ne 1 \text{ and } c = 5 \\ 2^{c-1}+1 & \text{if } \varepsilon(n) = 1, \text{ or } \varepsilon(n) = 0 \text{ and } n \equiv 3 \bmod 4 \\ 2^{c-2}+1 & \text{for all other cases.} \end{cases}$$

The results of Theorem 2.2 may be strengthened for cospherical elements of lower order by employing "KO-periodic" maps as in [**8**]. We shall not enumerate all such estimates obtainable in this manner from (2.1); some of the precise geometric dimensions are still unknown, despite the work of Davis, Gitler and Mahowald [**4**] which determined the "stable geometric dimension." Rather, we shall present a sample result by the following.

Proposition 2.3 *For every integer $h \ge 0$, we have*

$$12 \le gd\left(2^{11+4h}\xi_{32+8h}\right) \le 13 \text{ and } 8 \le gd\left(2^{12+4h}\xi_{32+8h}\right) \le 9\ .$$

Proof $gd(2^{11}\xi_{32}) \le 13$ by (2.2) with the cosphericity of $2^{11}\xi_{32}$ realizable via a map $g_0 : P^{32} \to S^{16}$. The restriction of g_0 to P^{16} must be null-homotopic; otherwise, g_0 would realize $2^7 x_{32}$. We can affirm that the restriction of g_0 to P^{18} is null-homotopic similarly. Thus g_0 induces a map $g_1 : P_{19}^{32} \to S^{16}$ whose restriction to S^{19} represents an element u of order ≤ 2 in $\pi_{19}(S^{16})$. If $u = 12\nu$, we would obtain a map $P_{20}^{32} \to S^{16} \cup_{12\nu} e^{20}$ of cohomological degree 1 in dimension 20. But this is incompatible with the Adams operation ψ^3 in KO-theory. Triviality of the 4-stem yields a map $g : P_{21}^{32} \to S^{16}$ realizing $2^{11}x_{32}$ when composed with $P^{32} \to P_{21}^{32}$. Since $gd(2\gamma) \le 9$ for the generator γ of $\widetilde{KO}(S^{16})$ by [**3**], (2.2) also gives $gd(2^{12}x_{32}) \le 9$. Using the following KO-periodic maps guaranteed by Theorem 3.3 of [**8**],

$$\cdots \longrightarrow P_{21+8h}^{32+8h} \longrightarrow \cdots \longrightarrow P_{29}^{40} \longrightarrow P_{21}^{32} \qquad (2.4)$$

one can compose with g and the obvious projection maps to conclude that $gd(2^{11+4h}\xi_{32+8h}) \leq 13$ and $gd(2^{12+4h}\xi_{32+8h}) \leq 9$. A strong nonsectioning result by Davis [**2**] affirms that $gd(2^{11+4h}\xi_{32+8h}) \geq 12$ and $gd(2^{12+4h}\xi_{32+8h}) \geq 8$. □

The techniques utilized in the proof of Theorem 2.1 together with Stiefel-Whitney class considerations determine $CS(P^m)$ for small m completely. We remark also that the subgroup of cospherical classes $\widetilde{K}(P^n)$ can be determined from Theorem 2.1.

Proposition 2.5 *$CS(P^m)$ is the subgroup of order 2 for $1 \leq m \leq 8$. The order of $CS(P^m)$ for any integer $8 < m < 32$ is given below, where $2^c \leq m < 2^{c+1}$.*

$$|CS(P^m)| = \begin{cases} 2^{c+1} & \text{if } \phi(m) \geq 2^{c-1}+c \\ 2^c & \text{if } \phi(m) = 2^{c-1}+c-1 \\ 2^{c-1} & \text{if } \phi(m) \leq 2^{c-1}+c-2 \end{cases}$$

3 Proof of Theorem 2.1

We first describe generators for $CS(P^n)$, which are produced via projectivity of the $Im\,J$ generator β_s in the stable s-stem. Now Theorem 7.1 of [**5**] affirms that $\beta_s : S^{8k+s} \to S^{8k}$ is a *projective* homotopy class on S^{8k}, provided $s \leq 8k-2$ and $\nu_2(8k) \leq \phi(s)$. Here $\nu_2(m)$ denotes the exponent of the highest power of 2 which divides m. Although a map $S^{16k-1} \to S^{8k}$ is not in the stable range, the inductive method in [**5**] still produces a projective class $\widehat{\beta} : S^{16k-1} \to S^{8k}$ with the same e-invariant as β_{8k-1}.

For $n \equiv 0, 1, 3$ or $7 \bmod 8$ with $2^c \leq n < 2^{c+1}$ and also $\phi(n) \geq 2^{c-1}+c$, projectivity of β_{n-2^c} (or $\widehat{\beta}$ in case $n = 2^{c+1}-1$) gives a map $g : P^n \to S^{2^c}$ such that the composite map

$$S^n \xrightarrow{\pi} P^n \xrightarrow{g} S^{2^c} \tag{3.1}$$

represents β_{n-2^c}. We shall prove that $g^*\gamma$ has order 2^{c+1} in $\widetilde{KO}(P^n)$ for any such map g in (3.1), where γ generates $\widetilde{KO}(S^{2^c})$. For $m \equiv 2, 4, 5$ or $6 \bmod 8$ with $2^c < m < 2^{c+1}$ and $\phi(m) \geq 2^{c-1}+c$, let n be the smallest integer $> m$ that is congruent to 3 or $7 \bmod 8$. Then any cospherical class of order 2^{c+1} in $\widetilde{KO}(P^n)$ produces the same in $\widetilde{KO}(P^m)$ through the inclusion $i : P^m \subset P^n$. For $2^c < m < 2^{c+1}$ with $\phi(m) = 2^{c-1}+c-1$, a generator for $CS(P^m)$ having order 2^c is given by

$$2^{\phi(m)-c}x_m = i^*2^{\phi(n)-(c+1)}x_n = i^*g^*\gamma \tag{3.2}$$

where $\phi(n) = 2^{c-1}+c$, and $g : P^n \to S^{2^c}$ yields a generator $g^*\gamma$ for $CS(P^n)$.

Let n be any integer with $n \equiv 0, 1, 3$ or $7 \bmod 8$ and $32 \leq 2^c \leq n < 2^{c+1}$ satisfying $\phi(n) \leq 2^{c-1}+c-1$. Set $s = n - 2^{c-1} - 2^{c-2}$. Theorem 7.1 of [**5**] affirms that the $Im\,J$ generator β_s is projective on the sphere of dimension $2^{c-1}+2^{c-2}$ since $3 \leq \nu_2(2^{c-1}+2^{c-2}) = c-2 \leq 2^{c-3} \leq \phi(s)$. Any map $g : P^n \to S^{2^{c-1}+2^{c-2}}$ such that $g \circ \pi$ represents β_s will produce a cospherical class $g^*\gamma$ of order 2^{c-1} in $\widetilde{KO}(P^n)$. Let m be any integer with $m \equiv 2, 4, 5$ or $6 \bmod 8$ and $32 \leq 2^c < m < 2^{c+1}$ satisfying $\phi(m) \leq 2^{c-1}+c-2$. Select the smallest integer $n > m$ that is congruent to 3 or $7 \bmod 8$. Then any cospherical class $g^*\gamma$ in $\widetilde{KO}(P^n)$ of order 2^{c-1} produces $i^*g^*\gamma$ of the same order in $\widetilde{KO}(P^m)$ where $i : P^m \subset P^n$.

In order to establish the orders of the cospherical classes produced via projectivity of the $Im\,J$ generators, we first note that J-theory produces an upper bound for the order of a cospherical class given by a map $g : P^n \to S^{8i}$ via the following diagram.

$$\begin{array}{ccc} \widetilde{KO}(P^n) & \xleftarrow{g^*} & \widetilde{KO}(S^{8i}) \\ \approx \big\downarrow J & & onto \big\downarrow J \\ J(P^n) & \xleftarrow[g^*]{} & J(S^{8i}). \end{array} \tag{3.3}$$

The 2-primary component of the finite cyclic group $J(S^{8i})$ has order $2^{\nu_2(8i)+1}$. If we use S^t in (3.1) for any t not divisible by 8, only cospherical classes whose order divides 8 could be produced by maps $g : P^n \to S^t$, since $\nu_2(t) + 1 \leq 3$. The order of $g^*\gamma$ divides 2^{c+1} for any map $g : P^n \to S^{2^c}$ by (3.3). In order to show that $g^*\gamma$ has order 2^{c+1} for the maps g constructed via projectivity of $\operatorname{Im} J$ generators with $\phi(n) \geq 2^{c-1} + c$, we first consider $n \equiv 7 \bmod 8$. Let $f = g \circ \pi : S^{8\ell-1} \to S^{8k}$ represent an $\operatorname{Im} J$ generator β_{n-2^c} where $8\ell - 1 = n$ and $8k = 2^c$. We consider the following diagram of cofibration sequences

$$\begin{array}{ccccccc} S^{8\ell-1} & \xrightarrow{f} & S^{8k} & \xrightarrow{i} & S^{8k} \cup_f e^{8\ell} & \xrightarrow{c} & S^{8\ell} \\ \| & & \big\uparrow g & & \big\uparrow G & & \| \\ S^{8\ell-1} & \xrightarrow{\pi} & P^{8\ell-1} & \longrightarrow & P^{8\ell} & \longrightarrow & S^{8\ell} \end{array} \tag{3.4}$$

Let α_{8k} and $\alpha_{8\ell}$ be generators for $\widetilde{KO}(S^{8k} \cup e^{8\ell}) \approx Z \oplus Z$ such that $i^*\alpha_{8k}$ generates $\widetilde{KO}(S^{8k})$, and $\alpha_{8\ell}$ belongs to $c^*\widetilde{KO}(S^{8\ell})$. We recall from [**1**] that the e-invariant of $\beta_{8(\ell-k)-1}$ is given by $\frac{odd}{m}$ in Q/Z where $\nu_2(m) = 4 + \nu_2(\ell - k)$. We claim that the integer t defined by the equation

$$\psi^3(\alpha_{8k}) = 3^{4k}\alpha_{8k} + t\alpha_{8\ell} \tag{3.5}$$

must be odd, as dictated by the e-invariant of f. For appropriate generators y_{8k} and $y_{8\ell}$ of $\widetilde{H}^*(S^{8k} \cup e^{8\ell}; Z)$, we have $ch(\alpha_{8k}) = y_{8k} + \lambda y_{8\ell}$ and $ch(\alpha_{8\ell}) = y_{8\ell}$. The rational number λ projects to $e(\beta_{8(\ell-k)-1})$ in Q/Z. By the following Chern character calculation as in [**6**], we obtain

$$3^{4k}y_{8k} + \lambda 3^{4\ell}y_{8\ell} = ch(\psi^3\alpha_{8k}) = 3^{4k}(y_{8k} + \lambda y_{8\ell}) + ty_{8\ell}\ . \tag{3.6}$$

So $t = \lambda 3^{4k}(3^{4(\ell-k)} - 1)$ must be *odd*, since $\nu_2(3^{4(\ell-k)} - 1) = \nu_2(\ell - k) + 4$. To determine the order of the cospherical class $rx_n = g^*\gamma = g^*(i^*\alpha_{8k})$, we deduce from (3.5) that

$$rx_{n+1} = G^*\alpha_{8k} = \psi^3(G^*\alpha_{8k}) = 3^{4k}(rx_{n+1}) + t2^{4\ell-1}x_{n+1}. \tag{3.7}$$

Thus rx_n has order $2^{\nu_2(k)+4}$ in $\widetilde{KO}(P^n)$.

The above argument can also be used for $n \equiv 0$, 1 or 3 mod 8. The integer t in (3.5) is again *odd* because of the e-invariant of an $\operatorname{Im} J$ generator, even though the relevant $\widetilde{KO}$ groups are now of the form $Z \oplus B$ where $B = Z$ or $Z/2$. Consequently, $rx_n = g^*\gamma$ has order $2^{\nu_2(k)+4}$ in $\widetilde{KO}(P^n)$ by (3.7). The cospherical classes produced via projectivity have order 2^{c+1} whenever $\phi(n) \geq 2^{c-1}+c$, and so must be generators for $CS(P^n)$ by (3.3).

Only $S^{2^{c-1}}$ or S^{2^c} can be spheres-of-origin for a cospherical class of order 2^c in $\widetilde{KO}(P^m)$ for $2^c \le m < 2^{c+1}$. For any integer n with $2^c \le n < 2^{c+1}$ and $\phi(n) < 2^{c-1} + c$, there can be no cospherical class of order 2^{c+1} in $\widetilde{KO}(P^n)$ for the following dimensional reason. Suppose there were a map $g : P^n \to S^{2^c}$ with $g^*\gamma = 2^{\phi(n)-(c+1)}x_n$. Since $\phi(n) - c - 1 < 2^{c-1} - 1$, the null-homotopic composite map $P^{2^c-1} \overset{i}{\subset} P^n \xrightarrow{g} S^{2^c}$ yields the contradiction that $(g \circ i)^*\gamma = 2^{\phi(n)-c-1}x_{2^c-1} \ne 0$. Consequently, $2^{c-1} \le |CS(P^n)| \le 2^c$ for $2^c \le n < 2^{c+1}$ with $\phi(n) < 2^{c-1} + c$.

Finally, we employ KO-isomorphisms constructed in [**8**] in order to produce cospherical classes of order 2^c in $\widetilde{KO}(P^n)$ for $2^c \le n < 2^{c+1}$ with $\phi(n) \le 2^{c-1}+c-2$ and $n \not\equiv 3 \bmod 4$. We assume $c \ge 6$. By the previous arguments, we have produced cospherical maps $f : P^{2^c-8} \to S^{2^{c-1}}$, $g : P^{2^c-6} \to S^{2^{c-1}}$ and $h : P^{2^c-2} \to S^{2^{c-1}}$ of order 2^c. These maps factor as follows:

$$\begin{aligned} f &: P^{2^c-8} \to P_\ell^{2^c-8} \xrightarrow{\hat{f}} S^{2^{c-1}} \\ g &: P^{2^c-6} \to P_\ell^{2^c-6} \xrightarrow{\hat{g}} S^{2^{c-1}} \\ h &: P^{2^c-2} \to P_\ell^{2^c-2} \xrightarrow{\hat{h}} S^{2^{c-1}} \end{aligned} \tag{3.8}$$

where $\ell = 2^{c-1} + 3$, or $2^{c-1} + 5$ for $\hat{h}$. By Theorem 3.3 of [**8**], we obtain KO-isomorphisms

$$P_{\ell+16}^{2^c+8} \to P_{\ell+8}^{2^c} \to P_\ell^{2^c-8}\,, \quad P_{\ell+8}^{2^c+2} \to P_\ell^{2^c-6}\,, \quad P_{\ell+8}^{2^c+6} \to P_\ell^{2^c-2}$$

whose compositions with $\hat{f}$, $\hat{g}$, and $\hat{h}$, respectively, and the obvious collapsing maps yield cospherical classes of order 2^c:

$$\begin{aligned} F &: P^{2^c} \to S^{2^{c-1}}\,, & G &: P^{2^c+2} \to S^{2^{c-1}}\,, \\ H &: P^{2^c+6} \to S^{2^{c-1}}\,, & \widetilde{F} &: P^{2^c+8} \to S^{2^{c-1}}. \end{aligned}$$

The restriction of H to P^{2^c+4} and also P^{2^c+5} produces cospherical classes of order 2^c. Employing the map $\phi : P_{\ell+8}^{2^c+1} \to P_{\ell+8}^{2^c}$ constructed in Theorem 4.1 of [**5**] yields a cospherical class of order 2^c via

$$P^{2^c+1} \to P_{\ell+8}^{2^c+1} \xrightarrow{\phi} P_{\ell+8}^{2^c} \to P_\ell^{2^c-8} \xrightarrow{\hat{f}} S^{2^{c-1}}\,.$$

Finally, we remark that restrictions of $\widetilde{F}$ to P^{2^c+7} and H to P^{2^c+3} yield only cospherical classes of order 2^{c-1}. In this manner, we proceed inductively to construct via KO-isomorphisms the desired cospherical classes of order 2^c for all $2^c \le n < 2^{c+1}$ with $\phi(n) \le 2^{c-1} + c - 2$ and $n \not\equiv 3 \bmod 4$.

Finally, we consider $c = 5$. We obtain a cospherical class of order 32 via $h : P_{19}^{31} \to S^{16}$ from projectivity of the $\operatorname{Im} J$ generator β_{15}. The composite map

$$P_{19}^{34} \xrightarrow{\phi} P_{19}^{33} \xrightarrow{\phi} P_{19}^{32} \xrightarrow{\phi} P_{19}^{31} \xrightarrow{h} S^{16}$$

produces cospherical classes of order 32 in $\widetilde{KO}(P^n)$ for $32 \le n \le 34$, where ϕ is the map of Theorem 4.1 of [**5**]. The composite

$$P_{27}^{38} \xrightarrow{g} P_{19}^{30} \subset P_{19}^{31} \xrightarrow{h} S^{16}$$

yields cospherical classes of order 32 for $36 \le n \le 38$, where g is a KO-equivalence. □

Remark It is regrettable that the undecided cases in Theorem 2.1 cannot be settled yet. We conjecture that $CS(P^n)$ has order 2^c whenever $\phi(n) < 2^{c-1} + c - 1$

and $n \equiv 3 \bmod 4$ in (2.1). This conjecture has been confirmed for some cases by delicate refinements of the techniques of [**5**] and [**8**].

Acknowledgement

We express our gratitude to the referee for valuable suggestions.

References

[1] Adams, J. F. [1966], *On the groups $J(X) - IV$*, Topology **5**, 21–71.
[2] Davis, D. M. [1984], *Desuspension of stunted projective spaces*, Pac. J. Math. **113**, 35–49.
[3] Davis, D. M. and Mahowald, M. [1989], *The $SO(n)$-of-origin*, Forum Math. **1**, 1–12.
[4] Davis, D. M., Gitler, S. and Mahowald, M. [1981], *The stable geometric dimension of vector bundles over real projective spaces*, Trans. Amer. Math. Soc. **268**, 39-61. Corrections **280** (1983), 841-843.
[5] Feder, S., Gitler, S. and Lam, K. Y. [1977], *Composition properties of projective homotopy classes*, Pacific J. Math. **68**, 47–61.
[6] Harris, B. [1969], *The Adams e-homomorphism and non-stable homotopy groups*, Topology **8**, 337–344.
[7] Hoffman, P. [1966], *On the unstable e-invariant*, Topology **4**, 343–349.
[8] Lam, K. Y. [1979], *KO-equivalences and existences of nonsingular bilinear maps*, Pacific J. Math. **82**, 145–154.
[9] Lam, K. Y. and Randall, D. [1995], *Geometric dimension of bundles on real projective spaces*, Contemporary Math. Vol. 188, Amer. Math. Soc., 137–160.
[10] Lam, K. Y. and Randall, D. [1992], *Low dimensional spinor representations, Adams maps and geometric dimension*, Adams Memorial Symposium on Algebraic Topology, London Math. Soc. Lecture Notes Series **195**, 89–102.

Fields Institute Communications
Volume **19**, 1998

Inert and Lazy n-cones

Jean-Michel Lemaire
Laboratoire de Mathématiques Jean Dieudonné
Unité Mixte de Recherche
CNRS Université de Nice Sophia-Antipolis
Parc Valrose, F-06108 Nice Cedex 2, France
Current address: Directeur adjoint
Département Sciences Physiques et Mathématiques
CNRS, 3 rue Michel-Ange
F-75794 Paris Cedex 16, France
jean-michel.lemaire@cnrs-dir.fr

To John Moore

Abstract. Let $f : Y \to X$ be a map between 1-connected spaces, both having finite-type homology with coefficients in a field κ. Let Z be the mapping cone of f, F be the fibre of the inclusion $i : X \to Z$, and $\delta : \Omega Z \to F$ be the connecting map. One says that f is inert (resp. lazy) when $H_*(\Omega i)$ is surjective (resp. $H_*(\delta)$ is trivial). An inert map is lazy. We show that the converse holds when X is a suspension, i.e., when Z is a 2–cone, and also when X is a 2–cone under a mild additional hypothesis. This follows from considering the 'small' spectral sequence for the path space fibration over an n–cone.

1 Introduction

A 1-connected space Z is a n-cone, $n \geq 1$ if there exists a sequence of cofibrations

$$Y_k \xrightarrow{f_k} Z_{k-1} \xrightarrow{i_k} Z_k, \quad i = 1, \dots, n$$

with $Z_0 = *, Y_k$ connected for all k and $Z_n \simeq Z$. By a result of O. Cornéa [**5**], one may assume without restriction that $Y_k \simeq \Sigma^{k-1} W_k$ is a $(k-1)$-fold suspension. In particular a space Z_2 is a 2-cone if it has the homotopy type of the mapping cone of a map $f : \Sigma W_2 \to \Sigma W_1$ between suspensions of connected spaces W_1 and W_2.

The loop space homology $H_*(\Omega Z_2; \kappa)$ of Z_2 is reasonably well understood, at least with coefficients in a field κ and under the usual finite type assumption ([**17, 9, 12**]). By the Bott-Samelson theorem [**4**], the Pontryagin algebra $H_*(\Omega\Sigma W_i; \kappa), i = 1, 2$ is isomorphic to the tensor algebra $T(U_i)$ on the graded κ-vector space $U_i = \tilde{H}_*(W_i)$, and the algebra $H_*(\Omega Z_2; \kappa)$ is isomorphic to the homology of the 'two-level' differential algebra $(T(U_1 \oplus sU_2), d)$ where $d|U_1 = 0$ and $d : sU_2 \to T(U_1)$ is

1991 *Mathematics Subject Classification.* 55P35, 55T10, 16P99.
The author is partially supported by CNRS.

(up to suspension) the map $\tilde{f}_* : \tilde{H}_*(W_2) \to H_*(\Omega\Sigma W_1)$ induced by the adjoint $\tilde{f}$ of f.

Actually $(T(U_1 \oplus sU_2), d)$ admits a bigrading, defined by putting $U_i, i = 1, 2$ in first degree 0, and giving the suspension operator s the bidegree $(1, 0)$, so that the differential d has bidegree $(-1, 0)$. Setting $H_{*,*} = H((T(U_1 \oplus sU_2), d)$ then $H_{0,*}$ is the quotient $T(U_1)/(dsU_2)$ of $T(_1)$ by the 2-sided ideal generated by dsU_2, in other words the algebra cokernel of the map $\Omega f_* : H_*(\Omega\Sigma W_2) \to H_*(\Omega\Sigma W_1)$. Moreover, $H_{*,*}$ is the tensor algebra over the algebra $H_{0,*}$ generated by $H_{1,*}$ [**3**]. For more details on the homology of the chain algebra $H((T(U_1 \oplus sU_2), d)$, see [**3, 9**].

A remarkable special case is when $H((TU_1 \oplus sU_2), d)$ is concentrated in first degree zero, in other words isomorphic to $T(U_1)/(dsU_2)$, or equivalently, when $H_*(\Omega i_2)$ is surjective. When this holds one says ([**14**]) that (dsU_2) is an *inert* ideal in $T(U_1)$.

More generally, let $f : Y \to X$ be any map, with X and Y simply connected, and let again $Z = X \cup_f CY$ be its mapping cone. Let F be the homotopy fibre of the inclusion $i : X \hookrightarrow Z$, and $\delta : \Omega Z \to F$ be the connecting map, i.e., the action of the loop space ΩZ on the fibre F, evaluated at the base point. From [**15**], we recall the following

Definition Let $f : Y \to X$ be as above.

(i) f is *inert* if $H_*(\Omega i) : H_*(\Omega X) \to H_*(\Omega Z)$ is surjective
(ii) The map f is *lazy* if $H_*(\delta) : H_*(\Omega Z) \to H_*(F)$ is the trivial map, i.e., $H_+(\delta) = 0$.

An inert map is lazy, but the converse is not true in general. In this note we wish to compare these two notions for the structure maps $f_k : \Sigma^{k-1}W_k \to Z_{k-1}$ of an n-cone. We shall show that f_2 is inert if and only if it is lazy, and that f_3 is inert if it is lazy and if in addition $H_*(i_2) : H_*(\Omega Z_2) \to H_*(\Omega Z_3)$ is injective on the algebra cokernel of $H_*(\Omega f_2)$.

Let us recall that the depth of a κ-algebra R is the least integer d such that $0 \neq \mathrm{Ext}^d_R(\kappa, R)$. If $f_k : \Sigma^{k-1}W_k \to Z_{k-1}$ is not lazy, the algebra $H_*(\Omega Z_k)$ has depth ≤ 1 ([**8, 15**]. Now a depth 1 algebra either grows exponentially, (in which case it contains a tensor algebra on two generators), or is linearly isomorphic to the tensor product of a polynomial algebra on a single generator and a finite dimensional algebra [**13, 6**]. Thus our results aim at understanding the growth of $H_*(\Omega Z_k)$.

In the case of 2-cones, this note answers a question Yves Félix raised during the 'Unstable Homotopy' session at The Fields Institute in May-June 1996. The author wishes to thank Yves Félix for asking this question and The Fields Institute for its hospitality.

2 The small spectral sequence of an n-cone

As in the introduction, we consider an n–cone $Z = Z_n$ defined by the sequence of cofibrations

$$\Sigma^{k-1}W_k \xrightarrow{f_k} Z_{k-1} \xrightarrow{i_k} Z_k, \quad i = 1, \dots, n$$

with $Z_0 = *$ and W_k connected for all k. All homology groups will have coefficients in a fixed field κ, which we shall omit in the notation. We denote by F_k the fibre of the inclusion $i_k : Z_{k-1} \to Z_k$, and by G_k the fibre of the inclusion $j_k : Z_{k-1} \to Z_n = Z$. We shall write $F = F_n = G_n$ for short. Note that $G_1 = \Omega Z$.

Let us consider the standard path space fibration $\Omega Z \to EZ \xrightarrow{p} Z$, where EZ is the space of paths in Z ending at the base point. Then G_k is the subspace of paths which start in Z_{k-1},

$$\begin{array}{ccccccccc} \Omega Z \subset & G_2 \subset & \cdots \subset & G_k & \subset \cdots \subset & F & \subset EZ \\ \downarrow & \downarrow p & & \downarrow p & & \downarrow p & \downarrow p \\ * \subset & Z_1 \subset & \cdots \subset & Z_{k-1} \subset & \cdots \subset & Z_{n-1} \subset & Z \end{array}$$

and the filtration of EZ by the G_k's gives rise to a spectral sequence, with

$$\begin{aligned} E^1_{0,l} &= H_l(\Omega Z) \\ E^1_{k,l} &= H_{k+l}(G_{k+1}, G_k), k = 1, \dots, n-1 \\ E^1_{n,l} &= H_{n+l}(EZ, F) \\ E^1_{k,l} &= 0, k < 0 \text{ and } k > n. \end{aligned}$$

and where the differential $d^1_{k,l} : E^1_{k,l} \to E^1_{k-1,l}$ is the connecting map in the homology exact sequence of the triple (G_{k+1}, G_k, G_{k-1}). We shall call this spectral sequence the 'small' spectral sequence for the loop space on the n–cone Z.

The right action of ΩZ on EZ restricts to the G_k's, and induces a structure of (right) $H_*(\Omega Z)$-module on this small spectral sequence.

Proposition 2.1 *For $k = 1 \dots n$, there are isomorphisms of right $H_*(\Omega Z)$–modules*

$$H_*(\Lambda_k) : H_*(W_k, *) \otimes H_*(\Omega Z) \cong E^1_{k,*}$$

Proof Let $\lambda_k : (C\Sigma^{k-1}W_k, \Sigma^{k-1}W_k) \to (G_{k+1}, G_k)$ map (t, y) to the path $s \mapsto (st, y)$. The map λ_k lifts the canonical map $(C\Sigma^{k-1}W_k, \Sigma^{k-1}W_k) \to (Z_k, Z_{k-1})$ through p. By a standard spectral sequence argument, the composite

$$\Lambda_k : (C\Sigma^{k-1}W_k, \Sigma^{k-1}W_k) \times \Omega Z \xrightarrow{\lambda_k \times \Omega Z} (G_{k+1}, G_k) \times \Omega Z \xrightarrow{\mu} (G_{k+1}, G_k)$$

where μ is the holonomy action, is a homology isomorphism (compare [**11**]). □

Note that $d^1_k : H_*(W_k, *) \otimes H_*(\Omega Z) \to H_*(W_{k-1}, *) \otimes H_*(\Omega Z)$ has (complementary) degree zero, and that d^r_k has degree $r - 1$.

Let us denote ${}^mE^r$ the small spectral sequence for the m-cone Z_m. The inclusion $i_{m,n} : Z_m \hookrightarrow Z_n = Z$ induces a morphism of spectral sequences ${}^mE^r_{*,*} \to {}^nE^r_{*,*}$.

Lemma 2.2 *For $k \leq m$, the diagram*

$$\begin{array}{ccc} {}^mE^1_{k,*} \otimes_{H_*(\Omega Z_m)} H_*(\Omega Z) & \xrightarrow{{}^md^1 \otimes id} & {}^mE^1_{k-1,*} \otimes_{H_*(\Omega Z_m)} H_*(\Omega Z) \\ \downarrow \cong & & \downarrow \cong \\ E^1_{k,*} & \xrightarrow{d^1} & E^1_{k-1,*} \end{array} \qquad (2.1)$$

where the vertical isomorphisms are induced by $i_{m,n}$ and the right action of $H_*(\Omega Z)$, is commutative.

Proof This follows from considering the diagram

$$\begin{array}{ccccc}
(C\Sigma^{k-1}W_k, \Sigma^{k-1}W_k) \times \Omega Z_m & \xrightarrow{\lambda_k \times \Omega Z_m} & ({}^mG_{k+1}, {}^mG_k) \times \Omega Z_m & \xrightarrow{\mu} & ({}^mG_{k+1}, {}^mG_k) \\
\downarrow{\scriptstyle id \times \Omega i_{m,n}} & & \downarrow & & \downarrow \\
(C\Sigma^{k-1}W_k, \Sigma^{k-1}W_k) \times \Omega Z & \xrightarrow{\lambda_k \times \Omega Z} & (G_{k+1}, G_k) \times \Omega Z & \xrightarrow{\mu} & (G_{k+1}, G_k)
\end{array} \tag{2.2}$$

which is obviously commutative. □

In particular, $d_1^1 : H_*(W_1,*) \otimes H_*(\Omega Z) \to H_*(\Omega Z)$ is the $H_*(\Omega Z)$-linear extension of the map $(\tilde{i}_{1,n})_*$ induced by the adjoint of $i_{1,n} : \Sigma W_1 \hookrightarrow Z$. Now in the small spectral sequence $E_0^2 = \operatorname{Coker} d_1^1$, therefore $E_0^2 = H_*(\Omega Z)/(\tilde{i}_{1,n})_* H_*(W_1,*).H_*(\Omega Z)$. Since $H_*(\Omega\Sigma W_1) \cong T(H_*(W_1,*))$, we obtain

Proposition 2.3 *There is an isomorphism of $H_*(\Omega Z)$-modules*

$$E_{0,*}^2 \cong \kappa \otimes_{H_*(\Omega\Sigma W_1)} H_*(\Omega Z).$$

□

Next, we observe that since the small spectral sequence $E_{k,l}^r$ is concentrated in the band $0 \leq k \leq n$, one has $E^{n+1} = E^\infty = \kappa$ concentrated in bidegree $(0,0)$. Moreover at the E^n level there is (at most) a single non-trivial $d_n^n : E_{n,*}^n \to E_{0,*}^n$, which is an isomorphism of degree $(n-1)$ onto $E_{0,+}^n$. Note that, since $d_k^n = 0$ for $k \neq n$, one has $E_{k,*}^n = E_{k,*}^{n+1} = 0$ for $k \neq 0, n$.

Finally, we have

Proposition 2.4 *There is an isomorphism*

$$E_{0,*}^n \cong \operatorname{Im}(H_*(\delta) : H_*(\Omega Z) \to H_*(F))$$

Proof Let $\{(E'^r, d'^r)\}$ be the small spectral sequence for the restriction of p to $F = p^{-1}Z_{n-1}$, defined by the filtration of Z_{n-1} by the Z_k's, $k \leq n-1$. Obviously, one has $E'^1_k = E^1_k$ and $d'1_k = d^1_k$ for $k < n$, and $E'^1_n = 0$. Thus one has $E'^r_k = E^r_k$ for $k \leq n-r$, and therefore $E'^n_0 = E^n_0$. But $E'^n = E'^\infty$ is the graded vector space associated to the filtration of $H_*(F)$, and the edge $H_*(\Omega Z) = E_0^1 \twoheadrightarrow E_0^n \subset H_*(F)$ coincides with $H_*(\delta)$ as usual. □

Corollary 2.5 *The map $f_n : \Sigma^{n-1}W_n \to Z_{n-1}$ is lazy if and only if $E_0^n = \kappa$, or equivalently $d_n^n = 0$, in the small spectral sequence for the n-cone Z_n.* □

3 Lazy 2-cones

If $n = 2$, the small spectral sequence has at most 6 non-zero terms, which we can display as follows

$$\begin{array}{ccccc}
\tilde{H}_*(W_2,*) \otimes H_*(\Omega Z_2) & \xrightarrow{d_2^1} & \tilde{H}_*(W_1,*) \otimes H_*(\Omega Z_2) & \xrightarrow{d_1^1} & H_*(\Omega Z_2) \\
\uparrow & & & & \downarrow \\
E_2^2 & & \xrightarrow[0]{d_2^2} & & \kappa \otimes_A H_*(\Omega Z_2) \\
& & & & \downarrow \\
& & & & \kappa
\end{array} \tag{3.1}$$

where A denotes the image of $H_*(\Omega\Sigma W_1)$ in $H_*(\Omega Z_2)$. This spectral sequence is used in [**16, 17**]. It is due to John Moore, who showed it to the author in 1968.

By Corollary 2.5, $f = \Sigma W_2 \to \Sigma W_1$ is lazy iff $E_0^2 = \kappa \otimes_A H_*(\Omega Z_2) = \kappa$, that is, if $A = H_*(\Omega Z_2)$ or in other words iff $H_*(\Omega\Sigma W_1) \xrightarrow{i} H_*(\Omega Z_2)$ is surjective. Thus f is lazy iff it is inert. Note that the argument does not actually depend upon the fact that the source of f is a suspension. This merely insures that E_2^1 is a free $H_*(\Omega Z_2)$–module, but we did not use this here. Thus we have proven

Proposition 3.1 *A map into a suspension is inert iff it is lazy.* □

Note that in the inert case the small spectral sequence reduces to the exact sequence

$$0 \to H_*(W_2,*) \otimes H_*(\Omega Z_2) \xrightarrow{d_2^1} H_*(W_1,*) \otimes H_*(\Omega Z_2) \xrightarrow{d_1^1} H_*(\Omega Z_2) \twoheadrightarrow \kappa \to 0$$

which is a projective resolution of κ as a $H_*(\Omega Z_2)$–module. Thus $\operatorname{gldim} H_*(\Omega Z_2) \leq 2$, $\operatorname{depth} H_*(\Omega Z_2) \leq 2$, and when $\operatorname{depth} H_*(\Omega Z_2) = 2$ the exact sequence above yields a non-zero element in $\operatorname{Ext}^2_{H_*(\Omega Z_2)}(\kappa, H_*(\Omega Z_2)) \otimes H_*(W_2,*)$.

In the non-inert case, one has $\operatorname{gldim} H_*(\Omega Z_2) = 1$. If we assume that $\Sigma W_2 = \bigvee_i S^{n_i}$ is a wedge of spheres (this is not a restriction if $\kappa = \mathbb{Q}$!), then one easily shows that there exists at least one index i such that the attachment

$$S^{n_i} \to \Sigma W_1 \cup \bigvee_{j \neq i} e^{n_j+1}$$

is not lazy, and therefore defines a non-zero element in $\operatorname{Ext}^1_{H_*(\Omega Z_2)}(\kappa, H_*(\Omega Z_2))$.

4 Lazy 3-cones

The space $Z = Z_3$ is now defined by the two cofibrations sequences

$$\Sigma W_2 \xrightarrow{f_2} \Sigma W_1 \xrightarrow{i_2} Z_2 \quad , \quad \Sigma^2 W_3 \xrightarrow{f_3} Z_2 \xrightarrow{i_3} Z$$

and the small spectral sequence has at most 11 non-zero terms :

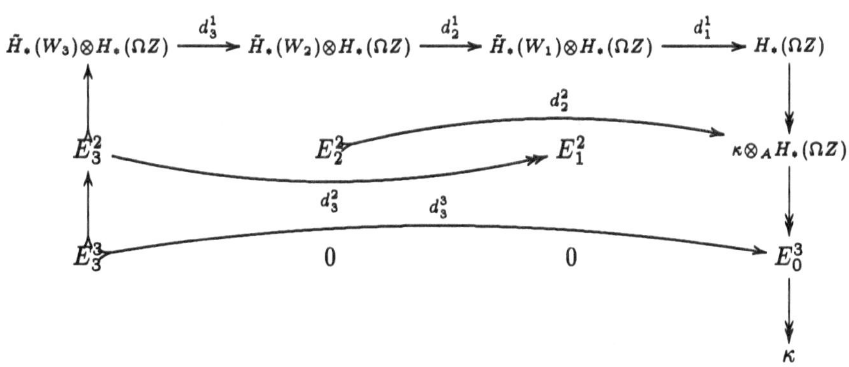

(4.1)

Let us denote again by A the image of $H_*(\Omega i_2) : H_*(\Omega\Sigma W_1) \to H_*(\Omega Z_2)$. We recall from the introduction that $H_*(\Omega Z_2)$ admits a bigrading, such that $A = H_{*,*}(\Omega Z_2)$. This bigrading induces a trigrading on the small spectral sequence ${}^2E^r$, such that

$${}^2E^1_{k,s,t} \cong \bigoplus_{t'+t''=t} \tilde{H}_{t'}(W_k) \otimes H_{s,t''}(\Omega Z_2) \quad , \quad k = 1, 2$$

such that ${}^2d^r$ has tridegree $(-r, r-1, 0)$. In particular, the component of second degree 0 in the ${}^2E^1$ complex reads

$$\tilde{H}_*(W_2) \otimes A \xrightarrow{d^1_{2,0,*}} \tilde{H}_*(W_1) \otimes A \xrightarrow{d^1_{1,0,*}} A \tag{4.2}$$

Since ${}^2E^2_1 = 0$, the sequence (4.2) is exact and the complex ${}^2E^1$ can be viewed as (4.2) tensored by $H_*(\Omega Z_2)$ over A. By Lemma 2.2, the sequence ${}^3E^1_2 \xrightarrow{{}^3d^1_2} {}^3E^1_1 \xrightarrow{{}^3d^1_1} {}^3E^1_0$ can in turn be viewed as (4.2) tensored by $H_*(\Omega Z)$ over A. In particular

Lemma 4.1 *The inclusion $i_{1,3}$ induces an injection $A \rightarrowtail H_*(\Omega Z)$ iff and only if ${}^3E^2_1 = 0$ in the small spectral sequence for $Z = Z_3$.*

Proof If A injects in $H_*(\Omega Z)$, the latter is a free A-module because the injection is a morphism of Hopf algebras [**18**]. Therefore $\cdot \otimes_A H_*(\Omega Z)$ is an exact functor. Conversely, consider the two sequences

$$\tilde{H}_*(W_2) \otimes A \to \tilde{H}_*(W_1) \otimes A \to A \twoheadrightarrow \kappa$$
$$\tilde{H}_*(W_2) \otimes H_*(\Omega Z) \to \tilde{H}_*(W_1) \otimes H_*(\Omega Z) \to H_*(\Omega Z) \twoheadrightarrow \kappa \otimes_A H_*(\Omega Z)$$

where the second is obtained by applying $\cdot \otimes_A H_*(\Omega Z)$ to the first. If they are exact, the change of ring morphism

$$\mathrm{Tor}^A_k(\kappa, \kappa) \to \mathrm{Tor}^{H_*(\Omega Z)}_k(\kappa \otimes_A H_*(\Omega Z), \kappa)$$

is bijective for $k = 1$ and surjective for $k = 2$. The change of ring spectral sequence

$$\mathrm{Tor}^{H_*(\Omega Z)}_p(\mathrm{Tor}^A_q(\kappa, H_*(\Omega Z)), \kappa) \Rightarrow \mathrm{Tor}^A_{p+q}(\kappa, \kappa)$$

then shows that $0 = E^2_{0,1} = \kappa \otimes_{H_*(\Omega Z)} \mathrm{Tor}^A_1(\kappa, H_*(\Omega Z))$, so $\mathrm{Tor}^A_1(\kappa, H_*(\Omega Z)) = 0$ and $H_*(\Omega Z)$ is A–free for reasons of connectivity. □

We can now prove

Proposition 4.2 *If $f_3 : \Sigma^2 W_3 \to Z_2$ is lazy and A injects in $H_*(\Omega Z)$, then f_3 is inert.*

Proof By Corollary (2.5) we have $E^3_0 = \kappa$ and $E^3_3 = 0$. By Lemma (4.1) we have $E^2_1 = 0$ and by the exactness of the sequence $0 \to E^3_3 \rightarrowtail E^2_3 \xrightarrow{d^2_3} E^2_1 \to 0$ we also get $E^2_3 = 0$. Thus the spectral sequence $E^r = {}^3E^r$ reduces to

$$\begin{array}{ccccccc}
\tilde{H}_*(W_3)\otimes H_*(\Omega Z) & \xrightarrow{d^1_3} & \tilde{H}_*(W_2)\otimes H_*(\Omega Z) & \xrightarrow{d^1_2} & \tilde{H}_*(W_1)\otimes H_*(\Omega Z) & \xrightarrow{d^1_1} & H_*(\Omega Z) \\
\uparrow & & & & & & \downarrow \\
0 & & E^2_2 & \xrightarrow{\quad d^2_2 \quad} & 0 & & \kappa\otimes_A H_*(\Omega Z) \\
 & & & & & & \downarrow \\
0 & & 0 & & 0 & & \kappa
\end{array} \tag{4.3}$$

We will now take the Poincaré series of the remaining terms. By a convenient abuse of notation, we will denote by $V = V(t) = \sum_{i=0}^{\infty} \dim(V_i)t^i$ the Poincaré series of the (positively) graded vector space $V = V_*$. Since A injects in $H_*(\Omega Z)$, there is an A–module isomorphism [**18**] $H_*(\Omega Z) \cong A \otimes (\kappa \otimes_A H_*(\Omega Z))$, from which we get $E_0^2 = H_*(\Omega Z).A^{-1}$. Now $E_0^1 - E_1^1 + E_2^1 - E_3^1 = E_2^2 + E_0^2$ and $tE_2^2 - E_0^2 + 1 = 0$ (d^2 has degree 1 !). Finally we get

$$H(\Omega Z)^{-1} = A^{-1}(1+t) - t(1 - \tilde{H}(W_1) + \tilde{H}(W_2) - \tilde{H}(W_3)),$$

while the same calculation performed on the small spectral sequence ${}^2E$ for Z_2 yields

$$H(\Omega Z_2)^{-1} = A^{-1}(1+t) - t(1 - \tilde{H}(W_1) + \tilde{H}(W_2)). \text{ (cf. [17])}$$

Thus

$$H(\Omega Z)^{-1} = H(\Omega Z_2)^{-1} + \tilde{H}(\Sigma W_3)$$

which insures by [**2**], 2.11 (see also [**14**] Theorem 2.5) that the algebra $H_*(\Omega Z)$ is the quotient of $H_*(\Omega Z_2)$ by the *inert* ideal generated by the image of $H_*(\tilde{f}_3)$: $\tilde{H}_*(\Sigma W_3) \to H_*(\Omega Z_2)$. Thus f_3 is inert. □

In plain words, Proposition 4.2 means that in a 3-level CW-complex, the attachment of the top cells is inert if it is lazy and if the top cells do not create new relations among the bottom cells.

In terms of Adams-Hilton models of the form $\mathcal{B} = (T(V_1 \oplus V_2 \oplus V_3), d)$, this condition will for instance be satisfied if there exists a grading on the differential algebra $\mathcal{B}$ such that the differential is homogeneous degree -1, V_1 lies in degree 0, V_2 in degree 1 and dV_3 lies in degrees > 0. This clearly insures that $A = H_0\mathcal{B}$ injects in $H_*\mathcal{B}$.

References

[1] Adams, J. F. and Hilton, P. J. [1955], *On the chain algebra of a loop space*, Comm. Math. Helv. **30**, 305–330.

[2] Anick, D. [1982], *Non-commutative graded algebras and their Hilbert series*, J. Algebra **78**, 120–140.

[3] Baues, H., Félix, Y. and Thomas, J.-C. [1992], *Presentations of algebras and the Whitehead Γ-functor for chain algebras*, J. Algebra **48**, 123–134.

[4] Bott, R. and Samelson, H. [1953], *On the Pontryagin product in the space of paths*, Comm. Math. Helv. **27**, 320–337.

[5] Cornea, O. [1995], *Strong LS category equals cone-length*, Topology **34** 2, 377–381.

[6] Bøgvad, R. and Jacobson, C. [1989], *Graded Lie algebras of depth one*, Manuscr. Math. **66**, 153–159.

[7] Félix, Y., Halperin, S., Lemaire, J.-M. and Thomas, J.-C. [1989], *Mod. p loop space homology*, Invent. Math. **95**, 247–262.

[8] Félix, Y. and Lemaire, J.-M. [1991], *On the Pontryagin algebra of the loops on a space with a cell attached*, Int. J. Math. **2**, 429–438.

[9] Félix, Y. and Lemaire, J.-M. [1992], *On the homology of two-level differential algebras*, Bull. Soc. Math. Belgique **44**, 1, Ser. B, 29–33.

[10] Félix, Y. and Thomas, J.-C. [1985], *Module d'holonomie d'une fibration*, Bull. Soc. Math. France **113**, 255–258.

[11] Félix, Y. and Thomas, J.-C. [1988], *The fibre-cofibre construction and its applications*, J. Pure Appl. Algebra **53**, 59–69.

[12] Félix, Y., Halperin, S. and Thomas, J.-C. [1990], *Loop space homology of spaces of LS category one and two*, Math. Ann. **287**, 377–386.

[13] Félix, Y., Halperin, S. and Thomas, J.-C. [1991], *Elliptic Hopf algebras*, Journ. London Math. Soc. (2)**43**, 545–555.

[14] Halperin, S. and Lemaire, J.-M. [1987], *Suites inertes dans les algèbres de Lie graduées ("Autopsie d'un meurtre II")*, Math. Scand. **61**, 39–67.

[15] Halperin, S. and Lemaire, J.-M. [1995], *The fibre of a cell attachment*, Proc. Edinburgh Math. Soc. **38**, 295–311.

[16] Lemaire, J.-M. [1969], *Un lemme sur l'homologie de certains espaces de lacets*, C. Rend. Acad. Sc. Paris **A 269**, 1122–1124.

[17] Lemaire, J.-M. [1974], *Algèbres Connexes et Homologie des Espaces de Lacets*, Lecture Notes in Mathematics **422**, Springer-Verlag.
[18] Milnor, J. W. and Moore, J. C. [1965], *On the structure of Hopf algebras*, Ann.Math. (2) **81**, 211–264.

Fields Institute Communications
Volume **19**, 1998

Mod 3 Truncated Polynomial Algebras over the Steenrod Algebra

James P. Lin
Department of Mathematics
University of California at San Diego
La Jolla, CA USA 92093–0112
jimlin@euclid.ucsd.edu

Abstract. Let X be a finite 3 torsion-free homotopy associative H-space. There exist nontrivial Steenrod operations connected algebra generators of $H^*(X;\mathbb{F}_3)$ if other Steenrod operations do not vanish.

Introduction

This paper is motivated by the following question:

Question: Is the mod 3 unstable action of the Steenrod algebra on the cohomology of a finite 3 torsion free, homotopy associative H-space the same as the unstable action on the cohomology of a finite 3 torsion free loop space?

Note that if X is a finite 3 torsion free homotopy associative H-space, then

$$H^*(X;\mathbb{F}_3) = \Lambda(x_{i_1},\dots,x_{i_k})$$

where the generators x_{i_ℓ} may be chosen to be primitive in odd degrees, by the Borel and Samelson Leray theorems [7]. We let degree $x_{i_\ell} = i_\ell$. Our result is

Main Theorem *Let X be a* mod 3 *homotopy associative finite H-space with*

$$H^*(X;\mathbb{F}_3) = \Lambda(x_{i_1},\dots,x_{i_k}).$$

Then if $j \geq 0$, $n > 0$ are integers

$$\mathcal{P}^{3^j} PH^{2(3^{j+1}n)+4(3^j)-1}(X;\mathbb{F}_3) = \mathcal{P}^{2(3^j)} PH^{2(3^{j+1}n)-1}(X;\mathbb{F}_3).$$

1991 *Mathematics Subject Classification.* Primary 55S20, 55S45, 55R35; Secondary 55D45, 55G20.

The unstable action of the Steenrod algebra on the odd generators of $H^*(X;\mathbb{F}_3)$ has been studied by several authors. Zabrodsky [12] proves there is no H-space X with

$$H^*(X;\mathbb{F}_3) = \Lambda(x_{6n+3}, \mathcal{P}^1 x_{6n+3}).$$

Hemmi [2] and Harper and Zabrodsky [1] have other theorems that describe under more technical assumptions, when odd generators always lie in the image of Steenrod operations.

The utility of our Main Theorem is that it holds for all homotopy associative mod 3 H-spaces with cohomology exterior. It proves there exist nontrivial Steenrod operations connecting odd degree algebra generators if other Steenrod operations do not vanish. For example for $j = 0$, the Main Theorem implies

$$\mathcal{P}^1 PH^{6n+3}(X;\mathbb{F}_3) = \mathcal{P}^2 PH^{6n-1}(X;\mathbb{F}_3).$$

We note that the assumption that some Steenrod operations not vanish on the generator is necessary. For example, the Lie group Spin $(3n+2)$ has mod 3 cohomology exterior and $H^*(\text{Spin }(3n+2);\mathbb{F}_3)$ has a primitive

$$x_{6n+3} \in PH^{6n+3}(\text{Spin }(3n+2);\mathbb{F}_3)$$

with

$$\mathcal{P}^i x_{6n+3} = 0 \quad \text{for all } i > 0$$

and x_{6n+3} does not lie in the image of any Steenrod operations. See [4, Appendix A]. By looking at the cohomology of Spin (k) it is possible to create generators in any odd degree in the cohomology of a mod 3 homotopy associative finite H-space that lie in the kernel of all Steenrod operations and also are not in the image of any Steenrod operations. For a complete list of the cohomology of the Lie groups, we refer the reader to [4, Appendix 2].

Our Main Theorem actually is a consequence of the following theorem about truncated polynomial algebras:

Theorem A *Let*

$$B = \frac{\mathbb{F}_3[y_{i_1}, \dots y_{i_k}]}{(y_{i_1}, \dots y_{i_k})^4}$$

be a finite generated polynomial algebra on generators of even degree truncated at height 4. *Suppose* B *admits an unstable action of the Steenrod algebra* $\mathcal{A}_3$. *Then if* $j \geq 0, n > 0$

$$\mathcal{P}^{3^j} QB^{2(3^{j+1}n)+4(3^j)} = \mathcal{P}^{2(3^j)} QB^{2(3^{j+1}n)}.$$

Theorem A is used to prove the Main Theorem in the following way. Hemmi [3, Theorem 1.1] proves if X satisfies the hypotheses of the Main Theorem, there is a space Y with

$$B \subseteq H^*(Y;\mathbb{F}_3).$$

Further, there is a map $\varepsilon : \Sigma X \to Y$. The induced map $\varepsilon^* : H^*(Y;\mathbb{F}_3) \to H^*(\Sigma X;\mathbb{F}_3)$ maps QB isomorphically onto $PH^*(X;\mathbb{F}_3)$ when $H^*(\Sigma X;\mathbb{F}_3)$ is identified with $H^{*-1}(X;\mathbb{F}_3)$. Thus the Steenrod action $PH^*(X;\mathbb{F}_3)$ is determined by the Steenrod action on QB.

We remark this is analogous to methods used by Wilkerson to classify finite loop spaces whose cohomology is exterior. In that case, the classifying space is a finitely generated polynomial algebra with an unstable Steenrod action. Via suspension, the algebra generators of the polynomial algebra are isomorphic to the primitives of the cohomology of the loop space.

Classification of the unstable action of the Steenrod algebra on full polynomial algebras is known. Wilkerson [11] and others provide a complete classification of all Noetherian, integrally closed integral domains with an unstable Steenrod action.

Notation: Throughout this paper $\mathcal{A}_3$ will denote the mod 3 Steenrod algebra and $\mathcal{A}_3^+$ will be the subHopf algebra of $\mathcal{A}_3$ generated by the symbols $\mathcal{P}^i$ for $i > 0$. The symbol B will be reserved for the truncated polynomial algebra described in the Main Theorem. All algebras will be connected and graded. Given C a connected graded algebra, let $I(C)$ denote the augmentation ideal. $QC = I(C)/I(C)^2$ will denote the module of indecomposables. If c is an element of C, $\bar{c}$ will denote its projection in QC. If C is an algebra over $\mathcal{A}_3^+$, the Cartan formula implies QC admits an action of the Steenrod algebra.

We assume the reader is familiar with the notation of the Steenrod algebra and various algebraic constructions. We follow the notation in [7, 9]. Occasionally we will use $\doteq$ to describe an Adem relation that holds modulo various constants.

1 $\mathcal{P}^1 QB^{6n+4} \subseteq im\mathcal{P}^2$

The proof of Theorem A begins with the case $j = 0$. Recall the Adem relations imply

$$\begin{aligned} &\mathcal{P}^{3n+2} \doteq \mathcal{P}^2\mathcal{P}^{3n} \doteq \mathcal{P}^1\mathcal{P}^{3n+1} \\ &\mathcal{P}^1\mathcal{P}^1\mathcal{P}^1 = 0, \quad \mathcal{P}^1\mathcal{P}^1 \doteq \mathcal{P}^2. \end{aligned} \tag{1.1}$$

Here the symbol $\doteq$ means the equations hold modulo constants. Note that $\mathcal{P}^1$ induces a map $\mathcal{P}^1 : QB^* \longrightarrow QB^*$. Let $im\mathcal{P}^i$ be the vector space image of QB^* under $\mathcal{P}^i$.

Further, we have the vector space isomorphisms

$$QB^* \cong \frac{(QB)^*}{im\mathcal{P}^1} \oplus \frac{im\mathcal{P}^1}{im\mathcal{P}^2} \oplus im\mathcal{P}^2. \tag{1.2}$$

There are the following maps induced by $\mathcal{P}^1$:

$$\begin{aligned} &\frac{(QB)^*}{im\mathcal{P}^1} \longrightarrow \frac{im\mathcal{P}^1}{im\mathcal{P}^2} \\ &\frac{im\mathcal{P}^1}{im\mathcal{P}^2} \longrightarrow im\mathcal{P}^2. \end{aligned} \tag{1.3}$$

Given $y \in B^{6n+4}$, let $\bar{y}$ be its projection in QB^{6n+4}. We wish to show $\mathcal{P}^1\bar{y} = \mathcal{P}^2\bar{z}$ for some $\bar{z} \in QB^*$.

For the remainder of the section we assume

$$\mathcal{P}^1\bar{y} \notin im\mathcal{P}^2, \quad \mathcal{P}^1\bar{y} \neq 0 \tag{1.4}$$

By (1.1) it follows that

$$\bar{y} \notin im\mathcal{P}^1 \tag{1.5}$$

Lemma 1.1 *There exists a basis $\{\bar{y}\}, \{\bar{v}_1\} \ldots \{\bar{v}_k\}$ for $\frac{QB^*}{im\mathcal{P}^1}$ and an integer $\ell \leq k$ such that $\{\mathcal{P}^1\bar{y}\}, \{\mathcal{P}^1\bar{v}_1\}, \ldots, \{\mathcal{P}^1\bar{v}_\ell\}$, is a basis for $\frac{im\mathcal{P}^1}{im\mathcal{P}^2}$.*

Proof By (1.3) $\{\mathcal{P}^1\bar{y}\}$ is nonzero in $\frac{im\mathcal{P}^1}{im\mathcal{P}^2}$. Extend $\{\mathcal{P}^1\bar{y}\}$ to $\{\mathcal{P}^1\bar{y}\}, \{\mathcal{P}^1\bar{v}_1\}, \ldots, \{\mathcal{P}^1\bar{v}_\ell\}$ a basis for $\frac{im\mathcal{P}^1}{im\mathcal{P}^2}$. Then extend $\{\bar{y}\}\{\bar{v}_1\}, \ldots, \{\bar{v}_\ell\}$ to a basis $\{\bar{y}\}\{\bar{v}_1\}, \ldots, \{\bar{v}_k\}$ for $\frac{QB^*}{im\mathcal{P}^1}$. □

Let $\mathcal{B}$ be a basis for QB^* formed by extending $\bar{y}, \bar{v}_1, \ldots, \bar{v}_k, \mathcal{P}^1\bar{y}, \mathcal{P}^1\bar{v}_1, \ldots, \mathcal{P}^1\bar{v}_\ell$ to a basis for QB^*. So $\mathcal{B}$ has the form

$$\mathcal{B} = \{\bar{y}, \bar{v}_1, \ldots, \bar{v}_k, \mathcal{P}^1\bar{y}, \mathcal{P}^1\bar{v}_1, \ldots, \mathcal{P}^1\bar{v}_\ell, \bar{w}_1, \ldots, \bar{w}_m\} \tag{1.6}$$

$$\hat{\mathcal{B}} = \{y, v_1, \ldots, v_k, \mathcal{P}^1 y, \mathcal{P}^1 v_1, \ldots, \mathcal{P}^1 v_\ell, w_1, \ldots, w_m\} \tag{1.7}$$

is a choice of representatives for $\mathcal{B}$. We have

$$B \cong \frac{\mathbb{F}_3[y, v_1 \ldots v_k, \mathcal{P}^1 y, \mathcal{P}^1 v_1, \ldots, \mathcal{P}^1 v_\ell, w_1 \ldots w_m]}{(y, v_1 \ldots v_k, \mathcal{P}^1 y, \mathcal{P}^1 v_1, \ldots, \mathcal{P}^1 v_\ell, w_1 \ldots w_m)^4} \tag{1.8}$$

Proposition 1.2 *There is no element $a \in B$ with*

$$P^1 a = y^2 + \sum a_i b_i$$

where $a_i, b_i \in \hat{\mathcal{B}}$ and not both a_i, b_i are equal to y.

Proof Suppose a exists and satisfies the above equation. Consider $B/I(B)^3$. There is an algebra map $\pi : B \longrightarrow B/I(B)^3$ and we have

$$\mathcal{P}^1\pi(a) = \pi(y)^2 + \sum \pi(a_i)\pi(b_i).$$

Further, $\mathcal{P}^2\mathcal{P}^1\pi(a) = 0 = \mathcal{P}^2[\pi(y)^2 + \sum \pi(a_i)\pi(b_i)]$

$$\begin{aligned} 0 =& [\mathcal{P}^1\pi(y)]^2 + 2\pi(y)(\mathcal{P}^2\pi(y)) \\ &+ \sum (\mathcal{P}^2\pi(a_i))\pi(b_i) + \mathcal{P}^1\pi(a_i)\mathcal{P}^1\pi(b_i) \\ &+ \sum \pi(a_i)\mathcal{P}^2\pi(b_i). \end{aligned} \tag{1.9}$$

By (1.4) $[\mathcal{P}^1\pi(y)]^2$ is nonzero in bidegree $(6n+8, 6n+8)$ and must be cancelled by some of the terms involving the a_i, b_i.

Equation (1.9) is an equation among decomposables of $B/I(B)^3$. The module of decomposables of $B/I(B)^3$ is isomorphic to $D[B/I(B)^3] \cong \frac{QB^* \otimes QB^*}{im(1-T^*)}$ where T^* is the twist function and by (1.2)

$$QB^* \otimes QB^* \cong \left(\frac{QB^*}{im\mathcal{P}^1} \oplus \frac{im\mathcal{P}^1}{im\mathcal{P}^2} \oplus im\mathcal{P}^2\right) \otimes \left(\frac{QB^*}{im\mathcal{P}^1} \oplus \frac{im\mathcal{P}^1}{im\mathcal{P}^2} \oplus im\mathcal{P}^2\right).$$

Via this isomorphism, $[\mathcal{P}^1\pi(y)]^2$ lies in

$$\frac{im\mathcal{P}^1}{im\mathcal{P}^2} \otimes \frac{im\mathcal{P}^1}{im\mathcal{P}^2}$$

and must be canceled by $\sum \mathcal{P}^1\pi(a_i)\mathcal{P}^1\pi(b_i)$. But by assumption a_i, b_i are not both equal to y. By the choice of basis, $[\mathcal{P}^1\pi(y)]^2$ and the $\mathcal{P}^1\pi(a_i)\mathcal{P}^1\pi(b_i)$ are linearly independent. This completes the proof. □

Theorem 1.3 $\mathcal{P}^1 QB^{6n+4} = \mathcal{P}^2 QB^{6n}$.

Proof Let $y \in B^{6n+4}$ with $\bar{y} \neq 0$ and suppose $\mathcal{P}^1\bar{y} \notin im\mathcal{P}^2$. Extend y to a basis $\hat{\mathcal{B}}$ of generators for the polynomial algebra B as in (1.7). Then by (1.1)

$$y^3 \doteq \mathcal{P}^1\mathcal{P}^{3n+1}y,$$

we will show

$$0 \neq \mathcal{P}^{3n+1}\bar{y} \notin im\mathcal{P}^2. \qquad (1.10)$$

Then $\mathcal{P}^{3n}\bar{y} \in QB^{18n+4}$ with $\mathcal{P}^1\mathcal{P}^{3n}\bar{y} \notin im\mathcal{P}^2$. So we will have shown that given any nonzero algebra generator $\bar{y} \in QB^{6n+4}$ with $\mathcal{P}^1\bar{y} \notin im\mathcal{P}^2$, one can produce another $0 \neq \mathcal{P}^{3n}\bar{y} \in QB^{18n+4}$ with $\mathcal{P}^1\mathcal{P}^{3n}\bar{y} \notin im\mathcal{P}^2$. By iterating this process, one can show B is not finitely generated as an algebra.

We now proceed to prove (1.10). Suppose $\mathcal{P}^{3n+1}\bar{y} = \mathcal{P}^2\bar{z}$. Then $\mathcal{P}^{3n+1}y = \mathcal{P}^2z + d$ where d is decomposable. Then by (1.1), $y^3 \doteq \mathcal{P}^1\mathcal{P}^{3n+1}y = \mathcal{P}^1 d$.

Expand d in terms of the basis $\hat{\mathcal{B}}$:

$$d = \sum a_i b_i + \sum c_j d_j e_j$$

$$a_i, b_i, c_j, d_j, e_j \in \hat{\mathcal{B}}.$$

So

$$y^3 = \mathcal{P}^1\left(\sum a_i b_i\right) + \mathcal{P}^1\left(\sum c_j d_j e_j\right). \qquad (1.11)$$

We claim $\mathcal{P}^1\left(\sum a_i b_i\right) \neq 0$. If not then $y^3 = \mathcal{P}^1\left(\sum c_j d_j e_j\right)$ implies $\{\bar{y}\} = 0$ in $\frac{QB^*}{im\mathcal{P}^1}$. By assumption $\{\bar{y}\} \neq 0$ in $\frac{QB^*}{im\mathcal{P}^1}$. It follows that either $\mathcal{P}^1 a_i$ or $\mathcal{P}^1 b_i$ must be decomposable and have y^2 as a summand when expanded in terms of the basis $\hat{\mathcal{B}}$. By Proposition 1.2 this is impossible. So we arrive at a contradiction. We conclude $\mathcal{P}^{3n+1}\bar{y} \notin im\mathcal{P}^2$. This proves (1.10). By induction we conclude the polynomial algebra is not finitely generated unless $\mathcal{P}^1\bar{y} \in im\mathcal{P}^2$. □

2 Tripletons

Let $\mathcal{A}_3^+$ be the subHopf algebra of $\mathcal{A}_3$ generated by the $\mathcal{P}^i$.

Recall there is a Hopf algebra map $\theta : \mathcal{A}_3^+ \longrightarrow \mathcal{A}_3^+$ that is dual to the cube map on the dual of the mod 3 Steenrod algebra. This map θ takes all elements of degree not divisible by three to zero, and

$$\theta(\mathcal{P}^{3^j}) = \mathcal{P}^{3^{j-1}}. \qquad (2.1)$$

Equation (2.1) can be used to define θ since the elements $\mathcal{P}^{3^j}$ generate $\mathcal{A}_3^+$ as an algebra. Note that θ takes admissible elements to admissible elements. The following definition is motivated by [10, Section 4].

Definition *An element $\gamma = \alpha_0\mathcal{P}^{3^k}\alpha_1\mathcal{P}^{3^k}\alpha_2\mathcal{P}^{3^k}\alpha_3$ is a tripleton of order k if $\deg\alpha_i \equiv 0 \bmod 3^{k+1}(4)$ for all i.*

Proposition 2.1 *Let $\delta \in \mathcal{A}_3^+$ be an element of degree $\equiv 0 \bmod 3^k(4)$. Then if $\delta \in \ker\theta^k$, δ is a linear combination of tripletons of order less than or equal to $k-1$.*

Proof By induction. If $k = 1$, then $\deg \delta \equiv 0 \bmod 12$. We claim δ lies in the subalgebra generated by $\mathcal{P}^1$. If not, then write δ as a sum of admissibles. Since θ is an algebra map, by (2.1), $\theta(\delta)$ is again a sum of admissibles, hence θ is monic on the admissibles that are not in the subalgebra generated by $\mathcal{P}^1$. So if $\delta \in \ker \theta$, δ lies in the subalgebra generated by $\mathcal{P}^1$.

Now since $\deg \delta \equiv 0 \bmod 12$, if we express δ as a sum of monomials, each monomial must have at least three $\mathcal{P}^1$s.

Hence δ is a sum of monomials that are tripletons of order zero. The inductive step is left to the reader. □

Lemma 2.2 *$QB^{6m} \cap im\mathcal{P}^2 = 0$ for $m > 2$.*

Proof Let $\bar{z} \in QB^{6m} \cap im\mathcal{P}^2$. Then $\bar{z} = \mathcal{P}^2\bar{w}$, degree $\bar{w} = 6m-8 = 6(m-2)+4$ where $m > 2$. By Theorem 1.3, $\mathcal{P}^1\bar{w} = \mathcal{P}^2\bar{w}_1$ so $\bar{z} \doteq \mathcal{P}^1\mathcal{P}^1\bar{w} = \mathcal{P}^1\mathcal{P}^2\bar{w}_1 = 0$ by (1.1). □

Proposition 2.3 *$\mathcal{P}^{3^j}\mathcal{P}^2 QB^{6m} \subseteq im\mathcal{P}^2$ for $m > 0$.*

Proof For $j = 1$, Adem relations imply

$$\begin{aligned}\mathcal{P}^3\mathcal{P}^2 &\doteq \mathcal{P}^4\mathcal{P}^1 + \mathcal{P}^5\\ &\doteq \mathcal{P}^1\mathcal{P}^3\mathcal{P}^1 + \mathcal{P}^2\mathcal{P}^3.\end{aligned}$$

By Theorem 1.3

$$\mathcal{P}^1\mathcal{P}^3\mathcal{P}^1 QB^{6m} \subseteq im\mathcal{P}^2$$

So $\mathcal{P}^3\mathcal{P}^2 QB^{6m} \subseteq im\mathcal{P}^2$.

For $j > 1$, the Adem relations imply

$$\begin{aligned}\mathcal{P}^{3^j}\mathcal{P}^2 &\doteq \mathcal{P}^6\mathcal{P}^{3^j-4} + \mathcal{P}^{3^j+2} + \mathcal{P}^{3^j+1}\mathcal{P}^1\\ &\doteq \mathcal{P}^6\mathcal{P}^2\mathcal{P}^{3^j-6} + \mathcal{P}^2\mathcal{P}^{3^j} + \mathcal{P}^1\mathcal{P}^{3^j}\mathcal{P}^1.\end{aligned} \tag{2.2}$$

By Theorem 1.3 $\mathcal{P}^1\mathcal{P}^{3^j}\mathcal{P}^1 QB^{6m} \subseteq im\mathcal{P}^2$, so by (2.2) it remains to prove $\mathcal{P}^6\mathcal{P}^2\mathcal{P}^{3^j-6} QB^{6m} \subseteq im\mathcal{P}^2$.

But

$$\begin{aligned}\mathcal{P}^6\mathcal{P}^2 \doteq \mathcal{P}^3\mathcal{P}^3\mathcal{P}^2 &\doteq \mathcal{P}^3(\mathcal{P}^5 + \mathcal{P}^4\mathcal{P}^1)\\ &\doteq \mathcal{P}^3(\mathcal{P}^2\mathcal{P}^3 + \mathcal{P}^1\mathcal{P}^3\mathcal{P}^1).\end{aligned} \tag{2.3}$$

If $\mathcal{P}^{3^j-6} QB^{6m} \subseteq QB^{6n}$, for some n, then

$$\begin{aligned}\mathcal{P}^6\mathcal{P}^2\mathcal{P}^{3^j-6} QB^{6m} &\subseteq \mathcal{P}^6\mathcal{P}^2 QB^{6n}\\ &\subseteq \mathcal{P}^3(\mathcal{P}^2\mathcal{P}^3 + \mathcal{P}^1\mathcal{P}^3\mathcal{P}^1) QB^{6n} \text{ by (2.3)}\\ &\subseteq im\mathcal{P}^3\mathcal{P}^2 QB^{6k} \text{ for some } k \text{ by the case } j = 1\\ &\subseteq im\mathcal{P}^2 \quad \text{Q.E.D.}\end{aligned}$$

□

Theorem 2.4 *A tripleton of order* 0 *annihilates QB^{6n} for $n > 0$.*

Proof Let $\alpha_0\mathcal{P}^1\alpha_1\mathcal{P}^1\alpha_2\mathcal{P}^1\alpha_3$ be a tripleton of order 0, $\deg\alpha_i \equiv 0 \bmod 12$. Then

$$\deg\alpha_2\mathcal{P}^1\alpha_3QB^{6n} \equiv 4 \mod 6$$

so

$$\mathcal{P}^1\alpha_2\mathcal{P}^1\alpha_3QB^{6n} \subseteq \mathcal{P}^2QB^{6n'}.$$

for some n' by Theorem 1.3.

Now α_1 is a sum of products of $\mathcal{P}^{3^j}s$ for $j \geq 1$. So by Proposition 2.3

$$\alpha_1\mathcal{P}^1\alpha_2\mathcal{P}^1\alpha_3QB^{6n} \subseteq im\mathcal{P}^2$$

so

$$\mathcal{P}^1\alpha_1\mathcal{P}^1\alpha_2\mathcal{P}^1\alpha_3QB^{6n} = 0$$

□

3 Contractions

We introduce the theory of contractions as first described in [10, Section 3]. This allows us to generalize Theorem 1.3.

Definition 3.1 *Let B and B' be algebras over $\mathcal{A}_3^+$. We say $f : B \longrightarrow B'$ is a contraction if*

(1) $f(B^{3\ell+k}) = 0$ for $0 < k < 3$
$f(B^{3\ell}) \subseteq (B')^\ell$;

(2) f is an algebra map.

We say $f : B \longrightarrow B'$ is a *contraction over θ* if the following diagram commutes

$$\begin{array}{ccc} \mathcal{A}_3^+ \otimes B & \xrightarrow{\phi} & B \\ {\scriptstyle\theta\otimes f}\downarrow & & \downarrow{\scriptstyle f} \\ \mathcal{A}_3^+ \otimes B' & \xrightarrow{\phi'} & B' \end{array} \tag{3.2}$$

where ϕ, ϕ' are the actions of $\mathcal{A}_3^+$ on B and B' respectively.

Let B^0 be the subalgebra of B generated by polynomial generators of degree not divisible by 6. Define $B^\#$ to be the quotient algebra of B by the ideal generated by B^0

$$B^\# = B//B^0 \tag{3.3}$$

Let $\mathcal{A}(1)$ be the subalgebra of $\mathcal{A}_3^+$ generated by all $\mathcal{P}^{3^j}$ for $j > 0$. Then since

$$\mathcal{P}^{3^j}(B_0) \subseteq I(B_0)B, \tag{3.4}$$

$B^\#$ admits an action of $\mathcal{A}(1)$.

Further $B^\#$ is concentrated in degrees congruent to zero mod 6.

We define

$$\mathcal{R}(B)^i = (B^\#)^{3i}.$$

Then $\mathcal{R}(B)$ is an algebra, isomorphic as an ungraded algebra to $B^\#$. Define ρ to be the composition

$$\rho : B \xrightarrow{\pi} B^\# \xrightarrow{i} \mathcal{R}(B) \tag{3.5}$$

where i is the isomorphism. Then ρ is a contraction. Further, one checks that ρ is a contraction over θ, and ρ is onto.

$$\mathcal{R}(B) \text{ is a finitely generated polynomial algebra truncated at height 4 which admits an action of } \mathcal{A}_3^+. \tag{3.6}$$

Proposition 3.1 $\mathcal{P}^3 QB^{18n+12} = \mathcal{P}^6 QB^{18n}$.

Proof Given $\mathcal{P}^3\bar{x} \in \mathcal{P}^3 QB^{18n+12}$ we have by (3.2) and (3.5)

$$\rho(\mathcal{P}^3\bar{x}) = \mathcal{P}^1\rho(\bar{x})$$

with $\rho(\bar{x}) \in Q\mathcal{R}(B)^{6n+4}$.

Since $\mathcal{R}(B)$ is again a truncated polynomial algebra, Theorem 1.3 applied to $\mathcal{R}(B)$ implies

$$\mathcal{P}^1\rho(\bar{x}) \in \mathcal{P}^2 Q\mathcal{R}(B)^{6n} = \mathcal{P}^2\rho(QB^{18n}).$$

But by (3.2)

$$\mathcal{P}^2\rho(QB^{18n}) = \rho(\mathcal{P}^6 QB^{18n}), \quad \mathcal{P}^1\rho(\bar{x}) = \rho(\mathcal{P}^3\bar{x}).$$

Hence

$$\mathcal{P}^3\bar{x} \in \mathcal{P}^6 QB^{18n} + \ker\rho.$$

By (3.5) $\ker\rho$ consists of the ideal generated by B^0, but because we are dealing with indecomposables in degrees congruent to zero mod 6, we have

$$\mathcal{P}^3\bar{x} \in \mathcal{P}^6 QB^{18n}$$

or

$$\mathcal{P}^3 QB^{18n+12} = \mathcal{P}^6 QB^{18n}$$

□

We now prove Theorem A:

Theorem A *For* $j \geq 0, n > 0$, $\mathcal{P}^{3^j} QB^{2(3^{j+1}n)+4(3^j)} = \mathcal{P}^{2(3^j)} QB^{2(3^{j+1}n)}$.

Proof The proof is by induction on j. The cases $j = 0, j = 1$ are Theorem 1.3 and Proposition 3.1. Assume for all finitely generated polynomial algebras truncated at height 4 that for $0 \leq j' < j$, $n > 0$

$$\mathcal{P}^{3^{j'}} QB^{2(3^{j'+1}n)+4(3^{j'})} = \mathcal{P}^{2(3^{j'})} QB^{2(3^{j'+1}n)}.$$

Then since $\mathcal{R}(B)$ is a finitely generated polynomial algebra truncated at height 4, we have by induction

$$\mathcal{P}^{3^{j-1}} Q\mathcal{R}(B)^{2(3^j n)+4(3^{j-1})} = \mathcal{P}^{2(3^{j-1})} Q\mathcal{R}(B)^{2(3^j n)} \tag{3.7}$$

But $\rho : B \longrightarrow \mathcal{R}(B)$ is onto by (3.5) so

$$\rho(QB)^{2(3^{j+1}n)+4(3^j)} = Q\mathcal{R}(B)^{2(3^j n)+4(3^{j-1})}$$

$$\rho(QB)^{2(3^{j+1}n)} = Q\mathcal{R}(B)^{2(3^j n)}$$

By (3.2) and (3.7)

$$\rho(\mathcal{P}^{3^j} QB^{2(3^{j+1}n)+4(3^j)}) = \rho(\mathcal{P}^{2(3^j)} QB^{2(3^{j+1}n)}).$$

Hence

$$\mathcal{P}^{3^j} QB^{2(3^{j+1}n)+4(3^j)} \equiv \mathcal{P}^{2(3^j)} QB^{2(3^{j+1}n)} \text{ modulo } \ker\rho. \tag{3.8}$$

But $\ker \rho$ in degrees divisible by 6 consists of decomposables. This completes the proof. □

Acknowledgement

The author would like to thank Yutaka Hemmi for many useful discussions related to this work.

References

[1] Harper, J. R. Harper and Zabrodsky, A. [1988] *Evaluating a p-th order cohomology operation* Publications Matematiques **32**, 61–78.

[2] Hemmi, Y. [1987] *Homotopy associative finite H-spaces and the mod 3 reduced power operations* RIMS, Kyoto University **23**, 1071–1084.

[3] Hemmi, Y. [1994] *On exterior A_n-spaces and modified projective spaces* Hiroshima Math. J. **24**, 583–605.

[4] Kane, R. [1988] *The homology of Hopf spaces*, North Holland.

[5] Lin, J. [1995] *H-spaces with finiteness conditions*, Chapter 22, Handbook of Algebraic Topology (I. James, ed.), North Holland.

[6] McGibbon, C. [1989] *Higher forms of homotopy commutativity and finite loop spaces* Math Z **201**, 363–374.

[7] Milnor, J. and Moore, J. C. [1965] *On the structure of Hopf algebras* Ann. Math. **81**, 211–264.

[8] Stasheff, J. [1963] *Homotopy associativity of H-spaces, I, II* Trans. Amer. Math. Soc. **108**, 275–292, 293–312.

[9] Steenrod, N. E. [1962] *Cohomology Operations*, Ann. Math. Studies no. 50, Princeton University Press.

[10] Thomas, E. [1965] *Steenrod squares and H-spaces II* Ann. Math. **81**, 473–495.

[11] Wilkerson, C. *Rings of invariants and inseparable forms of algebras over the Steenrod algebra, The Hopf archive.*

[12] Zabrodsky, A. [1979] *Some relations in the* mod 3 *cohomology of H-spaces* Israel J. of Math **33**, 59–72.

Fields Institute Communications
Volume **19**, 1998

A Differential in the Adams Spectral Sequence for Spheres

Wen-Hsiung Lin
Department of Mathematics
National Tsing Hua University
Hsinchu 30043, Taiwan
whlin@math.nthu.edu.tw

Abstract. $h_7^3 \in Ext_A^{3,380}(\mathbb{Z}_2, \mathbb{Z}_2)$ is shown not to detect homotopy elements in the Adams spectral sequence for spheres where A is the mod 2 Steenrod algebra.

1 Introduction

In a previous paper [**13**] we studied the differential

$$d_2 : Ext_A^{3,*} = Ext_A^{3,*}(\mathbb{Z}_2, \mathbb{Z}_2) \longrightarrow Ext_A^{5,*} = Ext_A^{5,*}(\mathbb{Z}_2, \mathbb{Z}_2)$$

in the mod 2 Adams spectral sequence for the 2-adic stable homotopy groups of spheres ${}_2\pi_*^s = {}_2\pi_*^s(S^0)$, where A denotes the mod 2 Steenrod algebra. In this paper we study a higher differential on a particular element in $Ext_A^{3,*}$. Let $h_i \in Ext_A^{1,2^i}$ be the class corresponding to $Sq^{2^i} \in A$. Adams proves in [**1**] that $h_i^2 \neq 0$ in $Ext_A^{2,2^{i+1}}$ and that $h_i^3 = h_{i-1}^2 h_{i+1} \neq 0$ in $Ext_A^{3,2^{i+1}+2^i}$ for all $i \geq 0$. The particular element referred to is h_7^3. We will show that $d_4(h_7^3) \neq 0$ in the ASS for S^0. Our method of proving this suggests that, for $i \geq 8$, $d_4(h_i^3)$ should also be non-zero, and in a similar fashion. Solid calculations have only been made for $i = 7$ and our purpose here is to demonstrate $d_4(h_7^3) \neq 0$ based on these calculations.

In order to describe what element in $Ext_A^{7,*}$ gets hit by $d_4(h_7^3)$ we first recall from May in [**18**] that $Ext_A^{4,24} \cong \mathbb{Z}_2$, generated by an element called g_1. Let $Sq^0 : Ext_A^{s,t} \to Ext_A^{s,2t}$ be the first squaring operation as in [**14**]. Let $(Sq^0)^j = \underbrace{Sq^0 \cdot Sq^0 \cdots Sq^0}_{j}$ ($(Sq^0)^0$ is the identity map). Tangora proves in [**20**] that $g_{j+1} = (Sq^0)^j (g_1) \neq 0$ in $Ext_A^{4,2^{j+4}+2^{j+3}}$ for all $j \geq 0$ (this is also proved in [**13**]), and also proves that $Ext_A^{5,67} \cong \mathbb{Z}_2 \oplus \mathbb{Z}_2 \oplus \mathbb{Z}_2$, generated by $h_5^2 h_0^3$, $D_3 h_1$ and H_1. We will give statements later in the paper about some elements in $Ext_A^{s,*}$ for $s = 5, 6, 7$ including

1991 *Mathematics Subject Classification.* 55Q45, 55T15.

The author was partially supported by N.S.C. of Taiwan and the Foundation for the Advancement of Outstanding Scholarship.

the results that $Ext_A^{5,2^7+2} \cong \mathbb{Z}_2$, generated by an element called K and that $g_5h_0^3 \neq 0$, $h_8h_6H_1 \neq 0$, $h_8h_0K \neq 0$ and are linearly independent in $Ext_A^{7,2^8+2^7+3}$.

Our main result is the following.

Theorem 1.1 *$d_4\left(h_7^3\right) = g_5h_0^3 + \epsilon_1 h_8h_6H_1 + \epsilon_2 h_8h_0K \neq 0$ in the Adams spectral sequence for ${}_2\pi_*^s$ where $\epsilon_i = 0$ or 1.*

Presumably ϵ_i should be zero, but we have not been able to prove this.

It is well known that h_i^3, for $0 \leq i \leq 3$, detect homotopy elements in ${}_2\pi_*^s$. Mahowald and Tangora prove in [**16**] that h_4^3 also persists to $E_\infty^{*,*}$ in the ASS for S^0. One reason for h_4^3 to detect homotopy elements is that $g_2h_0^3 = 0$ [**20**], and this is explained as follows. They have shown in [**16**] that ${}_2\pi_{30}^s \cong \mathbb{Z}_2$, generated by θ_4 which is detected by h_4^2. They also prove, together with M. G. Barratt, in [**2**] that $\sigma^2\theta_4$ is detected by $g_2h_0^2 \neq 0$ in $Ext_A^{6,50}$ where σ is the Hopf class in π_7^s detected by h_3. If $g_2h_0^3 \neq 0$ then it would follow that $d_4\left(h_4^3\right) = g_2h_0^3$ since, in the ASS for S^0, $d_2\left(h_i\right) = h_{i-1}^2h_0 \neq 0$ for $i \geq 4$, in particular, $d_2(h_4) = h_3^2h_0$ [**1**]. This can be extended to $i \geq 5$ as follows.

First consider the case $i = 5$. There exists a θ_5 detected by h_5^2 such that $2\theta_5 = 0$. This is due to Mahowald (see [**10**]). Mahowald and Tangora have recently also shown that $g_3h_0^3 \neq 0$ (private communication). Then $\theta_4\theta_5$ should be detected by $g_5h_0^2$, and one should have $d_4\left(h_5^3\right) = g_3h_0^3$. Efforts have yet to be made to give a formal proof of this. This case and the case $i = 6$ are different from the case $i \geq 7$ as to be explained in a moment and also in the next section.

It is possible to make calculations to show that $g_{i-2}h_0^3 \neq 0$ for $i \geq 6$. Solid calculations have been made to confirm this only for $i = 7$. For $i \geq 6$, it is not yet known whether there exist θ_i detected by h_i^2 such that $2\theta_i = 0$. If, for $i \geq 6$, there exist θ_{i-1} and θ_i with $2\theta_i = 0$ then $\theta_{i-1}\theta_i$ should be detected by $g_{i-2}h_0^2$ and we should have $d_4(h_i^3) = g_{i-2}h_0^3$.

When we say "should be" or "should have" above, it means "we conjecture". What we want to do here is to give an almost positive answer to this conjecture for $i = 7$ in the sense that we only get $d_4\left(h_7^3\right) = g_5h_0^3 + \epsilon_1 h_8h_6H_1 + \epsilon_2 h_8h_0K$. And we do this without the knowledge of θ_6 and θ_7.

There is a distinction between the case "$i \geq 7$" and the case "$i = 5, 6$" for $g_{i-2}h_0^3$. The distinction is that, for $i \geq 7$, the "algebraic" Hopf invariant of $g_{i-2}h_0^3$ is easier to recognize while those of $g_3h_0^3$ and $g_4h_0^3$ are not. We are not going to define here the notion of Hopf invariant for elements in $Ext_A^{*,*}$. Nor do we have a proof of what exactly the Hopf invariant of $g_{i-2}h_0^3$ is for $i \geq 7$. Nevertheless, we will point out in the next section what the Hopf invariant of $g_{i-2}h_0^3$ should be for $i \geq 7$ where we also show how Theorem 1.1 is proved based on this observation. Our method of proving Theorem 1.1 also suggests that $d_4(h_i^3)$ is non-zero for $i \geq 8$ and the value is $g_{i-2}h_0^3$ plus something else (the conjecture above is that this "something else" is zero). But because of the difficulties of the calculations involved, the case $i = 7$ is the only case we can do so far. The Hopf invariants of $g_3h_0^3$ and $g_4h_0^3$ do not have the pattern of those of $g_{i-2}h_0^2$ for $i \geq 7$. Indeed, they are more difficult to describe. Because of this we believe that more geometric arguments are needed in order to prove the conjectured $d_4(h_5^3)$ and $d_4(h_6^3)$.

2 Method of the proof

In this section we roughly describe our method of proving Theorem 1.1. Meanwhile we will also show how the method can be applied to $i \geq 8$ and point out what the difficulties are. The description, although rough, is quite lengthy. Our purpose here is to give the basic ideas of the proof.

Theorem 1.1 will be deduced from a corresponding result in the real projective space P^∞. Let $t : P^\infty \longrightarrow S^0$ be the transfer map as in [**9, 15**]. Consider the induced map

$$\begin{aligned} t_* : Ext_A^{s,t}(P^\infty) &= Ext_A^{s,t}(H^*(P^\infty), \mathbb{Z}_2) \longrightarrow \\ Ext_A^{s+1,t+1}(\mathbb{Z}_2, \mathbb{Z}_2) &= Ext_A^{s+1,t+1}(S^0) \end{aligned}$$

in the ASS for P^∞ and S^0. We recall [**6**] that $Ext_A^{0,2^i-1}(P^\infty) \cong \mathbb{Z}_2$, generated by an element called $\hat{h}_i$ $(i \geq 1)$, which corresponds to the non-zero class in $H_{2^i-1}(P^\infty) = \mathbb{Z}_2$ and that $t_*(\hat{h}_i) = h_i$. Thus $t_*\left(\hat{h}_{i-1}h_{i-1}h_{i+1}\right) = h_{i-1}^2 h_{i+1} = h_i^3$; in particular, $t_*\left(\hat{h}_6 h_6 h_8\right) = h_7^3$. We shall prove

(1) $d_2\left(\hat{h}_6 h_6 h_8\right) = 0$

(2) $d_3\left(\hat{h}_6 h_6 h_8\right) = 0$

(3) $d_4\left(\hat{h}_6 h_6 h_8\right) \neq 0$

in the ASS for P^∞. (1) is actually true for all i, that is, $d_2(\hat{h}_{i-1}h_{i-1}h_{i+1}) = 0$ for $i \geq 2$, and this is straightforward as will be seen from the differentials $d_2(h_i) = h_{i-1}^2 h_0$ $(i \geq 4)$ in S^0. (2) is also true for all i; the proof is less trivial. In this paper we will show it only for $i = 7$. Thus, by applying t_*, we may conclude that, in the ASS for S^0, if $d_r(h_i^3) \neq 0$ then $r \geq 4$.

The starting point of this work is the observation on (3) which will be described in a moment. We then proceed to evaluate $d_4\left(\hat{h}_6 h_6 h_8\right)$ which will be of the form

$$\begin{aligned} (4)\quad d_4\left(\hat{h}_6 h_6 h_8\right) = \overline{e_{111} h_4 h_7^2 h_0^3} + \varepsilon_1 \hat{h}_6 h_8 H_1 + \varepsilon_2 \overline{e_{47} h_0^2 h_4 h_6} h_8 h_0 + \varepsilon_3 \hat{h}_6 h_8 h_1 D_3 \\ + \varepsilon_4 \hat{h}_5 c_5 h_0^3 + \varepsilon_5 \overline{e_{17} h_3^2 h_5} h_0 \end{aligned}$$

where each term in the sum is a class in $Ext_A^{6,2^8+2^7+2}(P^\infty)$, $\varepsilon_i = 0$ or 1 and e_j is the non-zero class in $H_j(P^\infty)$. A detailed description of $\overline{e_{111} h_4 h_7^2 h_0^3}$ will be given later in this section. We will show that $t_*\left(\overline{e_{111} h_4 h_7^2 h_0^3}\right) = g_5 h_0^3$, $t_*\left(\hat{h}_6 h_8 H_1\right) = h_6 h_8 H_1$, $t_*\left(\overline{e_{47} h_0^2 h_4 h_6}\right) = \varepsilon K$, where $\varepsilon = 0$ or 1, and so $t_*\left(\overline{e_{47} h_0^2 h_4 h_6} h_8 h_0\right) = \varepsilon h_8 h_0 K$ and that $t_*(x) = 0$ for the remaining classes x in the sum (4). So we get $d_4(h_7^3) = g_5 h_0^3 + \varepsilon_1 h_8 h_6 H_1 + \varepsilon_2 h_8 h_0 K$ in S^0 as claimed. Of course we still need to show $g_5 h_0^3 + \varepsilon_1 h_8 h_6 H_1 + \varepsilon_2 h_8 h_0 K$ can not be hit by differentials d_2 and d_3 from elements in $Ext_A^{5,*}$ and $Ext_A^{4,*}$ respectively. This follows from the fact that $Ext_A^{4,2^8+2^7+1} \cong \mathbb{Z}_2$ and $Ext_A^{5,2^8+2^7+2} \cong \mathbb{Z}_2$, generated by $h_7^3 h_0$ and $h_7^3 h_0^2$ respectively as will be seen.

We will actually obtain (4) in the real projective space P^{113}. Indeed, (4) is true for any P^n with $n > 111$. For computational purpose we want $n(> 111)$ to be as close to 111 as possible and we find that P^{113} is an appropriate space to consider.

The class $\overline{e_{111} h_4 h_7^2 h_0^3}$ is the key term in (4). In order to see how this term comes up we suppose that $\hat{h}_6 h_6 h_8$ persists to E_∞ in the ASS for P^{113}. Let $k = 2^8 + 2^7 - 3$ and let $S^k \rightarrow P^{113}$ be a stable map detected by $\hat{h}_6 h_6 h_8$. Consider the collapsing maps $P^{113} \xrightarrow{q_1} P_{65}^{113} \xrightarrow{q_2} P_{97}^{113}$. $Ext_A^{0,95}\left(P_{65}^{113}\right) \cong \mathbb{Z}_2$ is generated by $\bar{e}_{95}$

and $Ext_A^{0,111}\left(P_{97}^{113}\right) \cong \mathbb{Z}_2$ is generated by $\bar{e}_{111}$, and we may consider the classes $\bar{e}_{95}h_5h_0h_8 \in Ext_A^{3,k+3}\left(P_{65}^{113}\right)$ and $\bar{e}_{111}h_4h_0^2h_8 \in Ext_A^{4,k+4}\left(P_{97}^{113}\right)$. Then it is not difficult to see that $S^k \xrightarrow{q_1 f} P_{65}^{113}$ is detected by $\bar{e}_{95}h_5h_0h_8$ and $S^k \xrightarrow{q_2q_1 f} P_{97}^{113}$ is detected by $\bar{e}_{111}h_4h_0^2h_8$. But the latter is impossible because $d_2\left(\bar{e}_{111}h_4h_0^2h_8\right) = \bar{e}_{111}h_4h_7^2h_0^3 \neq 0$ in the ASS for P_{97}^{113}. As will be shown later in the paper this implies $d_4\left(\hat{h}_6h_6h_8\right) \neq 0$ in the ASS for P^{113}. Here non-triviality of the class $\bar{e}_{111}h_4h_7^2h_0^3$ in $Ext_A^{6,k+5}\left(P_{97}^{113}\right)$ is crucial. As will be discussed in a moment, from $\bar{e}_{111}h_4h_7^2h_0^3 \neq 0$ we conjecture that the Hopf invariant of $g_5h_0^3$ should be $h_4h_7^2h_1^3 \neq 0$. The argument above can be carried over to any i with $i \geq 7$. More precisely, for any such i, let $a = 2^{i-1}+2^{i-2}+2^{i-3}-1$, $a_1 = 2^{i-1}+2^{i-2}-1$, $b = 2^{i-1}+2^{i-2}+1$, $b_1 = 2^{i-1}+1$ and consider $P^{a+2} \xrightarrow{q_1} P_{b_1}^{a+2} \xrightarrow{q_2} P_b^{a+2}$, $\bar{e}_{a_1}h_{i-2}h_0h_{i+1} \in Ext_A^{3,*}\left(P_{b_1}^{a+2}\right)$, $\bar{e}_a h_{i-3}h_0^2h_{i+1} \in Ext_A^{4,*}\left(P_b^{a+2}\right)$. Then $d_2\left(\bar{e}_a h_{i-3}h_0^2h_{i+1}\right) = \bar{e}_a h_{i-3}h_i^2h_0^3 \neq 0$ in P_b^{a+2} and this implies $d_4\left(\hat{h}_{i-1}h_{i-1}h_{i+1}\right) \neq 0$ in P^{a+2}. Thus $d_4\left(\hat{h}_{i-1}h_{i-1}h_{i+1}\right)$ is of the form

$$d_4\left(\hat{h}_{i-1}h_{i-1}h_{i+1}\right) = x_1 + x_2$$

where $x_1, x_2 \in Ext_A^{6,*}\left(P^{a+2}\right)$ with $(q_2q_1)_*(x_1) = \bar{e}_a h_{i-3}h_i^2h_0^3$ and $(q_2q_1)_*(x_2) = 0$. x_1 is not unique. We can find, for any $i \geq 7$, a canonical x_1 which is denoted by $\overline{e_a h_{i-3}h_i^2h_0^3}$. A detailed description of this will be given later. If x_2 can be evaluated so that we can show $t_*(x_2)$ and $g_{i-2}h_0^3$ are linearly independent then we would have $d_4\left(h_i^3\right) = g_{i-2}h_0^3 + t_*(x_2) \neq 0$ in the ASS for S^0. For $i = 7$ we can evaluate

$$x_2 = \varepsilon_1\hat{h}_6h_8H_1 + \cdots + \varepsilon_5\overline{e_{17}c_5h_3^2}h_0$$

as given in (4), due to the following two facts which will be proved in this paper.

(5) There exists a stable map $\varphi_1 : S^{62} \cup_{2\iota} e^{63} \to P^{113}$ such that

$$\varphi_1^* \neq 0 : H^{63}\left(P^{113}\right) = \mathbb{Z}_2 \to H^{63}\left(S^{62} \cup_{2\iota} e^{63}\right) = \mathbb{Z}_2.$$

(6) There exists a stable map $\varphi_2 : S^{94} \cup_{2\iota} e^{95} \to P_{65}^{113}$ such that

$$\varphi_2^* \neq 0 : H^{95}\left(P_{65}^{113}\right) = \mathbb{Z}_2 \to H^{95}\left(S^{94} \cup_{2\iota} e^{95}\right) = \mathbb{Z}_2.$$

From these facts the problem of evaluating x_2, for $i = 7$, will be reduced to the problem of computing many Ext groups including $Ext_A^{5,t}$ for $2^8+2^6+1 \leq t \leq 2^8+2^7$, $Ext_A^{s,t}(X) = Ext_A^{s,t}\left(H^*(X), \mathbb{Z}_2\right)$ for small s and certain t where X is either a stunted projective space P_m^n with $n \leq 113$ or a space derived from P_m^n by (5) and (6). Computing these *Ext* groups is an enormous and tedious work. The "solid calculations" referred to earlier means the calculations for these Ext groups for $i = 7$. It is impossible to write down here the details of these calculations as they are very complicated and very lengthy. We will state in Section 3 the results obtained about these Ext groups. And the main work of the paper is to organize a proof of Theorem 1.1 based on these *Ext* groups results. For $i \geq 8$, whether there exist analogues of the maps φ_1 and φ_2 is presently unknown. Without such maps it is still possible to evaluate x_2, but one needs to compute $Ext_A^{6,t}$ for $2^{i+1}+2^{i-2}+1 \leq t \leq 2^{i+1}+2^i$ in order to see what are the possible values for x_2. We have not been able to compute such *Ext* groups of high homological dimension and of large internal degrees t.

The relation $t_*\left(\overline{e_a h_{i-3}h_i^2h_0^3}\right) = g_{i-2}h_0^3$ for $a = 2^{i-1}+2^{i-2}+2^{i-3}-1$ with $i \geq 7$ is relevant to the Hopf invariant of $g_{i-2}h_0^3$. This is described as follows.

First we recall the lambda algebra Λ [**5**]. Λ is a bigraded differential algebra over $\mathbb{Z}_2$ generated by $\lambda_j \in \Lambda^{1,j}$ for $j \geq 0$ subject to the relations

(7) $\lambda_j \lambda_{2j+1+m} = \Sigma_{\nu \geq 0} \binom{m-\nu-1}{\nu} \lambda_{m+j-\nu} \lambda_{2j+1+\nu}$ for $m \geq 0$

with differential δ given on the generators by

(8) $\delta(\lambda_j) = \Sigma_{i \geq 0} \binom{j-i-1}{i+1} \lambda_{j-i-1} \lambda_i$.

Then $H^{s,t-s}(\Lambda) = Ext_A^{s,t}$ [**5**]. From (7) we see the set of admissible monomials $\{\lambda_{j_1} \cdots \lambda_{j_s} | 2j_i \geq j_{i+1}\}$ is a $\mathbb{Z}_2$-base for Λ. The class $h_i \in Ext_A^{1,2^i}$ is represented by λ_{2^i-1}. The relations $h_i h_{i+1} = 0$, $h_i^2 h_{i+2} = h_{i+1}^3$ and $h_i h_{i+2}^2 = 0$ in $Ext_A^{*,*}$ follow from the relations $\lambda_{2^i-1}\lambda_{2^{i+1}-1} = 0$, $\lambda_{2^i-1}^2 \lambda_{2^{i+2}-1} = \lambda_{2^{i+1}-1}^3$ and $\lambda_{2^i-1}\lambda_{2^{i+2}-1}^2 = 0$ in Λ. There is an operation $Sq^0 : \Lambda \to \Lambda$ given by

$$Sq^0(\lambda_{j_1} \cdots \lambda_{j_s}) = \lambda_{2j_1+1} \cdots \lambda_{2j_s+1}$$

which respects the relations (7) and commutes with the differential δ in (8). So Sq^0 induces a map $Sq^0 : H^{s,t-s}(\Lambda) \to H^{s,2t-s}(\Lambda)$ which coincides with the operation $Sq^0 : Ext_A^{s,t} \to Ext_A^{s,2t}$ in Section 1. Next we recall that g_1 has a Massey product representation $g_1 =< h_2, h_1, h_0, h_0 h_3^2 >$ (see [**15**]). The relation $h_0^2 h_3^2 = 0$ follows from the relation $\delta(\lambda_9 \lambda_3^2 + \lambda_3 \lambda_5 \lambda_7) = \lambda_0^2 \lambda_7^2$ in Λ. In Λ there are also the relations $\delta(\lambda_2) = \lambda_1 \lambda_0$, $\delta(\lambda_5) = \lambda_3 \lambda_1$, $\delta(\lambda_6) = \lambda_5 \lambda_0 + \lambda_3 \lambda_2$ and $\delta(\lambda_{11}\lambda_3^2 + \lambda_5 \lambda_9 \lambda_3) = \lambda_1(\lambda_9 \lambda_3^2 + \lambda_3 \lambda_5 \lambda_7) + \lambda_2 \lambda_0 \lambda_7^2$. So $g_1 =< h_2, h_1, h_0, h_0 h_3^2 >$ is represented by the cycle

$$\lambda_5 \lambda_9 \lambda_3^2 + \lambda_5 \lambda_3 \lambda_5 \lambda_7 + \lambda_6 \lambda_0 \lambda_7^2 + \lambda_3 \lambda_{11} \lambda_3^2 + \lambda_3 \lambda_5 \lambda_9 \lambda_3.$$

We have $\delta(\lambda_7 \lambda_{11} \lambda_3) = \lambda_7^2 \lambda_3^2 = \lambda_3 \lambda_{11} \lambda_3^2$. Then

$$g_1 = \{\lambda_6 \lambda_0 \lambda_7^2 + \lambda_5 \lambda_9 \lambda_3^2 + \lambda_5 \lambda_3 \lambda_5 \lambda_7 + \lambda_3 \lambda_5 \lambda_9 \lambda_3\}.$$

Applying $(Sq^0)^{i-3}$ and then multiplying respectively by h_0^3 and λ_0^3 we get

$$(9) \quad g_{i-2} h_0^3 = \left\{ \begin{array}{l} \lambda_a \lambda_{2^{i-3}-1} \lambda_{2^i-1}^2 \lambda_0^3 + \lambda_{a_1} \lambda_{2^i+2^{i-2}-1} \lambda_{2^{i-1}-1}^2 \lambda_0^3 \\ + \lambda_{a_1} \lambda_{2^{i-1}-1} \lambda_{a_1} \lambda_{2^i-1} \lambda_0^3 + \lambda_{2^{i-1}-1} \lambda_{a_1} \lambda_{2^i+2^{i-2}-1} \lambda_{2^{i-1}-1} \lambda_0^3 \end{array} \right\}$$

where $a = 2^{i-1} + 2^{i-2} + 2^{i-3} - 1$ and $a_1 = 2^{i-1} + 2^{i-2} - 1$. In particular, we have

$$g_5 h_0^3 = \{\lambda_{111} \lambda_{15} \lambda_{127}^2 \lambda_0^3 + \lambda_{95} \lambda_{159} \lambda_{63}^2 \lambda_0^3 + \lambda_{95} \lambda_{63} \lambda_{95} \lambda_{127} \lambda_0^3 + \lambda_{63} \lambda_{95} \lambda_{159} \lambda_{63} \lambda_0^3\}.$$

Note that $\lambda_{2^{i-3}-1} \lambda_{2^i-1}^2 \lambda_0^3$ represents $h_{i-3} h_i^2 \lambda_0^3$; in particular, $\lambda_{15} \lambda_{127}^2 \lambda_0^3$ represents $h_4 h_7^2 h_0^3$. We also recall that $Ext_A^{*,*}(P^\infty)$ can be computed from Λ as follows. Let e_j be the generator of $H_j(P^\infty) = \mathbb{Z}_2$ for $j \geq 1$. We have a complex $H_*(P^\infty) \otimes \Lambda$ which is a differential right Λ-module with differential δ given on the generators $e_j = e_j \otimes 1$ by

$$\delta(e_j) = \sum_{i \geq 0} \binom{j-i-1}{i+1} e_{j-i-1} \otimes \lambda_i.$$

Bigrade $H_*(P^\infty) \otimes \Lambda$ by $(H_*(P^\infty) \otimes \Lambda)^{s,t} = \Sigma_k H_k(P^\infty) \otimes \Lambda^{s,t-s-k}$. Then $H^{s,t}(H_*(P^\infty) \otimes \Lambda, \delta) = Ext_A^{s,t}(P^\infty)$. There is a chain map $\bar{t} : H_*(P^\infty) \otimes \Lambda \to \Lambda$ given by $\bar{t}(e_j \otimes \lambda_{j_1} \cdots \lambda_{j_s}) = \lambda_j \lambda_{j_1} \cdots \lambda_{j_s}$ which induces the induced map

$$t_* : Ext_A^{s,t}(P^\infty) \longrightarrow Ext_A^{s+1,t+1}(S^0)$$

of the transfer map $t : P^\infty \to S^0$ (see [**11**]). Consider the sum

(10) $e_a \lambda_{2^{i-3}-1} \lambda_{2^i-1}^2 \lambda_0^3 + e_{a_1} \lambda_{2^i+2^{i-2}-1} \lambda_{2^{i-1}-1}^2 \lambda_0^3$

$$+e_{a_1}\lambda_{2^{i-1}-1}\lambda_{a_1}\lambda_{2^i-1}\lambda_0^3+e_{2^{i-1}-1}\lambda_{a_1}\lambda_{2^i+2^{i-2}-1}\lambda_{2^{i-1}-1}\lambda_0^3$$

in $H_*(P^\infty)\otimes\Lambda$ which is obtained by replacing the leading λ_j in each term of the cycle (9) by the corresponding e_j. It is not difficult to show that (10) is a cycle in $H_*(P^\infty)\otimes\Lambda$, and so represents a class in $Ext_A^{6,*}(P^\infty)$. This class is denoted by $\overline{e_a h_{i-3}h_i^2h_0^3}$, and this is the canonical x_1 mentioned above. Then

$$t_*\left(\overline{e_a h_{i-3}h_i^2h_0^3}\right)=g_{i-2}h_0^3.$$

This explains the connection of $d_4(h_i^3)$, via $d_4\left(\hat{h}_{i-1}h_{i-1}h_{i+1}\right)$, to $g_{i-2}h_0^3$ as discussed above. In this connection the role of the Hopf invariant of $g_{i-2}h_0^3$, which is conjectured to be $h_1^3h_{i-3}h_i^2$ for $i\geq 7$, is described as follows. Since $h_{i-3}h_i^2h_0^3=h_0^3h_{i-3}h_i^2$ in $Ext_A^{*,*}$ there exists $L\in\Lambda^{5,*}$ such that $\delta(L)=\lambda_0^3\lambda_{2^{i-3}-1}\lambda_{2^i-1}^2+\lambda_{2^{i-3}-1}\lambda_{2^i-1}^2\lambda_0^3$. We have

$$\delta\left(\lambda_{a+1}\lambda_0^2+\lambda_{a-1}\lambda_2\lambda_0+\lambda_{a-2}\lambda_1\lambda_2\right)\equiv\lambda_a\lambda_0^3+\lambda_{a-3}\lambda_3\lambda_0^2 \mod \Lambda(a-4)$$

for $i\geq 4$ (recall $a=2^{i-1}+2^{i-2}+2^{i-3}-1$) where $\Lambda(k)$ is the $\mathbb{Z}_2$-submodule of Λ generated by admissible monomials $\lambda_{j_1}\lambda_{j_2}\cdots\lambda_{j_s}$ with $j_1\leq k$. These together with the cycle representation in (9) show that

(11) $g_{i-2}h_0^3=\left\{\lambda_{a-3}\lambda_3\lambda_0^2\lambda_{2^{i-3}-1}\lambda_{2^i-1}^2+\Gamma\right\}$ with $\Gamma\in\Lambda(a-4)$.

Let $F(k)\subset H_*(P^\infty)\otimes\Lambda$ be defined by $F(k)=\Sigma_{j\leq k}H_j(P^\infty)\otimes\Lambda$. A similar calculation shows that

(12) $\overline{e_a h_{i-3}h_i^2h_0^3}=\left\{e_{a-3}\lambda_3\lambda_0^2\lambda_{2^{i-3}-1}\lambda_{2^i-1}^2+\Gamma'\right\}$ with $\Gamma'\in F(a-4)$.

For $i\geq 7$, $h_1^3h_{i-3}h_i^2=h_2h_0^2h_{i-3}h_i^2=\left\{\lambda_3\lambda_0^2\lambda_{2^{i-3}-1}\lambda_{2^i-1}^2\right\}$ is non-zero in $Ext_A^{6,*}$. From (11) we see $g_{i-2}h_0^3$ can be represented by a cycle which lies in $\Lambda(a-3)$. We conjecture that $g_{i-2}h_0^3$, for $i\geq 7$, can not be represented by a cycle which lies in $\Lambda(a-4)$. This is equivalent to saying that the Hopf invariant of $g_{i-2}h_0^3$ is $h_1^3h_{i-3}h_i^2$. This conjecture is equivalent to the following.

(13) $\overline{e_a h_{i-3}h_i^2h_0^3}$, for $i\geq 7$, can not be represented by a cycle which lies in $F(a-4)$.

Anything that we said above about P^∞ can be applied to $P^{a+2}=P^{2^{i-1}+2^{i-2}+2^{i-3}+1}$. Thus we also have $F(k)$ for P^{a+2} with $k\leq a+2$, and $F(a+2)=H_*(P^{a+2})\otimes\Lambda$. In particular, the class $\overline{e_a h_{i-3}h_i^2h_0^3}$ in (12) can be considered as in $Ext_A^{6,*}(P^{a+2})$. One of our *Ext* groups results is that (13) is true if restricted to P^{a+2}. This implies $\bar{e}_a h_{i-3}h_i^2h_0^3\neq 0$ in $Ext_A^{6,*}(P_b^{a+2})$ for $i\geq 7$, in particular, $\bar{e}_{111}h_4h_7^2h_0^3\neq 0$, and this is the crucial point of our method as already mentioned above, since the assertion

(14) $d_4\left(\hat{h}_{i-1}h_{i-1}h_{i+1}\right)\neq 0$ in P^{a+2}

is recognized easily from this property. This property does not hold for $i=5,6$. In fact, $h_{i-3}h_i^2h_1^3=0$ in $Ext_A^{6,*}$, and also $\bar{e}_a h_{i-3}h_i^2h_0^3=0$ in $Ext_A^{6,*}(P_b^{a+2})$ for $i=5,6$. Nevertheless, since $g_3h_0^3\neq 0$, $g_4h_0^3\neq 0$, (14) should also hold for $i=5,6$. To deduce it and also to evaluate $d_4\left(\hat{h}_{i-1}h_{i-1}h_{i+1}\right)$ for these two cases depend heavily on the Hopf invariants of $g_3h_0^3$ and $g_4h_0^3$ which are not the same form as those for the cases $i\geq 7$ as just indicated. These two Hopf invariants will not be discussed here. As we have already said above, even if the Hopf invariants of $g_3h_0^3$

and $g_4h_0^3$ can be explicitly calculated, it seems that more geometric arguments are needed in order to show $d_4\left(h_i^3\right) = g_{i-2}h_0^3$ (plus something else) for $i = 5, 6$.

This concludes the description of our method.

This paper is organized as follows. The "solid calculations" mentioned above will be referred to hereafter as the **work of "solid calculations"**. This work computes two families of Ext groups. The first family is that of the Ext groups $Ext_A^{s,t} = Ext_A^{s,t}(\mathbb{Z}_2, \mathbb{Z}_2)$ for $s = 5, 6, 7$ and for some various t. We will also recall the known structure of $Ext_A^{s,*}$ for $1 \le s \le 4$. The results about these Ext groups will be stated in the first half of Section 3. The Ext groups in the second family are of the form $Ext_A^{s,t}(X) = Ext_A^{s,t}(H^*(X), \mathbb{Z}_2)$ for $0 \le s \le 6$ and for some various t where X is a stunted projective space P_ℓ^k or a space associated to one of the $P_\ell^{k's}$. These associated spaces are to be constructed from the topological results (5) and (6) above. The second half of Section 3 contains the statements of these Ext groups and the precise statements of (5) and (6). In addition we will illustrate our method in Section 3, from the work of "solid calculations", for proving some of the most difficult results about the groups $Ext_A^{s,t}(X)$. In Section 4 we compute the differential $d_2 : Ext_A^{s,t}(X) \to Ext_A^{s+2,t+1}(X)$ in the ASS for such X on some elements described in Section 3. Section 4 represents a preparatory work for the proof of Theorem 1.1 in Section 5. In Section 6 we prove the topological results (5) and (6), which will be stated as (3.8)(1), (2) in Section 3.

3 Results on some Ext groups over the mod 2 Steenrod algebra

We begin by stating some known results about $Ext_A^{s,t} = Ext_A^{s,t}(\mathbb{Z}_2, \mathbb{Z}_2)$ for small s. Recall that $h_i \in Ext_A^{1,2^i}$ is the class corresponding to the generator $Sq^{2^i} \in A$.

Theorem 3.1 [1, 21]

(1) *$\{h_i | i \ge 0\}$ is a $\mathbb{Z}_2$-base for $Ext_A^{1,*}$.*

(2) *$h_ih_{i+1} = 0$ and $\{h_ih_j | 0 \le i \le j, i \ne j-1\}$ is a $\mathbb{Z}_2$-base for $Ext_A^{2,*}$.*

(3) *The products $h_ih_jh_k$ in $Ext_A^{3,*}$ are subject only to the relations $h_ih_{i+2}^2 = 0$, $h_i^3 = h_{i-1}^2h_{i+1}$ and the relations induced by $h_ih_{i+1} = 0$.*

(4) *For each $i \ge 0$ there is a non-zero class $c_i \in Ext_A^{3,2^{i+3}+2^{i+1}+2^i}$ such that the set $\{c_i | i \ge 0\}$ together with a linearly independent set of the products $h_ih_jh_k$ form a $\mathbb{Z}_2$-base for $Ext_A^{3,*}$.*

We recall that h_i is represented in the lambda algebra Λ by λ_{2^i-1}. The indecomposable element c_i in Theorem (3.1) (4) has a cycle representation in Λ given by

$$c_i = \left\{\lambda_{2^{i+1}+2^i-1}\lambda_{2^{i+2}-1}^2\right\}.$$

In addition to the relations in Theorem (3.1)(3) among the products of the $h_i's$, May shows in [18] that there are further relations $h_i^2h_{i+3}^2 = 0$, $h_ih_0^{2^{i-1}} = 0$ and $h_i^2h_0^{2^{i-2}} = 0$.

Theorem 3.2 [13] *In $Ext_A^{s,*}$ for $s \le 5$, the products in the $h_i's$ are subject only to the relations $h_ih_{i+1} = 0$, $h_ih_{i+2}^2 = 0$, $h_i^3 = h_{i-1}^2h_{i+1}$, $h_i^2h_{i+3}^2 = 0$ and $h_0^4h_3 = 0$.*

Before stating the next result we recall from Tangora [20] the following indecomposable elements in $Ext_A^{4,*}$ and $Ext_A^{5,*}$:

(1) $d_0 \in Ext_A^{4,18}$
(2) $e_0 \in Ext_A^{4,21}$
(3) $f_0 \in Ext_A^{4,22}$
(4) $g_1 \in Ext_A^{4,24}$
(5) $p_0 \in Ext_A^{4,37}$
(6) $D_3 \in Ext_A^{4,65}$
(7) $p_0' \in Ext_A^{4,73}$
(8) $x \in Ext_A^{5,42}$
(9) $D_1 \in Ext_A^{5,57}$
(10) $H_1 \in Ext_A^{5,67}$.

Applying $(Sq^0)^i$ for each $i \geq 0$ to these elements we get the following elements:

(1)′ $d_i \in Ext_A^{4,2^{i+4}+2^{i+1}}, \quad i \geq 0.$
(2)′ $e_i \in Ext_A^{4,2^{i+4}+2^{i+2}+2^i}, \quad i \geq 0.$
(3)′ $f_i \in Ext_A^{4,2^{i+4}+2^{i+2}+2^{i+1}}, \quad i \geq 0.$
(4)′ $g_{i+1} \in Ext_A^{4,2^{i+4}+2^{i+3}}, \quad i \geq 0.$
(5)′ $p_i \in Ext_A^{4,2^{i+5}+2^{i+2}+2^i}, \quad i \geq 0.$
(6)′ $D_3(i) \in Ext_A^{4,2^{i+6}+2^i}, \quad i \geq 0.$
(7)′ $p_i' \in Ext_A^{4,2^{i+6}+2^{i+3}+2^i}, \quad i \geq 0.$
(8)′ $x(i) \in Ext_A^{5,2^{i+5}+2^{i+3}+2^{i+1}}, \quad i \geq 0.$
(9)′ $D_1(i) \in Ext_A^{5,2^{i+5}+2^{i+4}+2^{i+3}+2^i}, \quad i \geq 0.$
(10)′ $H_1(i) \in Ext_A^{5,2^{i+6}+2^{i+1}+2^i}, \quad i \geq 0.$

Theorem 3.3 [13]

(1) *The elements in (1)′ through (7)′ are all non-zero and are the only indecomposable elements in* $Ext_A^{4,*}$.
(2) *The products* $h_j c_i \in Ext_A^{4,*}$ *are subject only to the relations* $h_{i-1}c_i = 0$, $h_i c_i = 0$, $h_{i+2}c_i = 0$ *and* $h_{i+3}c_i = 0$.
(3) *The set of the indecomposable elements in (3.3)(1), the linearly independent set* $\{h_j c_i\}$ *(determined by (3.3)(2)) together with a linearly independent set* $\{h_i h_j h_k h_\ell\}$ *(determined by Theorem 3.2) form a* $\mathbb{Z}_2$*-base for* $Ext_A^{4,*}$.

The following results (3.4) and (3.5) about $Ext_A^{5,t}$ and $Ext_A^{6,t'}$ for certain t and t' are new. They are consequences of a part of the work of "solid calculations". These Ext groups and those above form the first family of Ext groups referred to at the end of Section 2. (3.4) describes completely $Ext_A^{5,t}$ and $Ext_A^{6,t'}$ for $t = 2^8 + 2^6 + \bar{t}$ with $1 \leq \bar{t} \leq 2^6$, for $t = 2^7 + 2, 2^8 + 2^5 + 2, 2^8 + 2^4 + 2, 2^8 + 2^7 + 2$ and for $t' = 2^8 + 2^6 + 3, 2^7 + 3$. (3.5) describes some monomials which are non-zero or zero in relevant $Ext_A^{5,*}$, $Ext_A^{6,*}$ and $Ext_A^{7,*}$.

Theorem 3.4

(1) *In (a), (b), (c), (d) below only* $Ext_A^{5,t}$ *for* $t = 2^8 + 2^6 + \bar{t}$ *with* $1 \leq \bar{t} \leq 2^6$ *are considered. In (a), if* $Ext_A^{5,2^8+2^6+\bar{t}} \cong \mathbb{Z}_2$ *and is generated by* y *then we write* $(\bar{t}, y)$ *to denote this relation. The notations* $(\bar{t}, y_1, y_2), (\bar{t}, y_1, y_2, y_3)$ *in (b) and (c) are similarly interpreted.*

(a) $Ext_A^{5,2^8+2^6+\bar{t}} \cong \mathbb{Z}_2$ *in the following cases:*

$(1, h_8D_3)$, $(3, h_8h_6h_0^3)$, $(5, h_8h_5^2h_2h_0)$,
$(6, h_8h_6h_2h_0^2)$, $(8, h_8d_2)$, $(9, h_8p_0')$,
$(11, h_8h_6c_0)$, $(12, h_8h_6h_3h_1^2)$, $(16, x(3))$,
$(21, h_8h_6h_4h_2h_0)$, $(22, h_8h_6c_1)$, $(25, h_8h_0c_3)$,
$(26, h_8h_1c_3)$, $(37, c_5h_2h_0)$, $(41, c_5h_3h_0)$,
$(42, c_5h_3h_1)$, $(44, h_8h_6c_2)$, $(48, f_4h_4 = e_4h_5 = c_5h_3^2)$.

(b) $Ext_A^{5,2^8+2^6+\bar{t}} \cong \mathbb{Z}_2 \oplus \mathbb{Z}_2$ *in the following cases:*

$(2, T_1, h_8h_5^2h_0^2)$, $(4, D_3(2)h_6, h_8h_5^2h_1^2)$,
$(10, h_8p_1, h_8h_6h_3h_0^2)$, $(17, e_4h_0, h_8h_6h_3^2h_0)$,
$(18, e_4h_1, h_8h_6h_4h_0^2)$, $(24, f_2h_8, h_8h_6h_4h_2^2)$,
$(32, h_8g_3, d_4h_6 = e_4h_4)$, $(33, f_4h_0, h_8h_6h_4^2h_0)$,
$(36, f_4h_2, c_5h_1^2)$,

(c) $Ext_A^{5,2^8+2^6+\bar{t}} \cong \mathbb{Z}_2 \oplus \mathbb{Z}_2 \oplus \mathbb{Z}_2$ *in the following cases:*

$(20, e_4h_2, e_2h_8, h_8h_6h_4h_1^2)$, $(34, f_4h_1, h_8h_6h_4^2h_1, c_5h_0^2)$
$(40, f_4h_3, c_5h_2^2, h_8h_4c_3)$.

(d) $Ext_A^{5,2^8+2^6+\bar{t}} = 0$ *for other* $\bar{t}$ $(1 \le \bar{t} \le 2^6)$.

(2) (a) $Ext_A^{5,2^7+2} \cong \mathbb{Z}_2$, *generated by an element called* K.
(b) $Ext_A^{5,2^8+2^5+2} \cong \mathbb{Z}_2 \oplus \mathbb{Z}_2$, *generated by* $h_8h_4^2h_0^2$ *and* d_4h_1.
(c) $Ext_A^{5,2^8+2^4+2} \cong \mathbb{Z}_2$, *generated by* $h_7^2h_4h_0^2$.
(d) $Ext_A^{5,2^8+2^7+2} \cong \mathbb{Z}_2$, *generated by* $h_7^3h_0^2$.

(3) (a) $Ext_A^{6,2^7+3} \cong \mathbb{Z}_2 \oplus \mathbb{Z}_2$, *generated by* h_6H_1 *and* Kh_0.
(b) $Ext_A^{6,2^8+2^6+3} \cong \mathbb{Z}_2 \oplus \mathbb{Z}_2 \oplus \mathbb{Z}_2$, *generated by* $h_8h_5^2h_0^3$, h_8H_1 *and* $h_8h_1D_3$.

Theorem 3.5 *In (1) (resp. (2)) below, only* $Ext_A^{6,2^8+2^6+\bar{t}}$ *(resp.* $Ext_A^{5,2^8+\bar{t}}$*) for certain* $\bar{t}$ *with* $1 \le \bar{t} \le 2^6$ *will be considered. If* y *is a non-zero element in* $Ext_A^{6,2^8+2^6+\bar{t}}$ *(resp.* $Ext_A^{5,2^8+\bar{t}}$*) then we write* $(y, \bar{t})$ *to denote this element. If* $y_1, \cdots, y_n$ *are linearly independent in* $Ext_A^{6,2^8+2^6+\bar{t}}$ *(resp.* $Ext_A^{5,2^8+\bar{t}}$*) then we write* $(y_1, \cdots, y_n, \bar{t})$ *to denote this linearly independent set. In (3) if* z *is a non-zero element in* $Ext_A^{5,t}$ *then we write* (z, t) *to denote this element.*

(1) *There are the following non-zero elements and linearly independent sets in* $Ext_A^{6,2^8+2^6+\bar{t}}$*:*

$(h_8h_1D_3, h_8h_5^2h_0^3, 3)$, $(h_8h_6h_0^4, 4)$, $(h_8h_5^2h_2h_0^2, 6)$,
$(h_8p_0'h_0, 10)$, $(h_8h_2d_2, 12)$, $(e_4h_0^2, h_1x(3), h_8h_6d_0, 18)$,
$(h_8h_6h_4h_2h_0^2, 22)$, $(h_8h_0^2c_3, h_8h_1f_2, h_8h_6h_2c_1, 26)$,
$(h_8h_6h_4c_0, 27)$, $(h_8f_2h_2 = h_8c_3h_1^2 = h_8e_2h_3, 28)$,
$(e_4h_4h_1, f_4h_0^2, h_8h_1g_3, h_8h_5^3h_0^2, 34)$, $(h_8g_3h_2, 36)$,
$(c_5h_2h_0^2, 38)$ $(f_4h_3h_1, c_5h_3h_0^2, 42)$, $(c_5h_2^3, 44)$,
$(e_4h_4^2, 48)$, $(h_8D_1(1), 50)$, $(h_8h_6h_3c_2, 52)$.

(2) *There are the following non-zero elements and linearly independent sets in* $Ext_A^{5,2^8+\bar{t}}$:

$(h_7^2h_4h_0^2, 18)$,	$(h_8h_4h_2h_0^2, 22)$,	$(h_7^2h_3^3, 24)$,
$(h_8h_2c_1, 26)$,	$(h_8h_4c_0, 27)$,	$(d_4h_1, h_8h_4^2h_0^2, 34)$,
$(h_8h_5h_2h_0^2, 38)$,	$(p_3h_1, h_8h_5h_3h_0^2, 42)$,	
$(h_8h_5h_2^3, 44)$,	$(h_8h_4^3h_0, 49)$,	$(h_8h_3c_2, 52)$.

(3) *There are the following non-zero elements in* $Ext_A^{5,t}$:

$(h_6c_0h_1, 77)$,	$(e_2h_0, 85)$,	$(f_2h_0, 89)$,	$(h_6h_2c_1, 90)$,
$(g_3h_0, 97)$,	$(h_4c_3h_0, 105)$,	$(h_4c_3h_1, 106)$,	$(h_6c_2h_0, 109)$.

(4) *(a)* $h_7^2h_4h_1^3 \neq 0$ *in* $Ext_A^{6,278}$, *(b)* $h_7^2h_4h_0^3 \neq 0$ *in* $Ext_A^{6,275}$.

(5) $g_5h_0^3, h_8h_6H_1, h_8h_0K$ *are linearly independent in* $Ext_A^{7,387}$.

(6) (a) $e_4h_4h_2 = 0$ *in* $Ext_A^{6,356}$,
(b) $h_8h_4h_1c_3 = 0$ *in* $Ext_A^{6,362}$,
(c) $T_1h_5 = 0$ *in* $Ext_A^{6,354}$,
(d) $h_6D_3 = 0$ *in* $Ext_A^{5,129}$,
(e) $D_3(1)h_0 = 0$ *in* $Ext_A^{5,131}$,
(f) $T_1h_0 = 0$ *in* $Ext_A^{6,323}$.

This completes the statements of Theorems 3.4 and 3.5.

Note that T_1 in Theorem 3.4(1)(b) and K in Theorem 3.4(2)(a) are new indecomposable elements in $Ext_A^{5,*}$. The work of "solid calculations" also contains the following.

$$(1)\ T_1 = \langle h_6, h_7, D_3(1) \rangle, \qquad (2)\ 0 \in \langle h_7, D_3(1), h_0 \rangle. \tag{3.6}$$

By Theorem 3.5(6)(d), $h_6D_3 = 0$ which implies $h_7D_3(1) = 0$ (by applying Sq^0) and by Theorem 3.5(6)(e), $D_3(1)h_0 = 0$. So the Massey products in (3.6) can be formed. The subscript "1" in T_1 comes from: if we let $T = T_0 = \langle h_5, h_6, D_3 \rangle$ then $Sq^0T_0 = T_1$. Using (3.6) the relations $h_5T_1 = 0$ and $T_1h_0 = 0$ in Theorem 3.5(6)(c), (f) follow easily from $h_5T_1 = h_5 \langle h_6, h_7, D_3(1) \rangle = \langle h_5, h_6, h_7 \rangle D_3(1) = 0$ and $T_1h_0 = \langle h_6, h_7, D_3(1) \rangle h_0 = h_6 \langle h_7, D_3(1), h_0 \rangle = 0$ since $h_6h_7 = 0$ and by (3.3), $Ext_A^{4,2^8+2} = \mathbb{Z}_2\left(h_7^2h_0^2\right)$. We also note that the relations $f_4h_4 = e_4h_5 = c_5h_3^2$, $d_4h_6 = e_4h_4$, $h_8f_2h_2 = h_8c_3h_1^2 = h_8e_2h_3$ and $h_8h_4c_3h_1 = 0$ in Theorem 3.4(1)(a), (b), Theorem 3.5(1) and Theorem 3.5(6)(b) are consequences of the known relations $f_0h_0 = e_0h_1$, $d_0h_2 = e_0h_0$, $f_1h_1 = c_2h_0^2$ and $h_3c_2h_0 = 0$ in $Ext_A^{*,*}$ **[2, 20]** by applying appropriate $\left(Sq^0\right)^i$.

We proceed to describe the second family of Ext groups which are of the form $Ext_A^{*,*}(X) = Ext_A^{*,*}\left(H^*(X), \mathbb{Z}_2\right)$ where X is either a stunted real projective space P_ℓ^k or a space associated to one of these P_ℓ^k, arising from the two topological results described in (5), (6) of Section 2. Some preliminaries are needed before we can state results about these *Ext* groups. These include the recalling of some facts about the cohomology $H^*\left(P_\ell^k\right)$ as a left module over the mod 2 Steenrod algebra A, about $Ext_A^{s,*}\left(P_\ell^k\right)$ for $s = 0, 1$ and also the precise statements of the two topological results.

We begin by recalling that $H^i\left(P_\ell^k\right) \cong \mathbb{Z}_2$, generated by x^i for $\ell \le i \le k$, and is zero for other i for which we let $x^i = 0$. The Steenrod algebra A acts on $H^*\left(P_\ell^k\right)$

by $Sq^j x^i = \binom{i}{j} x^{i+j}$. Let e_i be the dual element of x^i in $H_i\left(P_\ell^k\right)$. Transposing the left A-action on $H^*\left(P_\ell^k\right)$ we get a right A-action on the homology $H_*\left(P_\ell^k\right)$, and this action is given by

$$e_i Sq^j = \binom{i-j}{j} e_{i-j}. \tag{3.7}$$

For $\ell = 1$, only e_{2^i-1} with $2^i - 1 \le k$ have the property $e_{2^i-1}Sq^j = 0$ for all $j > 0$. These elements give rise to a $\mathbb{Z}_2$-base $\left\{\bar{e}_{2^i-1} | 1 \le 2^i - 1 \le k\right\}$ for $Ext_A^{0,*}\left(P_1^k = P^k\right)$. In [**6**], $\bar{e}_{2^i-1}$ is denoted by $\hat{h}_i$. For the remainder of this paper we will use the notation $\bar{e}_{2^i-1}$ to denote this class. If $k \le 2^j$ then, via the correspondence $e_{2^j+n} \to e_n$, $H_*\left(P_{2^j+1}^{2^j+k}\right)$ is isomorphic to $H_*\left(P^k\right)$ as a right A-module. So $Ext_A^{s,t}\left(P_{2^j+1}^{2^j+k}\right) \cong Ext_A^{s,t}\left(P^k\right)$ for all s, t. The corresponding non-zero elements in $Ext_A^{0,*}\left(P_{2^j+1}^{2^j+k}\right)$ will be denoted by $\bar{e}_{2^j+2^i-1}$ for $1 \le 2^i - 1 \le k$. In general, if e_n gives rise to a generator of $Ext_A^{0,n}\left(P_\ell^k\right) = \mathbb{Z}_2$ then this generator is denoted by $\bar{e}_n$. The stunted real projective spaces that we want to compute the Ext groups of are P^{53}, P^{64}, P^{113}, P_{63}^{113}, P_{65}^{113}, P_{97}^{113}, P_{65}^{96}, P_{54}^{56}. It is easy to see from (3.7) what the generators $\bar{e}_n$ in $Ext_A^{0,*}(\quad)$ of these spaces are. Recall that $Ext_A^{*,*}\left(P_\ell^k\right)$ is a right module over the cohomology algebra $Ext_A^{*,*}$. It is proved in [**6**] that, in $Ext_A^{1,*}\left(P^k\right)$, $\bar{e}_{2^i-1}h_j$ is zero for $j = i - 1$ and $2^i + 2^j - 1 \le k$ and is non-zero otherwise. In particular, $\bar{e}_{31}h_5 \ne 0$ in $Ext_A^{1,63}\left(P^k\right)$ for $k \ge 31$ and $\bar{e}_{79}h_4 \ne 0$ in $Ext_A^{1,95}\left(P_{65}^k\right)$ for $k \ge 79$.

The spaces associated to the P_ℓ^k to be considered will be constructed out of two maps in the following Proposition 3.8 which gives more precise statements on the two topological results referred to above. The proof of Proposition 3.8 will be given in Section 6.

Proposition 3.8 *(1) There exists a stable map* $\varphi_1 : S^{62} \cup_{2\iota} e^{63} \to P^{113}$ *such that* $\varphi_1^* \ne 0 : H^{63}(P^{113}) = \mathbb{Z}_2 \to H^{63}(S^{62} \cup_{2\iota} e^{63}) = \mathbb{Z}_2$ *and* $\varphi_1 | S^{62} : S^{62} \to P^{113}$ *is detected* $\bar{e}_{31}h_5$ *in the ASS for* P^{113}.
(2) There exists a stable map $\varphi_2 : S^{94} \cup_{2\iota} e^{95} \to P_{65}^{113}$ *such that* $\varphi_2^* \ne 0 : H^{95}(P_{65}^{113}) = \mathbb{Z}_2 \to H^{95}\left(S^{94} \cup_{2\iota} e^{95}\right) = \mathbb{Z}_2$ *and* $\varphi_2 | S^{94} : S^{94} \to P_{65}^{113}$ *is detected by* $\bar{e}_{79}h_4$ *in the ASS for* P_{65}^{113}.

The mapping cones C_{φ_1} and C_{φ_2} will be denoted by $\tilde{P}^{113}$ and $\hat{P}_{65}^{113}$ respectively. It is clear that $\tilde{P}^{113}$ (resp. $\hat{P}_{65}^{113}$) also has a stable cell e^n for each n with $1 \le n \le 113$ (resp. $65 \le n \le 113$); its k-skeleton will be denoted by $\tilde{P}^k$ (resp. $\hat{P}_{65}^k$). Note that $\tilde{P}^{k_1} = P^{k_1}$ and $\hat{P}_{65}^{k_2} = P_{65}^{k_2}$ for $k_1 \le 62$ and $k_2 \le 94$. $\tilde{P}^k / \tilde{P}^{\ell-1}$ (resp. $\hat{P}_{65}^k / \hat{P}_{65}^{\ell-1}$) is denoted by $\tilde{P}_\ell^k$ (resp. $\hat{P}_\ell^k$). Note that $\tilde{P}_{\ell_1}^{k_1} = P_{\ell_1}^{k_1}$ and $\hat{P}_{\ell_2}^{k_2} = P_{\ell_2}^{k_2}$ for $\ell_1 \ge 64$ and $\ell_2 \ge 96$. All of these are stable objects, but we will still refer to them as spaces. These spaces are called the associated stunted projective spaces.

The generator of $H_i(\tilde{P}^{113}) \cong \mathbb{Z}_2$ $(1 \le i \le 113)$ and $H_i\left(\hat{P}_{65}^{113}\right) \cong \mathbb{Z}_2$ $(65 \le i \le 113)$ will also be denoted by e_i except for two cases: that of $H_{63}(\tilde{P}^{113})$ is denoted by $\tilde{e}_{63}$ and that of $H_{95}(\hat{P}_{65}^{113})$ is denoted by $\hat{e}_{95}$. The right A-actions on these homology groups are given by the following rules which follow from (3.7) and the second properties of φ_1 and φ_2 in (3.8).

(1) For $H_*\left(\tilde{P}^{113}\right)$, $\tilde{e}_{63}Sq^j = \begin{cases} e_{31} & \text{for } j = 32 \\ 0 & \text{for } 0 < j \neq 32, \end{cases}$

$$e_iSq^j = \begin{cases} \binom{i-j}{j}e_{i-j} & \text{for } i-j \neq 63 \\ 0 & \text{for } i-j = 63. \end{cases}$$

(2) For $H_*(\hat{P}^{113}_{65})$, (3.9)

$$\hat{e}_{95}Sq^j = \begin{cases} e_{79} & \text{for } j = 16 \\ 0 & \text{for } 0 < j \neq 16, \end{cases}$$

$$e_iSq^j = \begin{cases} \binom{i-j}{j}e_{i-j} & \text{for } i-j \neq 95 \\ 0 & \text{for } i-j = 95. \end{cases}$$

The A-actions on $H_*(\tilde{P}^k_\ell)$, $H_*(\hat{P}^k_\ell)$ are also given by these rules. For $X = \tilde{P}^k_\ell$ or $\hat{P}^k_\ell$, if $x \in H_n(X)$ is an element that gives rise to a non-zero class in $Ext_A^{0,n}(X)$ then this class will be denoted by $\bar{x}$. The associated stunted projective spaces that we want to compute the Ext groups of are $\tilde{P}^{113}$, $\tilde{P}^{80}$, $\tilde{P}^{80}_{63}$, $\hat{P}^{113}_{65}$ and $\hat{P}^{96}_{65}$. It is not difficult to see from (3.9) what the generators in $Ext_A^{0,*}(\quad)$ of these spaces are. For example, $\{\bar{e}_1, \bar{e}_3, \bar{e}_7, \bar{e}_{15}, \bar{e}_{31}, \bar{e}_{95}\}$ is a $\mathbb{Z}_2$-base for $Ext_A^{0,*}\left(\tilde{P}^{113}\right)$ and is also a $\mathbb{Z}_2$-base for $Ext_A^{0,*}\left(\tilde{P}^{80}\right)$ if $\bar{e}_{95}$ is excluded.

In addition to the spaces described above we will also consider $\bar{P}^{113}_{62} = P^{113}_{62}/S^{63}$ which is described as follows. We have $P^{63}_{62} \simeq S^{62} \vee S^{63}$. Then $\bar{P}^{113}_{62}$ is the mapping cone of the composite $S^{63} \overset{i}{\hookrightarrow} S^{62} \vee S^{63} = P^{63}_{62} \overset{j}{\longrightarrow} P^{113}_{62}$ where i, j are inclusion maps. The A-action on $H_*(\bar{P}^{113}_{62})$ is also given by (3.7) with $e_{63} = 0$.

Recall the lambda algebra Λ and the complex $H_*(P^\infty) \otimes \Lambda$ in Section 2. We can consider a similar complex for computing $Ext_A^{*,*}(M) = Ext_A^{*,*}(M, \mathbb{Z}_2)$ for any connected locally finite graded left A-module M, and this is described as follows. Let $M_* = Hom_{\mathbb{Z}_2}(M, \mathbb{Z}_2)$. Transposing the left A-action on M is a right A-action on M_*. The complex is $M_* \otimes \Lambda$ which is a differential right Λ-module with differential δ given on the module generators $m_* = m_* \otimes 1$ by

$$\delta(m_*) = \sum_{j \geq 0} m_*Sq^{j+1} \otimes \lambda_j. \tag{3.10}$$

Bigrade $M_* \otimes \Lambda$ by $(M_* \otimes \Lambda)^{s,t} = \Sigma_k M_k \otimes \Lambda^{s,t-s-k}$. Then $H^{s,t}(M_* \otimes \Lambda, \delta) \cong Ext_A^{s,t}(M)$ [**5**]. Since $M_* \otimes \Lambda$ is a differential right Λ-module, $Ext_A^{*,*}(M)$ is a right $Ext_A^{*,*}$-module.

Applying $M_* \otimes \Lambda$ to $M_* = H_*(P^k_\ell)$, $H_*(\bar{P}^{113}_{62})$, $H_*\left(\tilde{P}^k_\ell\right)$ and $H_*\left(\hat{P}^k_\ell\right)$ we see, from (3.7), (3.9) and (3.10), that the differentials for these complexes are given by the following rules

(1) $\delta(e_i) = \sum_{j \geq 0} \binom{i-j-1}{j+1} e_{i-j-1} \otimes \lambda_j$ for $H_*(P^k_\ell) \otimes \Lambda$ and for $H_*(\bar{P}^{113}_{62}) \otimes \Lambda$.

(3.11)

$$\left.\begin{array}{ll}(2) & \delta(e_i) = \displaystyle\sum_{j\geq 0, i\neq j+64} \binom{i-j-1}{j+1} e_{i-j-1}\otimes\lambda_j \\ & \delta(\tilde{e}_{63}) = e_{31}\otimes\lambda_{31}\end{array}\right\} \text{ for } H_*(\tilde{P}^k_\ell)\otimes\Lambda.$$

$$\left.\begin{array}{ll}(3) & \delta(e_i) = \displaystyle\sum_{j\geq 0, i\neq j+96} \binom{i-j-1}{j+1} e_{i-j-1}\otimes\lambda_j \\ & \delta(\hat{e}_{95}) = e_{79}\otimes\lambda_{15}\end{array}\right\} \text{ for } H_*(\hat{P}^k_\ell)\otimes\Lambda.$$

The following are easy to see from (3.11).

$$\begin{array}{ll}(1) & \bar{e}_{2^i-1}h_{i-1} = 0 \text{ in } Ext_A^{1,2^i+2^{i-1}-1}(P^k) \text{ for } 2^i+2^{i-1}-1\leq k. \\ (2) & \bar{e}_{95}h_4 = 0 \text{ in } Ext_A^{1,111}(\tilde{P}^{113}) \text{ and also in } Ext_A^{1,111}(P^{113}_{65}). \\ (3) & \bar{e}_{79}h_3 = 0 \text{ in } Ext_A^{1,87}(P^{113}_{65}).\end{array} \qquad (3.12)$$

We describe some classes in $Ext_A^{s,t}(X)$ for small $s\geq 1$ where X is one of the spaces considered above. (3.13) below lists these classes. In the list, $\bar{x}=\{z\}$ means z is a cycle in $H_*(X)\otimes\Lambda$ for the X indicated and $\bar{x}$ is the class as named. To simplify the notation we let $m_*\lambda_I$ to denote $m_*\otimes\lambda_I$.

(1) $\overline{e_{79}h_0} = \{e_{79}\lambda_0 + e_{63}\lambda_{16} + e_{64}\lambda_{15}\}$ in $Ext_A^{1,80}\left(P^{113}_{63}\right)$.

(2) $\overline{e_{95}h_0} = \{e_{95}\lambda_0 + e_{63}\lambda_{32} + e_{64}\lambda_{31}\}$ in $Ext_A^{1,96}\left(P^{113}_{63}\right)$.

(3) $\overline{\tilde{e}_{63}h_6} = \{\tilde{e}_{63}\lambda_{63}\}$ in $Ext_A^{1,127}\left(\tilde{P}^{113}\right)$.

(4) $\overline{e_{79}h_4} = \{e_{79}\lambda_{15} + \tilde{e}_{63}\lambda_{31} + e_{47}\lambda_{47}\}$ in $Ext_A^{1,95}\left(\tilde{P}^{113}\right)$.

(5) $\overline{e_{103}h_3} = \{e_{103}\lambda_7 + \hat{e}_{95}\lambda_{15} + e_{87}\lambda_{23}\}$ in $Ext_A^{1,111}\left(\hat{P}^{113}_{65}\right)$.

(6) $\overline{\hat{e}_{95}h_5} = \{\hat{e}_{95}\lambda_{31}\}$ in $Ext_A^{1,127}\left(\hat{P}^{113}_{65}\right)$ and also in $Ext_A^{1,127}\left(\hat{P}^{96}_{65}\right)$.

(7) $\overline{\tilde{e}_{63}h_5^2} = \left\{\tilde{e}_{63}\lambda_{31}^2 + e_{47}\lambda_{15}\lambda_{63}\right\}$ in $Ext_A^{2,127}\left(\tilde{P}^{113}\right)$.

(8) $\overline{\hat{e}_{95}h_4^2} = \left\{\hat{e}_{95}\lambda_{15}^2 + e_{87}\lambda_7\lambda_{31}\right\}$ in $Ext_A^{2,127}\left(\hat{P}^{96}_{65}\right)$ and also in $Ext_A^{2,127}\left(\hat{P}^{113}_{65}\right)$.

(9) $\overline{e_{56}h_2^2} = \left\{e_{56}\lambda_3^2\right\}$ in $Ext_A^{2,64}\left(P^{56}_{54}\right)$.

(10) $\overline{e_{23}h_0h_3h_5} = \left\{e_{23}\lambda_0\lambda_7\lambda_{31} + e_{15}\lambda_8\lambda_7\lambda_{31} + e_{16}\lambda_7^2\lambda_{31}\right\}$ in $Ext_A^{3,64}\left(P^{113}\right)$.

(11) $\overline{e_{87}h_0h_3h_5} = \left\{e_{87}\lambda_0\lambda_7\lambda_{31} + e_{79}\lambda_8\lambda_7\lambda_{31} + e_{80}\lambda_7^2\lambda_{31}\right\}$ in $Ext_A^{3,128}\left(\hat{P}^{113}_{65}\right)$ and also in $Ext_A^{3,128}\left(\hat{P}^{96}_{65}\right)$.

(12) $\overline{e_{47}h_4h_6h_1} = \left\{\begin{array}{c} e_{47}\lambda_{15}\lambda_{63}\lambda_1 + e_{31}\left(\lambda_{15}^2\lambda_{65} + \lambda_{15}\lambda_{17}\lambda_{63} + \lambda_{17}\lambda_{15}\lambda_{63}\right) \\ + e_{33}\lambda_{15}^2\lambda_{63}\end{array}\right\}$ in $Ext_A^{3,129}\left(P^{113}\right)$ and also in $Ext_A^{3,129}\left(\tilde{P}^{113}\right)$.

(13) $\overline{\hat{e}_{111}h_4h_7^2} = \{e_{111}\lambda_{15}\lambda_{127}^2 + e_{95}\lambda_{159}\lambda_{63}^2 + e_{95}\lambda_{63}\lambda_{95}\lambda_{127} + e_{63}\lambda_{95}\lambda_{159}\lambda_{63}\}$ in $Ext_A^{3,383}\left(P^{113}\right)$. The notation $\overline{e_{111}h_4h_7^2}$ will also denote the class $\left\{e_{111}\lambda_{15}\lambda_{127}^2 + e_{95}\lambda_{159}\lambda_{63}^2 + e_{95}\lambda_{63}\lambda_{95}\lambda_{127}\right\}$ in $Ext_A^{3,383}\left(\tilde{P}^{113}\right)$ and also in $Ext_A^{3,383}\left(P^{113}_{65}\right)$.

(14) $\overline{e_{17}h_3^2c_5} = \left\{e_{17}\lambda_7^2\lambda_{95}\lambda_{127}^2\right\}$ in $Ext_A^{5,385}\left(P^{113}\right)$ and also in $Ext_A^{5,385}\left(\tilde{P}^{113}\right)$.

$$(15)\quad \overline{\tilde{e}_{63}T_1} = \{\tilde{e}_{63}L_1 + e_{31}L_2\} \text{ in } Ext_A^{5,385}\left(\tilde{P}^{113}\right). \tag{3.13}$$

Straightforward calculations show that the sums in the brackets { } in (1) through (14) are cycles. The cycle $\tilde{e}_{63}L_1 + e_{31}L_2$ in (15) is described as follows. L_1 is a cycle in $\Lambda^{5,*}$ representing T_1 where T_1 is the class as described in (3.6). From (3.11) (2) for $H_*\left(\tilde{P}^{113}\right)\otimes\Lambda$ we see $\delta\left(\tilde{e}_{63}L_1\right) = e_{31}\lambda_{31}L_1$. By (3.5)(6) (c), $h_5T_1 = 0$. So $\lambda_{31}L_1$ is a boundary, that is, $\delta(L_2) = \lambda_{31}L_1$ for some $L_2 \in \Lambda^{5,*}$. Then $\tilde{e}_{63}L_1 + e_{31}L_2$ is a cycle which represents a class called $\overline{\tilde{e}_{63}T_1}$. Note that the first cycle representation for $\overline{e_{111}h_4}h_7^2$ in (13) has already been described in Section 2.

We have the following relations.

$$\begin{aligned}
&(1)\quad \overline{e_{95}h_0}h_4 = 0 \text{ in } Ext_A^{2,112}\left(P_{63}^{113}\right).\\
&(2)\quad \overline{e_{79}h_4}h_5 = \overline{\tilde{e}_{63}h_5^2} \text{ in } Ext_A^{2,127}\left(\tilde{P}^{113}\right).\\
&(3)\quad \overline{\tilde{e}_{63}h_6}h_7 = 0 \text{ in } Ext_A^{2,255}\left(\tilde{P}^{113}\right).\\
&(4)\quad \overline{\tilde{e}_{63}T_1}h_0 = 0 \text{ in } Ext_A^{6,386}\left(\tilde{P}^{113}\right).\\
&(5)\quad \overline{e_{103}h_3}h_4 = \overline{\hat{e}_{95}h_4^2} \text{ in } Ext_A^{2,127}\left(\hat{P}_{65}^{113}\right).\\
&(6)\quad \overline{e_{87}h_0h_3h_5}h_0 = 0 \text{ in } Ext_A^{4,129}\left(\hat{P}_{65}^{113}\right).\\
&(7)\quad \overline{\hat{e}_{95}h_5}h_7^2 = 0 \text{ in } Ext_A^{3,383}\left(\hat{P}_{65}^{113}\right).
\end{aligned} \tag{3.14}$$

Except for (4) and (6), the remaining 5 relations are obtained by straightforward calculations. We illustrate the calculations for two cases of these which are (1) and (5). $\overline{e_{95}h_0}h_4$ is represented by $e_{95}\lambda_0\lambda_{15} + e_{63}\lambda_{32}\lambda_{15} + e_{64}\lambda_{31}\lambda_{15}$. We have

$$\delta\left(e_{95}\lambda_{16} + e_{111}\lambda_0 + e_{64}\lambda_{47} + e_{63}\lambda_{48}\right) = e_{95}\lambda_0\lambda_{15} + e_{63}\lambda_{32}\lambda_{15} + e_{64}\lambda_{31}\lambda_{15}$$

in $H_*\left(P_{63}^{113}\right)\otimes\Lambda$. So $\overline{e_{95}h_0}h_4 = 0$ proving (3.14)(1). For (3.14)(5) we note that $\overline{e_{103}h_3}h_4$ is represented by $e_{103}\lambda_7\lambda_{15} + e_{87}\lambda_{23}\lambda_{15} + \hat{e}_{95}\lambda_{15}^2 = \hat{e}_{95}\lambda_{15}^2 + e_{87}\lambda_7\lambda_{31}$ which also represents $\overline{\hat{e}_{95}h_4^2}$. (13.4) (6) is non-trivial; it is contained in the work of "solid calculations". (13.4)(4) follows from (3.6) and the proof goes as follows. From the description of the cycle representation of $\overline{\tilde{e}_{63}T_1}$ in (3.13) (15) it is not difficult to see that to prove $\overline{\tilde{e}_{63}T_1}h_0 = 0$ it suffices to show $0 \in \langle h_5, T_1, h_0 \rangle$. Recall that $T_1 = \langle h_6, h_7, D_3(1) \rangle$. Then

$$\begin{aligned}
\langle h_5, T_1, h_0 \rangle &= \langle h_5, \langle h_6, h_7, D_3(1) \rangle, h_0 \rangle\\
&= \langle\langle h_5, h_6, h_7 \rangle, D_3(1), h_0 \rangle + \langle h_5, h_6, \langle h_7, D_3(1), h_0 \rangle\rangle .
\end{aligned}$$

We have $\langle h_5, h_6, h_7 \rangle = 0$. By (3.6)(2), $0 \in \langle h_7, D_3(1), h_0 \rangle$. So $0 \in \langle h_5, T_1, h_0 \rangle$.

Now we state the results on the second family of Ext groups in (3.15) below. We write $Ext_A^{s,t}(X) \cong \mathbb{Z}_2(y_1)\oplus\cdots\oplus\mathbb{Z}_2(y_n)$ to mean that $Ext_A^{s,t}(X) \cong \underbrace{\mathbb{Z}_2\oplus\cdots\oplus\mathbb{Z}_2}_{n}$, generated by $y_1, \cdots, y_n$.

Theorem 3.15 *(1)* $Ext_A^{1,96}\left(P_{63}^{113}\right) \cong \mathbb{Z}_2\left(\overline{e_{95}h_0}\right)$.

(2) $Ext_A^{1,127}\left(\tilde{P}^{113}\right) \cong \mathbb{Z}_2\left(\overline{\tilde{e}_{63}h_6}\right) \oplus \mathbb{Z}_2\left(\bar{e}_{95}h_5\right)$.

(3) $Ext_A^{1,127}\left(\hat{P}_{65}^{113}\right) \cong \mathbb{Z}_2\left(\overline{\hat{e}_{95}h_5}\right) \oplus \mathbb{Z}_2\left(\bar{e}_{111}h_4\right)$.

(4) $Ext_A^{2,64}\left(P^n\right) \cong \mathbb{Z}_2\left(\bar{e}_{31}h_5h_0\right)$ *for* $n \geq 64$.

(5) $\bar{e}_{63}h_6h_8 \neq 0$ *in* $Ext_A^{2,383}\left(P^n\right)$ *for* $n \geq 64$.

(6) $\bar{e}_{95}h_5h_8 \neq 0$ *in* $Ext_A^{2,383}\left(P_{65}^{113}\right)$ *and in* $Ext_A^{2,383}\left(P_{65}^{96}\right)$.

(7) $Ext_A^{2,96}(P_{65}^{113}) \cong \mathbb{Z}_2(\bar{e}_{79}h_4h_0) \cong Ext_A^{2,96}(P_{65}^{96})$.

(8) $Ext_A^{2,112}(P_{97}^{113}) \cong \mathbb{Z}_2(\bar{e}_{103}h_3h_0)$.

(9) $Ext_A^{2,96}(\tilde{P}^{113}) \cong \mathbb{Z}_2(\overline{e_{79}h_4}h_0) \oplus \mathbb{Z}_2(\bar{e}_{31}h_6h_0)$.

(10) $Ext_A^{2,112}(\hat{P}_{65}^{113}) \cong \mathbb{Z}_2(\overline{e_{103}h_3}h_0) \oplus \mathbb{Z}_2(\bar{e}_{79}h_5h_0)$.

(11) $Ext_A^{2,96}(\bar{P}_{62}^{113}) \cong \mathbb{Z}_2(\bar{e}_{79}h_4h_0)$.

(12) $Ext_A^{2,127}(\hat{P}_{65}^{113}) \cong \mathbb{Z}_2(\overline{\hat{e}_{95}h_4^2}) \oplus \mathbb{Z}_2(\bar{e}_{111}h_3^2)$.

(13) $Ext_A^{3,64}(P^{113}) \cong \mathbb{Z}_2(\overline{e_{23}h_0h_3h_5})$.

(14) $Ext_A^{3,97}(P_{63}^{113}) \cong \mathbb{Z}_2(\overline{e_{79}h_0}h_4h_0)$

(15) $Ext_A^{3,384}(P_{63}^{113}) \cong \mathbb{Z}_2(\overline{e_{95}h_0}h_5h_8)$.

(16) $Ext_A^{3,384}(P_{65}^{113}) \cong \mathbb{Z}_2(\bar{e}_{95}h_0h_5h_8)$.

(17) $Ext_A^{3,97}(P_{65}^{113}) \cong \mathbb{Z}_2(\bar{e}_{79}h_4h_0^2)$.

(18) $Ext_A^{3,384}(P_{97}^{113}) \cong \mathbb{Z}_2(\bar{e}_{111}h_0h_4h_8)$.

(19) $Ext_A^{3,128}(P_{54}^{56}) \cong \mathbb{Z}_2(\bar{e}_{54}h_6h_3h_1) \oplus \mathbb{Z}_2(\overline{e_{56}h_2^2h_6})$.

(20) $Ext_A^{3,384}(\tilde{P}^{113}) \cong \mathbb{Z}_2(\bar{e}_{95}h_5h_8h_0) \oplus \mathbb{Z}_2(\overline{\tilde{e}_{63}h_6}h_8h_0)$.

(21) $Ext_A^{3,128}(\tilde{P}^{80}) \cong \mathbb{Z}_2(\overline{\tilde{e}_{63}h_5^2h_0})$.

(22) $Ext_A^{3,128}(\tilde{P}_{63}^{80}) \cong \mathbb{Z}_2(\bar{\tilde{e}}_{63}h_5^2h_0)$.

(23) $Ext_A^{3,128}(\hat{P}_{65}^{113}) \cong \mathbb{Z}_2(\overline{\hat{e}_{95}h_4^2h_0}) \oplus \mathbb{Z}_2(\overline{e_{87}h_0h_3h_5}) \oplus \mathbb{Z}_2(\bar{e}_{111}h_3^2h_0)$.

(24) $Ext_A^{3,128}(\hat{P}_{65}^{96}) \cong \mathbb{Z}_2(\overline{\hat{e}_{95}h_4^2h_0}) \oplus \mathbb{Z}_2(\overline{e_{87}h_0h_3h_5})$.

(25) $Ext_A^{3,384}(\hat{P}_{65}^{113}) \cong \mathbb{Z}_2(\overline{\hat{e}_{95}h_5}h_8h_0) \oplus \mathbb{Z}_2(\bar{e}_{111}h_0h_4h_8)$.

(26) $\bar{e}_{95}h_4^2h_0h_8 \neq 0$ *in* $Ext_A^{4,384}(P_{65}^{96})$.

(27) $Ext_A^{4,128}(P^{53}) = \mathbb{Z}_2(\overline{e_{40}h_6h_3^3}) \oplus \mathbb{Z}_2(\overline{e_{52}h_6h_2^3})$. *The classes* $\overline{e_{40}h_6h_3^3}$ *and* $\overline{e_{52}h_6h_2^3}$ *are defined to be the images of the classes* $\bar{e}_{54}h_1h_3h_6$, $\overline{e_{56}h_2^2h_6}$ *in* $Ext_A^{3,128}(P_{54}^{56})$ *(as given in (3.15) (19)) under the connecting homomorphism* $Ext_A^{3,128}(P_{54}^{56}) \xrightarrow{\delta} \to Ext_A^{4,128}(P^{53})$ *arising from the short exact sequence* $0 \to H_*(P^{53}) \to H_*(P^{56}) \to H_*(P_{54}^{56}) \to 0$.

(28) *There is a cycle of the form* $e_{47}\lambda_0^2\lambda_{15}\lambda_{63} + \Sigma_{i\leq 46} e_i\lambda_{I_i}$ *in* $H_*(P^{113}) \otimes \Lambda^{4,*}$ *and also in* $H_*(\tilde{P}^{113}) \otimes \Lambda^{4,*}$. $\overline{e_{47}h_0^2h_4h_6}$ *will denote the class* $\{e_{47}\lambda_0^2\lambda_{15}\lambda_{63} + \Sigma_{i\leq 46} e_i\lambda_{I_i}\}$ *in* $Ext_A^{4,129}(P^{113})$ *and also in* $Ext_A^{4,129}(\tilde{P}^{113})$.

(29) $Ext_A^{4,385}(\tilde{P}^{113}) \cong \mathbb{Z}_2(\overline{\tilde{e}_{63}h_5^2}h_8h_1) \oplus \mathbb{Z}_2(\overline{\tilde{e}_{63}h_6}h_8h_0^2) \oplus \mathbb{Z}_2(\bar{e}_{95}h_5h_8h_0^2)$ $\oplus \mathbb{Z}_2(\overline{e_{111}h_4h_7^2}h_1) \oplus \mathbb{Z}_2(\overline{e_{47}h_4h_6h_1}h_8) \oplus \mathbb{Z}_2(\bar{e}_{31}c_5h_1)$.

(30) $\bar{e}_{63}h_5^2h_0^2h_8 \neq 0$ *in* $Ext_A^{5,385}(P^{64})$.

(31) $\bar{e}_{63}T_1 \neq 0$ *in* $Ext_A^{5,385}(P^{113})$ *and also in* $Ext_A^{5,385}(P_{63}^{113})$.

(32) *ker* $\left(\bar{q}_* : Ext_A^{5,385}(\tilde{P}^{113}) \longrightarrow Ext_A^{5,385}(P_{65}^{113})\right) \cong \mathbb{Z}_2(\overline{\tilde{e}_{63}h_5^2}h_0^2h_8)$ $\oplus \mathbb{Z}_2(\overline{e_{47}h_0^2h_4h_6}h_8) \oplus \mathbb{Z}_2(\bar{e}_{31}f_4h_1) \oplus \mathbb{Z}_2(\bar{e}_{31}c_5h_0^2) \oplus \mathbb{Z}_2(\overline{e_{17}h_3^2c_5})$ $\oplus \mathbb{Z}_2(\overline{\tilde{e}_{63}T_1})$ *where* $\bar{q} : \tilde{P}^{113} \to \tilde{P}_{65}^{113} = P_{65}^{113}$ *is the collapsing map.*

(33) $\overline{e_{111}h_4h_7^2h_0^3} \neq 0$ *in* $Ext_A^{6,386}(P^{113})$, *in* $Ext_A^{6,386}\left(\tilde{P}^{113}\right)$ *and in* $Ext_A^{6,386}(P_{65}^{113})$, *and* $\bar{e}_{111}h_4h_7^2h_0^3 \neq 0$ *in* $Ext_A^{6,386}(P_{97}^{113})$.

This completes the statement of (3.15).

Most of the results in (3.15) are contained in the work of "solid calculations". Here we would like to illustrate the method, from this work, for proving some of the difficult results in (3.15) which include (3.15)(29), (31), (32). In the illustration one will see that the *Ext* groups in (3.1) through (3.5) are the foundation blocks on which the calculations for obtaining the results in (3.15) are based.

(3.15) is proved by making calculations with a spectral sequence. Let M_* denote $H_*(X)$ where X is one of the spaces considered in (3.15). Thus, for each k, M_k is either $\mathbb{Z}_2$ or 0. Filter the complex $M_*\otimes\Lambda$ by $F(i) = \Sigma_{k\le i}M_k\otimes\Lambda$. Then $F(i)/F(i-1)$ is either $\Sigma^i\Lambda$ or 0 according as $M_i = \mathbb{Z}_2$ or 0. This filtration gives rise to a spectral sequence $\left\{\bar{E}_r^{i,s,t}(M)\right\}_{r\ge1}$ with $\bar{E}_1^{i,s,t}(M) = H^{s,t}(F(i)/F(i-1)) \cong Ext_A^{s,t-i}$ or 0 according as $M_i = \mathbb{Z}_2$ or 0, and $\Sigma_i\bar{E}_\infty^{i,s,t}(M) \cong Ext_A^{s,t}(M)$ as a $\mathbb{Z}_2$-module. For each $r \ge 1$ the differential d_r goes from $\bar{E}_r^{i,s,t}(M)$ to $\bar{E}_r^{i-r,s+1,t}(M)$. If $M_i = \mathbb{Z}_2$ and if m_i is its generator then for $\alpha \in Ext_A^{s,t-i}$ the corresponding element in $\bar{E}_1^{i,s,t}(M)$ is denoted by $m_i\alpha$. If $d_r(m_i\alpha) = m_{i-r}\beta$ then we write $m_i\alpha \to m_{i-r}\beta$ to indicate this differential. If $m_i\alpha$ is an infinite cycle, that is, $d_r(m_i\alpha) = 0$ for all r, then we write $m_i\alpha \to 0$ to indicate this notion. It is clear that $m_i\alpha$ is an infinite cycle if and only if there is a cycle in $M_*\otimes\Lambda$ of the form m_iz+y where z is a cycle in Λ representing α and $y \in F(i-1)$. Furthermore, if $m_i\alpha$ is not a bouundary in the spectral sequence so that $m_i\alpha$ persists to $\bar{E}_\infty$ then $\{m_iz+y\} \neq 0$ in $Ext_A^{*,*}(M)$. The following is also clear which is a property of the spectral sequence.

> Suppose $\{m_{i_j}\alpha_j \mid 1 \le j \le n\}$ is a set of n elements all persisting to $\bar{E}_\infty$ such that the classes $\{m_{i_j}z_j + y_j\}$, $1 \le j \le n$, form a $\mathbb{Z}_2$-base for $Ext_A^{s,t}(M)$ where $\alpha_j = \{z_j\}$ and $y_j \in F(i_j - 1)$. If, for each j, $m_{i'_j}z'_j + y'_j$ is another cycle in $M_*\otimes\Lambda$ with $i'_j \ge i_j$, $y'_j \in F(i'_j - 1)$ such that
>
> $(*)\quad \delta(q_j) \equiv m_{i'_j}z'_j + y'_j + m_{i_j}z'_j \mod F(i_j-1)$ in $M_*\otimes\Lambda$ for some q_j
>
> then the classes $\{m_{i'_j}z'_j + y'_j\}, 1 \le j \le n$, also form a $\mathbb{Z}_2$-base for $Ext_A^{s,t}(M)$. Any class $\{m_{i'_j}z'_j + y'_j\}$ satisfying $(*)$ is said to be represented by $m_{i_j}\alpha_j$, to have filtration i_j and there it projects to $\alpha_j = \{z_j\}$.
>
> (3.16)

For $M_* = H_*(X)$ we let $\bar{E}_r^{i,s,t}(X)$ denote $\bar{E}_r^{i,s,t}(H^*(X))$ and $\bar{E}_r^{s,t}(X)$ denote $\Sigma_i\bar{E}_r^{i,s,t}(X)$.

This spectral sequence is first considered by Mahowald in [**15**] where $Ext_A^{s,t}(P_\ell^k)$ for $t-s \le 30$ are computed. Some extensive calculations with this spectral sequence are made in [**6**] to obtain a complete determination on $Ext_A^{s,t}(P_\ell^k)$ for $s = 0, 1, 2$ all t and any P_ℓ^k. In particular, the results (3.15) (1), (4), (5), (6), (7), (8), (11) are a part of these calculations. We illustrate in the following the use of this spectral sequence for proving five of the remaining results in (3.15). These five are (3.15)(9), (29), (31), (32) and (33). (3.15)(9) and (3.15)(33) are easier and the remaining three are the most difficult ones in the whole (3.15).

First, (3.15)(9). To compute $Ext_A^{2,96}\left(\tilde{P}^{113}\right)$ by this spectral sequence we have to consider $\bar{E}_1^{2,96}\left(\tilde{P}^{113}\right)$ and $\bar{E}_1^{1,96}\left(\tilde{P}^{113}\right)$ which are generated by the elements in the following (3.15)(9)*(1) and (2) respectively.

$$
\begin{array}{llllll}
(3.15)(9)^* & (1) & e_{94}h_0^2, & e_{92}h_1^2, & e_{91}h_2h_0, & e_{88}h_2^2, \quad e_{87}h_3h_0, \\
& & e_{86}h_3h_1, & e_{80}h_3^2, & e_{79}h_4h_0, & e_{78}h_4h_1, \quad e_{76}h_4h_2, \\
& & e_{64}h_4^2, & \tilde{e}_{63}h_5h_0, & e_{62}h_5h_1, & e_{60}h_5h_2, \quad e_{56}h_5h_3, \\
& & e_{32}h_5^2, & e_{31}h_6h_0, & e_{30}h_6h_1 & e_{28}h_6h_2, \quad e_{24}h_6h_3, \\
& & e_{16}h_6h_4 & & & \\
& (2) & e_{95}h_0, & e_{94}h_1, & e_{92}h_2, & e_{88}h_3, \quad e_{80}h_4, \\
& & e_{64}h_5, & e_{32}h_6. & &
\end{array}
$$

By straightforward calculations in the complex $H_*\left(\tilde{P}^{113}\right) \otimes \Lambda$ given by (3.11)(2) we get the following differentials in the spectral sequence.

$$
\begin{array}{lll}
(3.15)(9)^{**} & (1) & e_{94}h_0^2 \to e_{93}h_0^3, \qquad e_{92}h_1^2 \to e_{90}h_1^3, \\
& & e_{88}h_2^2 \to e_{84}h_2^3, \qquad e_{86}h_3h_1 \to e_{72}h_3^3, \\
& & e_{80}h_3^2 \to e_{79}h_3^2h_0, \qquad e_{76}h_4h_2 \to e_{75}h_4h_2h_0, \\
& & e_{64}h_4^2 \to e_{62}h_4^2h_1, \qquad \tilde{e}_{63}h_5h_0 \to e_{30}h_5^2h_1, \\
& & e_{60}h_5h_2 \to e_{59}h_5h_2h_0, \qquad e_{56}h_5h_3 \to e_{55}h_5h_3h_0, \\
& & e_{32}h_5^2 \to e_{31}h_5^2h_0, \qquad e_{28}h_6h_2 \to e_{27}h_6h_2h_0, \\
& & e_{24}h_6h_3 \to e_{23}h_6h_3h_0, \qquad e_{16}h_4^2 \to e_{15}h_4^2h_0, \\
& & \\
& (2) & e_{95}h_0 \to 0, \qquad e_{94}h_1 \to 0, \\
& & e_{92}h_2 \to e_{91}h_2h_0, \qquad e_{88}h_3 \to e_{87}h_3h_0, \\
& & e_{80}h_4 \to e_{79}h_4h_0, \qquad e_{64}h_5 \to e_{62}h_5h_1, \\
& & e_{32}h_6 \to e_{31}h_6h_0.
\end{array}
$$

The differential $e_{86}h_3h_1 \to e_{72}h_3^3$ is less trivial and is shown as follows. We have

$$
\begin{aligned}
\delta\left(e_{88}\lambda_7\right) \quad & \equiv e_{87}\lambda_0\lambda_7 + e_{86}\lambda_1\lambda_7 + \left(e_{84}\lambda_3\lambda_7 = 0\right) + e_{79}\lambda_8\lambda_7 \\
& +e_{78}\lambda_9\lambda_7 + e_{76}\lambda_{11}\lambda_7 \qquad \bmod \quad F(64).
\end{aligned}
$$

Let L be the sum of the last 2 terms. So $L \in F(78)$. Then

$$\delta\left(e_{86}\lambda_1\lambda_7 + L\right) \equiv \delta\left(e_{87}\lambda_0\lambda_7 + e_{79}\lambda_8\lambda_7\right) \equiv e_{79}\lambda_0\lambda_7^2 \quad \bmod \quad F(64).$$

We have $\delta\left(e_{80}\lambda_7^2\right) \equiv e_{79}\lambda_0\lambda_7^2 + e_{72}\lambda_7^3 \bmod F(64)$. So

$$\delta\left(e_{86}\lambda_1\lambda_7 + L + e_{80}\lambda_7^2\right) \equiv e_{72}h_7^3 \quad \bmod \quad F(64).$$

This proves $e_{86}h_1h_3 \to e_{72}h_3^3$. $e_{95}h_0 \to 0$ in (2) is clear. $e_{94}h_1 \to 0$ follows from $\delta\left(e_{96}\right) \equiv e_{95}\lambda_0 + e_{94}\lambda_1 \bmod F(93)$ and the fact that $e_{95}\lambda_0$ is a cycle in $H_*\left(\tilde{P}^{113}\right) \otimes \Lambda$.

From (3.15)(9)* and (3.15)(9)** we conclude that only $e_{78}h_4h_1$ and $e_{30}h_6h_1$ persist to $\bar{E}_\infty^{2,96}\left(\tilde{P}^{113}\right)$. We have

$$
\begin{aligned}
&\delta\left(e_{80}\lambda_{15} + e_{79}\lambda_{16}\right) \equiv e_{79}\lambda_{15}\lambda_0 + e_{78}\lambda_1\lambda_{15} \quad \bmod \quad F(77), \\
&\delta\left(e_{32}\lambda_{63}\right) \equiv e_{31}\lambda_0\lambda_{63} + e_{30}\lambda_1\lambda_{63} \quad \bmod \quad F(29).
\end{aligned}
$$

Then from (3.13)(4) and (3.16) we conclude

$$Ext_A^{2,96}\left(\tilde{P}^{113}\right) = \mathbb{Z}_2\left(\overline{e_{79}h_4}h_0\right) \oplus \mathbb{Z}_2\left(\bar{e}_{31}h_0h_6\right).$$

This proves (3.15)(9).

Next we prove (3.15)(33). First of all note that the classes in $Ext_A^{3,383}\left(P^{113}\right)$, $Ext_A^{3,383}\left(\tilde{P}^{113}\right)$ and $Ext_A^{3,383}\left(P_{65}^{113}\right)$ with the same name $\overline{e_{111}h_4h_7^2}$, as described in

(3.13)(13), are mapped to $\bar{e}_{111}h_4h_7^2 \in Ext_A^{3,383}\left(P_{97}^{113}\right)$ under the collapsing maps $P^{113} \to P_{97}^{113}$, $\tilde{P}^{113} \to P_{97}^{113}$, $P_{65}^{113} \to P_{97}^{113}$ respectively. Thus it suffices to show $\bar{e}_{111}h_0^3h_4h_7^2 \neq 0$ in $Ext_A^{6,386}\left(P_{97}^{113}\right)$. We have the relation

$$\begin{aligned}&\delta\left(e_{112}\lambda_0^2\lambda_{15}\lambda_{127}^2 + e_{110}\lambda_2\lambda_0\lambda_{15}\lambda_{127}^2 + e_{109}\lambda_1\lambda_2\lambda_{15}\lambda_{127}^2\right)\\ &\equiv e_{111}\lambda_0^3\lambda_{15}\lambda_{127}^2 + e_{108}\lambda_3\lambda_0^2\lambda_{15}\lambda_{127}^2 \mod F(107)\end{aligned}$$

in $H_*\left(P_{97}^{113}\right)\otimes\Lambda$. Note that $\bar{e}_{111}\lambda_0^3h_4h_7^2 = \left\{e_{111}\lambda_0^3\lambda_{15}\lambda_{127}^2\right\}$ and $h_1^3h_4h_7^2 = h_2h_0^2h_4h_7^2 = \left\{\lambda_3\lambda_0^2\lambda_{15}\lambda_{127}^2\right\}$. So if we can show $e_{108}h_1^3h_4h_7^2$ is not a boundary in the spectral sequence $\left\{\bar{E}_r^{i,s,t}\left(P_{97}^{113}\right)\right\}$ then $\bar{e}_{111}h_0^3h_4h_7^2 \neq 0$. In $H_*\left(P_{97}^{113}\right)\otimes\Lambda$, $\delta\left(e_{113}\right) \equiv e_{111}\lambda_1 + e_{109}\lambda_3$, $\delta\left(e_{112}\right) \equiv e_{111}\lambda_0 + e_{110}\lambda_1 + e_{108}\lambda_3$, $\delta\left(e_{111}\right) = 0$, $\delta\left(e_{110}\right) \equiv e_{109}\lambda_0$, $\delta\left(e_{109}\right) \equiv 0$ all mod $F(107)$. These imply that the only possible element in $\bar{E}_1^{5,386}\left(P_{97}^{113}\right)$ which might hit $e_{108}h_1^3h_4h_7^2$ in the spectral sequence comes from $\bar{E}_1^{112,5,386}\left(P_{97}^{113}\right) \cong Ext_A^{5,274}$. By (3.4)(2)(c), $Ext_A^{5,274} \cong \mathbb{Z}_2\left(h_0^2h_4h_7^2\right)$. So this possible element is $e_{112}h_0^2h_4h_7^2$. But $e_{112}h_0^2h_4h_7^2 \to e_{111}h_0^3h_4h_7^2$. This shows $e_{108}h_1^3h_4h_7^2$ is not a boundary proving (3.15)(33). In this proof we use the fact that $h_7^2h_4h_1^3 \neq 0$, $h_7^2h_4h_0^3 \neq 0$ in $Ext_A^{6,*}$ by (3.5)(4).

The proofs of (3.15)(29) and (3.15)(32) are to be given simultaneously and that of (3.15)(31) will be a part of these proofs. To prove (3.15)(29), (32) we have to consider $\bar{E}_1^{4,385}\left(\tilde{P}^{113}\right)$ and the $\mathbb{Z}_2$-submodule $\Sigma_{i\leq 64}\bar{E}_1^{i,5,385}\left(\tilde{P}^{113}\right)$ of $\bar{E}_1^{5,385}\left(\tilde{P}^{113}\right)$ which are generated by the elements in the following (3.15)(29)* and (3.15)(32)* respectively (by (3.3) and (3.4)(1)).

(3.15)(29)* $e_{113}h_7^2h_3^2$, $e_{112}h_8h_3^2h_0$, $e_{112}h_7^2h_4h_0$, $e_{111}h_8h_4h_0^2$,
$e_{111}h_7^2h_4h_1$, $e_{109}h_8h_4h_1^2$, $e_{109}h_7^2h_4h_2$, $e_{108}h_8h_4h_2h_0$,
$e_{107}h_8c_1$, $e_{105}h_8h_3^3$, $e_{97}d_4$, $e_{96}h_8h_4^2h_0$,
$e_{95}h_8h_5h_0^2$, $e_{95}h_8h_4^2h_1$, $e_{93}h_8h_5h_1^2$, $e_{93}p_2'$,
$e_{92}h_8h_5h_2h_0$, $e_{89}p_3$, $e_{89}h_8h_5h_2^2$, $e_{88}h_8h_5h_3h_0$,
$e_{87}h_8h_5h_3h_1$, $e_{85}h_8c_2$, $e_{81}h_5h_3^2h_8$, $e_{64}h_8h_5^2h_0$,
$\tilde{e}_{63}h_8h_5^2h_1$, $\tilde{e}_{63}h_8h_6h_0^2$, $e_{61}h_8h_5^2h_2$, $e_{61}h_8h_6h_1^2$,
$e_{60}h_8h_6h_2h_0$, $e_{57}h_8h_6h_2^2$, $e_{56}h_8h_6h_3h_0$, $e_{55}h_8h_6h_3h_1$,
$e_{49}h_8h_6h_3^2$, $e_{49}e_4$, $e_{48}h_8h_6h_4h_0$, $e_{47}h_8h_6h_4h_1$,
$e_{45}h_8h_6h_4h_2$, $e_{41}h_8c_3$, $e_{33}h_8h_6h_4^2$, $e_{33}f_4$,
$e_{32}c_5h_0$, $e_{31}c_5h_1$, $e_{29}c_5h_2$, $e_{25}c_5h_3$.

(3.15)(32)* $e_{64}h_8D_3$, $\tilde{e}_{63}h_8h_5^2h_0^2$, $\tilde{e}_{63}T_1$, $e_{62}h_8h_6h_0^3$,
$e_{61}h_8h_5^2h_1^2$, $e_{61}h_6D_3(2)$, $e_{60}h_8h_5^2h_2h_0$, $e_{59}h_8h_6h_2h_0^2$,
$e_{57}h_8d_2$, $e_{56}h_8p_0'$, $e_{55}h_8h_6h_3h_0^2$, $e_{55}h_8p_1$,
$e_{54}h_8h_6c_0$, $e_{53}h_8h_6h_3h_1^2$, $e_{49}x(3)$, $e_{48}h_8h_6h_3^2h_0$,
$e_{48}h_0e_4$, $e_{47}h_8h_6h_4h_0^2$, $e_{47}h_1e_4$, $e_{45}h_8h_6h_4h_1^2$,
$e_{45}h_8e_2$, $e_{45}h_2e_4$, $e_{44}h_8h_6h_4h_2h_0$, $e_{43}h_8h_6c_1$

$e_{41}h_8h_6h_3^3$, $e_{41}h_8f_2$, $e_{40}h_8h_0c_3$, $e_{39}h_8h_1c_3$,
$e_{33}h_4c_4$, $e_{33}h_8g_3$, $e_{32}f_4h_0$, $e_{32}h_8h_5^3h_0$,
$e_{31}c_5h_0^2$, $e_{31}f_4h_1$, $e_{31}h_8h_5^3h_1$, $e_{29}c_5h_1^2$,
$e_{29}f_4h_2$, $e_{28}c_5h_2h_0$, $e_{25}h_8h_4c_3$, $e_{25}c_5h_2^2$,
$e_{25}f_4h_3$ $e_{24}c_5h_3h_0$, $e_{23}c_5h_3h_1$, $e_{21}h_8h_6c_2$,
$e_{17}c_5h_3^2$.

We have the following differentials in the spectral sequence for $\tilde{P}^{113}$; elements of (3.15) (32)* which are targets of the differentials appear only in (ii).

(3.15)[(29), (32)]**

(*i*)

$$
\begin{aligned}
&e_{113}h_7^2h_3^2 \longrightarrow e_{105}h_7^2h_3^3\\
&e_{112}h_7^2h_4h_0 \longrightarrow e_{111}h_7^2h_4h_0^2,\\
&e_{112}h_8h_4h_0 \longrightarrow e_{111}h_8h_4h_0^2,\\
&e_{108}h_8h_4h_2h_0 \longrightarrow e_{107}h_8h_4h_2h_0^2,\\
&e_{113}h_8h_3^2 \longrightarrow e_{105}h_8h_3^3\\
&e_{96}h_8h_4^2h_0 \longrightarrow e_{95}h_8h_4^2h_0^2,\\
&e_{97}h_8h_4^2 \longrightarrow e_{95}h_8h_4^2h_1,\\
&e_{89}h_8h_5h_2^2 \longrightarrow e_{85}h_8h_5h_2^3,\\
&e_{88}h_8h_5h_3h_0 \longrightarrow e_{87}h_8h_5h_3h_0^2\\
&e_{85}h_8c_2 \longrightarrow e_{77}h_8c_2h_3,\\
&e_{65}h_8h_5^2 \longrightarrow e_{61}h_8h_5^2h_2,\\
&e_{57}h_8h_6h_3 \longrightarrow e_{55}h_8h_6h_3h_1\\
&e_{33}c_5 \longrightarrow e_{31}c_5h_1.
\end{aligned}
\qquad
\begin{aligned}
&e_{112}h_8h_3^2h_0 \longrightarrow e_{102}h_4c_0h_8,\\
&e_{113}h_7^2h_4 \longrightarrow e_{111}h_7^2h_4h_1,\\
&e_{109}h_8h_4h_1^2 \longrightarrow e_{80}h_8h_4^3h_0,\\
&e_{107}h_8c_1 \longrightarrow e_{103}h_8c_1h_2,\\
&e_{97}d_4 \longrightarrow e_{95}d_4h_1,\\
&e_{96}h_8h_5h_0 \longrightarrow e_{95}h_8h_5h_0^2,\\
&e_{92}h_8h_5h_2h_0 \longrightarrow e_{91}h_8h_5h_2h_0^2,\\
&e_{89}p_3 \longrightarrow e_{87}p_3h_1,\\
&e_{89}h_8h_5h_3 \longrightarrow e_{87}h_8h_5h_3h_1,\\
&e_{109}h_8h_4h_2 \longrightarrow e_{81}h_8h_5h_3^2,\\
&e_{64}h_8h_6h_0 \longrightarrow e_{61}h_8h_6h_1^2,\\
&e_{49}h_8h_6h_4 \longrightarrow e_{47}h_8h_6h_4h_1,
\end{aligned}
$$

(*ii*)

$$
\begin{aligned}
&e_{64}h_8D_3 \longrightarrow e_{62}h_8D_3h_1,\\
&e_{64}h_8h_5^2h_0 \longrightarrow e_{61}h_8h_5^2h_1^2,\\
&e_{60}h_8h_5^2h_2h_0 \longrightarrow e_{59}h_8h_5^2h_2h_0^2,\\
&e_{57}h_8d_2 \longrightarrow e_{53}h_8d_2h_2,\\
&e_{56}h_8h_6h_3h_0 \longrightarrow e_{55}h_8h_6h_3h_0^2\\
&e_{54}h_8h_6c_0 \longrightarrow e_{47}h_8h_6d_0,\\
&e_{49}x(3) \longrightarrow e_{47}x(3)h_1,\\
&e_{48}h_8h_6h_3^2h_0 \longrightarrow e_{38}h_8h_6h_4c_0,\\
&e_{49}e_4 \longrightarrow e_{47}h_1e_4,\\
&e_{45}h_2e_4 \longrightarrow e_{17}h_4^2e_4,\\
&e_{43}h_8h_6c_1 \longrightarrow e_{39}h_8h_6c_1h_2,\\
&e_{41}h_8f_2 \longrightarrow e_{39}h_8f_2h_1,\\
&e_{41}h_8c_3 \longrightarrow e_{39}h_8h_1c_3,\\
&e_{33}h_8g_3 \longrightarrow e_{31}h_1h_8g_3,\\
&e_{32}f_4h_0 \longrightarrow e_{31}f_4h_0^2,\\
&e_{33}f_4 \longrightarrow e_{31}f_4h_1,\\
&e_{28}c_5h_2h_0 \longrightarrow e_{27}c_5h_2h_0^2,\\
&e_{25}c_5h_2^2 \longrightarrow e_{21}c_5h_2^3,\\
&e_{24}c_5h_3h_0 \longrightarrow e_{23}c_5h_3h_0^2,\\
&e_{21}h_8h_6c_2 \longrightarrow e_{13}h_8h_6c_2h_3.
\end{aligned}
\qquad
\begin{aligned}
&e_{62}h_8h_6h_0^3 \longrightarrow e_{61}h_8h_6h_0^4,\\
&e_{93}p_2' \longrightarrow e_{61}h_6D_3(2),\\
&e_{60}h_8h_6h_2h_0 \longrightarrow e_{59}h_8h_6h_2h_0^2,\\
&e_{56}h_8p_0' \longrightarrow e_{55}h_8h_0p_0',\\
&e_{55}h_8p_1 \longrightarrow e_{29}h_8h_2g_3,\\
&e_{57}h_8h_6h_2^2 \longrightarrow e_{53}h_8h_6h_3h_1^2,\\
&e_{48}h_0e_4 \longrightarrow e_{47}h_0^2e_4,\\
&e_{48}h_8h_6h_4h_0 \longrightarrow e_{47}h_8h_6h_4h_0^2,\\
&e_{45}h_8e_2 \longrightarrow e_{37}h_8f_2h_2,\\
&e_{44}h_8h_6h_4h_2h_0 \longrightarrow e_{43}h_8h_6h_4h_2h_0^2,\\
&e_{49}h_8h_6h_3^2 \longrightarrow e_{41}h_8h_6h_4h_2^2,\\
&e_{40}h_8h_0c_3 \longrightarrow e_{39}h_8c_3h_0^2,\\
&e_{33}h_4e_4 \longrightarrow e_{31}h_1h_4e_4,\\
&e_{32}h_8h_5^3h_0 \longrightarrow e_{31}h_8h_5^3h_0^2,\\
&e_{32}c_5h_0 \longrightarrow e_{31}c_5h_0^2,\\
&e_{33}h_8h_5^3 \longrightarrow e_{31}h_8h_5^3h_1,\\
&e_{25}f_4h_3 \longrightarrow e_{23}f_4h_3h_1,\\
&e_{25}h_8h_4c_3 \longrightarrow e_{15}h_8D_1(1),\\
&e_{25}c_5h_3 \longrightarrow e_{23}c_5h_3h_1,
\end{aligned}
$$

Elements x in $Ext_A^{5,*}$ such that e_nx are target elements in (i) are non-zero or linearly independent by (3.5)(2) and elements y in $Ext_A^{6,*}$ such that e_ny are target elements in (ii) are non-zero or linearly independent by (3.5)(1).

The differentials in (3.15)[(29), (32)]** are the most difficult ones in the use of the spectral sequence $\{\bar{E}_r^{i,s,t}(X)\}$ for proving (3.15). These are contained in the work of "solid calculations".

From (3.15)(32)* and (3.15)[(29), (32)]** we see $\tilde{e}_{63}h_5^2h_8h_0^2, \tilde{e}_{63}T_1, e_{45}h_1^2h_4h_6h_8$, $e_{29}h_2f_4, e_{29}h_1^2c_5$ and $e_{17}h_3^2c_5$ are the only possible elements in $\Sigma_{i\leq 64}\bar{E}_1^{i,5,385}\left(\tilde{P}^{113}\right)$ to persist to $\bar{E}_\infty$. That these 6 elements are infinite cycles is seen as follows: the first two and the last one are clear by (3.13)(7), (14), (15), and the remaining three follow from (3.15)(28), the fact $\bar{e}_{31} = \{e_{31}\}$ and the relations $\delta(e_{33}) \equiv e_{31}\lambda_1 + e_{29}\lambda_3$ mod $F(28)$, $\delta(e_{32}\lambda_0 + e_{30}\lambda_2) \equiv e_{31}\lambda_0^2 + e_{29}\lambda_1^2$ mod $F(28)$, $\delta(e_{48}\lambda_0 + e_{46}\lambda_2) \equiv e_{47}\lambda_0^2 +$

$e_{45}\lambda_1^2$ mod $F(44)$ in $H_*\left(\tilde{P}^{113}\right)\otimes\Lambda$. Since they are not boundaries (by (3.15)(29)* and (3.15)[(29), (32)]*,*) from (3.16) we obtain the conclusion in (3.15)(32). From (3.15)(29)* and (3.15)[(29), (32)]** we also see that $e_{109}h_2h_4h_7^2$, $e_{93}h_1^2h_5h_8$, $\tilde{e}_{63}h_5^2h_8h_1$, $\tilde{e}_{63}h_6h_8h_0^2, e_{45}h_2h_4h_6h_8, e_{29}h_2c_5$ are the only possible elements to persist to $\bar{E}_\infty^{4,*}$. We claim they do. Similar simple calculations as above together with (3.13)(3), (7), (12), (13) and the facts $\bar{e}_{31}=\{e_{31}\}$, $\bar{e}_{95}=\{e_{95}\}$ show that they are infinite cycles. That they are not boundaries is seen as follows. That $\tilde{e}_{63}h_5^2h_8h_1$ and $\tilde{e}_{63}h_8h_6h_0^2$ are not boundaries follows from (3.11)(2). The basis elements in $\bar{E}_1^{3,385}\left(\tilde{P}^{113}\right)$ that appear in the domain of the differentials in (3.15)(29), (32)** (i) do not hit $e_{109}h_2h_4h_7^2$, $e_{93}h_1^2h_5h_8$, $e_{45}h_2h_4h_6h_8$, $e_{29}h_2c_5$ and the remaining basis elements are $e_{111}h_8h_4h_1$, $e_{95}h_8h_5h_1$, $e_{61}h_8h_6h_2$, $e_{93}h_8h_5h_2$ and $\tilde{e}_{63}h_8h_6h_1$. The first three are boundaries since $e_{113}h_8h_4\to e_{111}h_8h_4h_1$, $e_{97}h_8h_5\to e_{95}h_8h_5h_1$, $e_{65}h_8h_6\to e_{61}h_8h_6h_2$. The remaining $e_{93}h_8h_5h_2$ and $\tilde{e}_{63}h_8h_6h_1$ are infinite cycles by (3.13)(3) and by $\bar{e}_{95}=\{e_{95}\}$, $\delta(e_{97})\equiv e_{95}\lambda_1+e_{93}\lambda_3$ mod $F(92)$. So $e_{109}h_2h_4h_7^2$, $e_{93}h_1^2h_5h_8$, $e_{45}h_2h_4h_6h_8$, and $e_{29}h_2c_5$ are not boundaries. This proves the claim. Thus from (3.16) we obtain the conclusion (3.15)(29). Finally, (3.15)(31) is shown as follows. For all s,t the basis elements in $\bar{E}_1^{s,t}\left(\tilde{P}^{113}\right)$ are precisely the basis elements for $\bar{E}_1^{s,t}\left(P^{113}\right)$ with $\tilde{e}_{63}$ replaced by e_{63}. All the differentials in (3.15)[(29), (32)]** are unchanged for P^{113} except that $e_{65}h_8h_5^2\to e_{61}h_8h_5^2h_2$, $e_{64}h_8h_6h_0\to e_{61}h_8h_6h_1^2$, $e_{64}h_8h_5^2h_0\to e_{61}h_8h_5^2h_1^2$ are replaced by $e_{65}h_8h_5^2\to e_{63}h_8h_5^2h_1$, $e_{64}h_8h_6h_0\to e_{63}h_8h_6h_0^2$, $e_{64}h_8h_5^2h_0\to e_{63}h_8h_5^2h_0^2$. In particular, $e_{63}T_1$ is not a boundary, and since it is clearly also an infinite cycle we conclude $\bar{e}_{63}T_1\neq 0$ in $Ext_A^{5,385}\left(P^{113}\right)$. Clearly this argument also proves $\bar{e}_{63}T_1\neq 0$ in $Ext_A^{5,385}\left(P_{63}^{113}\right)$. This proves (3.15)(31).

This concludes our illustration of the proofs for (3.15). We would like to remark here that the non-zero elements in (3.5)(3) are needed in the spectral sequence $\left\{\bar{E}_r^{*,*}\left(P^{53}\right)\right\}$ for proving (3.15)(27) which is not included in this illustration of the proofs.

In concluding this section we prove two results concerning the induced maps in Ext groups from projective spaces P_ℓ^k to appropriate associated ones and also to S^0. For the first result let $f:P^{113}\to\tilde{P}^{113}$ and $f':P_{65}^{113}\to\hat{P}_{65}^{113}$ be the quotient maps.

Proposition 3.17

(1) $\overline{\tilde{e}_{63}h_5^2h_0^3h_8}\neq 0$ *in* $Ext_A^{6,386}\left(\tilde{P}^{113}\right)$ *which is not in the image of* $f_*:Ext_A^{6,386}\left(P^{113}\right)\to Ext_A^{6,386}\left(\tilde{P}^{113}\right)$.

(2) $\overline{\hat{e}_{95}h_4^2h_0^2h_8}$ *and* $\bar{e}_{111}h_4h_7^2h_0^2$ *are linearly independent in* $Ext_A^{5,385}\left(\hat{P}_{65}^{113}\right)$ *and* $\overline{\hat{e}_{95}h_4^2h_0^2h_8}\notin im\left(f'_*:Ext_A^{5,385}\left(P_{65}^{113}\right)\to Ext_A^{5,385}\left(\hat{P}_{65}^{113}\right)\right)$.

(3) $ker\left(Ext_A^{5,385}\left(P^{113}\right)\xrightarrow{f_*}Ext_A^{5,385}\left(\tilde{P}^{113}\right)\right)=\mathbb{Z}_2\left(\bar{e}_{63}T_1\right)$.

(4) $ker\left(Ext_A^{6,386}\left(P^{113}\right)\xrightarrow{f_*}Ext_A^{6,386}\left(\tilde{P}^{113}\right)\right)=\mathbb{Z}_2\left(\bar{e}_{63}h_8H_1\right)\oplus\mathbb{Z}_2\left(\bar{e}_{63}h_8h_1D_3\right)$

(5) $ker\left(Ext_A^{5,385}\left(P_{65}^{113}\right)\xrightarrow{f'_*}Ext_A^{5,385}\left(\hat{P}_{65}^{113}\right)\right)=\mathbb{Z}_2\left(\bar{e}_{95}d_4h_1\right)$.

Proof From (3.9) we see there are A-maps $H_*\left(\tilde{P}^{113}\right)\xrightarrow{g}\Sigma^{63}\mathbb{Z}_2$, $H_*\left(\hat{P}^{113}_{65}\right)\xrightarrow{g'}\Sigma^{95}\mathbb{Z}_2$ such that $g(\tilde{e}_{63})\neq 0$, $g'(\hat{e}_{95})\neq 0$ and the composites $H_*\left(P^{113}\right)\xrightarrow{f_*}H_*\left(\tilde{P}^{113}\right)\xrightarrow{g}\Sigma^{63}\mathbb{Z}_2$, $H_*\left(P^{113}_{65}\right)\xrightarrow{f'_*}H_*\left(\hat{P}^{113}_{65}\right)\xrightarrow{g'}\Sigma^{95}\mathbb{Z}_2$ are zero. Then $Ext_A^{6,386}\left(\tilde{P}^{113}\right)\xrightarrow{g_*}Ext_A^{6,386}\left(\Sigma^{63}\mathbb{Z}_2\right)=Ext_A^{6,323}$ has $g_*\left(\overline{\tilde{e}_{63}h_5^2h_0^3h_8}\right)=h_5^2h_0^3h_8\neq 0$ (by (3.5)(1)) and $Ext_A^{5,385}\left(\hat{P}^{113}_{65}\right)\xrightarrow{g'_*}Ext_A^{5,385}\left(\Sigma^{95}\mathbb{Z}_2\right)=Ext_A^{5,290}$ has $g'_*\left(\overline{\hat{e}_{95}h_4^2h_0^2h_8}\right)=h_4^2h_0^2h_8\neq 0$ (by (3.2)). By (3.15) (33), $\bar{e}_{111}h_4h_7^2h_0^2\neq 0$ in $Ext_A^{5,385}\left(\hat{P}^{113}_{65}\right)$ and lies in $im\left(Ext_A^{5,385}\left(P^{113}_{65}\right)\xrightarrow{f'_*}Ext_A^{5,385}\left(\hat{P}^{113}_{65}\right)\right)$. Since the composites

$$Ext_A^{6,386}\left(P^{113}\right)\xrightarrow{f_*}Ext_A^{6,386}\left(\tilde{P}^{113}\right)\xrightarrow{g_*}Ext_A^{6,323}$$
$$Ext_A^{5,385}\left(P^{113}_{65}\right)\xrightarrow{f_*}Ext_A^{5,385}\left(\hat{P}^{113}_{65}\right)\xrightarrow{g'_*}Ext_A^{5,290}$$

are zero, the conclusions of (1) and (2) follow.

Let $M=ker\left(g:H_*\left(\tilde{P}^{113}\right)\to\Sigma^{63}\mathbb{Z}_2\right)$. Then we have short exact sequences of A-modules

$$0\longrightarrow M\xrightarrow{i}H_*\left(\tilde{P}^{113}\right)\xrightarrow{g}\Sigma^{63}\mathbb{Z}_2\longrightarrow 0$$
$$0\longrightarrow\Sigma^{63}\mathbb{Z}_2\xrightarrow{j}H_*\left(P^{113}\right)\xrightarrow{\bar{f}_*}M\longrightarrow 0$$

where the composite $H_*\left(P^{113}\right)\xrightarrow{\bar{f}_*}M\xrightarrow{i}H_*\left(\tilde{P}^{113}\right)$ is f_* and e_{63} generates the second $\Sigma^{63}\mathbb{Z}_2$. Consider the following commutative diagram obtained by splicing together the long exact sequences in Ext groups arising from these two short exact sequences

$$\begin{array}{ccccc}
\cdots\longrightarrow Ext_A^{s,t}\left(\Sigma^{63}\mathbb{Z}_2\right)\xrightarrow{\delta_*} & Ext_A^{s+1,t}(M) & \xrightarrow{i*} & Ext_A^{s+1,t}\left(\tilde{P}^{113}\right)\xrightarrow{g_*}\cdots \\
 & \uparrow \bar{f}_* & & \| \\
 & Ext_A^{s+1,t}\left(P^{113}\right) & \xrightarrow{f_*} & Ext_A^{s+1,t}\left(\tilde{P}^{113}\right) \\
 & \uparrow j_* & & \\
 & Ext_A^{s+1,t}\left(\Sigma^{63}\mathbb{Z}_2\right). & &
\end{array}$$

From (3.9)(1), one sees that $Ext_A^{0,31}(M)=\mathbb{Z}_2\left(\bar{f}_*(\bar{e}_{31})\right)$. We have that $1\in Ext_A^{0,0}\cong Ext_A^{0,63}(\Sigma^{63}\mathbb{Z}_2)$ is mapped to $\bar{e}_{63}$ under j_* and is mapped to $\bar{f}_*(\bar{e}_{31})h_5$ under δ_*; the latter also follows from (3.9)(1). So $\delta_*(z)=\bar{f}_*(\bar{e}_{31})h_5z$ and $j_*(z)=\bar{e}_{63}z$ for any $z\in Ext_A^{*,*}$. By (3.3), $Ext_A^{4,322}\cong\mathbb{Z}_2\left(h_5^2h_8h_1\right)\oplus\mathbb{Z}_2\left(h_6h_8h_0^2\right)$ and, by (3.4)(1)(a), $Ext_A^{5,323}\cong\mathbb{Z}_2(h_6h_8h_0^3)$. Then $\delta_*(h_5^2h_8h_1)=\bar{f}_*(\bar{e}_{31})h_5^3h_8h_1=\bar{f}_*\left(\bar{e}_{31}h_5^3h_8h_1\right)=\bar{f}_*\left(\bar{e}_{31}h_5^3h_8h_1\right)=\bar{f}_*(0)=0$ (by (3.12)(1)) and $\delta_*(h_6h_8h_0^k)=\bar{f}_*(\bar{e}_{31})h_5h_6h_8h_0^k=0$ for $k=2,3$. Thus, for $(s+1,t)=(5,385)$ and $(6,386)$, $keri_*=im\delta_*=0$ and this implies $kerf_*=ker\bar{f}_*=imj_*$. By (3.4)(1)(b), $Ext_A^{5,385}\left(\Sigma^{63}\mathbb{Z}_2\right)\cong Ext_A^{5,322}\cong\mathbb{Z}_2(T_1)\oplus\mathbb{Z}_2(h_8h_5^2h_0^2)$, and by (3.4)(3)(b), $Ext_A^{6,386}(\Sigma^{63}\mathbb{Z}_2)\cong Ext_A^{6,323}\cong\mathbb{Z}_2(h_8H_1)\oplus\mathbb{Z}_2\left(h_8h_1D_3\right)\oplus\mathbb{Z}_2(h_8h_5^2h_0^3)$. Since

$\bar{e}_{63}h_5 = 0$, (by (3.12)(1)) it follows that $j_*(h_8h_5^2h_0^k) = \bar{e}_{63}h_5^2h_8h_0^k = 0$ for $k = 2, 3$. We conclude that $ker f_* = im j_* = \mathbb{Z}_2(\bar{e}_{63}T_1)$ for $(s+1,t) = (5, 385)$ and that $ker f_* = im j_* = \mathbb{Z}_2(\bar{e}_{63}H_1h_8) \oplus \mathbb{Z}_2(\bar{e}_{63}D_3h_1h_8)$ for $(s+1,t) = (6, 386)$. This proves (3) and (4). (5) is similarly proved by considering the short exact sequences

$$0 \longrightarrow M' \xrightarrow{i'} H_*\left(\hat{P}_{65}^{113}\right) \xrightarrow{g'} \Sigma^{95}\mathbb{Z}_2 \longrightarrow 0$$

$$0 \longrightarrow \Sigma^{95}\mathbb{Z}_2 \xrightarrow{j'} H_*\left(P_{65}^{113}\right) \xrightarrow{\bar{f}'_*} M' \longrightarrow 0$$

together with the facts: (i) $Ext_A^{4,290} \cong \mathbb{Z}_2\left(h_8h_4^2h_1\right) \oplus \mathbb{Z}_2\left(h_8h_5h_0^2\right)$ by (3.3), (ii) $Ext_A^{5,290} \cong \mathbb{Z}_2\left(h_8h_4^2h_0^2\right) \oplus \mathbb{Z}_2\left(d_4h_1\right)$ by (3.4)(2)(b) and (iii) $\bar{e}_{79}h_3 = 0$, $\bar{e}_{95}h_4 = 0$ in $Ext_A^{1,*}\left(P_{65}^{113}\right)$ by (3.12)(2), (3). This completes the proof of Proposition (3.17).

To state the second result recall the induced map $t_* : Ext_A^{s,t}\left(P^\infty\right) \to Ext_A^{s+1,t+1}(S^0)$ of the transfer $t : P^\infty \to S^0$ which is induced by the chain map $\bar{t} : H_*\left(P^\infty\right) \otimes \Lambda \to \Lambda$ with $\bar{t}\left(e_i \otimes \lambda_I\right) = \lambda_i\lambda_I$. We also recall the classes $\overline{e_{111}h_4h_7^2}$, $\overline{e_{17}h_3^2c_5}$ and $\overline{e_{47}h_0^2h_4h_6}$ in $Ext_A^{*,*}\left(P^{113}\right)$ as described in (3.13) and (3.15)(28). In the following we restrict t_* to $Ext_A^{s,t}\left(P^{113}\right)$.

Proposition 3.18

(1) $t_*\left(\overline{e_{111}h_4h_7^2h_0^3}\right) = g_5h_0^3$.

(2) $t_*\left(\bar{e}_{63}H_1h_8\right) = h_6h_8H_1$.

(3) $t_*\left(\overline{e_{47}h_0^2h_4h_6}\right) = \epsilon K$ *and so* $t_*\left(\overline{e_{47}h_0^2h_4h_6}h_8h_0\right) = \epsilon h_8h_0K$ *where* $\epsilon = 0$ *or* 1.

(4) $t_*\left(\bar{e}_{63}D_3h_8h_1\right) = 0$.

(5) $t_*\left(\bar{e}_{31}c_5h_0^3\right) = 0$.

(6) $t_*\left(\overline{e_{17}h_3^2c_5h_0}\right) = 0$.

Proof (1) has already been proved in Section 2. (2) is clear. By (3.4)(2)(a), $Ext_A^{5,130} \cong \mathbb{Z}_2(K)$. (3) follows from this since $t_*\left(\overline{e_{47}h_0^2h_4h_6}\right) \in Ext_A^{5,130}$. (4) follows from $h_6D_3 = 0$ by (3.5)(6)(d). (5) follows from $h_5c_5 = 0$ (by (3.2)). To see (6) note that $t_*\left(\overline{e_{17}h_3^2c_5h_0}\right) = h_0t_*\left(\overline{e_{17}h_3^2c_5}\right)$ which is represented in Λ by

$$\lambda_0\lambda_{17}\lambda_7^2\lambda_{95}\lambda_{127}^2 = \lambda_{15}\lambda_1\lambda_7^2\lambda_{95}\lambda_{127}^2 = 0.$$

This proves (3.18). □

4 Some first-stage differentials in the Adams spectral sequence for stunted real projective spaces and their associated spaces

For a spectrum Y consider the mod 2 Adams spectral sequence $\{E_r^{s,t}(Y)\}$ for $\pi_*^s(Y)$. So $E_2^{s,t}(Y) = Ext_A^{s,t}(Y)$. By the first-stage differentials in the ASS for Y we mean the differentials $d_2 : Ext_A^{s,t}(Y) \to Ext_A^{s+2,t+1}(Y)$. In this section we compute certain such differentials for some of the spaces in (3.15). We refer to (3.13), (3.14) and (3.15) for the elements in relevant Ext groups considered in statement (4.1) below and statement (4.2) later.

Proposition 4.1 *There are the following first-stage differentials in the Adams spectral sequence for the spaces specified.*

(1) $d_2\left(\bar{e}_{63}\right) = \bar{e}_{31}h_5h_0$ *in* P^n *for* $n \geq 64$.

(2) $d_2\left(\bar{e}_{95}\right) = \overline{e_{79}h_4h_0}$ *in* P_{65}^{113}.

(3) $d_2\left(\bar{e}_{95}\right) = \overline{e_{79}h_4h_0} + \epsilon\bar{e}_{31}h_6h_0$ *in* $\tilde{P}^{113}$ *where* $\epsilon = 0$ *or* 1.

(4) $d_2\left(\overline{e_{95}h_0}\right) = \overline{e_{79}h_0}h_4h_0$ *in* P_{63}^{113}.
(5) $d_2\left(\bar{e}_{111}\right) = \bar{e}_{103}h_3h_0$ *in* P_{97}^{113}.
(6) $d_2\left(\bar{e}_{111}\right) = \overline{e_{103}h_3}h_0 + \epsilon\bar{e}_{79}h_5h_0$ *in* $\hat{P}_{65}^{113}$ *where* $\epsilon = 0$ *or* 1.
(7) $d_2\left(\overline{\tilde{e}_{63}h_6}\right) = \overline{\tilde{e}_{63}h_5^2}h_0$ *in* $\tilde{P}^{113}$.
(8) $d_2\left(\overline{\hat{e}_{95}h_5}\right) = \overline{\hat{e}_{95}h_4^2}h_0 + \epsilon\overline{e_{87}h_0h_3h_5}$ *in* $\hat{P}_{65}^{113}$ *where* $\epsilon = 0$ *or* 1.
(9) $d_2\left(\overline{\tilde{e}_{63}h_5^2}\right) = 0$ *in* $\tilde{P}^{113}$.

We recall from [**1**] the Adams differentials $d_2(h_i) = h_{i-1}^2h_0 \neq 0$ for $i \geq 4$ in S^0. (4.1) (1) is deduced from $d_2(h_6) = h_5^2h_0$ in [**6**]. All other differentials in (4.1) are also deduced from the Adams differentials. Details go as follows.

First, (4.1) (2). Consider the cofibration sequence $S^{63} \xrightarrow{i} P_{62}^{113} \xrightarrow{q} \bar{P}_{62}^{113} \xrightarrow{\bar{t}} S^{64} = \Sigma S^{63}$ where $\bar{P}_{62}^{113}$ is defined by this sequence as in Section 3. It is easy to see that $Ext_A^{0,95}\left(\bar{P}_{62}^{113}\right) \cong \mathbb{Z}_2\left(\bar{e}_{95}\right)$. By (3.15) (11), $Ext_A^{2,96}\left(\bar{P}_{62}^{113}\right) \cong \mathbb{Z}_2\left(\bar{e}_{79}h_4h_0\right)$. $\bar{t}$ induces $Ext_A^{s,t}\left(\bar{P}_{62}^{113}\right) \xrightarrow{\bar{t}_*} Ext_A^{s+1,t+1}\left(S^{64}\right) = Ext_A^{s+1,t-63}(S^0)$ which maps $\bar{e}_{95}$ to h_5 and $\bar{e}_{79}h_4h_0$ to $h_4^2h_0$. Since $d_2(h_5) = h_4^2h_0$ in S^0 it follows that $d_2\left(\bar{e}_{95}\right) = \bar{e}_{79}h_4h_0$ in $\bar{P}_{62}^{113}$. By considering the collapsing map $\bar{P}_{62}^{113} \to P_{65}^{113}$ we obtain $d_2\left(\bar{e}_{95}\right) = \bar{e}_{79}h_4h_0$ in P_{65}^{113}. This proves (4.1) (2). A similar construction for P_{97}^{113} shows $d_2\left(\bar{e}_{111}\right) = \bar{e}_{103}h_3h_0$ in P_{97}^{113} by using $d_2(h_4) = h_3^2h_0$ and by (3.15) (8), and this is (4.1)(5). (4.1)(3),(4) are deduced from (4.1)(2) as follows. The induced maps in the correct Ext groups of the collapsing map $\tilde{P}^{113} \to P_{65}^{113}$ map $\bar{e}_{95}$ to $\bar{e}_{95}$, $\overline{e_{79}h_4}h_0$ to $\bar{e}_{79}h_4h_0$ and $\bar{e}_{31}h_6h_0$ to zero. By (3.15)(9), $Ext_A^{2,96}\left(\tilde{P}^{113}\right) \cong \mathbb{Z}_2\left(\overline{e_{79}h_4}h_0\right) \oplus \mathbb{Z}_2\left(\bar{e}_{31}h_6h_0\right)$. So $d_2\left(\bar{e}_{95}\right) = \overline{e_{79}h_4}h_0 + \epsilon\bar{e}_{31}h_6h_0$ in $\tilde{P}^{113}$ where $\epsilon = 0$ or 1. This proves (4.1) (3). The induced maps in the correct Ext groups of the collapsing map $P_{63}^{113} \to P_{65}^{113}$ map $\overline{e_{95}h_0}$ to $\bar{e}_{95}h_0$ and $\overline{e_{79}h_0}h_4h_0$ to $\bar{e}_{79}h_4h_0^2$. This implies $d_2\left(\overline{e_{95}h_0}\right) = \overline{e_{79}h_0}h_4h_0$ in P_{63}^{113} as $Ext_A^{3,97}\left(P_{63}^{113}\right) \cong \mathbb{Z}_2\left(\overline{e_{79}h_0}h_4h_0\right)$ by (3.15) (14). This proves (4.1) (4). Similarly, the differential (4.1) (6) is deduced from (4.1) (5) this way using $Ext_A^{2,112}\left(\hat{P}_{65}^{113}\right) \cong \mathbb{Z}_2\left(\overline{e_{103}h_3}h_0\right) \oplus \mathbb{Z}_2\left(\bar{e}_{79}h_5h_0\right)$ in (3.15)(10).

To prove (4.1) (7) it suffices to show $d_2\left(\overline{\tilde{e}_{63}h_6}\right) = \overline{\tilde{e}_{63}h_5^2}h_0$ in $\tilde{P}^{80}$. The induced maps in the correct Ext groups of the collapsing map $\tilde{P}^{80} \to \tilde{P}_{63}^{80}$ map $\overline{\tilde{e}_{63}h_6}$ to $\bar{\bar{e}}_{63}h_6$ and $\overline{\tilde{e}_{63}h_5^2}h_0$ to $\bar{\bar{e}}_{63}h_5^2h_0$. In $\tilde{P}_{63}^{80}$, $d_2\left(\bar{\bar{e}}_{63}h_6\right) = \bar{\bar{e}}_{63}h_5^2h_0 \neq 0$ (by (3.15) (22)) since the class $\bar{\bar{e}}_{63}$ is spherical. By (3.15) (21), $Ext_A^{3,128}\left(\tilde{P}^{80}\right) \cong \mathbb{Z}_2\left(\overline{\tilde{e}_{63}h_5^2}h_0\right)$. So $d_2\left(\overline{\tilde{e}_{63}h_6}\right) = \overline{\tilde{e}_{63}h_5^2}h_0$ proving (4.1)(7).

To prove (4.1) (8) it suffices to show $d_2\left(\overline{\hat{e}_{95}h_5}\right) = \overline{\hat{e}_{95}h_4^2}h_0 + \epsilon\overline{e_{87}h_0h_3h_5}$ in $\hat{P}_{65}^{96}$. The proof for this is similar to that for (4.1) (7) just given. Since $\bar{\hat{e}}_{95}$ is spherical in $\hat{P}_{95}^{96}$, by considering the collapsing map $\hat{P}_{65}^{96} \to \hat{P}_{95}^{96}$ we deduce $d_2\left(\overline{\hat{e}_{95}h_5}\right) = \overline{\hat{e}_{95}h_4^2}h_0 + \epsilon\overline{e_{87}h_0h_3h_5}$ in $\hat{P}_{65}^{96}$ where $\epsilon = 0$ or 1 since $Ext_A^{3,128}\left(\hat{P}_{65}^{96}\right) \cong \mathbb{Z}_2\left(\overline{\hat{e}_{95}h_4^2}h_0\right) \oplus \mathbb{Z}_2\left(\overline{e_{87}h_0h_3h_5}\right)$ by (3.15) (24). This proves (4.1) (8).

Finally we prove (4.1) (9) as follows. Recall that $\tilde{P}^{113}$ = mapping cone of $\varphi_1 : S^{62} \cup_{2\iota} e^{63} \to P^{113}$ where φ_1 is as in (3.8) (1). Let $\tilde{\theta}_5 = \varphi_1 | S^{62} : S^{62} \to P^{113}$ which is detected by $\bar{e}_{31}h_5$. Then the composite $S^{62} \xrightarrow{\tilde{\theta}_5} P^{113} \xrightarrow{t} S^0$, denoted by θ_5, is detected by h_5^2. Any suspension of θ_5 will also be denoted by θ_5. We want to show $d_2\left(\overline{\tilde{e}_{63}h_5^2}\right) = 0$ in $\tilde{P}^{113}$ for (4.1)(9). This is equivalent to proving the following.

(*) The composite $\psi : S^{124} \xrightarrow{\theta_5} S^{62} \xrightarrow{\tilde{\theta}_5} P^{113}$ has Adams filtration ≥ 5.

By dimensional reasons $\tilde{\theta}_5$ can be pulled back to $S^{62} \xrightarrow{\tilde{\theta}_5} P^{62}$. In fact, it can be further pulled back to $S^{62} \xrightarrow{\tilde{\theta}_5} P^{53}$ (see [**4**]). This yields a pull-back $S^{124} \xrightarrow{\psi} P^{53}$ of the ψ in (*). To prove (*) it suffices to show it for a suitable pull-back $S^{124} \xrightarrow{\psi} P^{53}$. Note that $S^{62} \xrightarrow{\tilde{\theta}_5} P^{53}$ is still detected by $\bar{e}_{31}h_5$. By (3.12)(1), $\bar{e}_{31}h_5^3 = \bar{e}_{31}h_4^2h_6 = 0$ in $Ext_A^{*,*}(P^{53})$. So $\psi = \tilde{\theta}_5\theta_5 : S^{124} \to P^{53}$ has Adams filtration ≥ 4. Suppose ψ has Adams filtration 4. Then from (3.15) (27) we see ψ is detected by $\overline{e_{40}h_6h_3^3}$ or $\overline{e_{52}h_6h_2^3}$ or $\overline{e_{40}h_6h_3^3} + \overline{e_{52}h_6h_2^3}$. Consider the cofibration sequence $P^{53} \to P^{56} \to P_{54}^{56} \xrightarrow{q} \Sigma P^{53}$. By (3.15) (19), $Ext_A^{3,128}\left(P_{54}^{56}\right) \cong \mathbb{Z}_2\left(\bar{e}_{54}h_6h_1h_3\right) \oplus \mathbb{Z}_2\left(\overline{e_{56}h_2^2h_6}\right)$. Since h_6h_1 detects homotopy elements in $\pi_{64}^s(S^0)$ ([**17**]), h_2^2 detects ν^2, $2\nu^2 = 0$, $\eta\nu^2 = 0$ and $< \nu^2, 2\iota, \theta_5 >$ is detected by $h_2^2h_6$ it follows that $\bar{e}_{54}h_6h_1h_3$ and $\overline{e_{56}h_2^2h_6}$ both are spherical classes. Let $\psi_1 : S^{125} \to P_{54}^{56}$, $\psi_2 : S^{125} \to P_{54}^{56}$ be stable maps detected by $\bar{e}_{54}h_6h_1h_3$ and $\overline{e_{56}h_2^2h_6}$ respectively. By the second conclusion of (3.15) (27) the composites $S^{125} \xrightarrow{\psi_1} P_{54}^{56} \xrightarrow{q} \Sigma P^{53}$ and $S^{125} \xrightarrow{\psi_2} P_{54}^{56} \xrightarrow{q} \Sigma P^{53}$ are detected by $\overline{e_{40}h_6h_3^3}$ and $\overline{e_{52}h_6h_2^3}$ respectively. This shows that $\Sigma\psi : S^{125} \to \Sigma P^{53}$ can be altered to have Adams filtration ≥ 5. This proves (*) and therefore (4.1) (9).

We have the following consequence of (4.1).

Corollary 4.2 *(1)* $d_2\left(\bar{e}_{63}h_6h_8\right) = \begin{cases} \bar{e}_{63}h_5^2h_0h_8 \neq 0 & in\ P^{64} \\ 0 & in\ P^{113}. \end{cases}$

(2) $d_2\left(\overline{e_{95}h_0}h_5h_8\right) = 0$ *in* P_{63}^{113}.

(3) $d_2\left(\bar{e}_{95}h_5h_8\right) = d_2\left(\overline{\tilde{e}_{63}h_6h_8}\right) = \overline{\tilde{e}_{63}h_5^2h_8h_0}$ *in* $\tilde{P}^{113}$.

(4) $d_2\left(\bar{e}_{95}h_5h_8\right) = \begin{cases} \bar{e}_{95}h_4^2h_0h_8 \neq 0 & in\ P_{65}^{96} \\ 0 & in\ P_{65}^{113}. \end{cases}$

(5) $d_2\left(\bar{e}_{111}h_4h_0h_8\right) = \bar{e}_{111}h_4h_7^2h_0^2 \neq 0$ *in* P_{97}^{113}.

(6) $d_2\left(\hat{e}_{95}h_5h_8h_0\right) = \hat{e}_{95}h_4^2h_0^2h_8 \neq 0$ *in* $\hat{P}_{65}^{113}$.

(7) $d_2\left(\bar{e}_{111}h_4h_0h_8\right) = \overline{\hat{e}_{95}h_4^2h_0^2h_8} + \bar{e}_{111}h_4h_7^2h_0^2 \neq 0$ *in* $\hat{P}_{65}^{113}$.

(4.2) follows from (4.1), (3.12), (3.14), (3.15) and the Adams differentials $d_2(h_i) = h_{i-1}^2h_0$. Details of the proof go as follows. First, (4.2) (1). We have

$$\begin{aligned} d_2\left(\bar{e}_{63}h_6h_8\right) &= d_2\left(\bar{e}_{63}\right)h_6h_8 + \bar{e}_{63}d_2\left(h_6\right)h_8 + \bar{e}_{63}h_6d_2\left(h_8\right) \\ &= \bar{e}_{31}h_5h_0h_6h_8 + \bar{e}_{63}h_5^2h_0h_8 + \bar{e}_{63}h_6h_7^2h_0 \quad \text{(by (4.1)(1))} \\ &= \bar{e}_{63}h_5^2h_0h_8 \end{aligned}$$

which is non-zero in P^{64} by (3.15) (30) and is zero in P^{113} by (3.12). Next, (4.2) (2). We have

$$\begin{aligned} &d_2\left(\overline{e_{95}h_0}h_5h_8\right) \\ &= \overline{e_{79}h_0}h_4h_0h_5h_8 + \overline{e_{95}h_0}h_4^2h_0h_8 + \overline{e_{95}h_0}h_5h_7^2h_0 \qquad \text{(by (4.1)(4))} \\ &= \overline{e_{95}h_0}h_4^2h_0h_8 = 0 \qquad \text{(by(3.14) (1))}. \end{aligned}$$

For (4.2)(3) we have

$$\begin{aligned}
&d_2\left(\overline{\bar{e}_{95}h_5h_8}\right)\\
&= \overline{e_{79}h_4}h_0h_5h_8 + \epsilon\bar{e}_{31}h_6h_0h_5h_8 + \bar{e}_{95}h_4^2h_0h_8 + \bar{e}_{95}h_5h_7^2h_0 \qquad \text{(by (4.1)(3))}\\
&= \overline{e_{79}h_4}h_5h_0h_8 + \bar{e}_{95}\overline{h_4^2}h_0h_8\\
&= \tilde{e}_{63}h_5^2h_0h_8 + 0 = \tilde{e}_{63}h_5^2h_0h_8 \qquad \text{(by (3.12), (3.14)(2))}
\end{aligned}$$

and

$$\begin{aligned}
d_2\left(\overline{\tilde{e}_{63}h_6}h_8\right) &= \overline{\tilde{e}_{63}h_5^2}h_0h_8 + \overline{\tilde{e}_{63}h_6}h_7^2h_0 \qquad (by (4.1)(7))\\
&= \overline{\tilde{e}_{63}h_5^2}h_0h_8 \qquad \text{(by (3.14)(3))}
\end{aligned}$$

which is non-zero by (3.17) (1). The proofs of (4.2) (4) and (4.2) (5) are similar to that of (4.2) (1) above which we leave to the reader. For (4.2) (6) we have

$$\begin{aligned}
&d_2\left(\overline{\hat{e}_{95}h_5}h_0h_8\right)\\
&= \overline{\hat{e}_{95}h_4^2}h_0^2h_8 + \epsilon\overline{e_{87}h_0h_3h_5}h_0h_8 + \overline{\hat{e}_{95}h_5}h_7^2h_0^2 \qquad \text{(by (4.1) (8))}\\
&= \overline{\hat{e}_{95}h_4^2}h_0^2h_8 \qquad \text{(by (3.14) (6), (3.14) (7))}
\end{aligned}$$

which is non-zero by (3.17) (2). For (4.2) (7) we have

$$\begin{aligned}
&d_2\left(\bar{e}_{111}h_4h_0h_8\right)\\
&= \overline{e_{103}h_3}h_4h_0^2h_8 + \epsilon\bar{e}_{79}h_5h_0^2h_4h_8 + \bar{e}_{111}h_3^2h_0^2h_8 + \bar{e}_{111}h_4h_7^2h_0^2 \quad \text{(by (4.1)(6))}\\
&= \hat{e}_{95}h_4^2h_0^2h_8 + \bar{e}_{111}h_4h_7^2h_0^2 \qquad \text{(by (3.14) (5))}
\end{aligned}$$

which is non-zero by (3.17) (2). This completes the proof of (4.2).

5 Proof of Theorem 1.1

We already have shown the following.

(4.2)(1) $\quad d_2\left(\bar{e}_{63}h_6h_8\right) = 0$ in the ASS for P^{113}.

We shall prove

(5.1) $\quad d_3\left(\bar{e}_{63}h_6h_8\right) = 0$

and

(5.2) $\quad d_4\left(\bar{e}_{63}h_6h_8\right) \neq 0$

and then estimate the value of $d_4\left(\bar{e}_{63}h_6h_8\right)$ as follows.

(5.3) $$\begin{aligned}d_4\left(\bar{e}_{63}h_6h_8\right) = \overline{e_{111}h_4h_7^2}h_0^3 + \epsilon_1\bar{e}_{63}h_8H_1 + \epsilon_2\bar{e}_{63}D_3h_8h_1\\ +\epsilon_3\overline{e_{47}h_0^2h_4h_6}h_8h_0 + \epsilon_4\bar{e}_{31}c_5h_0^3 + \epsilon_5\overline{e_{17}h_3^2c_5}h_0\end{aligned}$$

in the ASS for P^{113} where $\epsilon_i = 0$ or 1.

Theorem 1.1 is deduced from these as follows. Recall the induced map $t_* : Ext_A^{s,t}\left(P^{113}\right) \to Ext_A^{s+1,t+1}\left(S^0\right)$ of the transfer map $P^{113} \hookrightarrow P^\infty \xrightarrow{t} S^0$. We have $t_*\left(\bar{e}_{63}h_6h_8\right) = h_6^2h_8 = h_7^3$. By (3.18), $t_*\left(\overline{e_{111}h_4h_7^2h_0^3}\right) = g_5h_0^3$, $t_*\left(\bar{e}_{63}H_1h_8\right) = h_6H_1h_8$, $t_*\left(\bar{e}_{63}D_3h_8h_1\right) = 0$, $t_*\left(\overline{e_{47}h_0^2h_4h_6}h_8h_0\right) = \epsilon h_8h_0K$, $\epsilon = 0$ or 1, $t_*\left(\bar{e}_{31}c_5h_0^3\right) = 0$ and $t_*\left(\overline{e_{17}h_3^2c_5}h_0\right) = 0$. Since t_* commutes with d_4 in the ASS for P^{113} and S^0 we deduce that

$(*)$ $\quad d_4\left(h_7^3\right) = g_5h_0^3 + \epsilon_1h_8h_6H_1 + \epsilon_2h_8h_0K$ in the ASS for S^0 where $\epsilon_i = 0$ or 1.

By (3.5) (5) the classes $g_5h_0^3, h_8h_6H_1, h_8h_0K$ are linearly independent in $E_2^{7,387}\left(S^0\right)$

$= Ext_A^{7,387}$. By (3.3) and (3.4)(2)(d), $E_2^{4+s,385+s}(S^0) = Ext_A^{4+s,385+s} = \mathbb{Z}_2(h_7^3 h_0^{s+1})$, $s = 0, 1$. Since $t_*(\bar{e}_{63} h_6 h_8 h_0^{s+1}) = h_7^3 h_0^{s+1}$ and t_* commutes with d_2, d_3 in the ASS for P^{113} and S^0, from (4.2) (1), (5.1) we deduce $d_3(h_7^3 h_0) = 0$ and $d_2(h_7^3 h_0^2) = 0$. So $g_5 h_0^3, h_8 h_6 H_1, h_8 h_0 K$ are also linearly independent in $E_4^{7,387}(S^0)$. Thus the differential $(*)$ is a non-trivial one. This proves Theorem 1.1.

The remainder of this section is devoted to proving (5.1), (5.2) and (5.3).

To prove (5.1) we first prove the following (5.4). To state it we denote the image in $Ext_A^{*,*}(S^0 \cup_{2\iota} e^1)$ of any product $h_i h_j \cdots$ in $Ext_A^{*,*}(S^0)$ by the same notation.

Proposition 5.4 *$d_3(h_6 h_8) = 0$ in the ASS for $S^0 \cup_{2\iota} e^1$.*

Proof Let $\psi : S^{63} \to S^0 \cup_{2\iota} e^1$ be a stable map which is detected by $h_6 \in Ext_A^{1,64}(S^0 \cup_{2\iota} e^1)$. The existence of such a map is due to Mahowald (see [**3, 10**]). ψ has the property that the composite $S^{63} \xrightarrow{\psi} S^0 \cup_{2\iota} e^1 \xrightarrow{p} S^1$ is detected by h_5^2 where p is the pinching map. Mahowald also proves in [**17**] that there is a stable map $\Sigma^{320} P_{-33}^{-2} \xrightarrow{\psi_1} S^{63}$ which is detected by h_8 on the top cell $e^{318} = \Sigma^{320} e^{-2}$ and is detected by h_7^2 on the next top cell e^{317}. Here P_{-33}^{-2} is the spectrum $\Sigma^{-2^m} P_{2^m-33}^{2^m-2}$ for m large (see [**8**]).

From (3.1) we see $Ext_A^{s,t} = 0$ for $s \le 3$, $224 \le t - s \le 253$. So $\psi_1 | \Sigma^{320} P_{-33}^{-4}$ has Adams filtration ≥ 4 in the ASS for the stable groups $[\Sigma^{320} P_{-33}^{-4}, S^{63}]_*$. Since $h_7^2 h_6 = 0$ and $h_7^2 h_5^2 = 0$ the composite $\Sigma^{320} P_{-33}^{-2} \xrightarrow{\psi_2 = \psi\psi_1} S^0 \cup_{2\iota} e^1$ has the following properties:

(i) ψ_2 is detected on the top cell e^{318} by $h_6 h_8 \in Ext_A^{2,320}(S^0 \cup_{2\iota} e^1)$,

(ii) ψ_2 is detected on the next top cell e^{317} by some $\alpha \in Ext_A^{s,317+s}(S^0 \cup_{2\iota} e^1)$ with $4 \le s \le \infty$ which, if $s = 4$, comes from $Ext_A^{4,321}(S^0)$,

(iii) $\psi_2 | \Sigma^{320} P_{-33}^{-4}$ has Adams filtration ≥ 5 in the ASS for the stable groups $[\Sigma^{320} P_{-33}^{-4}, S^0 \cup_{2\iota} e^1]_*$.

Let $S^{317} \xrightarrow{g} \Sigma^{320} P_{-33}^{-3}$ be the attaching map for the top cell of $\Sigma^{320} P_{-33}^{-2}$. The composite $S^{317} \xrightarrow{g} \Sigma^{320} P_{-33}^{-3} \xrightarrow{q} S^{317}$ is a degree 2 map which is detected by h_0 where q is the collapsing map. It follows from this and (i), (ii), (iii) above that if the α in (ii) has Adams filtration $s \ge 5$ then we can conclude $d_3(h_6 h_8) = 0$ because then $\psi_2 | \Sigma^{320} P_{-33}^{-3}$ can be lifted to the fifth stage spectrum X_5 in any Adams resolution $\{X_i\}$ for $S^0 \cup_{2\iota} e^1$, and so the composite $S^{317} \xrightarrow{g} \Sigma^{320} P_{-33}^{-3} \longrightarrow X_5$ can be further lifted to X_6 and this implies $d_3(h_6 h_8) = 0$. By the same reasoning if α has Adams filtration 4 then $d_3(h_6 h_8) = \alpha h_0$. But $\alpha h_0 = 0$ in $Ext_A^{5,322}(S^0 \cup_{2\iota} e^1)$ since α comes from $Ext_A^{4,321}(S^0)$. So $d_3(h_6 h_8) = 0$. This completes the proof of Proposition (5.4).

An immediate consequence of (5.4) is the following.

Corollary 5.5 $d_3(\bar{e}_{63} h_6 h_8) = 0$ in the ASS for P_{63}^{113}.

Now we prove (5.1). We want to show for (5.1) that $d_3(\bar{e}_{63} h_6 h_8) = 0$ in the ASS for P^{113}. We prove this by contradiction. Suppose $d_3(\bar{e}_{63} h_6 h_8) \ne 0$ and let x be an element in $Ext_A^{5,385}(P^{113}) = E_2^{5,385}(P^{113})$ representing $d_3(\bar{e}_{63} h_6 h_8) \in E_3^{5,385}(P^{113})$. Let $f : P^{113} \to \tilde{P}^{113}$, $f_* : Ext_A^{5,385}(P^{113}) \to Ext_A^{5,385}(\tilde{P}^{113})$ be as in (3.17). We will show $f_*(x) = 0$ and $f_*(x) \ne 0$ both are impossible. If $f_*(x) = 0$ then, by (3.17)(3), $x = \bar{e}_{63} T_1$. Consider the collapsing map $p : P^{113} \to P_{63}^{113}$. By (3.15)(31), $p_*(\bar{e}_{63} T_1) = \bar{e}_{63} T_1$ is non-zero in $Ext_A^{5,385}(P_{63}^{113}) = E_2^{5,385}(P_{63}^{113})$

and is also non-zero in $E_3^{5,385}\left(P_{63}^{113}\right)$ because, by (3.15) (15), $Ext_A^{3,384}\left(P_{63}^{113}\right) \cong \mathbb{Z}_2\left(\overline{e_{95}h_0}h_5h_8\right)$ and, by (4.2) (2), $d_2\left(\overline{e_{95}h_0}h_5h_8\right) = 0$ in the ASS for P_{63}^{113}. Thus

$$\begin{aligned} d_3\left[p_*\left(\bar{e}_{63}h_6h_8\right)\right. &= \left.\bar{e}_{63}h_6h_8\right] \\ &= p_*\left(d_3\left(\bar{e}_{63}h_6h_8\right)\right) = p_*\left(x = \bar{e}_{63}T_1\right) = \bar{e}_{63}T_1 \neq 0 \end{aligned}$$

in the ASS for P_{63}^{113}. This contradicts (5.5). Thus $f_*(x) = 0$ is impossible. Suppose $f_*(x) \neq 0$ in $Ext_A^{5,385}\left(\tilde{P}^{113}\right) = E_2^{5,385}\left(\tilde{P}^{113}\right)$. Since $f_*\left(\bar{e}_{63}h_6h_8\right) = 0$ and $d_3\left(\bar{e}_{63}h_6h_8\right) = x$ it follows that $f_*(x) = 0$ in $E_3^{5,385}\left(\tilde{P}^{113}\right)$, that is, $d_2(v) = f_*(x)$ for some non-zero $v \in Ext_A^{3,384}\left(\tilde{P}^{113}\right)$ in the ASS for $\tilde{P}^{113}$. By (3.15) (20), $Ext_A^{3,384}\left(\tilde{P}^{113}\right) \cong \mathbb{Z}_2\left(\overline{\tilde{e}_{63}h_6}h_8h_0\right) \oplus \mathbb{Z}_2\left(\overline{e_{95}}h_5h_8h_0\right)$, and by (4.2) (3) and (3.17) (1), $d_2\left(\overline{\tilde{e}_{63}h_6}h_8h_0\right) = d_2\left(\bar{e}_{95}h_5h_8h_0\right) = \overline{\tilde{e}_{63}h_5^2}h_8h_0^2 \neq 0$ in $Ext_A^{5,385}\left(\tilde{P}^{113}\right)$ which is not in im f_*. This implies v is non-existent proving $f_*(x) \neq 0$ is also impossible. This completes the proof of (5.1).

We proceed to prove (5.2) and (5.3). To prove (5.2) we shall apply the following (5.6) that describes a well known property about the differentials in the Adams spectral sequence. To state it we suppose given a stable cofibration $X \xrightarrow{i} Y \xrightarrow{q} Z$ such that

$$0 \longrightarrow H_*(X) \xrightarrow{i_*} H_*(Y) \xrightarrow{q_*} H_*(Z) \longrightarrow 0$$

is an exact sequence of A-modules. This gives rise to a long exact sequence of Ext groups:

$$\cdots \xrightarrow{\delta_*} Ext_A^{s,t}(X) \xrightarrow{i_*} Ext_A^{s,t}(Y) \xrightarrow{q_*} Ext_A^{s,t}(Z) \xrightarrow{\delta_*} Ext_A^{s+1,t}(X) \longrightarrow \cdots.$$

Let $\alpha' \in Ext_A^{\bar{s},\bar{t}}(X)$ be a non-zero class such that $i_*(\alpha') = \alpha \neq 0$ in $Ext_A^{\bar{s},\bar{t}}(Y)$. We assume the following.

(i) $\beta = d_2(\alpha') \neq 0$ in $Ext_A^{\bar{s}+2,\bar{t}+1}(X)$ in the ASS for X.
(ii) $Ext_A^{\bar{s}+1,\bar{t}+1}(Z) \cong \mathbb{Z}_2(\gamma)$ with $\delta_*(\gamma) = \beta$ (this implies $i_*(\beta) = 0$).
(iii) $d_{\bar{r}}(\gamma) \neq 0$ in the ASS for Z for some $\bar{r} \geq 2$, and $d_{\bar{r}}(\gamma)$ has a representative z in $Ext_A^{\bar{s}+\bar{r}+1,\bar{t}+\bar{r}}(Z)$ such that $z \in imq_*$.
(iv) $d_{r'}(\alpha) = 0$ in the ASS for Y for $2 \leq r' \leq \bar{r}$.

Proposition 5.6 *Under the above assumptions, $d_{\bar{r}+1}(\alpha) \neq 0$ in the ASS for Y and $q_*\left(d_{\bar{r}+1}(\alpha)\right) = d_{\bar{r}}(\gamma)$, that is, if $d_{\bar{r}+1}(\alpha)$ is represented by $y \in Ext_A^{\bar{s}+\bar{r}+1,\bar{t}+\bar{r}}(Y)$ then $q_*(y)$ is hit by γ via $d_{\bar{r}}$ in the ASS for Z.*

The following will be proved later.

Proposition 5.7 $d_3\left(\bar{e}_{95}h_5h_8\right) = \overline{e_{111}h_4h_7^2}h_0^2 + \epsilon\bar{e}_{95}d_4h_1$ *in the ASS for* P_{65}^{113} *where* $\epsilon = 0$ *or* 1.

This implies the following.

Corollary 5.8 $d_3\left(\bar{e}_{95}h_5h_8h_0\right) = \overline{e_{111}h_4h_7^2}h_0^3 \neq 0$ in the ASS for P_{65}^{113}.

We deduce (5.2) from (5.6) and (5.8) as follows. Consider the cofibration $P^{64} \xrightarrow{i} P^{113} \xrightarrow{q} P_{65}^{113}$ and the resulting long exact sequence of Ext groups:

$$\cdots \xrightarrow{\delta_*} Ext_A^{s,t}(P^{64}) \xrightarrow{i_*} Ext_A^{s,t}(P^{113}) \xrightarrow{q_*} Ext_A^{s,t}(P_{65}^{113}) \xrightarrow{\delta_*} Ext_A^{s+1,t}(P^{64}) \longrightarrow \cdots.$$

Let $\alpha' = \bar{e}_{63}h_6h_8 \in Ext_A^{2,383}\left(P^{64}\right)$ and $\alpha = i_*(\alpha') = \bar{e}_{63}h_6h_8 \in Ext_A^{2,383}\left(P^{113}\right)$. By (4.2)(1), $\beta = d_2(\alpha') = d_2\left(\bar{e}_{63}h_6h_8\right) = \bar{e}_{63}h_5^2h_0h_8 \neq 0$ in the ASS for P^{64}. By (3.15) (16), $Ext_A^{3,384}\left(P_{65}^{113}\right) \cong \mathbb{Z}_2\left(\bar{e}_{95}h_5h_8h_0\right)$. Let $\gamma = \bar{e}_{95}h_5h_8h_0$. We have

$\delta_*(\gamma = \bar{e}_{95}h_5h_8h_0) = \bar{e}_{63}h_5^2h_0h_8 = \beta$. By (5.8), $d_3(\gamma = \bar{e}_{95}h_5h_8h_0) = \overline{e_{111}h_4h_7^2h_0^3} \neq 0$ in the ASS for P_{65}^{113}. $\overline{e_{111}h_4h_7^2h_0^3} \in Ext_A^{5,385}(P_{65}^{113})$ lies in im q_* since $q_*\left(\overline{e_{111}h_4h_7^2h_0^3}\right) = \overline{e_{111}h_4h_7^2h_0^3}$ (see (3.13)(13)). By (4.2)(1) and (5.1), $d_r(\alpha = \bar{e}_{63}h_6h_8) = 0$ for $r = 2, 3$ in the ASS for P^{113}. Applying (5.6) we conclude that $d_4(\alpha = \bar{e}_{63}h_6h_8) \neq 0$ in the ASS for P^{113} and that

$$(a) \qquad q_*(d_4(\bar{e}_{63}h_6h_8)) = d_3(\bar{e}_{95}h_5h_8h_0) = \overline{e_{111}h_4h_7^2h_0^3} \neq 0 \quad \text{in} \quad P_{65}^{113}.$$

This proves (5.2).

Equation (a) means that $d_4(\bar{e}_{63}h_6h_8)$ has a representative y in $E_2^{6,386}(P^{113}) = Ext_A^{6,386}(P^{113})$ such that $q_*(y)$ is non-zero in $E_2^{6,386}(P_{65}^{113}) = Ext_A^{6,386}(P_{65}^{113})$, continues to be non-zero in $E_3^{6,386}(P_{65}^{113})$ and gets hit by $\bar{e}_{95}h_5h_8h_0$ via d_3 in the ASS for P_{65}^{113}. We want to find the estimated value of y as described in (5.3). Note that $P^{113} \xrightarrow{q} P_{65}^{113}$ has a factorization

$$q : P^{113} \xrightarrow{f} \tilde{P}^{113} \xrightarrow{\bar{q}} P_{65}^{113}$$

where f is as in (3.17) and $\bar{q}$ is the collapsing map. To estimate the value of y we have to estimate the value of $f_*(y)$ in $Ext_A^{6,386}\left(\tilde{P}^{113}\right)$ first.

Since $q_*(y) = \bar{q}_*(f_*(y))$ and $q_*(y) \neq 0$ in $E_3^{6,386}(P_{65}^{113})$ it follows that $f_*(y) \neq 0$ in $E_3^{6,386}\left(\tilde{P}^{113}\right)$. Since $d_4(\bar{e}_{63}h_6h_8) = y$ and $f_*(\bar{e}_{63}h_6h_8) = 0$ it follows that $f_*(y) = 0$ in $E_4^{6,386}\left(\tilde{P}^{113}\right)$, that is, there is some non-zero $u \in E_3^{3,384}\left(\tilde{P}^{113}\right)$ such that $d_3(u) = f_*(y) \neq 0$ in $E_3^{6,386}\left(\tilde{P}^{113}\right)$. By (3.15) (20), $E_2^{3,384}\left(\tilde{P}^{113}\right) \cong \mathbb{Z}_2\left(\tilde{e}_{63}\overline{h_6}h_8h_0\right) \oplus \mathbb{Z}_2(\bar{e}_{95}h_5h_8h_0)$, and by (4.2)(3) and (3.17) (1), $d_2\left(\tilde{e}_{63}\overline{h_6}h_8\right) = d_2(\bar{e}_{95}h_5h_8) = \tilde{e}_{63}h_5^2h_8h_0 \neq 0$ in $E_2^{4,384}\left(\tilde{P}^{113}\right)$ which implies $d_2\left(\tilde{e}_{63}\overline{h_6}h_8h_0\right) = d_2(\bar{e}_{95}h_5h_8h_0) = \tilde{e}_{63}h_5^2h_8h_0^2 \neq 0$ in $E_2^{5,385}\left(\tilde{P}^{113}\right)$. So $E_3^{3,384}\left(\tilde{P}^{113}\right) \cong \mathbb{Z}_2\left(\tilde{e}_{63}\overline{h_6}h_8h_0 + \bar{e}_{95}h_5h_8h_0\right)$. Thus this non-zero u must be $u = \tilde{e}_{63}\overline{h_6}h_8h_0 + \bar{e}_{95}h_5h_8h_0$. Then

$$\begin{aligned} f_*(y) = d_3(u) &= d_3\left(\overline{\tilde{e}_{63}h_6}h_8h_0 + \bar{e}_{95}h_5h_8h_0\right) \\ &= \left[d_3\left(\overline{\tilde{e}_{63}h_6}h_8 + \bar{e}_{95}h_5h_8\right)\right]h_0 \end{aligned}$$

where we have also just shown above that $\overline{\tilde{e}_{63}h_6}h_8 + \bar{e}_{95}h_5h_8 \in E_3^{2,383}\left(\tilde{P}^{113}\right)$. To estimate the value of $f_*(y) \in E_2^{6,386}\left(\tilde{P}^{113}\right)$ it suffices to estimate the value of $d_3\left(\overline{\tilde{e}_{63}h_6}h_8 + \bar{e}_{95}h_5h_8\right)$.

As noted above, $E_2^{3,384}(P_{65}^{113}) \cong \mathbb{Z}_2(\bar{e}_{95}h_5h_8h_0)$ (by (3.15)(16)). By (4.2) (4) $d_2(\bar{e}_{95}h_5h_8h_0) = d_2(\bar{e}_{95}h_5h_8)h_0 = 0$ in $E_2^{5,385}(P_{65}^{113})$. This implies

(b) $E_3^{5,385}(P_{65}^{113}) \subset E_2^{5,385}(P_{65}^{113})$

We have

$$\begin{aligned} (c)\ \bar{q}_*\left[d_3\left(\overline{\tilde{e}_{63}h_6}h_8 + \bar{e}_{95}h_5h_8\right)\right] &= d_3\left[\bar{q}_*\left(\overline{\tilde{e}_{63}h_6}h_8 + \bar{e}_{95}h_5h_8\right)\right] \\ &= d_3(\bar{e}_{95}h_5h_8) \qquad \text{(since } \bar{q}_*\left(\overline{\tilde{e}_{63}h_6}h_8\right) = 0) \\ &\doteq \overline{e_{111}h_4h_7^2h_0^2} + \epsilon\bar{e}_{95}d_4h_1 \text{ in } E_3^{5,385}(P_{65}^{113}) \text{ (by (5.7)).} \end{aligned}$$

We also have

$$(d) \qquad \bar{q}_*\left(\overline{e_{111}h_4h_7^2h_0^2}\right) = \overline{e_{111}h_4h_7^2h_0^2} \quad \text{and} \quad \bar{q}_*(\bar{e}_{95}d_4h_1) = \bar{e}_{95}d_4h_1.$$

From (b), (c), (d) we see if $\tilde{y}$ is a representative of $d_3\left(\tilde{e}_{63}\overline{h_6}h_8 + \bar{e}_{95}h_5h_8\right)$ in $E_2^{5,385}\left(\tilde{P}^{113}\right)$ then $\tilde{y}$ is of the form

$$\tilde{y} = \overline{e_{111}h_4h_7^2}h_0^2 + \epsilon\bar{e}_{95}d_4h_1 + \tilde{y}_1$$

with $\tilde{y}_1$ lying in $ker\left(\bar{q}_* : E_2^{5,385}\left(\tilde{P}^{113}\right) \to E_2^{5,385}\left(P_{65}^{113}\right)\right)$. By (3.15) (32), $ker\bar{q}_*$ is generated by $\overline{\tilde{e}_{63}h_5^2}h_0^2h_8$, $\overline{e_{47}h_0^2h_4h_6}h_8$, $\bar{e}_{31}f_4h_1$, $\bar{e}_{31}c_5h_0^2$, $\overline{e_{17}h_3^2c_5}$ and $\overline{\tilde{e}_{63}T_1}$. Since, by (4.2) (3), $d_2\left(\tilde{e}_{63}\overline{h_6}h_8h_0\right) = \overline{\tilde{e}_{63}h_5^2}h_0^2h_8$ in $E_2^{5,385}\left(\tilde{P}^{113}\right)$, we see $\tilde{y}$ can be chosen to be of the form

$$\begin{aligned}\tilde{y} = \ & \overline{e_{111}h_4h_7^2}h_0^2 + \epsilon\bar{e}_{95}d_4h_1 + \epsilon_3\overline{e_{47}h_0^2h_4h_6}h_8 + \epsilon_4\bar{e}_{31}c_5h_0^2 \\ & +\epsilon_5\overline{e_{17}h_3^2c_5} + \epsilon_6\bar{e}_{31}f_4h_1 + \epsilon_7\overline{\tilde{e}_{63}T_1}\end{aligned}$$

where $\epsilon, \epsilon_i = 0$ or 1. Why ϵ_i are indexed beginning with 3 will be clear later. Note that $\bar{e}_{31}f_4h_1h_0 = 0 = \bar{e}_{95}d_4h_1h_0$, and by (3.14)(4), $\overline{\tilde{e}_{63}T_1}h_0 = 0$. Let

$$\begin{aligned} w \ &= \overline{e_{111}h_4h_7^2}h_0^3 + \epsilon_3\overline{e_{47}h_0^2h_4h_6}h_8h_0 + \epsilon_4\bar{e}_{31}c_5h_0^3 + \epsilon_5\overline{e_{17}h_3^2c_5}h_0 \in E_2^{6,386}\left(P^{113}\right) \\ \tilde{w} \ &= \overline{e_{111}h_4h_7^2}h_0^3 + \epsilon_3\overline{e_{47}h_0^2h_4h_6}h_8h_0 + \epsilon_4\bar{e}_{31}c_5h_0^3 + \epsilon_5\overline{e_{17}h_3^2c_5}h_0 \in E_2^{6,386}\left(\tilde{P}^{113}\right).\end{aligned}$$

Then

$$\begin{aligned} f_*(y) &= \left[d_3\left(\tilde{e}_{63}\overline{h_6}h_8 + \bar{e}_{95}h_5h_8\right)\right]h_0 \\ &= \tilde{y}h_0 = \tilde{w} = f_*(w),\end{aligned}$$

that is,

(e) $$f_*(y + w) = 0.$$

This is an equation in $E_3^{6,386}\left(\tilde{P}^{113}\right)$. We will prove the following in a moment.

Proposition 5.9 *For* $z \in E_2^{6,386}\left(P^{113}\right)$ *if* $f_*(z) = 0$ *in* $E_3^{6,386}\left(\tilde{P}^{113}\right)$ *then either (i)* $f_*(z) = 0$ *in* $E_2^{6,386}\left(\tilde{P}^{113}\right)$
or (ii) there exists a $z_1 \in E_2^{6,386}\left(P^{113}\right)$ *with* $z_1 = 0$ *in* $E_3^{6,386}\left(P^{113}\right)$ *such that* $f_*(z + z_1) = 0$ *in* $E_2^{6,386}\left(\tilde{P}^{113}\right)$.

Let $Q = ker\left(f_* : E_2^{6,386}\left(P^{113}\right) \to E_2^{6,386}\left(\tilde{P}^{113}\right)\right)$. Applying (5.9) to $z = y+w$ in (e) we see either $y + w \in Q$ or there is some $z_1 \in E_2^{6,386}\left(P^{113}\right)$ with $z_1 = 0$ in $E_3^{6,386}\left(P^{113}\right)$ such that $y + w + z_1 \in Q$. By (3.17) (4), $Q = \mathbb{Z}_2\left(\bar{e}_{63}H_1h_8\right) \oplus \mathbb{Z}_2\left(\bar{e}_{63}D_3h_1h_8\right)$. So $y = w + \epsilon_1\bar{e}_{63}H_1h_8 + \epsilon_2\bar{e}_{63}D_3h_1h_8 + \epsilon' z_1$ where ϵ', $\epsilon_i = 0$ or 1. Recall that y is a representative of $d_4\left(\bar{e}_{63}h_6h_8\right)$ in $E_2^{6,386}\left(P^{113}\right)$. Since z_1 is zero in $E_3^{6,386}\left(P^{113}\right)$ it is also zero in $E_4^{6,386}\left(P^{113}\right)$. So we may assume $\epsilon' = 0$. Then

$$\begin{aligned} d_4\left(\bar{e}_{63}h_6h_8\right) = y \ &= w + \epsilon_1\bar{e}_{63}H_1h_8 + \epsilon_2\bar{e}_{63}D_3h_1h_8 \\ &= \overline{e_{111}h_4h_7^2}h_0^3 + \epsilon_1\bar{e}_{63}H_1h_8 + \epsilon_2\bar{e}_{63}D_3h_1h_8 \\ &\quad +\epsilon_3\overline{e_{47}h_0^2h_4h_6}h_8h_0 + \epsilon_4\bar{e}_{31}c_5h_0^3 + \epsilon_5\overline{e_{17}h_3^2c_5}h_0\end{aligned}$$

in the ASS for P^{113}. This prove (5.3).

(5.9) is proved as follows. Suppose $f_*(z) \neq 0$ in $E_2^{6,386}\left(\tilde{P}^{113}\right)$. Since $f_*(z) = 0$ in $E_3^{6,386}\left(\tilde{P}^{113}\right)$, $d_2(v) = f_*(z)$ for some $v \in E_2^{4,385}\left(\tilde{P}^{113}\right)$. By (3.15)(29),

$$\begin{aligned} &E_2^{4,385}\left(\tilde{P}^{113}\right) \\ &\quad\cong \mathbb{Z}_2\left(\overline{\tilde{e}_{63}h_5^2}h_8h_1\right) \oplus \mathbb{Z}_2\left(\overline{\tilde{e}_{63}h_6}h_8h_0^2\right) \oplus \mathbb{Z}_2\left(\bar{e}_{95}h_5h_8h_0^2\right) \oplus \mathbb{Z}_2\left(\overline{e_{111}h_4h_7^2h_1}\right) \\ &\qquad \oplus \mathbb{Z}_2\left(\overline{e_{47}h_4h_6h_1}h_8\right) \oplus \left(\bar{e}_{31}c_5h_1\right). \end{aligned}$$

So

$$\begin{aligned} (f)\quad f_*(z) = d_2\Big(&\epsilon_1'\overline{\tilde{e}_{63}h_5^2}h_8h_1 + \epsilon_2'\overline{\tilde{e}_{63}h_6}h_8h_0^2 + \epsilon_3'\bar{e}_{95}h_5h_8h_0^2 + \epsilon_4'\overline{e_{111}h_4h_7^2h_1} \\ &+\epsilon_5'\overline{e_{47}h_4h_6h_1}h_8 + \epsilon_6'\bar{e}_{31}c_5h_1\Big) \end{aligned}$$

where $\epsilon_i' = 0$ or 1. Note that $f_*\left(\overline{e_{111}h_4h_7^2h_1}\right) = \overline{e_{111}h_4h_7^2h_1}$, $f_*\left(\overline{e_{47}h_4h_6h_1}h_8\right) = \overline{e_{47}h_4h_6h_1}h_8$ and $f_*\left(\bar{e}_{31}c_5h_1\right) = \bar{e}_{31}c_5h_1$. By (4.2)(3), $d_2\left(\overline{\tilde{e}_{63}h_6}h_8h_0^2\right) = d_2\left(\bar{e}_{95}h_5h_8h_0^2\right) = \overline{\tilde{e}_{63}h_5^2}h_0^3h_8$. By (4.1) (9), $d_2\left(\overline{\tilde{e}_{63}h_5^2}\right) = 0$ which implies $d_2\left(\overline{\tilde{e}_{63}h_5^2}h_8h_1\right) = 0$ since $d_2\left(h_8h_1\right) = 0$. Let $z_1 = d_2(\epsilon_4'\overline{e_{111}h_4h_7^2h_1} + \epsilon_5'\overline{e_{47}h_4h_6h_1}h_8 + \epsilon_6'\bar{e}_{31}c_5h_1)$ in $E_2^{6,386}\left(P^{113}\right)$. Then

$$\begin{aligned} f_*\left(z+z_1\right) &= f_*(z) + f_*\left(z_1 = d_2\left(\epsilon_4'\overline{e_{111}h_4h_7^2h_1} + \epsilon_5'\overline{e_{47}h_4h_6h_1}h_8 + \epsilon_6'\bar{e}_{31}c_5h_1\right)\right) \\ &= f_*(z) + \epsilon_4'd_2\left(f_*\left(\overline{e_{111}h_4h_7^2h_1}\right)\right) + \epsilon_5'd_2\left(f_*\left(\overline{e_{47}h_4h_6h_1}h_8\right)\right) + \epsilon_6'd_2\left(f_*\left(\bar{e}_{31}c_5h_1\right)\right) \\ &= f_*(z) + \epsilon_4'd_2\left(\overline{e_{111}h_4h_7^2h_1}\right) + \epsilon_5'd_2\left(\overline{e_{47}h_4h_6h_1}h_8\right) + \epsilon_6'd_2\left(\bar{e}_{31}c_5h_1\right) \\ &= \epsilon_1'd_2\left(\overline{\tilde{e}_{63}h_5^2}h_8h_1\right) + \epsilon_2'd_2\left(\overline{\tilde{e}_{63}h_6}h_8h_0^2\right) + \epsilon_3'd_2\left(\bar{e}_{95}h_5h_8h_0^2\right) \quad \text{(by (f))} \\ &= \left(\epsilon_2' + \epsilon_3'\right)\overline{\tilde{e}_{63}h_5^2}h_0^3h_8 \end{aligned}$$

By (3.17) (1), $\overline{\tilde{e}_{63}h_5^2}h_0^3h_8 \notin imf_*$. So $f_*(z+z_1) = 0$. Since $z_1 = 0$ in $E_3^{6,386}\left(P^{113}\right)$, (5.9) is proved.

It remains to prove (5.7) and (5.8). We already have shown the following.

(4.2)(5) $\quad d_2\left(\bar{e}_{111}h_4h_0h_8\right) = \bar{e}_{111}h_4h_7^2h_0^2 \neq 0$ in the ASS for P_{97}^{113}.

We shall deduce (5.7) from (4.2)(5) and (5.6). The deduction is similar to that for (5.2), (5.3) from (5.6), (5.7) and (5.8) above. Details go as follows.

Consider the long exect sequence

$$\cdots \xrightarrow{\delta_*} Ext_A^{s,t}(P_{65}^{96}) \xrightarrow{i_*} Ext_A^{s,t}(P_{65}^{113}) \xrightarrow{q_*'} Ext_A^{s,t}(P_{97}^{113}) \xrightarrow{\delta_*} Ext_A^{s+1,t}(P_{65}^{96}) \longrightarrow \cdots.$$

corresponding to the cofibration $P_{65}^{96} \xrightarrow{i} P_{65}^{113} \xrightarrow{q'} P_{97}^{113}$. Let $\alpha' = \bar{e}_{95}h_5h_8 \in Ext_A^{2,383}\left(P_{65}^{96}\right)$ and $\alpha = i_*(\alpha') = \bar{e}_{95}h_5h_8 \in Ext_A^{2,383}\left(P_{65}^{113}\right)$. By (4.2) (4), $\beta = d_2(\alpha') = d_2\left(\bar{e}_{95}h_5h_8\right) = \bar{e}_{95}h_4^2h_0h_8 \neq 0$ in the ASS for P_{65}^{96}. By (3.15) (18), $Ext_A^{3,384}\left(P_{97}^{113}\right) = \mathbb{Z}_2\left(\bar{e}_{111}h_0h_4h_8\right)$. Let $\gamma = \bar{e}_{111}h_0h_4h_8$. We have $\delta_*(\gamma = \bar{e}_{111}h_0h_4h_8) = \bar{e}_{95}h_4^2h_0h_8 = \beta$. By (4.2)(5), $d_2\left(\gamma = \bar{e}_{111}h_4h_0h_8\right) = \bar{e}_{111}h_4h_7^2h_0^2 \neq 0$ in the ASS for P_{97}^{113}. $\bar{e}_{111}h_4h_7^2h_0^2 \in Ext_A^{5,385}\left(P_{97}^{113}\right)$ lies in q_*' since $q_*'\left(\overline{e_{111}h_4h_7^2h_0^2}\right) = \bar{e}_{111}h_4h_7^2h_0^2$. By (4.2)(4), $d_2\left(\alpha = \bar{e}_{95}h_5h_8\right) = 0$ in the ASS for P_{65}^{113}. Applying (5.6) we conclude that $d_3\left(\alpha = \bar{e}_{95}h_5h_8\right) \neq 0$ in the ASS for P_{65}^{113} and that

$$q_*'\left(d_3\left(\bar{e}_{95}h_5h_8\right)\right) = d_2\left(\bar{e}_{111}h_4h_0h_8\right) = \bar{e}_{111}h_4h_7^2h_0^2 \neq 0 \quad \text{in } P_{97}^{113}.$$

This equation means that $d_3(\bar{e}_{95}h_5h_8)$ has a representative y' in $E_2^{5,385}(P_{65}^{113})$ such that $q'_*(y') = \bar{e}_{111}h_4h_7^2h_0^2 \neq 0$ in $E_2^{5,385}(P_{97}^{113})$. $P_{65}^{113} \xrightarrow{q'} P_{97}^{113}$ has a factorization

$$q' : P_{65}^{113} \xrightarrow{f'} \hat{P}_{65}^{113} \xrightarrow{\hat{q}'} P_{97}^{113}$$

where f' is as in (3.17) and $\hat{q}'$ is the collapsing map. Since $q'_*(y') = \hat{q}'_*(f'_*(y')) \neq 0$ it follows that $f'_*(y') \neq 0$ in $E_2^{5,385}\left(\hat{P}_{65}^{113}\right)$. Since $d_3(\bar{e}_{95}h_5h_8) = y'$ in $E_3^{5,385}(P_{65}^{113})$ and $f'_*(\bar{e}_{95}h_5h_8) = 0$ it follows that $f'_*(y') = 0$ in $E_3^{5,385}\left(\hat{P}_{65}^{113}\right)$, that is, there is some non-zero $u \in E_2^{3,384}\left(\hat{P}_{65}^{113}\right)$ such that $d_2(u) = f'_*(y') \neq 0$ in $E_2^{5,385}\left(\hat{P}_{65}^{113}\right)$. By (3.15)(25), $E_2^{3,384}\left(\hat{P}_{65}^{113}\right) = \mathbb{Z}_2\left(\overline{\hat{e}_{95}h_5}h_8h_0\right) \oplus \mathbb{Z}_2(\bar{e}_{111}h_8h_4h_0)$. By (4.2)(6), (7), we have

$$d_2\left(\overline{\hat{e}_{95}h_5}h_8h_0\right) = \overline{\hat{e}_{95}h_4^2}h_0^2h_8 \neq 0$$

and
$$d_2(\bar{e}_{111}h_8h_4h_0) = \overline{\hat{e}_{95}h_4^2}h_0^2h_8 + \bar{e}_{111}h_4h_7^2h_0^2 \neq 0$$

in the ASS for $\hat{P}_{65}^{113}$. By (3.17) (2), $\overline{\hat{e}_{95}h_4^2}h_0^2h_8 \notin im f'_*$. So u must be $\overline{\hat{e}_{95}h_5}h_8h_0 + \bar{e}_{111}h_4h_8h_0$. Thus

$$\begin{aligned} f'_*(y') &= d_2\left(\overline{\hat{e}_{95}h_5}h_8h_0 + \bar{e}_{111}h_4h_8h_0\right) \\ &= \bar{e}_{111}h_4h_7^2h_0^2 = f'_*\left(\overline{e_{111}h_4h_7^2h_0^2}\right), \end{aligned}$$

that is, $y' + \overline{e_{111}h_4h_7^2h_0^2} \in ker\left(f'_* : E_2^{5,385}(P_{65}^{113}) \to E_2^{5,385}\left(\hat{P}_{65}^{113}\right)\right)$. By (3.17) (5) this $ker f'_*$ is generated by $\bar{e}_{95}d_4h_1$. So

$$d_3(\bar{e}_{95}h_5h_8) = y' = \overline{e_{111}h_4h_7^2h_0^2} + \epsilon\bar{e}_{95}d_4h_1$$

in the ASS for P_{65}^{113} where $\epsilon = 0$ or 1. This completes the proof of (5.7).

Finally we prove that the differential $d_3(\bar{e}_{95}h_5h_8h_0) = \overline{e_{111}h_4h_7^2h_0^3}$ for P_{65}^{113} in (5.8), which is immediate from (5.7), is a non-trivial one. The proof of (3.15) (33) in Section 3 shows that $\overline{e_{111}h_4h_7^2h_0^3}$ has filtration 108 in the spectral sequence $\left\{\bar{E}_r^{i,s,t}(P_{65}^{113})\right\}$. This implies that if $\tilde{q} : P_{65}^{113} \to P_{108}^{113}$ is the collapsing map then $\tilde{q}_*\left(\overline{e_{111}h_4h_7^2h_0^3}\right) = \bar{e}_{111}h_4h_7^2h_0^3$ is non-zero in $Ext_A^{6,386}(P_{108}^{113})$.

From the differentials for $\left\{\tilde{E}_r^{i,s,t}\left(\tilde{P}^{113}\right)\right\}$ exhibited in (3.15) [(29), (32)]** one easily calculates that

$$Ext_A^{4,385}(P_{65}^{113}) \cong \mathbb{Z}_2\left(\overline{e_{93}p'_2}\right) \oplus \mathbb{Z}_2(\bar{e}_{95}h_5h_8h_0^2) \oplus \mathbb{Z}_2\left(\overline{e_{111}h_4h_7^2h_1}\right).$$

We have $\tilde{q}_*\left(\overline{e_{93}p'_2}\right) = \tilde{q}_*(\bar{e}_{95}h_5h_8h_0^2) = 0$ and $\tilde{q}_*\left(\overline{e_{111}h_4h_7^2h_1}\right) = \bar{e}_{111}h_4h_7^2h_1$. Since, in P_{108}^{113}, $\bar{e}_{111}$ is a spherical class and $d_2\left(h_4h_7^2h_1\right) = 0$ it follows that $d_2(\bar{e}_{111}h_4h_7^2h_1) = 0$ in the ASS for P_{108}^{113}. These together with $\tilde{q}_*\left(\overline{e_{111}h_4h_7^2h_0^3}\right) = \bar{e}_{111}h_4h_7^2h_0^3 \neq 0$ in $E_2^{6,386}(P_{108}^{113})$ imply $d_2(v) = \overline{e_{111}h_4h_7^2h_0^3}$ cannot happen in P_{65}^{113}, that is, $\overline{e_{111}h_4h_7^2h_0^3} \neq 0$ in $E_3^{6,386}(P_{65}^{113})$. This proves (5.8).

6 Proof of Proposition 3.8

Concerning (3.8) (1) we will prove the following more general result. If $X \xrightarrow{g} Y$ is a map of spaces or spectra with $H^n(X) \cong H^n(Y) \cong \mathbb{Z}_2$ and $g^* : H^n(Y) \longrightarrow H^n(X)$ is non-trivial then we say $g^* \neq 0$ in dimension n.

Proposition 6.1 *For $i \geq 3$ if there is a stable map $\theta_i : S^{2^{i+1}-2} \to S^0$ which is detected by h_i^2 and has order 2 then there is a stable map $\bar{\theta}_i : S^{2^{i+1}-2} \cup_{2\iota} e^{2^{i+1}-1} \longrightarrow P^\infty$ such that $\bar{\theta}_i^* \neq 0$ in dimension $2^{i+1}-1$ and $\bar{\theta}_i | S^{2^{i+1}-2}$ is detected by $\bar{e}_{2^i-1} h_i$.*

Proof The assumption on θ_i implies the existence of a stable map

$$\hat{\theta}_i : S^{2^{i+1}-2} \cup_{2\iota} e^{2^{i+1}-1} \longrightarrow S^0$$

which is detected on the top cell by h_{i+1} and has $\hat{\theta}_i | S^{2^{i+1}-2} = \theta_i$. This follows from Adams differential $d_2(h_{i+1}) = h_i^2 h_0$ in S^0. Recall the transfer map $t : P^\infty \to S^0$ and the induced map $Ext_A^{s,t}(P^\infty) \xrightarrow{t_*} Ext_A^{s+1,t+1}(S^0)$. It suffices to show $\hat{\theta}_i$ has a lifting $\bar{\theta}_i : S^{2^{i+1}-2} \cup_{2\iota} e^{2^{i+1}-1} \longrightarrow P^\infty$ with respect to t because then $\bar{\theta}_i | S^{2^{i+1}-2}$ is detected by $\bar{e}_{2^i-1} h_i$, since $t_*(\bar{e}_{2^i-1} h_i) = h_i^2$, and $\bar{\theta}_i^* \neq 0$ in dimension $2^{i+1}-1$ since $t_*(\bar{e}_{2^{i+1}-1}) = h_{i+1}$. The existence of such a lifting $\bar{\theta}_i$ follows from Kahn-Priddy Theorem [**9**], and this is described as follows. The stable objects $S^{2^{i+1}-2} \cup_{2\iota} e^{2^{i+1}-1}$, P^∞ and S^0 are actually the suspension spectra $\Sigma^\infty \left(S^{2^{i+1}-2} \cup_{2\iota} e^{2^{i+1}-1}\right)$, $\Sigma^\infty P^\infty$ and $\Sigma^\infty S^0$ where the latter S^0 denotes the space of two points. So the transfer map t should be $\Sigma^\infty P^\infty \xrightarrow{t} \Sigma^\infty S^0$, and $\hat{\theta}_i$ should be $\Sigma^\infty \left(S^{2^{i+1}-2} \cup_{2\iota} e^{2^{i+1}-1}\right) \xrightarrow{\hat{\theta}_i} \Sigma^\infty S^0$. Let $Q = \Omega^\infty \Sigma^\infty$ be the infinite loop space functor on spaces. Let $P^\infty \xrightarrow{t^*} QS^0$ and $S^{2^{i+2}-2} \cup_{2\iota} e^{2^{i+1}-1} \xrightarrow{\hat{\theta}_i^*} QS^0$ be the adjoint maps of t and $\hat{\theta}_i$ respectively. Let $\bar{t}^* : QP^\infty \longrightarrow QS^0$ be the infinite loop map induced by t^*. We may assume $\bar{t}^*$ maps QP^∞ to Q_0S^0, the 0th component. Then Kahn-Priddy Theorem says that there is a map $Q_0S^0 \xrightarrow{j} QP^\infty$ such that, at prime 2, the composite $\bar{t}^* j \simeq id$. We may also assume $\hat{\theta}_i^*$ maps $S^{2^{i+1}-2} \cup_{2\iota} e^{2^{i+1}-1}$ to Q_0S^0. The adjoint map $\Sigma^\infty \left(S^{2^{i+1}-2} \cup_{2\iota} e^{2^{i+1}-1}\right) \xrightarrow{\bar{\theta}_i} \Sigma^\infty P^\infty$ of the composite $S^{2^{i+1}-2} \cup_{2\iota} e^{2^{i+1}-1} \xrightarrow{\hat{\theta}_i^*} Q_0S^0 \xrightarrow{j} QP^\infty$ is then a (homotopy) lifting of $\hat{\theta}_i$ with respect to $\Sigma^\infty P^\infty \xrightarrow{t} \Sigma^\infty S^0$. This proves (6.1).

For $i = 5$ there exists a θ_5 with $2\theta_5 = 0$. This result is due to Mahowald [**3, 9**]. So, by (6.1), there exists a map $\varphi_1 = \bar{\theta}_5 : S^{62} \cup_{2\iota} e^{63} \to P^\infty$ with the properties described. This proves (3.8) (1) since φ_1 clearly can be pulled back to P^{113}.

By dimensional reasons, $S^{62} \cup_{2\iota} e^{63} \xrightarrow{\varphi_1} P^{113}$ can be pulled back up to P^{63}. The resulting map is still denoted by $\varphi_1 : S^{62} \cup_{2\iota} e^{63} \xrightarrow{\varphi_1} P^{63}$. Clearly this φ_1 still has $\varphi_1^* \neq 0$ in dimension 63. Mahowald and Tangora prove in [**16**] that there is a θ_4 with $2\theta_4 = 0$. Again, by (6.1), this implies there is a stable map $\varphi_1' : S^{30} \cup_{2\iota} e^{31} \to P^{31}$ such that $\varphi_1'^* \neq 0$ in dimension 31. The smash product

$$\left(S^{62} \cup_{2\iota} e^{63}\right) \wedge \left(S^{30} \cup_{2\iota} e^{31}\right) \xrightarrow{\varphi_1 \wedge \varphi_1'} P^{63} \wedge P^{31}$$

therefore has $(\varphi_1 \wedge \varphi_1')^* \neq 0$ in dimension 94. We will show the following.

(6.2) There is a stable map $\Sigma P^{63} \wedge P^{31} \xrightarrow{f} P_{65}^{95}$ such that $f^* \neq 0$ in dimension 95.

(6.3) ${}_2\pi_{93}^s\left(P_{65}^{95}\right) = 0$.

Granting these, consider the composite

$$\bar{\varphi}_2 : Y = \left(S^{63} \cup_{2\iota} e^{64}\right) \wedge \left(S^{30} \cup_{2\iota} e^{31}\right) \xrightarrow{\Sigma\varphi_1 \wedge \varphi_1'} \Sigma P^{63} \wedge P^{31} \xrightarrow{f} P_{65}^{95}$$

which has $\bar{\varphi}_2^* \neq 0$ in dimension 95. Note that $Y/S^{93} \simeq S^{94} \vee \left(S^{94} \cup_{2\iota} e^{95}\right)$ where S^{93} is the bottom sphere of Y. By (6.3), $\bar{\varphi}_2 | S^{93} \simeq 0$. So $\bar{\varphi}_2$ factorizes through Y/S^{93}

and the restriction of the factorization map to $S^{94}\cup_{2\iota}e^{95}$ is a map $\tilde{\varphi}_2 : S^{94}\cup_{2\iota}e^{95} \to P_{65}^{95}$ with $\tilde{\varphi}_2^* \neq 0$ in dimension 95. Let $\varphi_2 : S^{94}\cup_{2\iota}e^{95} \xrightarrow{\tilde{\varphi}_2} P_{65}^{95} \xhookrightarrow{i} P_{65}^{113}$. φ_2 still has $\varphi_2^* \neq 0$ in dimension 95 and this implies $\varphi_2|S^{94}$ is detected by $\bar{e}_{79}h_4$ since we already have shown in (4.1) (2) that $d_2(\bar{e}_{95}) = \bar{e}_{79}h_4h_0$ in the ASS for P_{65}^{113}. This proves (3.8)(2).

(6.2) is obtained from a construction of I. M. James on the real Stiefel manifolds $V_{n,k}$ [**8**]. Recall that $V_{n,k}$ is the space of orthogonal k-frames in $\mathbb{R}^n$. For $1 \leq \ell < k$ there is a projection map $p : V_{n,k} \to V_{n,\ell}$ obtained by taking the last ℓ vectors in each k-frame. In particular, we have $p : V_{n,k} \to V_{n,1} = S^{n-1}$. We also recall [**8, 19**] that there is a standard inclusion map $P_{n-k}^{n-1} \xrightarrow{i} V_{n,k}$ such that

(i) $\left(V_{n,k}, P_{n-k}^{n-1}\right)$ is $(2n-2k)$-connected,

(ii) the composite $g : P_{n-k}^{n-1} \xrightarrow{i} V_{n,k} \xrightarrow{p} S^{n-1}$ has $g^* \neq 0$ in dimension $n-1$,

(iii) P_{n-k}^{n-1} is a stable retract of $V_{n,k}$, that is, there is a stable map $V_{n,k} \xrightarrow{r} P_{n-k}^{n-1}$ such that $ri \simeq id$.

Now consider the join $V_{m,k} * V_{n,k}$. James constructs in [**8**] a map

$$h : V_{m,k} * V_{n,k} \longrightarrow V_{m+n,k}$$

which he calls the intrinsic map and has the following properties:

(iv) h is functorial in k; in particular, there is a commutative diagram

$$\begin{array}{ccc} V_{m,k} * V_{n,k} & \xrightarrow{h} & V_{m+n,k} \\ p*p \downarrow & & \downarrow p \\ S^{m-1} * S^{n-1} & \xrightarrow{h} & S^{m+n-1} \end{array}$$

(v) $S^{m-1} * S^{n-1} \xrightarrow{h} S^{m+n-1}$ is a homeomorphism.

Recall that $X * Y$ has the homotopy type of $\Sigma X \wedge Y$. The map we want for (6.2) is the composite of the stable maps

$$\begin{aligned} f : \Sigma P^{63} \wedge P^{31} &\xrightarrow{q\wedge 1} \Sigma P_{33}^{63} \wedge P^{31} \simeq P_{33}^{63} * P^{31} \\ &\xrightarrow{i*i} V_{64,31} * V_{32,31} \xrightarrow{h} V_{96,31} \xrightarrow{r} P_{65}^{95} \end{aligned}$$

where q is the collapsing map. It follows from (i), (ii), (iii), (iv) and (v) above that $f^* \neq 0$ in dimension 95. This proves (6.2).

To prove (6.3) we first note that $H_*\left(P_{65}^{95}\right)$ is isomorphic to $H_*\left(P^{31}\right)$ as an A-module. So $Ext_A^{s,t+64}\left(P_{65}^{95}\right) \cong Ext_A^{s,t}\left(P^{31}\right)$ all s,t. For $x \in Ext_A^{s,t}\left(P^{31}\right)$ the corresponding element in $Ext_A^{s,t+64}\left(P_{65}^{95}\right)$ is also denoted by x. By Mahowald [**15**] we have the following.

(1) $Ext_A^{s,s+29}\left(P^{31}\right) \cong Ext_A^{s,s+93}\left(P_{65}^{95}\right)$ is $\mathbb{Z}_2$, generated by $\hat{k}h_0^i$ for $s = 6+i, 0 \leq i \leq 2$ and is zero for other s. And $\hat{k}h_1 = 0$.

(2) $Ext_A^{s,s+28}\left(P^{31}\right) \cong Ext_A^{s,s+92}\left(P_{65}^{95}\right)$ is $\mathbb{Z}_2$, generated by $p^1\hat{g}h_0^i$ for $s = 7 + i, 0 \leq i \leq 2$ and is zero for other s. And $p^1\hat{g}h_1 = \hat{k}h_0^2$.

(3) $Ext_A^{5,35}\left(P^{31}\right) \cong Ext_A^{5,99}\left(P_{65}^{95}\right) \cong \mathbb{Z}_2\left(\hat{r}\right)$.

We shall prove the following.

(6.4) In the ASS for P_{65}^{95}, $d_2\left(\hat{k}h_0^i\right) = p^1\hat{g}h_0^{i+1}$ for $i = 0, 1$ and $d_3(\hat{r}) = \hat{k}h_0^2$.

This implies $E_\infty^{s,t}\left(P_{65}^{95}\right) = 0$ for $t - s = 93$. So ${}_2\pi_{93}^s\left(P_{65}^{95}\right) = 0$ proving (6.3).

The result (6.4) is true for P^{31} ([15]). But $P_{65}^{95} \not\simeq P^{31}$ [**7**]. So we can not simply deduce (6.4) from the result for P^{31}. To prove (6.4) consider P_{64}^{95} and the map $P_{64}^{95} \xrightarrow{\psi} S^{64}$ with $C_\psi = \Sigma P_{63}^{95}$. As an A-module, $H_*\left(P_{64}^{95}\right) \cong H_*\left(S^{64}\right) \oplus H_*\left(P_{65}^{95}\right)$. So

$$Ext_A^{s,t}\left(P_{64}^{95}\right) \cong Ext_A^{s,t}\left(S^{64}\right) \oplus Ext_A^{s,t}\left(P_{65}^{95}\right).$$

The Ext groups $Ext_A^{s,s+93}(S^{64}) = Ext_A^{s,s+29}(S^0)$, $Ext_A^{s,s+92}\left(S^{64}\right) = Ext_A^{s,s+28}(S^0)$ and $Ext_A^{6,100}(S^{64}) = Ext_A^{6,36}(S^0)$ are isomorphic respectively to $Ext_A^{s-1,s+28}(P^{31})$, $Ext_A^{s-1,s+27}\left(P^{31}\right)$ and $Ext_A^{5,35}\left(P^{31}\right)$ [**15, 20**]. The corresponding elements in $Ext_A^{*,*}$ of the non-zero classes in (1), (2), (3) above are denoted by $kh_0^i, p^1gh_0^i$ and r. We also have that $kh_1 = 0$ and $p^1gh_1 = kh_0^2$, and that $d_2\left(kh_0^i\right) = p^1gh_0^{i+1}$ for $i = 0, 1$ and $d_3(r) = kh_0^2$ in the ASS for S^0 [**16, 20**]. The map $P_{64}^{95} \xrightarrow{\psi} S^{64}$ induces a map

$$\psi_* : Ext_A^{s,t}\left(P_{64}^{95}\right) \cong Ext_A^{s,t}\left(S^{64}\right) \oplus Ext_A^{s,t}\left(P_{65}^{95}\right) \longrightarrow Ext_A^{s+1,t+1}\left(S^{64}\right)$$

in E_2 terms of the Adams spectral sequences which, when restricted to $Ext_A^{*,*}\left(S^{64}\right)$, is the multiplication by h_0 map and, when restricted to $Ext_A^{*,*}\left(P_{65}^{95}\right)$, has $\psi_*(\hat{r}) = r$, $\psi_*\left(\hat{k}h_0^i\right) = kh_0^i$ and $\psi_*\left(p^1\hat{g}h_0^i\right) = p^1gh_0^i$. Since $d_2(k) = p^1gh_0 \neq 0$, $\psi_*\left(\hat{k}\right) = k$ and ψ_* commutes with d_2 it follows that, in the ASS for P_{64}^{95}, $d_2\left(\hat{k}\right) \neq 0$ in $E_2^{8,100}\left(P_{64}^{95}\right) \cong \mathbb{Z}_2(p^1g) \oplus \mathbb{Z}_2\left(p^1\hat{g}h_0\right)$. So $d_2\left(\hat{k}\right) = \epsilon_1 p^1 g + \epsilon_2 p^1\hat{g}h_0 \neq 0$ where $\epsilon_j = 0$ or 1. Then either $\epsilon_1 = 1, \epsilon_2 = 0$ or $\epsilon_1 = 0, \epsilon_2 = 1$ since $\psi_*\left(p^1g\right) = \psi_*\left(p^1\hat{g}h_0\right) = p^1gh_0$ and $d_2(k) = p^1gh_0 \neq 0$. The case $d_2\left(\hat{k}\right) = p^1g$ is impossible because $\hat{k}h_1 = 0$ and $p^1gh_1 = kh_0^2 \neq 0$. Thus we have the following.

(∗) In the ASS for P_{64}^{95}, $d_2(\hat{k}) = p^1\hat{g}h_0$ which implies $d_2(\hat{k}h_0) = p^1\hat{g}h_0^2$.

We also note that, in the ASS for P_{64}^{95}, $d_2(\hat{r}) = 0$ as $d_2(\hat{r}) \in E_2^{7,100}\left(P_{64}^{95}\right) \cong \mathbb{Z}_2(k) \oplus \mathbb{Z}_2(\hat{k}h_0)$ and the set $\{d_2(k) = p^1gh_0, d_2(\hat{k}h_0) = p^1\hat{g}h_0^2\}$ is linearally independent in $E_2^{9,101}\left(P_{64}^{95}\right)$. Since $d_3(r) = kh_0^2$, $\psi_*(\hat{r}) = r$ and ψ_* commutes with d_3 it follows that, in the ASS for P_{64}^{95}, $d_3(\hat{r}) \neq 0$. Since $d_2(kh_0) = p^1gh_0^2$ it follows that $E_3^{8,101}\left(P_{64}^{95}\right) = \mathbb{Z}_2(\hat{k}h_0^2)$. Therefore we conclude the following.

(∗∗) In the ASS for $P_{64}^{95}, d_3(\hat{r}) = \hat{k}h_0^2$.

By considering the collapsing map $P_{64}^{95} \to P_{65}^{95}$ we see (∗) and (∗∗) also hold in P_{65}^{95}. This proves (6.4).

Note added in proof. (3.4)(2)(d) should be $Ext_A^{5,2^8+2^7+2} \cong \mathbb{Z}_2 \oplus \mathbb{Z}_2 \oplus \mathbb{Z}_2$, generated by $h_7^3h_0^2, g_5h_1$ and $h_8D_3(1)$. It is not difficult to show that $d_2(g_5h_1) = d_2(h_8D_3(1)) = 0$ in S^0. These are neglected in the proof on Page 21. So $g_5h_0^3$, $h_8h_6H_1$ and h_8h_0K are indeed linearly independent.

References

[1] Adams, J. F. [1960], *On the non-existence of elements of Hopf invariant one*, Ann. of Math. **72**, 20-104.

[2] Barratt, M. G., Mahowald, M. E. and Tangora, M. C. [1970], *Some differntials in the Adams sppectral sequence II*, Topology **9**, 309-316.

[3] Barratt, M. G., Jones, J. D. S. and Mahowald, M. E. [1984], *Relations amongst Toda brackets and the Kervaire invariant in dimension 62*, J. London Math. Soc. **30**, 533-550.

[4] Barratt, M. G., Jones, J. D. S. and Mahowald, M. E. [1987], *The Kervaire invariant and the Hopf invariant, (H. R. Miller, D. C. Ravenel, eds.)*, Lect. Notes in Math. **1286**, Springer-Verlag, 135-173.

[5] Bousfield, A. K. and Curtis, E. B. [1970], *A spectral sequence for the homotopy of nice spaces*, J. Trans. AMS. **151**, 457-479.

[6] Cohen, R. L., Lin, W. H. and Mahowald, M. E. [1988], *The Adams spectral sequence of the real projective spaces*, Pacific J. Math. **134** No.1, 27-55.

[7] Davis, D. M. and Mahowald, M. E. [1986], *Classification of stunted real projective space*, Pacific J. Math. **125**, 335-345.

[8] James, I. M. [1976], *The Topology of Stiefel manifolds, London Math. Soc. Lecture Notes 24*, Cambridge University Press.

[9] Kahn, D. S. and Priddy, S. B. [1972], *Applications of the transfer to the stable homotopy theory*, Bull. Amer. Math. Soc. **78**, 981-991.

[10] Kochman, S. L. [1990], *Stable Homotopy groups of spheres; A computer-assisted approach*, Lect. Notes in Math. **1423**, Springer-Verlag.

[11] Lin, W. H. [1981], *Algebraic Kahn-Priddy theorem*, Pacific J. Math. **96**, 435-455.

[12] Lin, W. H. [1990], *A conjecture of May on the E_2-term of the May spectral sequence for the cohomology of the Steenrod algebra*, in: Homotopy Theory and Related Topics (M. Mimura, ed.), Lect. Notes in Math. **1418**, Springer-Verlag, 15-52.

[13] Lin, W. H., *Some differentials in Adams spectral sequence for spheres*, Trans. AMS, to appear.

[14] Liulevicius, A. [1962], *The factorization of cyclic reduced powers by secondary cohomology operations*, Mem. AMS. **42**.

[15] Mahowald, M. E. [1967], *The metastable homotopy of S^n*, Memo. AMS. **72**.

[16] Mahowald, M. E. and Tangora, M. C. [1967], *Some differential, in the Adams spectral sequence*, Topology **6**, 349-369.

[17] Mahowald, M. E. [1977], *A new infinite family in ${}_2\pi_*^s$*, Topology **16**, 249-256.

[18] May, J. P. [1964], *The cohomology of restricted Lie algebras and Hopf algebras, applications to the Steenrod algebra*, Ph.D. Thesis, Princeton University.

[19] Steenrod, N. E. (with D. B. A. Epstein) [1962], *Cohomology operations*, Ann. Math. Study. **50**, Princeton University Press.

[20] Tangora, M. C. [1970], *On the cohomology of Steenrod algebra*, Math. Z. **116**, 18-64.

[21] Wang, J. S. P. [1967], *On the cohomology of the mod 2 Steenrod algebra and the nonexistence of element of Hopf invariant one III*, J. Math. **11**, 480-490.

Fields Institute Communications
Volume **19**, 1998

Some Problems about Phantom Maps

C. A. McGibbon
Mathematics Department
Wayne State University
Detroit, Michigan 48202 USA
mcgibbon@math.wayne.edu

Abstract. This article consists of 21 research problems about phantom maps - mostly from an unstable homotopy point of view. Most of these problems are still open. The rest have been solved only recently and their solutions have yet to appear in the literature.

Introduction

This is an updated version of the problem list that appeared in my survey paper on phantom maps in the *Handbook of Algebraic Topology*. Seven of the original fourteen problems have been essentially solved. Some of their solutions are included here – the others will (hopefully) appear in more detail elsewhere. The original problems are listed first in the same order as in [**14**]. Seven new problems have been added at the end of the list.

In this note a phantom map will mean a based map f out of a CW-complex X whose restriction to each n-skeleton X_n is null homotopic. Thus, for our purposes, the only phantom out of a finite dimensional complex is, up to homotopy, the trivial map. The universal phantom map out of X is a map $\Theta : X \to \bigvee_{n=1}^{\infty} \Sigma X_n$ through which all other phantom maps out of X factor, [**9**]. Hence if this map is null homotopic, then all phantom maps out of X are trivial. This is the case, for example, when $X = \Omega S^n$. However, this case appears to be exceptional when one considers the iterated loop spaces of the sphere.

Question 1. *Is the universal phantom map out of $\Omega^k S^n$ essential when localized at each prime p, for each $k \geq 2$, and for each $n \geq 2$?*

It is worth noting that some phantom maps are globally essential and yet locally trivial at every prime. This occurs, for example, among maps $\Omega^2 S^5 \to S^4$. See [**12**] where such maps were called *clones* of the trivial map. It is also known that there are no essential phantom maps from $\Omega^k S^n$ into p-local targets of finite type.[1] Thus

1991 *Mathematics Subject Classification.* 55P99, 55Q05.
[1]See Theorem 2 of [**11**] and note that the nilpotent restriction stated there was never needed in the proof.

Question 1 is a rather delicate problem which necessarily involves spaces not of finite type.

The answer to Question 1 is yes; its solution is joint work with Fred Cohen and Le Minh Ha, [**4**]. To give some of the details, we first need a definition. A space X is said to be *finitely controlled* if there is some natural number n such that any self-map of X which is homotopic to the identity map through dimension n is, in fact, a homotopy equivalence. This notion is less restrictive than that of an *atomic space* (where the number $n = conn(X) + 1$ and every n-equivalence is a homotopy equivalence). An important ingredient in the solution to Question 1 is the following

Lemma *If X is a finitely controlled, p-local, double loop space of finite type, whose homology $H_*(X; \mathbb{Z}/p)$ is not finitely generated as an algebra, then the universal phantom map out of X is essential.*

The homology of $\Omega^k S^n$ is well known for $k \leq n$ and for $k \geq 2$ it is not finitely generated. Moreover, in this range $\Omega^k S^n$ is often atomic at each prime p and those cases where it is not are well understood, by the results in [**3**]. The answer to Question 1 follows from these facts for $2 \leq k \leq n$.

The situation when S^n is looped past its connectivity is quite different. Very little is known about the homology of $\Omega^k S^n$ in this case; even less is known about mod p decompositions of its path components. It is known that every proper connected retract of $\Omega^k S^n$ must have infinitely many nonzero homotopy groups (by the Sullivan conjecture) and infinitely many nonzero homology groups (by standard results about finite H-spaces). It is tempting to speculate that no such retracts exist when n is odd but at present there is little evidence to support this. The answer to Question 1 when $k > n$ involves showing that the irreducible retract of $\Omega^k S^n_{(p)}$ whose homotopy groups contain the unstable image of J must have mod p homology which is too big to be dominated by any finitely generated algebra.

The next four questions deal with phantom maps out of the first loop space of a finite complex K. Under such conditions ΩK is usually an infinite dimensional space. It would be unreasonable to expect the universal phantom map out of ΩK to always be null homotopic as it is in the special case when $K = S^n$ or, more generally, when K is a finite suspension, [**9**]. In fact, it is known that the universal phantom map out of ΩK is essential for certain compact Lie groups K, *ibid.* For all examples currently known, the answer to the following question is yes.

Question 2. *If K is a finite 1-connected complex, does the universal phantom map out of ΩK become trivial when localized at sufficiently large primes?*

This is equivalent to asking if $\Sigma\Omega K$ is homotopy equivalent to a bouquet of finite complexes at all sufficiently large primes by Theorem 3.9 of [**14**]. The answer is known to be yes when K is rationally elliptic; that is, when K has only a finite number of nonzero rational homotopy groups by [**21**]. The answer is also yes when K has the rational homotopy type of a suspension e.g., see [**14**], Example 3.5. David Anick showed this is also true if K has Lusternik-Schnirelmann category two and $H_*(\Omega K; \mathbb{Z})$ has p-torsion for at most a finite number of primes, [**1**].

The following problem was raised about ten years ago in a paper with Clarence Wilkerson, [**21**]. It is still open. An answer of yes would lend some credibility to the previous question.

Question 3. *If K is a 1-connected finite complex, do the Steenrod reduced powers $\mathcal{P}^t$ act trivially on $H^*(\Omega K; \mathbb{Z}/p)$ for all primes sufficiently large?*

The answer is yes for all of the examples mentioned after Question 2.

Question 4. *Does there exist a finite complex K and an essential phantom map from ΩK to a target of finite type?*

As mentioned earlier, if K is a compact Lie group, then the universal phantom map out of ΩK is often essential. However, it is also known that in this special case there are no essential phantom maps from ΩK into targets of finite type, [**9**].

Question 5. *Let G/H be a homogeneous space where G is a compact Lie group and H is a closed subgroup. Is there a map*

$$\Omega(G/H) \longrightarrow \prod_{\alpha} S^{2n_\alpha - 1} \times \prod_{\beta} \Omega S^{2n_\beta + 1}$$

which induces an isomorphism in rational homology?

Question 5 is a special case of the one that preceded it, by Theorem 8.7 in [**14**]. However, the homogeneous spaces provide good test cases. It is possible that techniques needed to handle the homogeneous spaces might also handle the rationally elliptic spaces at small primes. In particular, what is the answer when $G/H = U(2n)/Sp(n)$?

Question 6. *Suppose that X is an arbitrary CW-complex and that Y is group-like. Is the subgroup of phantom maps in $[X, Y]$ necessarily divisible?*

The original version of this question contained a regrettable blunder and asked if the subgroup of phantom maps is abelian under the conditions stated. It is abelian; this is evident from the $\varprojlim^1$ description. This group is also known to be divisible when X and Y have finite type, [**14**], Corollary 5.3. I will show here it is not necessarily divisible when the finite type hypothesis on the target Y is dropped. To this end, let p be a fixed odd prime. It suffices to produce a p-local complex X, of finite type such that
1) The suspension ΣX is not homotopy equivalent to a bouquet of finite complexes, and
2) The space X is a double suspension and, with respect to this co-H-structure, the identity map on X has order p.

The first condition implies that the universal phantom map out of X is essential, by Theorem 3.3 of [**9**]. The second condition forces the functor $[X, \ ]$ to take values in elementary abelian groups. Together the two conditions imply that the group $Ph(X, \vee\Sigma X_n)$ is a nontrivial elementary abelian p-group and hence is not divisible.

To simplify notation, let $W = \Sigma^3 B\mathbb{Z}/p$ and let X denote the cofiber of the pth power map on W with respect to the suspension co-H-structure. This power map induces the trivial homomorphism in mod p cohomology and thus $H^*(X) \approx H^*(W \vee \Sigma W)$, even as modules over the Steenrod algebra. Thus from the $\mathcal{A}_p$ module structure of $H^*(B\mathbb{Z}/p)$ it follows that X satisfies condition 1.

Although this power map on W looks trivial in cohomology it is not null; in fact, the suspension order of the identity on W is infinite. One way to see this is to use the Kahn-Priddy theorem to see that the stable homotopy of $B\mathbb{Z}/p$ has no exponent. However, the pth power map on the cofiber X is null. This can be verified by noticing that

$$X \simeq P^4(p) \wedge B\mathbb{Z}/p$$

where the first factor is a mod p Moore space with top cell in dimension 4. The pth power map on X is homotopic to the pth power map on the Moore space smashed with the identity on the second factor. Since the identity on $P^4(p)$ has suspension order p, the result follows.

Question 7. *Suppose that X and Y are simply connected finite complexes. Is it true that every map from $Q(X)$ to Y is a phantom map?*

As usual, $Q(X) = \mathrm{colim}\Omega^n\Sigma^n X$. The answer to this question turned out to be yes. More generally, it is shown in [**15**] that if E is a connected infinite loop space with a torsion fundamental group, then the function space of based maps $map_*(E, \widehat{Y})$ is weakly contractible. Here $\widehat{Y}$ denotes the profinite completion of Y. The result is equivalent to saying that every map from E, or any iterated suspension of E, to Y is a phantom map of the second kind; that is, one whose restriction to any finite subcomplex of E is null homotopic. When E has finite type the two notions of phantom map agree.

Given a space Y, let $Y^{(n)}$ denote its Postnikov approximation through dimension n. Thus the homotopy groups of $Y^{(n)}$ are naturally isomorphic to those of Y in dimensions $\leq n$ and are zero in higher dimensions.

Question 8. *Given finite type spaces X and Y, define*

$$AP(X,Y) \quad = \quad colim_n Hom([X, \Omega Y^{(n)}], \mathbb{Z}).$$

Then $AP(X,Y)$ is a countable torsion-free abelian group. Can every such group be obtained in this manner?

The group $AP(X,Y)$ is interesting for two reasons. First, this abelian group is free if and only if there are no essential phantom maps from X to Y, [**19**]. Indeed, in certain cases one has a bijection $\mathrm{Ph}(X,Y) \approx \mathrm{Ext}(AP(X,Y),\mathbb{Z})$. Second, it seems like an impossible problem to characterize, even up to pro-isomorphism, those towers which occur as $\{[X, \Omega Y^{(n)}]\}$ for some spaces X and Y. Question 8 is a weaker version of this realization problem.

Richard Steiner and I have some partial results on this problem, [**20**]. We can show that every additive subgroup of $\mathbb{Q}\oplus\mathbb{Q}$ can be realized as $AP(X,Y)$ for suitable X and Y. We let $Y = QS^n$ for some fixed $n \geq 6$. The space X, of course, depends very much on the abelian subgroup in question. When the subgroup has rank two and is not free, we express it as the direct limit of a sequence of free abelian groups of rank two. The space X is then constructed so that the tower $\{[X_k, QS^n]\}$ is pro-isomorphic to the dual of the direct sequence which defined the subgroup. The space X is then an infinite dimensional complex with the rational homotopy type of $S^n \vee S^n$. It is interesting to note (but probably irrelevant to Question 8) that subgroups of $\mathbb{Q}^n$ have yet to be classified up to isomorphism, for $n \geq 3$.

Question 9. *Suppose that X and X' are two finite type domains that are homotopy equivalent after one suspension. Is it then true, for any target Y, that $Ph(X,Y) = *$ if and only if $Ph(X',Y) = *$?*

This problem arose in work with Joe Roitberg, [**18**]. We can show the answer is yes in many special cases. The following problem is the algebraic version of Question 9. It deals with the extent to which $\underleftarrow{\lim}^1 G$ depends upon the actual group structures in the tower. To be more precise, suppose that $\phi : G \to G'$ is a bijection between two towers of groups. Thus for each n there is a commutative diagram

$$\begin{array}{ccc} G_{n+1} & \xrightarrow{\phi_{n+1}} & G'_{n+1} \\ \downarrow{\scriptstyle \pi_n} & & \downarrow{\scriptstyle \pi'_n} \\ G_n & \xrightarrow{\phi_n} & G_n \end{array}$$

where the horizontal maps are bijections, but not necessarily homomorphisms. Assume also that for each n

i) ϕ_{n+1} restricts to an isomorphism between the abelian kernels, $\ker(\pi_n)$ and $\ker(\pi'_n)$, and

ii) ϕ_n induces an isomorphism between the abelian cokernels, $\mathrm{coker}(\pi_n)$ and $\mathrm{coker}(\pi'_n)$.
Here is the problem.

Question 10. *Does it follow that* $\varprojlim^1 G = *$ *if and only if* $\varprojlim^1 G' = *$?

When two spaces X and X' become homotopy equivalent after one suspension, the homotopy equivalence between the suspensions will induce a bijection between the towers $\{[\Sigma X, Y^{(n)}]\}$ and $\{[\Sigma X', Y^{(n)}]\}$, with the two properties just mentioned. This bijection is not necessarily a group isomorphism however. Thus Question 10 is the algebraic version of the problem which preceded it.

Recently, Ioannis Emmanouil has shown that the answer to Question 10 is no, [**7**]. Whether or not the towers he constructs can be realized in the above sense is not known. His towers necessarily consist of uncountably large groups. If the groups G_n are countable then the answer to Question 10 is yes. This can be shown using the Mittag-Leffler condition. Recall that this is a set-theoretic condition which, for towers of countable groups, is equivalent to the vanishing of their $\varprojlim^1$ terms. Bijections between towers clearly preserve this property. Incidentally, Emmanouil has some remarkable new results on the Mittag-Leffler property for towers in [**6**].

Question 11. *Does there exist a tower of finitely generated nilpotent groups* $\{G_n\}$ *with the property that* $\varprojlim^1_n \varprojlim_m G_n^{(m)} \neq *$ *or such that* $\varprojlim^1_c \varprojlim_n G_n^{[c]} \neq *$?

Here $G_n^{(m)}$ denotes the image of G_{m+n} in G_n while $H^{[c]}$ denotes the maximal nilpotent quotient of H of class $\leq c$. This is a problem that Steiner and I encountered in [**19**]. Simply put, we know very little about nonabelian towers and are unsure how complicated they can be.

Recently, Emmanouil has answered the first part of Question 11 affirmatively, [**7**], even among towers of finitely generated abelian groups.

Let G again denote a tower of finitely generated nilpotent groups and consider the function

$$\delta_* : \varprojlim^1 G \longrightarrow \varprojlim^1 \check{G},$$

where $\check{G}$ denotes the local expansion $\prod_p G_{(p)}$. The map δ_* is induced by the diagonal map and localization at each prime. In his thesis, [**25**], Steiner showed that δ_* is surjective under the conditions stated. The main question left open is this:

Question 12. *If the kernel of δ_* is nontrivial, does it follow that $\delta_*^{-1}(y)$ is an infinite set for each y in $\varprojlim^1 \check{G}$?*

The answer is yes when G is a tower of finitely generated abelian groups. Indeed, in this case the function δ_* is a homomorphism whose kernel is a divisible subgroup. See the proof of Theorem 4 of [**12**]. The answer to Question 12 is also known to be yes in certain special nonabelian cases considered in [**19**].

Question 13. *If $Ph(X,Y)$ has a natural abelian group structure, does it follow that the torsion in $Ph(X,Y)$ lies in the kernel of $\delta_* : Ph(X,Y) \to \Pi_p Ph(X, Y_{(p)})$?*

There is an abelian group structure on $\mathrm{Ph}(X,Y)$, for example, when either the domain X is a co-H-space or the target Y is an H-space, and this group structure is natural with respect to a certain class of maps. Theorem 4 of [**12**] shows this is true more generally when X is rationally co-H or Y is rationally an H-space. So under conditions such as these Question 13 asks if a phantom map of finite order necessarily vanishes when localized at any prime.

There were enough simple-minded examples around to suggest the answer might be yes. Indeed, when $\mathrm{colim}_n \mathrm{Hom}([\Sigma X, Y^{(n)}], \mathbb{Z}])$ is isomorphic to a direct sum of rank 1 groups, the answer is yes. However in general the answer is no. Steiner and I show in [**20**] that there exist realizable towers G with p-torsion which maps nontrivially under the map induced by p-localization $\varprojlim^1 G \to \varprojlim^1 G_{(p)}$.

The next question deals with towers of large abelian groups and their $\varprojlim^1$ terms. First recall the finitely generated case. Jensen has shown that when G is a tower of finitely generated abelian groups, $\varprojlim^1 G \approx \mathrm{Ext}(A, \mathbf{Z})$ where A is a countable torsion-free abelian group. The connection between G and A is given by $A = \mathrm{colimHom}(G_n, \mathbf{Z})$. His result shows that the possible values of $\varprojlim^1 G$ are quite restricted. Indeed, $\mathrm{Ext}(A, \mathbf{Z})$ if nontrivial, is a divisible group with a torsion-free summand of cardinality $2^{\aleph_0}$ and, for each prime p, has p-torsion isomorphic to $\oplus \mathbf{Z}/p^\infty$ where the number of summands can either be finite or $2^{\aleph_0}$. Jensen also showed that for any countable torsion-free abelian group A, the group $\mathrm{Ext}(A, \mathbf{Z})$ can be realized as the $\varprojlim^1$ term of an appropriate tower of finitely generated abelian groups, [**10**]. Roitberg has shown that each of these same Ext groups can be realized as groups of phantom maps $\mathrm{Ph}(X,Y)$ for appropriate choices of spaces X and Y, [**24**].

Suppose that the condition that each group in the tower is finitely generated abelian is relaxed to the point that each group is only required to be countable abelian. It is still true that the $\varprojlim^1$ term of such a tower is either zero or uncountably large. However, a nontrivial $\varprojlim^1$ term need no longer be divisible nor must it have a nontrivial torsion-free summand. For an example, consider an $\mathbb{F}_2$-vector space with a countable infinite basis. Take the tower of inclusions whose n-th term is the subspace of vectors whose first $n-1$ coordinates are zero. The $\varprojlim^1$ term in this case is $\Pi^\infty \mathbb{F}_2 / \oplus^\infty \mathbb{F}_2$. This prompts the following problem.

Question 14. *How does one characterize $\varprojlim^1 G$ when all the groups in the abelian tower are countable?*

When all restrictions (except commutativity) are removed from the groups in the tower there is a very nice characterization of the possible $\varprojlim^1$ terms. Warfield and Huber have shown in this case that $\varprojlim^1 G$ is a cotorsion group and conversely that every cotorsion group can be realized as the $\varprojlim^1$ term of a tower of abelian groups, [**26**]. Recall that an abelian group B is cotorsion if and only if $\mathrm{Ext}(\mathbb{Q}, B) = 0$. Thus every finite abelian group is cotorsion as is every countable bounded abelian group; see Fuchs's book, [**8**] Chapter 9. Of course, the simplest example of a group which is *not* cotorsion is the integer $\mathbb{Z}$.

Question 15. *Let X be a space of finite type with $\pi_q X \neq 0$ for some $q > 4$. Does it follow that the universal phantom map out of $\Omega^3 X$ must be essential?*

The reason for stating this problem in terms of three-fold loop spaces is evident if one considers the examples $\Omega^1 S^n$ and $\Omega^2 \mathbb{H}P^\infty$. The techniques used in answering Question 1 suggest that the answer to Question 15 may be yes.

Question 16. *Let G be a 1-connected compact Lie group. Is the universal phantom map out of ΩG essential at each prime p iff G is not p-equivalent to a product of spheres?*

An answer of yes seems likely. We know, for example, that this is true when $G = SU(n)$, [**4**]. However some problems remain in proving this for certain exceptional Lie groups at small primes. Speaking of Lie groups, the following problem was pointed out to me by Haynes Miller.

Question 17. *Let G and H be compact Lie groups. Are there any essential stable phantom maps from BG to BH?*

Thus this question deals with maps $\Sigma^n \underline{BG} \to \underline{BH}$ where $\underline{X}$ denotes the suspension spectrum of X and the degree n could be any integer. To avoid trivialities, assume neither group is finite. Since the rational homologies of BG and BH are concentrated in even degrees, it follows that there are no essential phantom maps of even degrees (positive or negative) from the suspension spectrum of BG to that of BH. This is an easy $\varprojlim^1$ argument. But what about odd degrees? The first nontrivial case is when $H = U(1)$. By a theorem of Graeme Segal, $Q(\mathbb{C}P^\infty) = BU \times F$, where F is a space with finite homotopy groups. This splitting allows one to recast the problem in terms of the K-theory of BG and its skelata. The Mittag-Leffler property[2] of the tower $\{K^*(B_n G)\}$ implies that there are no essential phantom maps of positive degree from the suspension spectrum of BG to that of $\mathbb{C}P^\infty$.

Added later: From this special case it follows that there are no essential stable phantom maps from BG to BH in positive degrees for any compact Lie group H. The complete answer is given by the following:

Theorem *For any two compact Lie groups G and H, the group $Ph(\Sigma^d \underline{BG}, \underline{BH})$, of stable phantom maps of degree d, is zero if d is even, or if $d+1 \geq 0$, or if either group is finite. However, if d is odd with $d+1 < 0$, and if G and H have nonzero rank, then this group of stable phantom maps is isomorphic to a rational vector space of uncountably large rank.*

Details will appear in [**17**].

Question 18. *Is the stable suspension order of the universal phantom map out of X finite when X is a $K(\pi, n)$ with π finite? What about when $X = \Omega Sp(2)$?*

---

[2] Here $B_n G$ denotes the nth term in the Milnor filtration on BG. The Mittag-Leffler property for such towers was demonstrated by Atiyah and Segal, [**2**].

The spaces in question represent some basic examples where the universal phantom map is known to be essential and yet all phantoms from these domains into finite type targets are trivial. I can show that the answer is no for a nontrivial $K(\pi,1)$. In other words, the universal phantom map out of $\Sigma^k B\mathbb{Z}/p$, for example, has infinite order for each $k \geq 1$. This is an easy consequence of the Segal conjecture.

Haynes Miller's solution to the Sullivan conjecture has been a key ingredient in many results about phantom maps. Let us call a discrete group G a Miller group if the function space $map_*(BG,Y)$ is weakly contractible for all finite dimensional based CW-complexes Y.

Problem 19. *Find necessary and sufficient conditions for a group G to be a Miller group.*

Miller's celebrated theorem could be rephrased as saying that locally finite groups are Miller groups. Recall that a group is locally finite if each of its finite subsets generates a finite subgroup. This property is not a necessary condition however. Indeed, if G and H are Miller groups then it is clear that $map_*(BG \vee BH, Y)$ is also weakly contractible for the same class of targets. But $BG \vee BH$ is the classifying space for the free product $G \star H$ by a theorem of J. H. C. Whitehead. So the class of Miller groups is closed under free products. On the other hand, the free product of two groups is never locally finite if both factors are nontrivial.

Since the Eilenberg-MacLane space $K(\mathbb{Q}, 2n+1)$ can be built as an infinite telescope using self maps of the sphere S^{2n+1}, it is a finite dimensional space and hence a target worth considering. This implies that for G to be a Miller group it is necessary that BG have trivial rational homology. While it is easy to see that this condition is not sufficient – use the Kan-Thurston theorem to get a homology equivalence $K(\pi,1) \to M$ where M is a mod p Moore space – it nevertheless raises the interesting

Problem 20. *Find necessary and sufficient conditions on a group G such that $H_*(BG;\mathbb{Q})$ is trivial.*

Obviously, the answer will have to take into account the class of acyclic groups. Just having some nontrivial sufficient conditions would be interesting. I once thought that the existence of a finite exponent for G might be sufficient in this problem, but I have heard rumors that there exist certain free Burnside groups with nontrivial rational homology! I have yet to find any mention of this in the literature, however.

Let us say that a space Y is a *Miller space* if $map_*(B\mathbb{Z}/p, Y)$ is weakly contractible. Thus Miller's theorem could be rephrased as saying that all finite dimensional spaces and their iterated loop spaces are Miller spaces.

Question 21. *Is the localization of a Miller space necessarily a Miller space?*

The answer depends upon the sort of localization used. Here are two examples where the answer is no. The first involves the Postnikov approximation $Y^{(n)}$ of a finite complex Y through dimension n. This can be viewed as the localization of Y with respect to the constant map $f: S^{n+1} \to *$, in the sense of Dror-Farjoun, [**5**]. There are a number of ways to see that $Y^{(n)}$ is not a Miller space in general. One of the most elegant methods for checking this is due to Neisendorfer, [**23**]. He defined a localization functor $\mathcal{L}_p$ with the with the property that $\mathcal{L}_p(Y) = Y$ whenever Y is a p-complete Miller space, and then showed that $\mathcal{L}_p(Y^{(n)}) = *$ whenever Y

is a connected simple space with a finite fundamental group. Later I showed that $\mathcal{L}_p(E) = *$ whenever E is an infinite loop space with a torsion fundamental group, [**16**]. This brings me to the second example which involves the Bousfield localization of a space with respect to p-local K-theory. Here p is an odd prime. Take Y to be the basepoint component of $\Omega^{2n+1}S^{2n+1}$. Then this is a Miller space whose localization Y_K can be taken to be the homotopy fiber of the power map p^n on the p-primary summand of $\mathrm{Im}\, J$. In particular this Y_K is an infinite loop space, and so an application of Neisendorfer localization shows that it is not a Miller space.

Acknowledgements

I thank the referee for his careful reading of the manuscript and for his suggestions which improved it. The revisions were made at the University of Queensland in Brisbane, Australia. I thank my hosts for their hospitality in such beautiful surroundings.

References

[1] Anick, D. [1989], *Homotopy exponents for spaces of category two*, Springer Lecture Notes in Math. **1370**, 24–52.

[2] Atiyah, M. F. and Segal, G. B. [1969], *Equivariant K-theory and completion*, J. Diff. Geometry **3**, 1–18.

[3] Campbell, H. E. A., Peterson, F. P. and Selick, P. S. [1986], *Self-maps of loop spaces, I*, Trans. Amer. Math. Soc. **293**, 1–39.

[4] Cohen, F. R., Ha, Le Minh and McGibbon, C. A., *Phantom maps out of double loop spaces*, in preparation.

[5] Dror-Farjoun, E. [1996], *Cellular Spaces, Null spaces and Homotopy Localization*, Lecture Notes in Math. **1622**, Springer Verlag, Berlin – New York.

[6] Emmanouil, I. [1996], *Mittag-Leffler condition and the vanishing of* $\varprojlim^1$, Topology **35**, 267–271.

[7] Emmanouil, I. [1996], *Some remarks on inverse systems*, unpublished manuscript June 96.

[8] Fuchs, L. [1970], *Infinite Abelian Groups*, Pure and Applied Math. Vol. 36-I, Academic Press, New York.

[9] Gray, B. and McGibbon, C. A. [1993], *Universal phantom maps*, Topology **32**, 371-394.

[10] Jensen, C. U. [1972], *Les foncteurs dérivés de* $\varprojlim$ *et leurs applications en théorie des modules*, Lecture Notes in Math. **254**, Springer Verlag, Berlin – New York.

[11] McGibbon, C. A. [1993], *Loop spaces and phantom maps*, Contemporary Math. **146**, 297–308.

[12] McGibbon, C. A. [1993], *Clones of spaces and maps in homotopy theory*, Comment. Math. Helv. **68**, 263–277.

[13] McGibbon, C. A. [1994], *Wilson spaces and stable splittings of* BT^n, Glasgow Math. J. **36**, 287–290.

[14] McGibbon, C. A. [1995], *Phantom maps*, Chapter 25 in The Handbook of Algebraic Topology, North Holland, Amsterdam.

[15] McGibbon, C. A. [1996], *A note on Miller's theorem about maps out of classifying spaces*, Proc. Amer. Math. Soc. **124**, 3241–3245.

[16] McGibbon, C. A. [1997], *Infinite loop spaces and Neisendorfer localization*, Proc. Amer. Math. Soc. **125**, 309–313.

[17] McGibbon, C. A. [1997], *Stable phantom maps between classifying spaces of compact Lie groups*, submitted March.

[18] McGibbon, C. A. and Roitberg, J. [1992], *Phantom maps and rational equivalences II*, Boletin Soc. Mat. Mexico **37**, 367–381.

[19] McGibbon, C. A. and Steiner, R. [1995], *Some questions about the first derived functor of the inverse limit*, J. Pure and Applied Algebra **103**, 325–340.

[20] McGibbon, C. A. and Steiner, R., *Phantom maps and the towers that determine them*, J. London Math Soc., to appear.

[21] McGibbon, C. A. and Wilkerson, C. W. [1986], *Loop spaces of finite complexes at large primes*, Proc. Amer. Math. Soc. **96**, 698–702.

[22] Miller, H. [1984], *The Sullivan fixed point conjecture on maps from classifying spaces*, Annals of Math. **120**, 39–87.
[23] Neisendorfer, J. A. [1995], *Localization and connected covers of finite complexes*, Contemporary Math. **181**, 385–390.
[23] Roitberg, J. [1994], *Phantom maps and torsion*, Topology and its applications **59**, 261–271.
[24] Steiner, R. J. [1977], *Localization, completion and infinite complexes*, Mathematika **24**, 1–15.
[25] Warfield, Jr., R. B. and Huber, M. [1979], *On the values of the functor* $\varprojlim^1$, Archiv der Math. **33**, 430–436.

Fields Institute Communications
Volume **19**, 1998

On some BP_*BP-Primitive Elements Related to the Kervaire Invariant Problem

Norihiko Minami
Mathematics Department
The University of Alabama
Box 870350
Tuscaloosa, AL USA 35487–0350

In [8], we showed there is no BP_*BP-primitive element in $BP_*B(\mathbb{Z}/2 \oplus \mathbb{Z}/2)$, which corresponds to a Kervaire invariant one element θ_j [1, 2, 5, 6], for $j \geq 5$. With this result in mind, it is natural to wonder whether similar results hold for other classifying spaces like $B\mathbb{Z}/2$, BD_8, and $BO(2)$. In fact, Kahn-Priddy theorem [4], Priddy's theorem [9] (see also [7]), and Snaith-Tornehave theorem [13] always guarantee appropriate lifts of Kervaire invariant one elements to these classifying spaces, whenever they exist:

Lifting Theorem *The following claims are equivalent:*
(i) $h_j^2 \in \operatorname{Ext}_{\mathcal{A}_*}^{2,2^{j+1}}(\mathbb{Z}/2, \mathbb{Z}/2) \cong \mathbb{Z}/2$ *is a permanent cycle in the classical Adams spectral sequence for* S^0.
(ii) *The nontrivial element in* $\operatorname{Ext}_{\mathcal{A}_*}^{1,2^{j+1}-1}(\mathbb{Z}/2, H_*(\mathbb{R}P_+^\infty)) \cong \mathbb{Z}/2$ *is a permanent cycle in the classical Adams spectral sequence for* $\mathbb{R}P_+^\infty$.
(iii) *Let* $t_{2^j-1} \in H_{2^j-1}(\mathbb{R}P_+^\infty)$ *be the generator, and let* $t_{2^j-1} \otimes t_{2^j-1}$ *denote both the corresponding obvious element in* $H_{2^{j+1}-2}(B(\mathbb{Z}/2 \oplus \mathbb{Z}/2)$ *and its image in* $H_{2^{j+1}-2}(BD_8)$, *under a group level embedding* $\mathbb{Z}/2 \oplus \mathbb{Z}/2 \to D_8$. *Then* $t_{2^j-1} \otimes t_{2^j-1} \in H_{2^{j+1}-2}(BD_8)$ *is in the mod-2 stable Hurewicz image.*
(iv) *Let* $t_{2^j-1} * t_{2^j-1}$ *denote the image of* $t_{2^j-1} \otimes t_{2^j-1}$ *in* $H_{2^{j+1}-2}(BO(2))$, *under the group level diagonal embedding* $\mathbb{Z}/2 \oplus \mathbb{Z}/2 \to O(2)$. *Then* $t_{2^j-1} * t_{2^j-1} \in H_{2^{j+1}-2}(BO(2))$ *is in the mod-2 stable Hurewicz image.* □

Here, $\mathcal{A}_*$ is the 2 primary dual Steenrod algebra and $H_*(X)$ always stands for the mod 2 homology of X.

Recently, attempts have been made to settle the Kervaire invariant problem by showing (iii) in the above Lifting Theorem is impossible through BP_*BP calculations. Now the purpose of this note is to show there *do exist* certain $\operatorname{Ext}_{BP_*BP}$

1991 *Mathematics Subject Classification.* Primary 55Q10, 55Q45, 55T15; Secondary 55R12, 55R35, 55N22, 57R77.

elements for these classifying spaces, which correspond to Kervaire invariant one elements. This indicates that we cannot disprove the existence of a Kervaire invariant one element θ_j by such a naive BP_*BP calculation alone.

The key ingredient is the following result of Ravenel [10], Lemma 6:

Lemma R *Let $b_n \in BP_{2n}(\mathbb{C}P_+^\infty)$ be the canonical generator, and let $x_{i,1} = b_{2^i}\prod_{0\le t<i}\left(b_{2^t} - v_1^{2^t}\right) \in BP_{2^{i+2}-2}(\mathbb{C}P_+^\infty)$. Then the following two properties hold:*

1. *Write $BP_*\mathbb{C}P_+^\infty/(2)$ for the mod 2 reduction $BP_*\mathbb{C}P_+^\infty/2BP_*\mathbb{C}P_+^\infty$. Then $x_{i,1}$, considered as an element of $BP_{2^{i+2}-2}\mathbb{C}P_+^\infty/(2)$, is BP_*BP-primitive.*
2. *$x_{i,1}$ projects to the nonzero element in $H_{2^{i+2}-2}(\mathbb{C}P_+^\infty)$.* □

From this, we immediately obtain our main result for $B\mathbb{Z}/2$:

Corollary 1 *For $i \ge 2$, The following composite sends $x_{i,1}$ to h_{i+1}^2 :*

$$\mathrm{Ext}_{BP_*BP}^{0,2^{i+2}-2}(BP_*, BP_*\mathbb{C}P_+^\infty/(2)) \cong \mathrm{Ext}_{BP_*BP}^{0,2^{i+2}-1}(BP_*, BP_*\Sigma\mathbb{C}P_+^\infty/(2)) \xrightarrow{\beta}$$

$$\mathrm{Ext}_{BP_*BP}^{1,2^{i+2}-1}(BP_*, BP_*\Sigma\mathbb{C}P_+^\infty) \xrightarrow{t_1} \mathrm{Ext}_{BP_*BP}^{1,2^{i+2}-1}(BP_*, BP_*\mathbb{R}P_+^\infty) \xrightarrow{t_2}$$

$$\mathrm{Ext}_{BP_*BP}^{2,2^{i+2}}(BP_*, BP_*) \to \mathrm{Ext}_{\mathcal{A}_*}^{2,2^{i+2}}(\mathbb{Z}/2, \mathbb{Z}/2) \cong \mathbb{Z}/2\{h_{j+1}^2\},$$

where the first map $\cong$ is the suspension isomorphism, the second map β is the Bockstein, the third map t_1 is induced by the S^1-transfer associated to the S^1-fiber bundle $\mathbb{R}P^\infty \to \mathbb{C}P^\infty$, the fourth map t_2 is induced by the Kahn-Priddy map, and the fifth map is the Thom reduction.

Proof Applying the geometric boundary theorem [3], we see the composite $t_2 \circ t_1 \circ \beta$ is essentially the composite

$$\mathrm{Ext}_{BP_*BP}^{0,2^{i+2}-1}(BP_*, BP_*\Sigma\mathbb{C}P_+^\infty/(2)) \xrightarrow{t_2\circ t_1} \mathrm{Ext}_{BP_*BP}^{1,2^{i+2}}(BP_*, BP_*/(2)) \xrightarrow{\beta} \mathrm{Ext}_{BP_*BP}^{2,2^{i+2}}(BP_*, BP_*).$$

From the property (2) of $x_{i,1}$ in Lemma R, we see $t_2 \circ t_1(x_{i,1})$ is a Hopf invariant one element in $\mathrm{Ext}_{BP_*BP}^{1,2^{i+2}}(BP_*, BP_*/(2))$. However, for $i \ge 2$, there is no Hopf invariant one element in $\mathrm{Ext}_{BP_*BP}^{1,2^{i+2}}(BP_*, BP_*)$, and so any Hopf invariant one element in $\mathrm{Ext}_{BP_*BP}^{1,2^{i+2}}(BP_*, BP_*/(2))$ is sent to a nontrivial element in $\mathrm{Ext}_{BP_*BP}^{2,2^{i+2}}(BP_*, BP_*)$. This nontrivial element turns out to be a Kervaire invariant one element, i.e., whose Thom reduction image is h_{i+1}^2 (cf. [1, 11, 12]). This completes the proof. □

We now turn our attention to the cases of BD_8 and $BO(2)$.

Lemma 2 *For any $x, y \in BP_{odd}(\Sigma\mathbb{C}P_+^\infty)$, we have $x \otimes y + y \otimes x = 0$ in $BP_{even}(S_+^1 \wedge_{\mathbb{Z}/2} (\Sigma\mathbb{C}P_+^\infty \wedge \Sigma\mathbb{C}P_+^\infty))$.*

Proof Let σ generate $\mathbb{Z}/2$. Then we have a cofiber sequence

$$\Sigma\mathbb{C}P^\infty_+ \wedge \Sigma\mathbb{C}P^\infty_+ \xrightarrow{1-\sigma} \Sigma\mathbb{C}P^\infty_+ \wedge \Sigma\mathbb{C}P^\infty_+ \to S^1_+ \wedge_{\mathbb{Z}/2} \left(\Sigma\mathbb{C}P^\infty_+ \wedge \Sigma\mathbb{C}P^\infty_+\right).$$

So the claim follows from $(1-\sigma)_*(x\otimes y) = x\otimes y-(-1)^{|x||y|}y\otimes x = x\otimes y+y\otimes x$. □

Proposition 3 *The element*

$$\Theta_i := (\Sigma x_{i-1,1}) \otimes (\Sigma x_{i-1,1}) \in BP_{2^{i+1}-2}(S^1_+ \wedge_{\mathbb{Z}/2} (\Sigma\mathbb{C}P^\infty_+ \wedge \Sigma\mathbb{C}P^\infty_+))$$

is an order 2 BP_*BP *primitive element.*

Proof For a spectrum X, let $\psi_X : BP_*(X) \to BP_*(X) \otimes_{BP_*} BP_*BP$ be its right coaction. Then, by Lemma R, we have $\psi_{\Sigma\mathbb{C}P^\infty_+}(\Sigma x_{i-1,1}) = \Sigma x_{i-1,1} \otimes 1 + 2\sum_\lambda p_\lambda \otimes u_\lambda$, where p_λ is an odd dimensional element in $BP_*(\Sigma\mathbb{C}P^\infty_+)$ and u_λ is a positive even dimensional element in BP_*BP. From this, we immediately get

$$\psi_{S^1_+\wedge_{\mathbb{Z}/2}(\Sigma\mathbb{C}P^\infty_+\wedge\Sigma\mathbb{C}P^\infty_+)}(\Theta_i) - \Theta_i \otimes 1 = 2\sum_\lambda(\Sigma x_{i-1,1} \otimes p_\lambda + p_\lambda \otimes \Sigma x_{i-1,1}) \otimes u_\lambda +$$
$$4\sum_{\lambda\neq\mu}(p_\lambda \otimes p_\mu + p_\mu \otimes p_\lambda) \otimes u_\lambda u_\mu + 2\sum_\lambda(2p_\lambda \otimes p_\lambda) \otimes u_\lambda^2.$$

However, this is certainly 0 by Lemma 2, and so Θ_i is primitive. Of course, $2\Theta_i = 0$ by Lemma 2. So the claim follows. □

Now we are ready to prove our main result for BD_8 and $BO(2)$:

Corollary 4 (i) *For any one of the following spectra, there exists an order* 2 *BP_*BP-primitive element which corresponds to a Kervaire invariant one element*

$$\begin{array}{ccccc} S^1_+ \wedge_{\mathbb{Z}/2} \left(\Sigma\mathbb{C}P^\infty_+ \wedge \Sigma\mathbb{C}P^\infty_+\right) & \xrightarrow{1\wedge t_1\wedge t_1} & S^1_+ \wedge_{\mathbb{Z}/2} \left(\mathbb{R}P^\infty_+ \wedge \mathbb{R}P^\infty_+\right) & & \\ \downarrow & & \downarrow & & \\ D_2(\Sigma\mathbb{C}P^\infty_+) & \xrightarrow{D_2(t_1)} & D_2(\mathbb{R}P^\infty_+) = BD_{8+} & \xrightarrow{Bf} & BO(2)_+, \end{array}$$

where all the unnamed maps are obvious ones, and f is chosen so that the composite $B(\mathbb{Z}/2 \oplus \mathbb{Z}/2)_+ = \mathbb{R}P^\infty_+ \wedge \mathbb{R}P^\infty_+ \to S^1_+ \wedge_{\mathbb{Z}/2} \left(\mathbb{R}P^\infty_+ \wedge \mathbb{R}P^\infty_+\right) \to D_2(\mathbb{R}P^\infty_+) = BD_{8+} \xrightarrow{Bf} BO(2)_+$ is induced by the diagonal embedding $\mathbb{Z}/2 \oplus \mathbb{Z}/2 \to O(2)$.
(ii) *Let Y be any one of the above spectra, and let $Y^{(2^{i+1}-2)}$ be its ($2^{i+1} - 2$)-th skeleton. Then the BP_*BP-primitive element constructed in* (i) *always lifts to a BP_*BP-primitive element of $BP_{2^{i+1}-2}(Y^{(2^{i+1}-2)})$.*

Proof (i) This follows immediately from Lifting Theorem and Proposition 3. (ii) Of course, a lift always exists. To see why any lift is BP_*BP primitive, just observe that the map $BP_*(Y^{(2^{i+1}-2)}) \to BP_*(Y)$ is an isomorphism for $* \leq 2^{i+1} - 2 - 1$. □

To conclude, it now looks much more difficult to disprove the existence of a Kervaire invariant one element θ_j (if that were ever possible) than when the author wrote [8].

References

[1] Barratt, M. G., Jones, J. D. S. and Mahowald, M. E. [1983] *The Kervaire invariant problem* Contemp. Math. Amer. Math. Soc. **19**, 9–23.

[2] Browder, W. [1969] *The Kervaire invariant of a framed manifold and its generalization* Annals of Math. **90**, 157–186.

[3] Johnson, D. C., Miller, H. R., Wilson, W. S. and Zahler, R. S. [1975] *Boundary homomorphisms in the generalized Adams spectral sequence and the non-triviality of infinitely many γ_t in stable homotopy*, Reunion sobre teorie de homotopia, Soc. Mat. Mexicana, pp. 47–59.

[4] Kahn, D. S. and Priddy S. B. [1972] *Applications of the transfer to stable homotopy theory* Bull. Amer. Math. Soc. **78**, 981–987.

[5] Kervaire, M. A. [1960] *A manifold which does not admit any differentiable structure* Comment. Math. Helv. **34**, 256–270.

[6] Kervaire, M. A. and Milnor, J. W. [1963] *Groups of homotopy spheres-I* Ann. of Math. **77**, 504–537.

[7] Minami, N. [1991] *Multiplicative homology operations and transfer* Memoirs. Amer. Soc. **457**.

[8] Minami, N. [1995] *The Kervaire invariant one element and the double transfer* Topology **34**, 481-488.

[9] Priddy, S. B. [1978] *Homotopy splittings of G and G/O* Comm. Math. Helv. **53**, 470–484.

[10] Ravenel, D. C. [1982] *On primitives in $BP_*\mathbb{C}P^\infty/I_n$*, Canadian Mathematical Society Conference Proceedings, vol. 2, part 1, pp. 245–256.

[11] Ravenel, D. C. [1986] *Complex Cobordism and Stable Homotopy Groups of Spheres*, Academic Press, Orlando, Florida.

[12] Shimomura, K. [1981] *Novikov's* Ext^2 *at the prime* 2. Hiroshima Math. J. **11**, 499-513.

[13] Snaith, V. and Tornehave, J. [1982] *On $\pi_*^s(BO)$ and the Arf invariant of framed manifolds* Contemp. Math. **12**, 299–313.

Fields Institute Communications
Volume **19**, 1998

Deterministic p-Compact Groups

Jesper M. Møller
Matematisk Institut
Universitet København
Universitetsparken 5
DK–2100 København Ø Danmark
moller@math.ku.dk

Abstract. We investigate the class of N-determined p-compact groups and the class of p-compact groups with N-determined automorphisms. A p-compact group is said to be N-determined if it is determined up to isomorphism by the normalizer of a maximal torus. The automorphisms of a p-compact group are said to be N-determined if they are determined by their restrictions to this maximal torus normalizer.

1 Introduction

The p-compact groups, introduced by Dwyer and Wilkerson [**5**] as homotopy theoretic Lie groups, have indeed turned out to possess an interesting and rich structure [**17, 7, 8, 18**]. For instance, any p-compact group is equipped with a maximal torus supporting a Weyl group action and the Borel construction of this action provides the normalizer of the maximal torus. The theme of this paper is the interrelation between this maximal torus normalizer and the p-compact group itself. Specifically, I'd like to discuss whether

- automorphisms of p-compact groups are determined by their restrictions to the maximal torus normalizer.
- p-compact groups are determined up to isomorphism by their maximal torus normalizer.

The first of these two problems will be reduced to a computation of two obstruction groups. The second problem, cf. [**13**, 5.2], [**20**, 5.20], will be related to the computation of yet another two obstruction groups. The main technical tools are the induction principle based on the homology decomposition theorem of Dwyer and Wilkerson [**7**, 8.1, 9.2] and the preferred lifts of [**16**] of monomorphisms of elementary abelian p-groups into p-compact groups.

To be more explicit, let X be a p-compact group (for the sake of this introduction assumed to be connected) and let $N \longrightarrow X$ be the normalizer of a maximal torus.

1991 *Mathematics Subject Classification.* 55P35, 55P10, 55P15.

Consider the groups $\mathrm{Out}(X)$ of homotopy classes of self-homotopy equivalences of BX and $\mathrm{Out}(N)$ of homotopy classes of self-homotopy equivalences of BN. There is a homomorphism

$$N\colon \mathrm{Out}(X) \to \mathrm{Out}(N) \tag{1.1}$$

defined by restricting automorphisms of X to N. This homomorphism is set up in Section 3. If (1.1) is injective, we say that X has N-determined automorphisms. Section 4 contains information about the class of p-compact groups with N-determined automorphisms. It is shown, for instance, that if the adjoint form $X/Z(X)$ has N-determined automorphisms, so does X itself. The conjecture that all p-compact groups have N-determined automorphisms is reduced to the computation of the two Wojtkowiak obstruction groups of (2.1).

Now let X' be another connected p-compact group and assume that also X' admits a monomorphism $N \longrightarrow X'$ that is conjugate to the normalizer of a maximal torus. Does it follow that X and X' are isomorphic? If this is so for all choices of X', we say that X is N-determined. I think of this as an alternative and perhaps more geometric version of uniqueness than the cohomological uniqueness of e.g., [**3, 4, 19, 1**]. Section 5 contains a discussion of determinism (at primes $p > 2$). It is shown, for instance, that if the adjoint form $X/Z(X)$ is N-determined, so is X itself. The conjecture that in fact all p-compact groups are N-determined is related to the computation of the two Wojtkowiak obstruction groups from (2.2).

This paper should clarify the very last sentence of my Research Report [**18**].

2 Obstruction theory

Let X be a p-compact group and $\mathbb{A}(X)$ the category whose objects are pairs (V, ν) where V is a nontrivial elementary abelian p-group and $\nu\colon V \to X$ a conjugacy class of monomorphisms. A morphism $g\colon (V_1, \nu_1) \to (V_2, \nu_2)$ in this category is an injection $g\colon V_1 \to V_2$ with $\nu_2 \circ g$ conjugate to ν_1.

Consider the functor BZ from $\mathbb{A}(X)$ to topological spaces given by

$$BZ(V, \nu) = \mathrm{map}(BC_X(\nu), BX)_{Be(\nu)}$$

where $Be(\nu)\colon BC_X(\nu) = \mathrm{map}(BV, BX)_{B\nu} \to BX$ is the evaluation monomorphism. By centricity [**2**], the space $BZ(V, \nu)$ is homotopy equivalent [**7**, 1.3] to the classifying space $\mathrm{map}(BC_X(\nu), BC_X(\nu))_{B1}$ of the center $Z(C_X(\nu))$ of the centralizer $C_X(\nu)$.

The $H\mathbb{F}_p$-equivalence $\mathrm{hocolim}_{\mathbb{A}(X)} BC_X(\nu) \longrightarrow BX$ of the Homology Decomposition Theorem [**7**, 8.1] induces for any p-compact group X' a map

$$[BX, BX'] \longrightarrow \lim^0_{\mathbb{A}(X)}[BC_X(\nu), BX']$$

which, however, may be neither injective nor surjective. Nonetheless, Wojtkowiak's obstruction theory [**22**] provides some information.

Lemma 2.1 *Let $f\colon X \to X$ be an endomorphism of the p-compact group X such that $f \circ e(\nu)$ and $e(\nu)$ are conjugate homomorphisms $C_X(\nu) \longrightarrow X$ for each object (V, ν) of $\mathbb{A}(X)$. If*

$$\lim^1_{\mathbb{A}(X)}\pi_1(BZ(V, \nu)) = 0 = \lim^2_{\mathbb{A}(X)}\pi_2(BZ(V, \nu))$$

then f is conjugate to the identity.

Lemma 2.2 *Let* $(Bf(\nu))_{\nu\in\mathbb{A}(X)} \in \lim^0[BC_X(\nu), BX']$ *be a collection of centric morphisms. If*

$$\lim^2_{\mathbb{A}(X)} \pi_1(BZ(V,\nu)) = 0 = \lim^3_{\mathbb{A}(X)} \pi_2(BZ(V,\nu))$$

then there exists a morphism $f\colon X \to X'$ *such that* $f \circ e(\nu) \simeq f(\nu)$ *for all objects* (V,ν) *of* $\mathbb{A}(X)$.

The morphism $f(\nu)\colon C_X(\nu) \to X'$ is centric if composition with $Bf(\nu)$ is a homotopy equivalence $\mathrm{map}(BC_X(\nu), BC_X(\nu))_{B1} \longrightarrow \mathrm{map}(BC_X(\nu), BX')_{Bf(\nu)}$.

3 Normalizer homomorphisms

The aim of this section is to relate the automorphism group of a p-compact group to that of the normalizer of a maximal torus.

For any loop space BG, let $\mathrm{Aut}(G)$ denote the group of invertible elements of the monoid $[BG, *; BG]$ of based homotopy classes of self-maps and let $\mathrm{Out}(G)$ denote the group of invertible elements of the monoid $[BG; BG]$ of free homotopy classes of self-maps of BG. There is an exact sequence

$$\pi_0(G) \longrightarrow \mathrm{Aut}(G) \longrightarrow \mathrm{Out}(G) \longrightarrow 1$$

reducing to an isomorphism $\mathrm{Aut}(G) \cong \mathrm{Out}(G)$ if G is connected.

Define $BZ(G) := \mathrm{map}(BG, BG)_{B1}$, the *center* of G. Any morphism $f\colon G \to H$ of loop spaces induces post- and precompostion maps

$$BZ(G) = \mathrm{map}(BG, BG)_{B1} \xrightarrow{Bf} \mathrm{map}(BG, BH)_{Bf} \xleftarrow{\overline{Bf}} \mathrm{map}(BH, BH)_{B1} = BZ(H)$$

between mapping spaces. Provided f is *centric*, i.e., $\underline{Bf}$ is a homotopy equivalence, f determines a homomorphism of centers

$$Z(f)\colon Z(H) \to Z(G)$$

given by $\underline{Bf} \circ BZ(f) \simeq \overline{Bf}$. If the square of centric maps

$$\begin{array}{ccc} G & \xrightarrow{g} & G \\ {\scriptstyle f}\downarrow & & \downarrow{\scriptstyle f} \\ H & \xrightarrow[h]{} & H \end{array}$$

commutes up to conjugacy, then the induced square of centers

$$\begin{array}{ccc} Z(G) & \xleftarrow{Z(g)} & Z(G) \\ {\scriptstyle Z(f)}\uparrow & & \uparrow{\scriptstyle Z(f)} \\ Z(H) & \xleftarrow[Z(h)]{} & Z(H) \end{array}$$

also commutes up to conjugacy. In particular, we obtain when $H = G$ an antihomomorphism

$$Z\colon \mathrm{Out}(G) \to \mathrm{Out}(Z(G))$$

between groups of outer automorphisms.

Suppose that X is a *connected* p-compact group. Let $j\colon N \to X$ be the normalizer [7, 9.8] of a maximal torus $i\colon T \to X$ such that $Bj\colon BN \to BX$ is a fibration.

Let $f \in \mathrm{Out}(X)$ be an outer automorphism and consider the space $(X/N)^{hN}$ of lifts

$$\begin{array}{ccccc} & & & & BN \\ & & & \nearrow & \downarrow{\scriptstyle Bj} \\ BN & \xrightarrow[Bj]{} & BX & \xrightarrow[Bf]{} & BX \end{array}$$

of $Bf \circ Bj$ over Bj.

Lemma 3.1 [**14**, 5.1] *$(X/N)^{hN}$ is contractible.*

According to this lemma, there exists a homomorphism

$$N\colon \mathrm{Out}(X) \to \mathrm{Out}(N) \tag{3.2}$$

which to any $f \in \mathrm{Out}(X)$ associates the unique outer automorphism $N(f) \in \mathrm{Out}(N)$ that makes

$$\begin{array}{ccc} N & \xrightarrow{N(f)} & N \\ {\scriptstyle j}\downarrow & & \downarrow{\scriptstyle j} \\ X & \xrightarrow[f]{} & X \end{array} \tag{3.3}$$

commute up to conjugacy. Note also that

$$\underline{Bj}\colon \mathrm{map}(BN, BN)_{BN(f)} \to \mathrm{map}(BN, BX)_{Bf \circ Bj}$$

is a homotopy equivalence and, in particular, that j is centric and induces a homomorphism $Z(j)$ between centers.

Lemma 3.4 [**7**, 7.1] *$Z(j)\colon Z(X) \to Z(N)$ is an isomorphism if p is odd.*

Proof It follows from Shapiro's lemma,

$$\mathrm{map}(BN, BN) = \mathrm{map}(BT_{hW_T(X)}, BN) = \mathrm{map}(BT, BN)^{hW_T(X)},$$

and the homotopy equivalence [**7**, 10.6]

$$BC_T(T)^{hW_T(X)} \longrightarrow BC_N(T)^{hW_T(X)} = \mathrm{map}(BT, BN)_{Bi}^{hW_T(X)}$$

induced from the $W_T(X)$-equivariant homotopy equivalence $BC_T(T) \longrightarrow BC_N(T)$ [**16**, 2.6], [**5**, 9.1] that $BZ(N) = \mathrm{map}(BN, BN)_{B1}$ is homotopy equivalent to the identity component of $BC_T(T)^{hW_T(X)}$. Thus (the loop space of) the identity component of $BC_{\check{T}}(\check{T})^{hW_T(X)}$, where $\check{T} \longrightarrow T$ is a discrete approximation [**5**, 6.4], is a discrete approximation to $Z(N)$ and also to $Z(X)$ by [**7**, Section 7]. □

We now turn to the case of a possibly *nonconnected* p-compact group.

Let π be a finite group and assume that G is a π-loop space, i.e., a fibration $BG \longrightarrow BG_{h\pi} \longrightarrow B\pi$ of based spaces over $B\pi$ with fibre BG. A π-automorphism of the π-loop space G is a fibre homotopy equivalence of the form

$$\begin{array}{ccc} BG_{h\pi} & \xrightarrow{Bf_{h\chi}} & BG_{h\pi} \\ \downarrow & & \downarrow \\ B\pi & \xrightarrow[B\chi]{} & B\pi \end{array}$$

where χ is a group automorphism of π. We say that two such fibre maps, $(Bf_{h\chi}, B\chi)$ and $(Bf'_{h\chi'}, B\chi')$, are homotopic if $\chi = \chi'$ and $Bf_{h\chi}$ and $Bf'_{h\chi'}$ are homotopic over $B\chi = B\chi'$. Let $\mathrm{Aut}_{h\pi}(G)$ denote the group of homotopy classes of π-automorphisms of the π-loop space G.

This new automorphism group is related to the ones previously introduced.

Remark 3.5 Let G be a π-loop space.

- $BZ(G) := \mathrm{map}(BG, BG)_{B1}$ is a π-space and there is, cf. [**15**, 5.1, 5.2], a short exact sequence of groups

$$1 \longrightarrow \pi_0(BZ(G)^{h\pi}) \longrightarrow \underset{h\pi}{\mathrm{Aut}}(G) \longrightarrow \mathrm{Aut}(\pi) \times \mathrm{Out}(G)$$

  where $\pi_0(BZ(G))^{h\pi}$ is the group of fibre homotopy classes of self-maps of $BG_{h\pi}$ over $B\pi$ and under self-maps homotopic to the identity of BG. In case G is a p-compact group or an extended p-compact torus and $\check{Z}(G) \longrightarrow Z(G)$ a discrete approximation, this group identifies to the cohomology group $H^1(\pi; \check{Z}(G))$.
- There exists a commutative diagram

$$\begin{array}{ccccccccc}
1 & \longrightarrow & \pi_0(G) & \longrightarrow & \pi_0(G_{h\pi}) & \longrightarrow & \pi & \longrightarrow & 1 \\
 & & \| & & \downarrow & & \downarrow & & \\
 & & \pi_0(G) & \longrightarrow & \mathrm{Aut}(G_{h\pi}) & \longrightarrow & \mathrm{Aut}_{h\pi}(G) & \longrightarrow & 1 \\
 & & & & \downarrow & & \downarrow & & \\
 & & & & \mathrm{Out}(G_{h\pi}) & = & \mathrm{Out}(G_{h\pi}) & & \\
 & & & & \downarrow & & \downarrow & & \\
 & & & & 1 & & 1 & &
\end{array}$$

  with exact rows and columns. In particular, $\mathrm{Aut}(G_{h\pi}) \cong \mathrm{Aut}_{h\pi}(G)$ if G is connected.

Consider a p-compact group $X_{h\pi}$ with identity component X and component group π. The normalizer $N_{h\pi}$ of the maximal torus $T \longrightarrow X \longrightarrow X_{h\pi}$ is related to the normalizer N of the maximal torus $T \longrightarrow X$ by the commmutative diagram

$$\begin{array}{ccccc}
N & \longrightarrow & N_{h\pi} & \longrightarrow & \pi \\
{\scriptstyle j}\downarrow & & \downarrow{\scriptstyle j_{h\pi}} & & \| \\
X & \longrightarrow & X_{h\pi} & \longrightarrow & \pi
\end{array}$$

with exact rows. Thus both N and X are π-loop spaces and $j \colon N \to X$ is a π-loop space homomorphism.

Let $f_{h\chi} \in \mathrm{Aut}_{h\pi}(X)$ be a π-loop space automorphism and $(X_{h\pi}/N_{h\pi})^{hN_{h\pi}}$ the space of lifts

$$\begin{array}{ccccc}
 & & & \dashrightarrow & BN_{h\pi} \\
 & & & & \downarrow{\scriptstyle Bj_{h\pi}} \\
BN_{h\pi} & \xrightarrow[Bj_{h\pi}]{} & BX_{h\pi} & \xrightarrow[Bf_{h\chi}]{} & BX_{h\pi}
\end{array}$$

of $Bf_{h\chi} \circ Bj_{h\pi}$ over $Bj_{h\pi}$.

Lemma 3.6 *$(X_{h\pi}/N_{h\pi})^{hN_{h\pi}}$ is contractible.*

Proof We have

$$(X_{h\pi}/N_{h\pi})^{hN_{h\pi}} \simeq (X_{h\pi}/N_{h\pi})^{hN})^{h\pi} \simeq ((X/N)^{hN})^{h\pi} \simeq *$$

since $(X/N)^{hN}$ is contractible by (3.1). □

Lemma 3.6 implies that there exists a homomorphism

$$N_{h\pi}\colon \operatorname{Aut}_{h\pi}(X) \to \operatorname{Aut}_{h\pi}(N) \tag{3.7}$$

that to any $f_{hX} \in \operatorname{Aut}_{h\pi}(X)$ associates the unique $N(f)_{hX} \in \operatorname{Aut}_{h\pi}(N)$ such that

$$\begin{array}{ccc} BN_{h\pi} & \xrightarrow{BN(f)_{hX}} & BN_{h\pi} \\ {\scriptstyle Bj_{h\pi}}\downarrow & & \downarrow{\scriptstyle Bj_{h\pi}} \\ BX_{h\pi} & \xrightarrow[Bf_{hX}]{} & BX_{h\pi} \end{array}$$

commutes up to vertical homotopy. In (3.7), $\operatorname{Aut}_{h\pi}(X) \cong \operatorname{Aut}(X_{h\pi})$ and $\operatorname{Aut}_{h\pi}(N) \cong W_T(X_0)\backslash \operatorname{Aut}(N_{h\pi})$.

Ignoring the π-structure, (3.6) also determines a homomorphism

$$N\colon \operatorname{Out}(X_{h\pi}) \to \operatorname{Out}(N_{h\pi}) \tag{3.8}$$

such that $N(f)$ for any $f \in \operatorname{Out}(X_{h\pi})$ is the unique outer automorphism of $N_{h\pi}$ that makes

$$\begin{array}{ccc} BN_{h\pi} & \xrightarrow{BN(f)} & BN_{h\pi} \\ {\scriptstyle Bj_{h\pi}}\downarrow & & \downarrow{\scriptstyle Bj_{h\pi}} \\ BX_{h\pi} & \xrightarrow[Bf]{} & BX_{h\pi} \end{array}$$

commutative up to homotopy. The homomorphisms (3.7) and (3.8) are related by a commutative diagram

$$\begin{array}{ccccccc} \pi & \longrightarrow & \operatorname{Aut}_{h\pi}(X) & \longrightarrow & \operatorname{Out}(X_{h\pi}) & \longrightarrow & 1 \\ \| & & {\scriptstyle N_{h\pi}}\downarrow & & {\scriptstyle N}\downarrow & & \\ \pi & \longrightarrow & \operatorname{Aut}_{h\pi}(N) & \longrightarrow & \operatorname{Out}(N_{h\pi}) & \longrightarrow & 1 \end{array} \tag{3.9}$$

with exact rows. In particular, $N_{h\pi}$ mono $\Rightarrow N$ mono and N epi $\Rightarrow N_{h\pi}$ epi.

Definition 3.10 *The p-compact group $X_{h\pi}$ has N-determined automorphisms if the morphism $N_{h\pi}$ of (3.7) is injective.*

If the component group π is trivial, i.e., $X_{h\pi} = X$ is connected, then $N_{h\pi}$ and N are the same map.

There are p-compact groups with N-determined automorphisms.

Proposition 3.11 *If the Weyl group order of X is prime to p, then X has N-determined automorphisms.*

This follows from

Lemma 3.12 *$H^*(Bj,\mathbb{F}_p)\colon H^*(BX;\mathbb{F}_p) \to H^*(BN;\mathbb{F}_p)$ is an isomorphism if p does not divide $|\pi_0(N)|$.*

Proof Since BX is p-complete, the map Bj factors through a map $(BN)^\wedge_p \longrightarrow BX$. The domain of this map is a connected (Clark-Ewing) p-compact group with maximal torus normalizer N [**16**, 1.2] and the map itself is an isomorphism of p-compact groups since [**17**, 3.7] it is monomorphism (as the restriction to the p-normalizer of the maximal torus is) and a rational equivalence [**5**, 9.7]. □

4 N-determined automorphisms of p-compact groups

First let $X_{h\pi}$ be a p-compact group with connected component X and component group π. We shall compare the two homomorphisms

$$N_{h\pi}\colon \mathrm{Aut}_{h\pi}(X) \to \mathrm{Aut}_{h\pi}(N)$$
$$N\colon \mathrm{Out}(X) \to \mathrm{Out}(N)$$

of automorphism groups. Let $\rho\colon \pi \to \mathrm{Out}(X)$ denote the monodromy action and

$$H^1_{Z^{-1}\rho}(\pi; \check{Z}(X)) \longrightarrow H^1_{Z^{-1}N\rho}(\pi; Z(\check{N})) \tag{4.1}$$

the map induced on cohomology by the discrete approximation to $Z(N(i))$.

Proposition 4.2 *Let $X_{h\pi}$ be a p-compact group with identity component X and component group π. Suppose that X has N-determined automorphisms and that (4.1) is a monomorphism. Then $X_{h\pi}$ has N-determined automorphisms.*

Proof The homomorphism $N_{h\pi}$ fits into a commutative diagram (3.5) with exact rows

$$\begin{array}{ccccccc}
0 \longrightarrow & H^1_{Z^{-1}\rho}(\pi; \check{Z}(X)) & \longrightarrow & \mathrm{Aut}_{h\pi}(X) & \longrightarrow & \mathrm{Aut}(\pi) \times \mathrm{Out}(X) \\
& \downarrow & & \downarrow{\scriptstyle N_{h\pi}} & & \downarrow{\scriptstyle 1\times N} \\
0 \longrightarrow & H^1_{Z^{-1}N\rho}(\pi; Z(\check{N})) & \longrightarrow & \mathrm{Aut}_{h\pi}(N) & \longrightarrow & \mathrm{Aut}(\pi) \times \mathrm{Out}(N)
\end{array}$$

where the outer vertical homomorphisms are injective. □

Recall (3.4) that at odd primes, (4.1) automatically is an isomorphism.

Example 4.3 If the order of the Weyl group of the identity component X is prime to p and (4.1) is an isomorphism (e.g. if $p > 2$), then $X_{h\pi}$ has N-determined automorphisms. Obviously, all p-compact toral groups have N-determined automorphisms.

Next, let X be a connected p-compact group with center $Z \longrightarrow X$ and adjoint form X/Z. If $i\colon T \to X$ is a maximal torus for X with normalizer $j\colon N \to X$, then $i/Z\colon T/Z \to X/Z$ is a maximal torus for X/Z [**17**, 4.6] with normalizer $j/Z\colon N/Z \to X/Z$ [**16**, 3.8].

We shall, for the benefit of the prime $p = 2$, need a slight refinement of the homomorphism (3.2). The homomorphism

$$Z = Z(X) \xrightarrow{Z(j)} Z(N) \longrightarrow N \tag{4.4}$$

is a central monomorphism of extended p-compact toral groups [**7**, 7.7] (that equals the center of N if p is odd (3.4)). Define $\mathrm{Out}(N, Z)$ to be the subgroup of $\mathrm{Out}(N)$ consisting of those conjugacy classes of automorphisms f of N that preserve the central monomorphism (4.4) in the sense that there exists some automorphism of Z such that

$$\begin{array}{ccc} Z & \longrightarrow & Z \\ \downarrow & & \downarrow \\ N & \xrightarrow[f]{} & N \end{array}$$

commutes up to conjugacy. ($\mathrm{Out}(N, Z) = \mathrm{Out}(N)$ if p is odd.) Note, for instance by using discrete approximations [**7**, 3.13], that the automorphism of Z, if it exists, is uniquely determined by f, so that we have a restriction homomorphism $|Z\colon \mathrm{Out}(N, Z) \to \mathrm{Out}(Z)$. Now note that the homomorphism (3.2) actually takes values in the subgroup $\mathrm{Out}(N, Z)$ of $\mathrm{Out}(N)$ and that $N(f)|Z = Z(f)^{-1}$ for all $f \in \mathrm{Out}(X)$.

By naturality [**5**, 8.3] of the short exact sequence of p-compact groups

$$Z \longrightarrow X \longrightarrow X/Z$$

there is a homomorphism

$$(Z^{-1}, /Z)\colon \mathrm{Out}(X) \to \mathrm{Out}(Z) \times \mathrm{Out}(X/Z) \tag{4.5}$$

taking $f \in \mathrm{Out}(X)$ to $(Z(f)^{-1}, f/Z)$ where f/Z is the automorphism of the adjoint form X/Z induced by $Z(f)^{-1}$ and f. Observe that [**14**, 4.3] implies that (4.5) is injective.

Similarly, naturality of the short exact sequence of extended p-compact toral groups

$$Z \longrightarrow N \longrightarrow N/Z$$

allows us to define a group homomorphism

$$(|Z, /Z)\colon \mathrm{Out}(N, Z) \to \mathrm{Out}(Z) \times \mathrm{Out}(N/Z) \tag{4.6}$$

taking $f \in \mathrm{Out}(N, Z)$ to the pair consisting of restriction $f|Z$ of f to Z and the quotient automorphism f/Z as above.

The group homomorphisms (4.5) and (4.6) are related by a commutative diagram

$$\begin{array}{ccc} \mathrm{Out}(X) & \xrightarrow{\quad N \quad} & \mathrm{Out}(N, Z) \\ {\scriptstyle (Z^{-1},/Z)}\downarrow & & \downarrow{\scriptstyle (|Z,/Z)} \\ \mathrm{Out}(Z) \times \mathrm{Out}(X/Z) & \xrightarrow{1 \times N/Z} & \mathrm{Out}(Z) \times \mathrm{Out}(N/Z) \end{array} \tag{4.7}$$

and as left the vertical homomorphism (4.5) is injective, the next proposition is immediate.

Proposition 4.8 *Let X be a connected p-compact group with adjoint form X/Z. If X/Z has N-determined automorphisms, so does X.*

We now invoke the Homology Decomposition Theorem [**7**].

Theorem 4.9 *Let X be a connected p-compact group. Assume that*

1. *$C_X(\nu)$ has N-determined automorphisms for each object (V, ν) of $\mathbb{A}(X)$.*
2. $\lim^1_{\mathbb{A}(X)} \pi_1(BZ(V, \nu)) = 0 = \lim^2_{\mathbb{A}(X)} \pi_2(BZ(V, \nu))$.

Then X has N-determined automorphisms.

Proof Let $f \in \mathrm{Out}(X)$ be an outer automorphism with $N(f) = 1$. Let $\nu\colon V \to X$ be a monomorphism of a nontrivial elementary abelian p-group V to X with [16] preferred lift $\mu\colon V \to N$ to the normalizer of the maximal torus.

Since $f \circ \nu = f \circ j \circ \mu = j \circ N(f) \circ \mu = j \circ \mu = \nu$, f restricts to an automorphism $C_f(\nu)$ of $C_X(\nu)$ such that $f \circ e(\nu) = e(\nu) \circ C_f(\nu)$ and such that the diagram

$$\begin{array}{ccc} & C_N(\mu) & \\ \swarrow & & \searrow \\ C_X(\nu) & \xrightarrow{\;\;C_f(\nu)\;\;} & C_X(\nu) \end{array}$$

commutes up to conjugacy. By assumption (1), $C_f(\nu)$ is conjugate to the identity of $C_X(\nu)$ and $f \circ e(\nu) \simeq e(\nu)$. By assumption (2) and obstruction theory (2.1), f is conjugate to the identity of X. □

Remark 4.10 If it can be shown that the obstruction groups of 4.9.(2) vanish for any connected, centerfree p-compact group X, then (4.2, 4.8, 4.9) will, by the induction principle of [7, Section 9], imply that all p-compact groups, $p > 2$, have N-determined automorphisms.

Example 4.11 Let $X = \mathrm{PU}(3)$ and $p = 3$. There are, up to isomorphism, two conjugacy classes, L_1 and L_2, of monomorphisms $\mathbb{Z}/3 \rightarrowtail X$, and two conjugacy classes, V_1 and V_2, of monomorphisms $(\mathbb{Z}/3)^2 \rightarrowtail X$. Their centralizers are

$$C_X(L_1) = \frac{\mathrm{U}(2) \times \mathrm{U}(1)}{\mathrm{U}(1)} \qquad C_X(L_2) = T \rtimes \langle \sigma \rangle$$
$$C_X(V_1) = T \qquad\qquad C_X(V_2) = V_2$$

where T is the maximal torus, V_1 consists of the elements of order 3 in T, σ is the permutation matrix of the 3-cycle $\sigma = (123)$, and $V_2 = \langle \sigma, \mathrm{diag}(1, \zeta, \zeta^2) \rangle$ where ζ is a primitive third root of unity. Note that these centralizers all have N-determined automorphisms (4.3).

Of the rank two elementary abelian p-groups in X, only V_2 has a centralizer with disconnected center and the automorphism group of V_2 as an object of the category $\mathbb{A}(X)$ can be shown to be the special linear group $\mathrm{SL}(2,3)$. Oliver's cochain complex [21] for the computation of the derived functors of the inverse limit functor now shows that $\lim^1_{\mathbb{A}(X)} \pi_1(BZ(V,\nu)) = 0$ since there are no nontrivial $\mathrm{SL}(2,3)$-homomorphisms from the Steinberg representation $\mathrm{St}(V_2)$ to V_2. Also $\lim^2_{\mathbb{A}(X)} \pi_2(BZ(V,\nu)) = 0$ since there are no monomorphisms from a rank three elementary abelian p-group to X.

Theorem 4.9 now implies that $\mathrm{PU}(3)$ (and hence also $\mathrm{U}(3)$ and $\mathrm{SU}(3)$) has N-determined automorphisms at $p = 3$ (and in fact at all odd primes). (Alternatively, this follows from [10]; see Section 5)

For later use we note that Lemma 3.4 can be extended to nonconnected p-compact groups whose identity component has N-determined automorphisms.

Corollary 4.12 *Assume that X has N-determined automorphisms and that p is odd. Then the maps*

$$\mathrm{map}(BN_{h\pi}, BN_{h\pi})_{B1} \longrightarrow \mathrm{map}(BN_{h\pi}, BX_{h\pi})_{Bj_{h\pi}} \longleftarrow \mathrm{map}(BX_{h\pi}, BX_{h\pi})_{B1}$$

are homotopy equivalences.

Proof The left map is a homotopy equivalence by Lemma 3.6. To see that also the right map is a homotopy equivalence, it suffices by Shapiro's lemma to see that the π-map $\mathrm{map}(BN, BX_{h\pi})_{Bj} \longrightarrow \mathrm{map}(BX, BX_{h\pi})_{Bj}$, where j denotes an appropriate inclusion homomorphism, is a homotopy equivalence.

Since X has N-determined automorphisms, the orbit $\pi \cdot B1 \subset \mathrm{Out}(X)$ of the π-action on the identity map of BX is taken bijectively onto the orbit $\pi \cdot Bj \subset [BN, BX]$ of the π-action on Bj. This means that the fibre map

$$\begin{array}{ccc} \mathrm{map}(BN, BX_{h\pi})_{Bj} & \longleftarrow & \mathrm{map}(BX, BX_{h\pi})_{Bj} \\ \downarrow & & \downarrow \\ \mathrm{map}(BN, B\pi)_{B0} & \xleftarrow{\simeq} & \mathrm{map}(BX, B\pi)_{B0} \end{array} \tag{4.13}$$

restricts to a bijection on the set of path components of the fibre. Since also the restriction to a typical component of the fibre, $\mathrm{map}(BN, BX)_{Bj} \longleftarrow \mathrm{map}(BX, BX)_{B1}$, is a homotopy equivalence by Lemma 3.4, we may conclude that (4.13) is a (fibre) homotopy equivalence. □

5 The Lie case

The purpose of this section is to verify that compact Lie groups have N-determined automorphisms (and that they do not contradict (7.2)).

Let G be any compact connected Lie group and let $\hat{G}$ denote G viewed as a p-compact group, i.e. $B\hat{G} = (BG)^{\wedge}_{p}$, where the prime p as in the previous section, is assumed to be *odd*. Let $N \longrightarrow \hat{G}$ be the normalizer of a maximal torus $T \longrightarrow \hat{G}$. (This maximal torus as well as its normalizer can be obtained from Lie group constructions.)

Proposition 5.1 [**9**, 3.5] $N\colon \mathrm{Out}(\hat{G}) \to \mathrm{Out}(N)$ *is an isomorphism.*

The proof of this proposition amounts to an interpretation of the fundamental results obtained by Jackowski, McClure, and Oliver.

The Weyl group $W = \pi_0(N)$ of the Lie group G or, what is the same, the p-compact group $\hat{G}$ acts on the short exact sequence

$$0 \longrightarrow \pi_1(T) \longrightarrow \pi_1(T) \otimes \mathbb{Q} \xrightarrow{\varepsilon} \breve{T} \longrightarrow 0$$

where $\breve{T}$ is the discrete approximation to T. We start by computing the first cohomology group of $\breve{T}$ as a W-module.

Lemma 5.2 $H^1(W; \breve{T}) = 0.$

Proof [**10**, 3.5] The Weyl group has a presentation

$$W = \langle R_1, \ldots, R_n | R_i^2 = 1,\ (R_iR_j)^{m_{ij}} = 1,\ i \neq j \rangle$$

as a Coxeter group for certain integers m_{ij}. The equations

$$0 = (1 + R_i)(1 - R_i), \quad 1 = \frac{1}{2}(1 + R_i) + \frac{1}{2}(1 - R_i),$$

which hold in the group ring $\mathbb{Q}_p[W]$, imply that $\mathrm{im}(1 - R_i) = \ker(1 + R_i)$ for all i in any $\mathbb{Q}_p[W]$-module ($p > 2$).

Now let $f\colon W \to \check{T}$ be a 1-cocycle. It suffices to lift f to a 1-cocycle $W \longrightarrow \pi_1(T) \otimes \mathbb{Q}$, i.e., to find vectors $v_1, \dots, v_n \in \pi_1(T) \otimes \mathbb{Q}$ such that $\varepsilon(v_i) = f(R_i)$ and

$$(1 + R_i)v_i = 0 \tag{5.3}$$

$$(1 + (R_iR_j) + \cdots + (R_iR_j)^{m_{ij}-1})(v_i + R_iv_j) = 0 \tag{5.4}$$

for all i and all $i \neq j$, respectively.

Since p is odd and the reflection R_i has order two, the cohomology group $H^1(\langle R_i\rangle; \check{T})$ is trivial. This means that f restricts to a 1-coboundary on the subgroup $\langle R_i\rangle$, i.e.,

$$f(R_i) \in \mathrm{im}(1 - R_i) = \mathrm{im}(1 - R_i)\varepsilon = \varepsilon(\mathrm{im}(1 - R_i)) = \varepsilon(\ker(1 + R_i))$$

so that $\varepsilon(v_i) = f(R_i)$ for some $v_i \in \ker(1 + R_i)$. Thus condition (5.3) is satisfied.

For $i \neq j$, the composition R_iR_j acts as rotation on the 2-dimensional subspace $\ker(1 + R_i) + \ker(1 + R_j)$ of $\pi_1(T) \otimes \mathbb{Q}$. Since

$$v_i + R_iv_j = v_i + v_j - (1 - R_i)v_j$$

belongs to this subspace, also condition (5.4) is satisfied. □

Proof of Proposition 5.1 Let $\mathrm{Aut}(W, \check{T})$ denote the subgroup of $\mathrm{Aut}(W) \times \mathrm{Aut}(\check{T})$ consisting of pairs (χ, ϕ) such that ϕ is χ-equivariant. The natural map

$$\mathrm{Aut}(N) \longrightarrow \mathrm{Aut}(W, \check{T})$$

is injective since its kernel $H^1(W, \check{T}) = 0$ by (5.2), cf. [**15**, Section 5]. Thus also the induced map of quotient groups

$$\mathrm{Out}(N) = \frac{\mathrm{Aut}(N)}{W} \longrightarrow \frac{\mathrm{Aut}(W, \check{T})}{W}$$

is injective.

According to Jackowski, McClure, and Oliver [**9**, 3.5], the composition

$$\mathrm{Out}(\hat{G}) \xrightarrow{N} \mathrm{Out}(N) \rightarrowtail \frac{\mathrm{Aut}(W, \check{T})}{W}$$

is an isomorphism. Hence both maps in the above diagram are isomorphisms. □

6 Centralizers

This section contains preparatory material for the proof of Proposition 7.10 and Proposition 7.17.

Let X be a p-compact group, N an extended p-compact torus, and $j\colon N \to X$ a monomorphism conjugate to the normalizer of a maximal torus $T \longrightarrow X$. If V is a nontrivial elementary abelian p-group and $\nu\colon V \to X$ a conjugacy class of monomorphisms, then a *preferred lift* of ν is a conjugacy class of monomorphisms $\mu\colon V \to N$ such that $j \circ \mu$ and ν are conjugate and the morphism $C_N(\mu) \longrightarrow C_X(\nu)$ induced by j is conjugate to the normalizer of a maximal torus.

For a preferred lift $\mu\colon V \to N$ of a monomorphism $\nu\colon V \to X$, let $A(\mu)$ denote the representation

$$V \xrightarrow{\pi_0(\mu)} \pi_0(N) \longrightarrow \mathrm{Aut}(\pi_1(N))$$

of V in the free $\mathbb{Z}_p$-module $\pi_1(N) = \pi_1(T)$.

Definition 6.1 *The preferred lift μ is a* special *preferred lift of ν if the representation $A(\mu)$ fails to be faithful.*

Note the following properties of special preferred lifts [**16**]:

- If $C_X(\nu)$ has maximal rank [**7**, Section 4], the unique preferred lift of ν is special (indeed, $A(\mu)$ is the trivial representation) [**16**, 4.8].
- If X is connected, the unique preferred lift of any monomorphism $\nu\colon \mathbb{Z}/p \to X$ is special [**16**, 4.6, 4.9]
- If X is connected, any ν has a special preferred lift [**16**, Proof of 1.3].
- If X is connected, there exists for any monomorphism $f\colon V_1 \to V_2$ between nontrivial elementary abelian p-groups a special preferred lift μ_2 of ν_2 such that $\mu_2 f$ is a special preferred lift of $\nu_2 f$ [**16**, Proof of 1.3].

It is possible to enumerate the preferred lifts in the rank two case.

Proposition 6.2 *Let $\nu\colon (\mathbb{Z}/p)^2 \to X$ be a monomorphism of a rank two elementary abelian p-group into a connected p-compact group X.*

1. *If the centralizer $C_X(\nu)$ has maximal rank, then ν admits a unique (special) preferred lift.*
2. *If the centralizer $C_X(\nu)$ is not of maximal rank, then the map $\mu \longrightarrow \ker A(\mu)$ determines a bijection between the set of N-conjugacy classes of special preferred lifts μ of ν and the set of nontrivial, proper subgroups of V.*

Proof The uniqueness of (special) preferred lifts in the maximal rank case is a general fact [**16**, 4.8].

Assume now that the rank of the centralizer $C_X(\nu)$ is less than the rank of X. Let μ_1 and μ_2 be two special preferred lifts of ν. The associated representations $A(\mu_1)$ and $A(\mu_2)$ have kernels of rank one since they are nontrivial and nonfaithful.

Suppose that $\ker A(\mu_1)$ and $\ker A(\mu_2)$ both equal the rank one subgroup $L < V$. Then the restrictions $\mu_1|L$ and $\mu_2|L$ are conjugate for they are both preferred lifts of $\nu|L$; put $\mu|L = \mu_1|L = \mu_2|L$. Write $(\mathbb{Z}/p)^2 = L \oplus L^\perp$ as a direct sum of L and a complement $L^\perp$ and let $\nu^\perp\colon L^\perp \to C_X(L)$, $\mu_1^\perp, \mu_2^\perp\colon L^\perp \to C_N(\mu|L)$ denote the monomorphisms adjoint to ν, μ_1, μ_2, respectively, relative to this splitting of $(\mathbb{Z}/p)^2$. Then $C_N(\mu|L)$ is the normalizer of a maximal torus of $C_X(\nu|L)$ and $\mu_1^\perp$ and $\mu_2^\perp$ are preferred lifts of $\nu|L$. Since preferred lifts are unique in the rank one case, $\mu_1^\perp$ and $\mu_2^\perp$ are conjugate in $C_N(\mu|L)$ and, therefore, μ_1 and μ_2 are conjugate in N.

Given any nontrivial and proper subgroup $L < (\mathbb{Z}/p)^2$ there exists a preferred lift μ of ν such that $\mu|L$ is a preferred lift of $\nu|L$. We then have $\ker A(\mu) = L$. □

Proposition 6.2 shows that, unless the centralizer is of maximal rank, any monomorphism of a rank two elementary abelian p-group into a connected p-compact group has exactly $p+1$ special preferred lifts.

Now follows a variation on the theme of centricity.

For any elementary abelian p-group V and any loop space Y, the mapping space $\mathrm{map}(BV, BY)$ as well as any of its components $\mathrm{map}(BV, BY)_{B\nu} = BC_Y(\nu)$ are spaces over BY by the evaluation maps. We say that a homotopy class of maps $BC_Y(\nu) \longrightarrow \mathrm{map}(BV, BY)$ is a homotopy class over BY if

$$\begin{array}{ccc} BC_Y(\nu) & \longrightarrow & \mathrm{map}(BV, BY) \\ & \searrow \quad \swarrow & \\ & BY & \end{array}$$

commutes up to homotopy.

Lemma 6.3 *Let V be a nontrivial elementary abelian p-group and $\nu\colon V \to X$ a monomorphism with preferred lift $\mu\colon V \to N$. Suppose that p is odd and that*

$C_X(\nu)$ has N-determined automorphisms. Then there is bijective correspondence between homotopy classes $BC_N(\mu) \longrightarrow \mathrm{map}(BV, BN)$ over BN and homotopy classes $BC_X(\nu) \longrightarrow \mathrm{map}(BV, BX)$ over BX.

The homotopy class $BC_N(\mu) \longrightarrow \mathrm{map}(BV, BN)$ over BN and the homotopy class $BC_X(\nu) \longrightarrow \mathrm{map}(BV, BX)$ over BX correspond to each other if and only if the diagram

$$\begin{array}{ccc} BC_N(\mu) & \longrightarrow & \mathrm{map}(BV, BN) \\ \downarrow & & \downarrow \\ BC_X(\nu) & \longrightarrow & \mathrm{map}(BV, BX) \end{array}$$

commutes up to homotopy.

Proof The map $BC_Y(\lambda) \longrightarrow \mathrm{map}(BV, BY)$ is a map over BY up to homotopy if and only if its adjoint $BV \longrightarrow \mathrm{map}(BC_Y(\lambda), BY)_{Be}$ maps into the connected component of the evaluation map $Be\colon BC_Y(\lambda) = \mathrm{map}(BV, BY)_{B\lambda} \to BY$, $(Y, \lambda) = (N, \mu)$, (X, ν). Thus we want to compare homotopy classes of maps $BV \longrightarrow \mathrm{map}(BC_N(\mu), BN)_{Be}$ to homotopy classes of maps $BV \longrightarrow \mathrm{map}(BC_X(\nu), BX)_{Be}$.

By centricity [**2**], there are homotopy equivalences

$$\mathrm{map}(BC_N(\mu), BC_N(\mu))_{B1} \longrightarrow \mathrm{map}(BC_N(\mu), BN)_{Be}$$
$$\mathrm{map}(BC_X(\nu), BC_X(\nu))_{B1} \longrightarrow \mathrm{map}(BC_X(\nu), BX)_{Be}$$

and coupled with the general homotopy equivalences of (4.12) we see that $\mathrm{map}(BC_N(\mu), BN)_{Be}$ and $\mathrm{map}(BC_X(\nu), BX)_{Be}$ are homotopy equivalent in such a way so as to induce the asserted the correspondence. □

For the following lemma, let V_1 and V_2 be nontrivial elementary abelian p-groups and $\nu_1\colon V_1 \to X$, $\nu_2\colon V_2 \to X$ monomorphisms with preferred lifts $\mu_1\colon V_1 \to N$, $\mu_2\colon V_2 \to N$. Suppose that there exists a monomorphism $g\colon V_1 \to V_2$ such that $\mu_2 \circ g$ is conjugate to μ_1. Then also $\nu_2 \circ g$ is conjugate to ν_1 and g induces morphisms $g^*\colon C_N(\mu_2) \to C_N(\mu_1)$ and $g^*\colon C_X(\nu_2) \to C_X(\nu_1)$.

Lemma 6.4 *Suppose that p is odd and that $C_X(\nu_2)$ has N-determined automorphisms. Let $\gamma\colon C_X(\nu_2) \to C_X(\nu_1)$ be any morphism such that the diagram*

$$\begin{array}{ccc} BC_N(\mu_2) & \xrightarrow{Bg^*} & BC_N(\mu_1) \\ \downarrow & & \downarrow \\ BC_X(\nu_2) & \xrightarrow[B\gamma]{} & BC_X(\nu_1) \end{array}$$

commutes up to homotopy. Then $B\gamma$ is homotopic to $Bg^*\colon BC_X(\nu_2) \to BC_X(\nu_1)$ induced by Bg.

Proof By (6.3) it suffices to show that $B\gamma$ is a map over BX.

Let $\overline{\nu}_2\colon V_2 \to C_X(\nu_2)$ denote the canonical central factorization of ν_2 through its centralizer [**5**, 8.2]. Using that $Bg^*\colon BC_N(\mu_2) \to BC_N(\mu_1)$ obviously is a map over BN it follows that

$$\begin{array}{ccccc} & & BV_2 & & \\ & \swarrow^{B\overline{\nu}_2} & & \searrow^{B\nu_2} & \\ BC_X(\nu_2) & & \xrightarrow[Be \circ B\gamma]{} & & BX \end{array}$$

commutes up to homotopy, i.e. that $Be \circ B\gamma$ is a map under BV_2. Viewing $BC_X(\nu_2)$ as the total space of the fibration

$$BV_2 \xrightarrow{B\overline{\nu}_2} BC_X(\nu_2) \longrightarrow B(C_X(\nu_2)/V_2),$$

the Shapiro lemma tells us that

$$\operatorname{map}(BC_X(\nu_2), BY) = \operatorname{map}((BV_2)_{h(C_X(V_2)/V_2)}, BY) = \operatorname{map}(BV_2, BY)^{h(C_X(V_2)/V_2)}$$

with, for instance, $Y = C_X(\nu_2)$ or $Y = X$. Noting that the homotopy equivalence $\operatorname{map}(BV_2, BC_X(\nu_2))_{B\overline{\nu}_2} \longrightarrow \operatorname{map}(BV_2, BX)_{B\nu_2}$ induces a homotopy equivalence

$$\operatorname{map}(BV_2, BC_X(\nu_2))_{B\overline{\nu}_2}^{h(C_X(\nu_2)/V_2)} \longrightarrow \operatorname{map}(BV_2, BX)_{B\nu_2}^{h(C_X(\nu_2)/V_2)}.$$

of homotopy fixed point spaces, we see that there exists a uniquely determined endomorphism $\overline{\gamma}$ of $C_X(\nu_2)$ such that

$$\begin{array}{ccc} BC_X(\nu_2) & \xrightarrow{B\overline{\gamma}} & BC_X(\nu_2) \\ {\scriptstyle B\gamma}\downarrow & & \downarrow{\scriptstyle Be} \\ BC_X(\nu_1) & \xrightarrow[Be]{} & BX \end{array}$$

commutes up to homotopy.

Similarly, there exists a uniquely determined endomorphism of $C_N(\mu_2)$ such that

$$\begin{array}{ccc} BC_N(\mu_2) & \longrightarrow & BC_N(\mu_2) \\ {\scriptstyle Bg^{\bullet}}\downarrow & & \downarrow{\scriptstyle Be} \\ BC_N(\mu_1) & \xrightarrow[Be]{} & BN \end{array}$$

commutes up to homotopy; in fact, this endomorphism of $C_N(\mu_2)$ can only be the identity.

By naturality of this construction, it follows that $\overline{\gamma}$ is an automorphism of $C_X(\nu_2)$ that is covered by the identity map of $C_N(\mu_2)$ and, since $C_X(\nu_2)$ is assumed to have N-determined automorphisms, γ itself must be conjugate to the identity. This shows that $B\gamma$ is a map over BX up to homotopy. □

We now observe that for a connected p-compact group, the nontrivial center of the p-normalizer of the maximal torus is contained in the maximal torus.

Lemma 6.5 *Suppose that X is connected and let N_p be the p-normalizer of the maximal torus $T \longrightarrow X$. Then the center $Z(N_p)$ of N_p has the form $Z(N_p) \longrightarrow T \longrightarrow N_p$ and $Z(N_p)$ is a nontrivial abelian p-compact group.*

Proof The center of N_p is nontrivial [**6**, 1.3] and the evaluation map $BZ(N_p) = \operatorname{map}(BN_p, BN_p)_{B1} \longrightarrow BN_p$ factors through $\operatorname{map}(BT, BN_p)_{Bi} = BC_{N_p}(T) = BT$ [**17**, 4.1], [**16**, 2.2]. □

Finally a version of what is known as the "Zabrodsky lemma".

Lemma 6.6 *In the commutative diagram*

$$\begin{array}{ccc} E_0 & \xrightarrow{u} & E_1 \\ & {\scriptstyle f_0}\searrow \quad \swarrow{\scriptstyle f_1} & \\ & B & \end{array}$$

suppose that B is connected and that f_0 and f_1 are fibrations with fibres F_0 and F_1, respectively. If $u|F_0\colon F_0 \to F_1$ is null homotopic and if constant maps provide a homotopy equivalence $F_1 \longrightarrow \mathrm{map}(F_0, F_1)_0$, then there exists a section v of f_1 such that $v \circ f_0$ and u are vertically homotopic.

Proof Let $M \longrightarrow B$ be the fibration whose fibre over any point $b \in B$ is $\mathrm{map}(f_0^{-1}(b), f_1^{-1}(b))_0$. Constant maps provide a fibre homotopy equivalence

$$\begin{array}{ccc} E_1 & \xrightarrow{\simeq} & M \\ & {\scriptstyle f_1}\searrow \quad \swarrow & \\ & B & \end{array}$$

so, since the adjoint of the lift u is a section of $M \longrightarrow B$, f_1 admits a corresponding section, v. □

7 N-determined p-compact groups

This section contains an investigation of the class of p-compact groups that are determined by the normalizer of the maximal torus. *It is a standing assumption throughout the section that the prime p is odd.*

Definition 7.1 *The p-compact group X is* N-determined *if for any diagram of morphisms of the form*

$$X \xleftarrow{j} N \xrightarrow{j'} X'$$

where

- *X' is a p-compact group and N an extended p-compact torus*
- *j is conjugate to the normalizer of a maximal torus of X*
- *j' is conjugate to the normalizer of a maximal torus of X'*

there exists an isomorphism $f\colon X \to X'$ such that $fj \simeq j'$.

A p-compact group is totally N-determined *if it has N-determined automorphisms (3.10) and is N-determined.*

Note that if X has N-determined automorphisms, then the isomorphism f, if it exists, is unique up to conjugacy.

Total N-determinacy has strong consequences for automorphism groups.

Corollary 7.2 *If the p-compact group $X_{h\pi}$ with component group π and identity component is totally N-determined then the maps*

$$N_{h\pi}\colon \mathrm{Aut}_{h\pi}(X) \to \mathrm{Aut}_{h\pi}(N)$$
$$N\colon \mathrm{Out}(X_{h\pi}) \to \mathrm{Out}(N_{h\pi})$$

are isomorphisms.

Proof It remains (3.9) to show surjectivity of the homomorphism N of (3.8). But that follows immediately from Definition 7.1 since any square of the form

$$\begin{array}{ccc} N_{h\pi} & \xrightarrow{\cong} & N_{h\pi} \\ \downarrow & & \downarrow \\ X_{h\pi} & \dashrightarrow & X_{h\pi} \end{array}$$

can be completed by an isomorphism. □

As shown in Section 5, compact connected Lie groups satisfy the conclusion of Corollary 7.2. Thus Lie groups have N-determined automorphisms and might be N-determined.

Now a few words about component groups to ensure that the π_0-image of the diagram of Definition 7.1 always can be completed.

Consider the Weyl representation $w\colon \pi_0(N) \to \mathrm{Aut}(\pi_1(N)\otimes\mathbb{Q})$ of the component group $\pi_0(N) = W_T(X)$ in the $\mathbb{Q}_p$-vector space $\pi_1(N)\otimes\mathbb{Q} = \pi_1(T)\otimes\mathbb{Q}$. Recall that an element $s \in \pi_0(N)$ is a *reflection* if $w(s)$ is nontrivial and pointwise fixes a hyperplane.

We want to characterize the Weyl group $\ker \pi_0(j) = W_T(X_0)$ of [**17**, 3.8] the identity component X_0 of X.

Lemma 7.3 *Let $s \in \pi_0(N)$ be a reflection. Then*

$$s \in \ker \pi_0(j) \Leftrightarrow \mathrm{ord}(s) = \mathrm{ord}(w(s)).$$

Proof One direction follows from the fact [**5**, 9.7] that the Weyl representation is faithful on the Weyl group of the identity component.

Suppose, conversely, that $\mathrm{ord}(s) = \mathrm{ord}(w(s))$. Since s is a reflection (and p is odd) the order of $w(s)$ is prime to p. Hence s is mapped to the unit of the finite p-group $\pi_0(X)$. □

Corollary 7.4 *The subgroup* $\ker \pi_0(j)$ *of* $\pi_0(N)$ *is generated by the set of reflections* $s \in \pi_0(N)$ *with* $\mathrm{ord}(s) = \mathrm{ord}(w(s))$.

Proof By [**5**, 9.7], the subgroup $\ker \pi_0(j)$, which is the Weyl group of the identity component, is generated by the reflections that it contains. □

Remark 7.5 Let $\nu\colon V \to X$ be a monomorphism which admits a factorization $\mu\colon V \to T$ through the maximal torus. The component group

$$\pi_0(C_N(\mu)) = W^\mu$$

is the isotropy subgroup of $B\mu$ for the W-action on $[BV, BT] = \mathrm{Rep}(V,T)$, and by (7.3), the component group $\pi_0(C_X(\nu))$ is isomorphic to the quotient of W^μ by the normal subgroup generated by the set of reflections it contains. In particular, $C_X(\nu)$ is connected if and only if W^μ is a reflection group.

It follows from (7.4) that $\ker \pi_0(j) = \ker \pi_0(j')$ so that in the situation of Definition 7.1 there exists a unique isomorphism

$$\varphi\colon \pi_0(X) = \pi_0(N)/\ker \pi_0(j) \to \pi_0(X') = \pi_0(N)/\ker \pi_0(j') \tag{7.6}$$

such that $\varphi \circ \pi_0(j) = \pi_0(j')$.

Note also that the class of N-determined p-compact groups is nonempty.

Proposition 7.7 *If p does not the Weyl group order of X, in particular if X is a p-compact torus, then X is N-determined.*

Proof If the Weyl group order of X is prime to p, the p-completions of Bj and Bj' are homotopy equivalences (3.12). □

Proposition 7.8 *Suppose that the identity component X_0 of the p-compact group X is totally N-determined. Then X itself is totally N-determined.*

Proof It remains (4.2) to show that X is N-determined.

Let BX_0, BN_0, BX_0' be the fibres of $BX \longrightarrow B\pi_0(X)$, $BN \longrightarrow B\pi_0(N) \longrightarrow B\pi_0(X)$, $BX' \longrightarrow B\pi_0(X')$, respectively, and let $X_0 \xleftarrow{j_0} N_0 \xrightarrow{j_0'} X_0'$ be the restrictions of j, j'. Since X_0 is assumed to be N-determined, there exists an isomorphism $f_0\colon X_0 \to X_0'$ with $Bf_0 \circ Bj_0 \simeq Bj_0'$.

The map Bf_0 is a $\pi_0(X)$-map in the sense that $Bf_0 \circ \xi \simeq \varphi(\xi) \circ Bf_0$ for any $\xi \in \pi_0(X)$ acting by monodromy on BX_0. To see this, note that the restriction map $[BX_0, BX_0'] \longrightarrow [BN_0, BX_0']$ is injective on the subset of isomorphisms since X_0 is assumed to have N-determined automorphisms and that $Bf_0 \circ \xi \circ Bj_0 \simeq Bf_0 \circ Bj_0 \circ \xi \simeq Bj_0' \circ \xi \simeq \varphi(\xi) \circ Bf_0 \circ Bj_0$.

Because Bf_0 is a $\pi_0(X)$-map, the mapping space $\mathrm{map}(BX_0, BX_0')_{Bf_0}$ is a $\pi_0(X)$ space. Composition with Bj induces a homotopy equivalence (3.4)

$$\mathrm{map}(BX_0, BX_0')_{Bf_0}^{h\pi_0(X)} \longrightarrow \mathrm{map}(BN_0, BX_0')_{Bj_0'}^{h\pi_0(X)}$$

and since the homotopy fixed point space to the right is nonempty – it contains Bj' – the homotopy fixed point space to the left is also nonempty: It contains a fibre homotopy equivalence $Bf\colon BX \to BX'$ over $B\pi$ such that $Bf \circ Bj$ is vertically homotopic to Bj'. □

Example 7.9 If the order of the Weyl group of the identity component X of $X_{h\pi}$ is prime to $p > 2$, then X is totally N-determined (3.11,7.7), so (7.8) $X_{h\pi}$ is also totally N-determined and (7.2) $\mathrm{Aut}_{h\pi}(X) \cong \mathrm{Aut}_{h\pi}(N)$ and $\mathrm{Out}(X_{h\pi}) \cong \mathrm{Out}(N_{h\pi})$ All p-compact toral groups, in particular, are totally N-determined.

Proposition 7.10 *Let X be a connected p-compact group with center $Z(X) \longrightarrow X$. If the adjoint form $X/Z(X)$ is N-determined, X itself is N-determined.*

Proof Let N be an extended p-compact torus, X' a connected p-compact group with center $Z(X') \longrightarrow X'$ and suppose that there exist morphisms, $j\colon N \to X$ and $j'\colon N \to X'$, that are conjugate to normalizers of maximal tori. The induced morphisms of centers, $Z(j)\colon Z(X) \to Z(N)$ and $Z(j')\colon Z(X') \to Z(N)$, are isomorphisms since p is assumed to be odd. In the induced commutative diagram

$$\begin{array}{ccccc}
Z(X) & \xleftarrow[\cong]{Z(j)^{-1}} & Z(N) & \xrightarrow[\cong]{Z(j')^{-1}} & Z(X') \\
\downarrow & & \downarrow & & \downarrow \\
X & \xleftarrow{\quad j \quad} & N & \xrightarrow{\quad j' \quad} & X' \\
\downarrow & & \downarrow & & \downarrow \\
X/Z(X) & \xleftarrow[j/Z]{} & N/Z(N) & \xrightarrow[j'/Z]{} & X'/Z(X')
\end{array} \qquad (7.11)$$

the morphisms j/Z and j'/Z are conjugate to normalizer of maximal tori so since $X/Z(X)$ is assumed to be N-determined there exists an isomorphism $f/Z\colon X/Z(X) \to X'/Z(X')$ such that $B(f/Z) \circ B(j/Z) \simeq B(j'/Z)$.

Our aim is to complete the diagram

$$\begin{array}{ccc} Z(X) & \xrightarrow[\cong]{Z(f)^{-1}} & Z(X') \\ \downarrow & & \downarrow \\ X & \overset{f}{\dashrightarrow} & X' \\ \downarrow & & \downarrow \\ X/Z(X) & \xrightarrow[f/Z]{\cong} & X'/Z(X') \end{array}$$

by an isomorphism f such that $Bf \circ Bj \simeq Bj'$; here $Z(f)^{-1} := Z(j')^{-1} \circ Z(j)$.

We shall consider the following diagram of fibered mapping spaces derived from diagram (7.11)

$$\begin{array}{ccccc} \operatorname{map}(BZ(X), BZ(X'))_{BZ(f)} & \longrightarrow & BZ(X')_{h(X/Z(X))} & \longrightarrow & B(X/Z(X)) \\ \| & & \uparrow & & \uparrow B(j/Z) \\ \operatorname{map}(BZ(X), BZ(X'))_{BZ(f)} & \longrightarrow & BZ(X')_{h(N/Z(N))} & \longrightarrow & B(N/Z(N)) \\ \simeq\downarrow & & \downarrow & & \| \\ \operatorname{map}(BZ(N), BZ(X'))_{BZ(j')^{-1}} & \longrightarrow & BZ(X')_{h(N/Z(N))} & \longrightarrow & B(N/Z(N)) \end{array}$$

where

- in the bottom horizontal fibration, the fibre over any point $b \in B(N/Z(N))$ is $\operatorname{map}(BN_b, BX'_{B(j'/Z)(b)})_{BZ(j')^{-1}}$
- in the middle horizontal fibration, the fibre over any point $b \in B(N/Z(N))$ is $\operatorname{map}(BX_{B(j/Z)(b)}, BX'_{B(j'/Z)(b)})_{BZ(f)}$
- in the top horizontal fibration, the fibre over any point $b \in B(X/Z(X))$ is $\operatorname{map}(BX_b, BX'_{B(f/Z)(b)})_{BZ(f)}$
- the map of the middle fibration to the bottom fibration is induced by Bj
- the map of the middle fibration to the top fibration is induced by B(j/Z).

Note that the bottom fibration admits a section adjoint to the fibre map Bj'. Thus also the middle fibration admits a section. By (6.6) this section will be the pull back of a section of the top fibration provided

1. $X/N \longrightarrow \operatorname{map}(BZ(X), BZ(X'))_{BZ(f)}$ is null homotopic
2. $X/N \longrightarrow \operatorname{map}(X/N, BZ(X'))_0$ is a homotopy equivalence.

The first condition is satisfied because the restriction to the fibre $X/N \subseteq B(N/Z(N))$ of the middle fibration is trivial and the section corresponds to the constant map of X/N into $\operatorname{map}(BZ(X), BZ(X'))_{BZ(f)}$. To see that the second condition is satisfied, recall that $\operatorname{map}(BZ(X), BZ(X'))_{BZ(f)} \simeq BZ(X') \simeq K(A,1) \times K(\mathbb{Z}_p, 2)^r$ for some finite abelian p-group A and that $\pi_1(X/N) \cong W_T(X)$ is finite.

This section of the top fibration is adjoint to a map $Bf\colon BX \to BX'$ over $B(f/Z)$ and under $BZ(f)$ such that $Bf \circ Bj \simeq Bj'$. □

From now on, let X and X' be two connected, centerfree p-compact groups and N an extended p-compact torus with monomorphisms

$$X \xleftarrow{j} N \xrightarrow{j'} X' \tag{7.12}$$

conjugate to normalizers of maximal tori. Also, make the induction hypothesis that all centralizers $C_X(\nu)$, $(V,\nu) \in \mathrm{Ob}(\mathbb{A}(X))$, have N-determined automorphisms and are N-determined.

We shall first see that the special preferred lifts are the same for X and X'. (The proof was communicated to me by Bill Dwyer.)

Lemma 7.13 *Let $\mu\colon V \to N$ be a monomorphism of a nontrivial elementary abelian p-group V into N. Then μ is a special preferred lift of $j\mu$ if and only if μ is a special preferred lift of $j'\mu$.*

Proof Put $\nu = j\mu$ and $\nu' = j'\mu$ and assume that μ is a special preferred lift of ν. Let $L < V$ denote the (nontrivial) kernel of the representation $A(\mu)$. Then $\mu|L$ is the unique preferred lift of $\nu|L$ and of $\nu'|L$. Let (if $L \neq V$) $L^\perp$ be a complement to L and $\mu^\perp\colon L^\perp \to C_N(\mu|L)$, $\nu^\perp\colon L^\perp \to C_X(\nu|L)$, $(\nu')^\perp\colon L^\perp \to C_{X'}(\nu'|L)$ the monomorphisms that are adjoint to μ, ν, ν', respectively. By the above induction hypothesis there exists an isomorphism $C_X(\nu|L) \longrightarrow C_{X'}(\nu'|L)$ under $C_N(\mu|L)$ inducing an isomorphism

$$C_X(\nu) = C_{C_X(\nu|L)}(\nu^\perp) \longrightarrow C_{C_{X'}(\nu'|L)}((\nu')^\perp) = C_{X'}(\nu')$$

under $C_{C_N(\mu|L)}(\mu^\perp) = C_N(\mu)$. Thus μ is also a preferred lift of ν'. □

For any object (V,ν) of $\mathbb{A}(X)$, let $\mathrm{SPL}(V,\nu) \subset [BV, BN]$ denote the set of conjugacy classes of special preferred lifts of (V,ν). Note that the automorphism group $\mathrm{Aut}(V,\nu)$ of (V,ν), consisting of those automorphisms α of V for which $\nu\alpha$ and ν are conjugate, acts on the set $\mathrm{SPL}(V,\nu)$.

Let $\mathbb{A}(N)$ be the category whose objects (V,μ) are conjugacy classes of monomorphisms $\mu\colon V \to N$ of nontrivial elementary abelian p-groups into N such that μ is a special preferred lift of $j\mu$ and $j'\mu$. A morphism $\alpha\colon (V_1,\mu_1) \to (V_2,\mu_2)$ in $\mathbb{A}(N)$ is a monomorphism $\alpha\colon V_1 \to V_2$ such that μ_1 and $\mu_2\alpha$ are conjugate. The morphisms j and j' induce full functors $\mathbb{A}(X) \longleftarrow \mathbb{A}(N) \longrightarrow \mathbb{A}(X')$ that restrict (see the remarks immediately after Definition 6.1) to isomorphisms on the full subcategories $\mathbb{A}_1(X) \cong \mathbb{A}_1(N) \cong \mathbb{A}_1(X')$ of rank one objects.

Lemma 7.14 *Let $\alpha\colon (V_1,\mu_1) \to (V_2,\mu_2)$ be a morphism in $\mathbb{A}(N)$. Then the diagram*

$$\begin{array}{ccc} C_X(j\mu_2) & \xrightarrow{\alpha^*} & C_X(j\mu_1) \\ {\scriptstyle f(\mu_2)}\downarrow & & \downarrow{\scriptstyle f(\mu_1)} \\ C_{X'}(j'\mu_2) & \xrightarrow[\alpha^*]{} & C_{X'}(j'\mu_1) \end{array}$$

commutes up to conjugacy where $f(\mu_1)\colon C_X(j\mu_1) \to C_{X'}(j'\mu_1)$ is the isomorphism under $C_N(\mu_1)$ and $f(\mu_2)\colon C_X(j\mu_2) \to C_{X'}(j'\mu_2)$ the isomorphism under $C_N(\mu_2)$.

Proof As the diagram

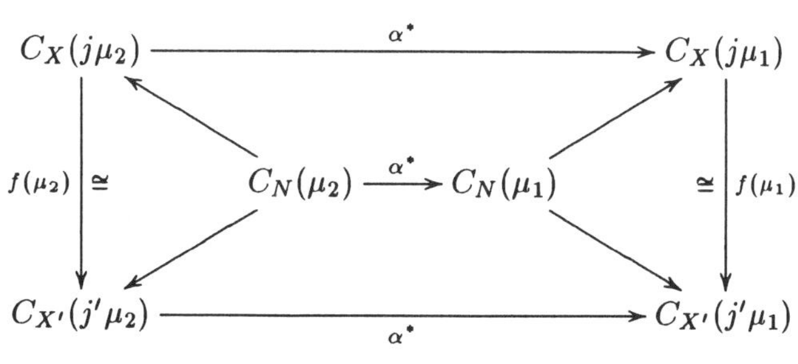

shows that $f(\mu_1)^{-1} \circ \alpha^* \circ f(\mu_2)$ is covered by α^*, we infer (6.4) that $\alpha^* \circ f(\mu_2)$ and $f(\mu_1) \circ \alpha^*$ are conjugate morphisms. □

We shall now employ the fact that the cohomology of BX is determined by $\mathbb{A}_{\leq 2}(X)$ to find conditions under which the cohomological image of the diagram of Definition 7.1 can be completed.

Proposition 7.15 *In the situation of (7.12), assume (1) and either (2) or (3) from the list*

1. $\mathrm{Aut}(V,\nu)$ *acts transitively on* $\mathrm{SPL}(V,\nu)$ *for all* $(V,\nu) \in \mathrm{Ob}(\mathbb{A}_2(X))$.
2. $\mathrm{Aut}(V,\nu')$ *acts transitively on* $\mathrm{SPL}(V,\nu')$ *for all* $(V,\nu') \in \mathrm{Ob}(\mathbb{A}_2(X'))$.
3. *Any two objects of* $\mathbb{A}_2(X)$ *with isomorphic centralizers of less than maximal rank are isomorphic.*

Then $H^*(Bj)H^*(BX;\mathbb{F}_p) = H^*(Bj')H^*(BX';\mathbb{F}_p)$ *in* $H^*(BN;\mathbb{F}_p)$.

Proof Assume first that hypotheses (1) and (2) hold. We are going to apply Oliver's cochain complex [**21**] to the covariant functor $H(X)$ from $\mathbb{A}(X)$ to the category of $\mathbb{F}_p$-modules given by $H(X)(V,\nu) = H^*(BC_X(V,\nu))$.

As noted above, there are bijective correspondences between the sets $\varepsilon_1(X) = \varepsilon_1(N) = \varepsilon_1(X')$ of isomorphism classes of objects of the rank one subcategories. Choose one representing object (L,μ) in each isomorphism class of $\mathbb{A}_1(N)$ and let $(L,\phi) = (L, j\mu)$ and $(L,\phi') = (L, j'\mu)$ be the corresponding objects of $\mathbb{A}_1(X)$ and $\mathbb{A}_1(X')$. The isomorphisms $f(\mu)\colon C_X(L,\phi) \to C_{X'}(L,\phi')$ under $C_N(L,\mu)$ induce an isomorphism

$$\Phi = \prod f(\mu)^* \colon \prod_{(L,\phi')\in\varepsilon_1(X')} H(X')(L,\phi') \longrightarrow \prod_{(L,\phi)\in\varepsilon_1(X)} H(X)(L,\phi)$$

which in turn induces an isomorphism of 0-cochains

$$\prod_{(L,\phi')\in\varepsilon_1(X')} \mathrm{Hom}(\mathrm{St}(L), H(X')(L,\phi')) \longrightarrow \prod_{(L,\phi)\in\varepsilon_1(X)} \mathrm{Hom}(\mathrm{St}(L), H(X)(L,\phi))$$

where $\mathrm{St}(L)$ is the Steinberg representation (in this case the trivial representation) and Hom is taken in the category of $\mathrm{Aut}(L,\phi) = \mathrm{Aut}(L,\phi')$ representations.

The morphisms j and j' induce full functors $\mathbb{A}_2(X) \longleftarrow \mathbb{A}_2(N) \longrightarrow \mathbb{A}_2(X')$ but these full subcategories of rank two objects are not necessarily isomorphic since in the rank two case special preferred lifts are not in general uniqely determined (6.2). However, by hypotheses (1) and (2), special preferred lifts are unique up to isomorphism and hence j and j' induce bijections $\varepsilon_2(X) = \varepsilon_2(N) = \varepsilon_2(X')$ of isomorphism classes of rank two objects in these three Quillen categories.

Oliver defines in [**21**, Proposition 5] for each isomorphism class $(E,\psi) \in \varepsilon_2(X)$ the coboundary map

$$\delta(E,\psi)\colon \prod_{(L,\phi)\in\varepsilon_1(X)} \mathrm{Hom}(\mathrm{St}(L), H(X)(L,\phi)) \longrightarrow \mathrm{Hom}(\mathrm{St}(E), H(X)(E,\psi))$$

as the homomorphism that to a collection $(c(L,\phi))$, $(L,\phi)\in\varepsilon_1(X)$, of $\mathrm{Aut}(L,\phi)$-homomorphisms of the domain associates the homomorphism

$$\mathrm{St}(E) \xrightarrow{R_E} \oplus_{[E:A]=p} \mathrm{St}(A) \xrightarrow{\oplus c(A,\psi|A)} \oplus_{[E:A]=p} H(X)(A,\psi|A) \xrightarrow{\oplus H(X)(i_A)} H(X)(E,\psi)$$

where R_E is a certain connecting homomorphism (which is surjective in this case where E has rank two) and $i_A\colon A \to E$ is the inclusion. I claim that

$\Phi(\ker\delta(E,\psi')) = \ker\delta(E,\psi)$, where (E,ψ) and (E,ψ') correspond under the identification $j'j^{-1}\colon \varepsilon_2(X) = \varepsilon_2(X')$.

Consider first the case where the centralizer $C_X(E,\psi)$ has maximal rank. Then (E,ψ) has a unique preferred lift (E,μ) and for any rank one subgroup A of E, $(A,\mu|A)$ is the preferred lift of $(A,\psi|A)$. According to (7.14), the diagram

$$\begin{array}{ccc}
\oplus H(X)(A,\psi|A) & \xrightarrow{\oplus H(X)(i_A)} & \oplus H(X)(E,\psi) \\
{\scriptstyle \oplus f(\mu|A)^*}\uparrow{\scriptstyle\cong} & & {\scriptstyle\cong}\uparrow{\scriptstyle \oplus f(\mu)^*} \\
\oplus H(X')(A,\psi'|A) & \xrightarrow[\oplus H(X')(i_A)]{} & \oplus H(X')(E,\psi')
\end{array}$$

commutes. It follows that $\ker\delta(E,\psi)$ and $\ker\delta(E,\psi')$ correspond under the isomorphism Φ in this case.

Next, consider the case where the rank of the centralizer $C_X(E,\psi)$ is less than maximal. The $p+1$ special preferred lifts μ_A of ψ are indexed (6.2) by the rank one subgroups $A < E$. By hypothesis (1), $\mu_B = \mu_A\beta$ for some automorphism β of (E,ψ). Then $\beta(B) = A$, $(B,\mu_B|B) = (B,\mu_A\beta|B)$ is isomorphic to $(A,\mu_A|A)$, and $(B,j\mu_B) = (B,\psi|B)$ is isomorphic to $(A,j\mu_A) = (A,\psi|A)$. Hence the restrictions $(A,\psi|A)$ and $(B,\psi|B)$ are isomorphic objects for all rank one subgroups $A,B < E$. Let (L,ϕ_E) be this common isomorphism class and $f_A\colon (L,\phi_E)\to(A,\psi|A)$ an isomorphism. The commutative diagram

$$\begin{array}{ccccc}
\oplus\,\mathrm{St}(A) & \xrightarrow{\oplus c(A,\psi|A)} & \oplus H(X)(A,\psi|A) & \xrightarrow{\oplus H(X)(i_A)} & \oplus H(X)(E,\psi) \\
{\scriptstyle\cong}\uparrow{\scriptstyle \oplus\,\mathrm{St}(f_A)} & & {\scriptstyle\cong}\uparrow{\scriptstyle \oplus H(X)(f_A)} & & \| \\
\oplus\,\mathrm{St}(L) & \xrightarrow[\oplus c(L,\phi_E)]{} & \oplus H(X)(L,\phi_E) & \xrightarrow[H(X)(f_A i_A)]{} & \oplus H(X)(E,\psi)
\end{array}$$

shows that $\ker\delta(E,\psi)$ consists of those

$$(c(L,\phi)) \in \prod_{(L,\phi)\in\varepsilon_1(X)} \operatorname{Hom}(\underset{L}{\mathrm{St}}, H(X)(L,\phi))$$

for which the coordinate $c(L,\phi_E) = 0$. Similarly, $\ker\delta(E,\psi')$ is described by the condition that $c(L,\phi'_E) = 0$ where (L,ϕ_E) corresponds to (L,ϕ'_E) under the identification $j'j^{-1}\colon \varepsilon_1(X) = \varepsilon_1(X')$. Therefore, $\Phi(\ker\delta(E,\psi')) = \ker\delta(E,\psi)$ also in this case of centralizers of less than maximal rank.

From Oliver [**21**] and Dwyer and Wilkerson [**6**, 1.2] we know that the intersection of the kernels $\ker\delta(E,\psi)$ as (E,ψ) ranges over $\varepsilon_2(X)$ equals

$$\lim\nolimits^0_{\mathbb{A}(X)} H(X) = H^*(BX),$$

and similarly for X', and hence we obtain a commutative diagram of monomorphisms

$$\begin{array}{ccccc}
H^*(BX) & \xrightarrow{H^*(Bj)} & H^*(BN) & \xleftarrow{H^*(Bj')} & H^*(BX') \\
\downarrow & & \downarrow & & \downarrow \\
\prod_{(L,\phi)\in\varepsilon_1(X)} H(X)(L,\phi) & \longrightarrow & \prod_{(L,\mu)\in\varepsilon_1(N)} H(N)(L,\mu) & \longleftarrow & \prod_{(L,\phi')\in\varepsilon_1(X')} H(X')(L,\phi') \\
 & & \xleftarrow[\Phi]{\cong} & &
\end{array}$$

where the isomorphism Φ restricts to an isomorphism between $H^*(BX')$ and $H^*(BX)$. (Use (6.5) to get injectivity of the middle vertical arrow.)

Assume next that hypotheses (1) and (3) hold. As above, hypothesis (1) shows that the full functor $j\colon \mathbb{A}_2(N) \to \mathbb{A}_2(X)$ induces a bijection $j\colon \varepsilon_2(N) \to \varepsilon_2(X)$ of objects and we obtain a surjection

$$j'j^{-1}\colon \varepsilon_2(X) \twoheadrightarrow \varepsilon_2(X') \tag{7.16}$$

by composing j^{-1} with the surjection $j'\colon \varepsilon_2(N) \to \varepsilon_2(X')$. Since $C_X(V,\nu)$ is assumed to be N-determined, $C_X(V,\nu)$ is isomorphic to $C_{X'}(j'j^{-1}(V,\nu))$ for all $(V,\nu) \in \mathrm{Ob}(\mathbb{A}_2(X))$. In particular, the surjection (7.16) takes objects with centralizers of less than full rank to objects of the same kind. By hypothesis (3), (7.16) is injective on this subset of $\varepsilon_2(X)$. But (7.16) is also injective on the set of isomorphism classes of objects with full rank centralizers since special preferred lifts are unique in this case. Hence, (7.16) is in fact a bijection.

Let (E,ψ) be an object of $\mathbb{A}_2(X)$ and (E,ψ') an object of $\mathbb{A}(X')$ representing isomorphism classes that correspond under the bijection (7.16). Suppose that the common rank of the centralizers $C_X(E,\psi)$ and $C_{X'}(E,\psi')$ is less then maximal. Let $A < E$ be any rank one subgroup of E. As before, the isomorphism class of $(A,\psi|A)$ is independent of the choice of A and since $(A,\psi'|A)$ corresponds to $(A,\psi|A)$ under the bijection $\varepsilon_1(X) = \varepsilon_1(X')$, also the restriction $(A,\psi'|A)$ is independent, up to isomorphism, of $A < E$.

Taking these observations into account, the above argument under the assumption of (1) and (2) also applies under the assumption of (1) and (3). $\square$

It can be shown that the p-compact group $BX = B\,\mathrm{PU}(n)^\wedge_p$, $n \geq 2$, $p > 2$, satisfies 7.15.(1) and 7.15.(3). In fact, there does not exist any monomorphism $(\mathbb{Z}/p)^2 \rightarrowtail X$ with centralizer of rank $< n-1$ unless p divides n in which case there exists an essentially unique such monomorphism and its automorphism group in $\mathbb{A}(X)$ is $\mathrm{SL}(2,p)$ which acts transitively on the set (6.2) of special preferred lifts.

When $H^*(BX)$ and $H^*(BX')$ have the same image in $H^*(BN)$ there exists an isomorphism Φ of algebras over the Steenrod algebra that makes the diagram

$$\begin{array}{ccc} & H^*(BN) & \\ {\scriptstyle H^*(Bj)}\nearrow & & \nwarrow{\scriptstyle H^*(Bj')} \\ H^*(BX) & \xleftarrow[\Phi]{\cong} & H^*(BX') \end{array}$$

commute. Since Quillen categories, by Lannes theory **[12]**, only depend on cohomological information, this isomorphism induces an isomorphism $\Phi\colon \mathbb{A}(X) \to \mathbb{A}(X')$ of categories. On objects, $\Phi(V, j\mu) = (V, j'\mu)$ for all $(V,\mu) \in \mathrm{Ob}(\mathbb{A}(N))$, and $\Phi(\alpha) = \alpha$ for any morphism.

Proposition 7.17 *In the situation of (7.12), assume that* $H^*(Bj)H^*(BX) = H^*(Bj')H^*(BX')$ *in* $H^*(BN)$ *and that*

1. $\mathrm{Aut}(V,\nu)$ *acts transitively on* $\mathrm{SPL}(V,\nu)$ *for all objects* (V,ν) *of* $\mathbb{A}(X)$.
2. *The isomorphism* $f(\mu)\colon C_X(j\mu) \to C_{X'}(j'\mu)$ *under* $C_N(\mu)$ *is* $\mathrm{Aut}(j\mu)$-*equivariant for any object* (V,μ) *of* $\mathbb{A}(N)$ *with centralizer of less than maximal rank.*
3. $\lim^2_{\mathbb{A}(X)} \pi_1(BZ(V,\nu)) = 0 = \lim^3_{\mathbb{A}(X)} \pi_2(BZ(V,\nu))$.

Then there exists an isomorphism $f\colon X \to X'$ *such that* $f \circ j$ *is conjugate to* j'.

Proof Hypotheses (1) and (2) imply that all choices of special preferred lift (V,μ) of (V,ν) lead to the same isomorphism $f(\mu)\colon C_X(V,\nu) \to C_{X'}(\Phi(V,\nu))$: By hypothesis (1), all special preferred lifts of (V,ν) have the form $(V,\mu\alpha)$ for some $\alpha \in \operatorname{Aut}(V,\nu) = \operatorname{Aut}(\Phi(V,\nu))$ and (7.14) $f(\mu\alpha) = \alpha^* \circ f(\mu) \circ (\alpha^*)^{-1} = f(\mu)$ if we assume, as in hypothesis (2), that $f(\mu)$ is equivariant. Let $f(V,\nu)\colon C_X(V,\nu) \to C_{X'}(\Phi(V,\nu))$ denote $f(\mu)$ for any special preferred lift μ.

Then (7.14) the centric morphisms

$$C_X(V,\nu) \stackrel{f(V,\nu)}{\longrightarrow} C_{X'}(\Phi(V,\nu)) \stackrel{e(\Phi(V,\nu))}{\longrightarrow} X'$$

define an element of $\lim^0_{\mathbb{A}(X)}[BC_X(V,\nu), BX']$. By hypothesis (3), see also (2.2), there exists an isomorphism $f\colon X \to X'$ such that $f \circ e(V,\nu) = e(\Phi(V,\nu)) \circ f(V,\nu)$ for all objects (V,ν) of $\mathbb{A}(X)$.

It remains to show that $f \circ j \simeq j'$. By construction, $f \circ j|C_N(\mu) \simeq j'|C_N(\mu)$ for any special preferred lift $\mu\colon V \to N$. Take $V = \mathbb{Z}/p$ and take $\mu\colon V \to T$ to be central in the Sylow p-subgroup N_p (6.5). It then follows that $f \circ j|N_p \simeq j'|N_p$ and, in particular, that $f \circ j|T\colon T \to X'$ is the maximal torus of X'.

The set of homotopy classes of maps $BN \longrightarrow BX'$ under BT,

$$\pi_0(\operatorname{map}(BT, BX')^{hW_T(X')}_{Bi'}) = H^2(W_T(X'); \pi_2(BT)),$$

injects into the set of homotopy classes of maps $BN_p \longrightarrow BX'$ under BT,

$$\pi_0(\operatorname{map}(BT, BX')^{hW_T(X')_p}_{Bi'}) = H^2(W_T(X')_p; \pi_2(BT)).$$

In particular, $f \circ j$ and j' must be conjugate since their restrictons to N_p are. □

It is clear that more work has to be done here before we can claim that the vanishing of the obstruction groups in condition 7.17.(3) for any centerfree, connected p-compact group will imply that all p-compact groups are N-determined.

Remark 7.18 By the Splitting Theorem of [**8**], it represents no loss of generality to assume that X and X' are simple p-compact groups in Proposition 7.17. (This is also necessary for hypothesis (1).) Note also that all assumptions (1), (2), and (3) of (7.15) as well as (1) and (2) of (7.17) are satisfied in case $H^*(BX)$ injects into $H^*(BT)$ for then all centralizers of elementary abelian p-groups have maximal rank and preferred lifts are unique.

Example 7.19 Let $X = \mathrm{PU}(3)$ and $p = 3$. All objects of $\mathbb{A}(X)$ have centralizers that have N-determined automorphisms and are N-determined since they are p-compact toral groups or (4.3,7.9) their Weyl groups have order prime to p. The automorphism group, $\mathrm{SL}(2,3)$, of $V_2 \in \mathrm{Ob}(\mathbb{A}_2(X))$, the only isomorphism class with a centralizer of less then maximal rank, acts transitively on the set $\mathrm{SPL}(V_2)$ of (6.2) rank one subgroups of V_2. Therefore (7.15), $H^*(BX)$ and $H^*(BX')$ have the same image in $H^*(BN)$ in the situation of (7.12). Any special preferred lift $\mu\colon V_2 \to N$ of the centric morphism $V_2 \longrightarrow X$ must also be centric. Using this, the diagram $C_X(V_2) \longleftarrow C_N(\mu) \longrightarrow C_{X'}(V_2)$ identifies to a diagram $V_2 \longleftarrow V_2 \longrightarrow V_2$ of $\mathrm{SL}(2,3)$-equivariant isomorphisms. Thus also $f(\mu)\colon C_X(V_2) \to C_{X'}(V_2)$ is $\mathrm{SL}(2,3)$-equivariant up to homotopy. Finally, the limits $\lim^2_{\mathbb{A}(X)} \pi_1(BZ(V,\nu)) = 0 = \lim^3_{\mathbb{A}(X)} \pi_2(BZ(V,\nu))$ for general reasons [**21**]. Hence $\mathrm{PU}(3)$ is N-determined at $p = 3$. (See [**1**] for a related result.)

When combined with (4.8, 4.11, 7.10) this example shows, independently of [**10, 11**], that U(3), SU(3), and PU(3) are totally N-determined at the prime $p = 3$;

it follows (7.2) for instance that the automorphism group of PU(3) is Out(PU(3)) = $\mathbb{Z}_3^*$.

References

[1] Broto, C. and Viruel, A. [1995], *Homotopy uniqueness of $BPU(3)$*, preprint.

[2] Dwyer, W. G. and Kan, D. M. [1992], *Centric maps and realizations of diagrams in the homotopy category*, Proc. Amer. Math. Soc. **114**, 575–584.

[3] Dwyer, W. G., Miller, H. R. and Wilkerson, C. W. [1987], *The homotopy uniqueness of BS^3*, in Algebraic Topology (J. Aguadé and R. Kane, eds.), Barcelona 1986. Lecture Notes in Mathematics, vol. 1298, pp. 90–105, Springer-Verlag, Berlin–Heidelberg–New York–London–Paris–Tokyo.

[4] Dwyer, W. G., Miller, H. R. and Wilkerson, C. W. [1992], *Homotopical uniqueness of classifying spaces*, Topology **31**, 29–45.

[5] Dwyer, W. G. and Wilkerson, C. W. [1994], *Homotopy fixed point methods for Lie groups and finite loop spaces*, Ann. of Math. (2) **139**, 395–442.

[6] Dwyer, W. G. and Wilkerson C. W. [1992], *A cohomology decomposition theorem*, Topology 31, 433–443.

[7] Dwyer, W. G. and Wilkerson, C. W. [1995], *The center of a p-compact group*, in The Čech Centennial (M. Cenkl and H. Miller, eds.). Contemporary Mathematics, vol. 181, pp. 119–157, American Mathematical Society, Providence, R. I.

[8] Dwyer, W. G. and Wilkerson, C. W. [1995], *Product splittings for p-compact groups*, Fund. Math. **147**, 279–300.

[9] Jackowski, S., McClure, J. and Oliver, R. [1995], *Self Homotopy Equivalences of classifying spaces of compact connected Lie groups*, Fund. Math. **147**, 99–126.

[10] Jackowski, S., McClure, J. and Oliver, R. [1992], *Homotopy classification of self-maps of BG via G-actions, Part I*, Ann. of Math. **135** (2), 183–226.

[11] Jackowski, S., McClure, J. and Oliver, R. [1992], *Homotopy classification of self-maps of BG via G-actions, Part II*, Ann. of Math. **135** (2), 227–270.

[12] Lannes, J. [1992], *Sur les espaces fonctionnels dont la source est le classifiant d'un p-group abélien élémentaire*, Publ. I.H.E.S **75**, 135–244.

[13] Lannes, J. [1995], *Theorie homotopique des groupes de Lie (d'aprés W.G. Dwyer and C.W. Wilkerson)*, Astérisque **227**, 21–45, (Séminaire Bourbaki, Vol. 1993/94, Exp. no. 776).

[14] Møller, J. M. [1996], *Rational Isomorphisms of p-compact Groups*, Topology **35**, 201–225.

[15] Møller, J. M. [1996], *Extensions of p-compact groups*, in Algebraic Topology: New Trends in Localization and Periodicity (C. Broto, C. Casacuberta and G. Mislin, eds.), Progress in Mathematics, vol. 136", pp. 307–327, Birkhäuser, Basel–Boston–Berlin.

[16] Møller, J. M. [1995], *Normalizers of maximal tori*, preprint.

[17] Møller, J. M. and Notbohm, D. [1994], *Centers and Finite Coverings of Finite Loop Spaces*, J. reine angew. Math. **456**, 99–133.

[18] Møller, J. M. [1995], *Homotopy Lie groups*, Bull. Amer. Math. Soc. **32**, 413–428.

[19] Notbohm, D. [1994], *Homotopy uniqueness of classifying spaces of compact connected Lie groups at primes dividing the order of the Weyl group*, Topology **33**, 271–330.

[20] Notbohm, D. [1995], *Classifying spaces of compact Lie groups*, in Handbook of Algebraic Topology (I. M. James, ed.), pp. 1049–1094, North-Holland, Amsterdam-Lausanne-New York-Oxford-Shannon-Tokyo.

[21] Oliver, B. [1994], *Higher limits via Steinberg representations*, Comm. Algebra **22**, 1381-1393.

[22] Wojtkowiak, Z. [1987], *On maps from* holim F *to* Z, in Algebraic Topology (J. Aguadé and R. Kane, eds.), Barcelona 1986. Lecture Notes in Mathematics, vol. 1298, pp. 227–236, Springer Verlag, Berlin–Heidelberg–New York–London–Paris–Tokyo.

Fields Institute Communications
Volume **19**, 1998

Space Exponents for Loop Spaces of Spheres

Paul Selick
Department of Mathematics
University of Toronto
Toronto, Ontario, Canada M5T 1A1
selick@math.utoronto.ca

Abstract. Let $S^{2n+1}\langle 2n+1\rangle$ denote the $(2n+1)$-connected cover of S^{2n+1}. For odd primes it is known that in the H-space $\Omega^{2n}S^{2n+1}\langle 2n+1\rangle_{(p)}$ the p^nth power map $(x \mapsto x^{p^n})$ is null homotopic, but that in $\Omega^{2n-2}S^{2n+1}\langle 2n+1\rangle_{(p)}$ the kth power map is essential for any k. The main purpose of this paper is to plug the gap by using a theorem of Anick to show that the p^nth power map is null homotopic on $\Omega^{2n-1}S^{2n+1}\langle 2n+1\rangle_{(p)}$.

Introduction

An integer M is called an exponent for the torsion of an abelian group G if $M * (\text{torsion of } G) = 0$. We say that M is a homotopy exponent for a space X if M is an exponent for $\pi_k(X)$ for all k. If X is an associative H-space with homotopy exponent M, one can ask if X satisfies the stronger condition that the Mth power map, $x \to x^M$, is null homotopic; if it is then we say that M is an H-space exponent for X. If p is a prime then for each of these concepts the statement that M is a mod p exponent means that M becomes an exponent after localization at p.

In this author's thesis and follow-up papers ([**12**] and [**13**]) it was shown that p is a mod p homotopy exponent for S^3 when p is an odd prime. In fact, after localization at p, a homotopy decomposition $\Omega^2 S^{2p+1}\{p\} \approx \Omega^2 S^3\langle 3\rangle \times C(p)$ was given, where $X\{p\}$ denotes the homotopy-theoretic fibre of the pth power map of an associative H-space X, $\Omega^2 S^3\langle 3\rangle$ is the universal covering space of $\Omega^2 S^3$, and $C(n)$ denotes the homotopy theoretic fibre of the double suspension map $E^2 : S^{2n-1} \to \Omega^2 S^{2n+1}$. A consequence of this is that $\Omega^2 S^3\langle 3\rangle$ has mod p H-space exponent p for such p.

A sequence of papers by Cohen, Moore, and Neisendorfer around 1980 ([**4**], [**5**], [**6**]) established for arbitrary n that p^n is the best possible (i.e., least) mod p homotopy exponent for S^{2n+1} when $p \geq 5$ and this was extended by Neisendorfer [**9**] to cover the case $p = 3$ as well.

For $k \leq 2n+1$, it is not possible for $\Omega^k S^{2n+1}{}_{(p)}$ to have H-space exponent M for any integer M because its bottom homotopy group is $\mathbb{Z}_{(p)}$. If one knocks off this

1991 *Mathematics Subject Classification.* 55P35, 55Q40.
Research partially supported by a grant from NSERC.

copy of the integers then the remaining homotopy groups are torsion, so the possibility of a mod p homotopy exponent exists. Let $S^n\langle n\rangle$, called the n-connected cover of S^n, denote the homotopy-theoretic fibre of a map $S^n \to K(\mathbb{Z}, n)$ which represents a of generator of $\pi_n(S^n)$. Cohen, Moore, and Neisendorfer for $p \geq 5$ and Neisendorfer for $p = 3$ actually showed that p^n is an H-space exponent for $\Omega^k S^{2n+1}\langle 2n+1\rangle_{(p)}$ when $k \geq 2n$. Neisendorfer and Selick [**10**] showed $\Omega^k S^{2n+1}\langle 2n+1\rangle_{(p)}$ has no H-space exponent when $k < 2n-1$, leaving open the case $k = 2n-1$.

The main purpose of this paper is to use the work of Anick [**1**] to deloop the decomposition of $\Omega^2 S^{2p+1}\{p\}$ given in [**13**] and use this to show that $\Omega^{2n-1} S^{2n+1}\langle 2n+1\rangle_{(p)}$ has H-space exponent p^n when $p \geq 5$, plugging the hole between the positive exponent statements of ([**4**], [**5**], [**6**]), and the negative exponent statements of [**10**]. Note that it was shown in [**14**] that such a homotopy decomposition of $\Omega S^{2n+1}\{p\}$ cannot exist for $n \neq 1$ or p.

Although the corresponding questions at the prime 3 are left open, I have no reason to suppose that the results of this paper or indeed the major theorems of [**1**] do not hold in that case. For $p = 2$, the homotopy exponent 2^{2n} for S^{2n+1} was first shown by James and reformulated as an H-space exponent for $\Omega^{2n} S^{2n+1}\langle 2n+1\rangle_{(2)}$ by Moore (unpublished course notes). For this prime, it is known that 2^n cannot be a homotopy exponent for S^{2n+1} in general, since $\pi_6(S^3)$ is a counterexample, however the best possible exponents are not known. The best exponent known is $2^{3n/2+1}$, given in [**15**]. Although the results of [**15**] were given only in terms of homotopy exponents, they can be reformulated in terms of H-space exponents for the spaces $\Omega^{2n} S^{2n+1}\langle 2n+1\rangle_{(2)}$ as demonstrated by Richter [**11**]. (See also Cohen [**3**].) The question of whether $\Omega^{2n-1} S^{2n+1}\langle 2n+1\rangle_{(2)}$ has any H-space exponent remains open.

Added in proof. In his recent thesis, Steve Theriault [**16**] has shown that Anick's fibration is valid for the prime 3 and consequently the results of this paper are also valid when $p = 3$.

Another proof which emerged from discussion of this author with Neisendorfer some 10 years ago (with the main input coming from his side, as I recall) of the negative exponent results mentioned from [**10**] is also given in this paper. Although that result is already known, the new proof, an application of the Sullivan conjecture proved by Miller, is so slick that it deserves to be recorded into the literature.

1 Existence of Space Exponents

Throughout this section, let p be a prime greater than 3 and assume that all spaces and maps have been localized at p. In particular, S^{2n+1} becomes a homotopy associative H-space under these assumptions.

For each integer n Anick constructs a space $T_\infty^{2n-1}\{p\}$ together with a map $T_\infty^{2n-1}\{p\} \to \Omega S^{2n+1}\{p\}$ (subsequently shown by Anick and Gray to be an H-map [**2**]) which fits in a commutative diagram

$$\begin{array}{ccccc} C(n) & \longrightarrow & T_\infty^{2n-1}\{p\} & \longrightarrow & \Omega S^{2n+1}\{p\} \\ \downarrow & & \| & & \downarrow \\ S^{2n-1} & \longrightarrow & T_\infty^{2n-1}\{p\} & \longrightarrow & \Omega S^{2n+1} \end{array}$$

in which the rows are homotopy fibrations.

The space $S^3\langle 3\rangle$ is a loop space with a delooping $B\left(S^3\langle 3\rangle\right)$ which is the homotopy fibre of a map $BS^3 \to K(\mathbb{Z},4)$ which represents a generator of $H^4(BS^3)$. Let $\alpha : S^{2p+1} \to B\left(S^3\langle 3\rangle\right)$ represent a generator of $\pi_{2p+1}\left(B(S^3\langle 3\rangle)\right)$, which is the least nonvanishing homotopy group of $B\left(S^3\langle 3\rangle\right)$. As is well known, and can be verified by calculating $H_*(S^3\langle 3\rangle)$, the order of α is p. Let $\bar{\alpha}$ be a map of homotopy-theoretic fibres induced from the homotopy commutative square at the bottom of the diagram

$$\begin{array}{ccc} S^{2p+1}\{p\} & \xrightarrow{\bar{\alpha}} & S^3\langle 3\rangle \\ \downarrow & & \downarrow \\ S^{2p+1} & \xrightarrow{*} & PB\left(S^3\langle 3\rangle\right) \\ \downarrow{\scriptstyle p} & & \downarrow \\ S^{2p+1} & \xrightarrow{\alpha} & B\left(S^3\langle 3\rangle\right). \end{array}$$

In general the homotopy class of $\bar{\alpha}$ would not be uniquely determined by such a diagram, however for any choice of $\bar{\alpha}$ the composite $\Omega S^{2p+1} \xrightarrow{\partial} S^{2p+1}\{p\} \xrightarrow{\bar{\alpha}} S^3\langle 3\rangle$ is homotopic to $\Omega\alpha$, where ∂ is induced from the homotopy fibration $S^{2p+1}\{p\} \to S^{2p+1} \xrightarrow{p} S^{2p+1}$. Since $\pi_{2p+1}\left(S^3\langle 3\rangle\right) = 0$, in this particular case the diagram does in fact uniquely determine the homotopy class of $\bar{\alpha}$, although we will not need to use this.

Theorem 1.1 $\Omega S^{2p+1}\{p\} \approx \Omega S^3\langle 3\rangle \times B(p)$, *where* $B(p)$ *is a delooping of* $C(p)$.

Proof Let F be the homotopy-theoretic fibre of $\bar{\alpha}$. Then $\Omega^2 F \approx C(p)$ as shown in [**13**]. As Hopf algebras, $H_*(T_\infty^{2p-1}\{p\};\mathbb{Z}/p\mathbb{Z}) \cong H_*(\Omega S^3\langle 3\rangle;\mathbb{Z}/p\mathbb{Z}) \cong (\mathbb{Z}/p\mathbb{Z})[x] \otimes E[\beta x]$, where $|x| = 2p$ and $E[x]$ denotes the exterior algebra on the single generator x. The map induced on $H_*(\ ;\mathbb{Z}/p\mathbb{Z})$ by the composition $T_\infty^{2p-1}\{p\} \longrightarrow \Omega S^{2p+1}\{p\} \xrightarrow{\Omega\bar{\alpha}} \Omega S^3\langle 3\rangle$ is an isomorphism on $H_{2p-1}(\ ;\mathbb{Z}/p\mathbb{Z})$ and so commutativity with the Bocksteins (including higher Bocksteins) and coalgebra structure forces it to be a homology isomorphism. (Using commutativity with the coalgebra structure is sufficient to prove the isomorphism, but of course it is even easier if the aforementioned result of Anick-Gray is used, since it implies that the composition induces a homomorphism of Hopf algebras.) Therefore the composition is a homotopy equivalence and the theorem follows. □

Remark The fact that $T_\infty^{2p-1}\{p\} \approx \Omega S^3\langle 3\rangle$ was certainly known to Anick, although his statement, "It is true that $T_\infty^{2p-1}\{p\} \approx \Omega S^3\langle 3\rangle$ but the proof of this fact is beyond the scope of this book." suggests that he had a more complicated proof in mind.

Theorem 1.2 $\Omega S^3\langle 3\rangle$ *has* H*-space exponent* p.

Proof This follows from the fact that $\Omega S^{2p+1}\{p\}$ has H-space exponent p and that $\Omega S^{2p+1}\{p\} \xrightarrow{\Omega\bar{\alpha}} \Omega S^3\langle 3\rangle$ is an H-map with a right homotopy splitting. □

Corollary 1.3 $\Omega^{2n-1}S^{2n+1}\langle 2n+1\rangle$ *has* H*-space exponent* p^n.

Proof Cohen-Moore-Neisendorfer construct maps $\bar{\pi} : \Omega^2 S^{2n+1}\langle 2n+1\rangle \to S^{2n-1}\langle 2n-1\rangle$ having the property that

$$\overline{E^2} \circ \bar{\pi} = p : \Omega^2 S^{2n+1}\langle 2n+1\rangle \to \Omega^2 S^{2n+1}\langle 2n+1\rangle$$

where $\overline{E^2}$ is induced from the double suspension map $E^2 : S^{2n-1} \to \Omega^2 S^{2n+1}$. Therefore $\Omega^{2n-1} S^{2n+1}\langle 2n+1\rangle$ has H-space exponent p^n by Theorem 1.2 and induction on n. □

2 Nonexistence of Space Exponents

In this section, the results hold after localization at any prime, including 2 and 3. $S^{2n+1}\langle 2n+1\rangle$ denotes the homotopy fibre of a generator $S^{2n+1} \to K(\mathbb{Z}, 2n+1)$ of $\pi_{2n+1}\big(K(\mathbb{Z}, 2n+1)\big)$.

Let G be a locally finite group. Then Miller's theorem (Sullivan conjecture) says that $\mathrm{Map}_*(BG, X) \approx *$ for any finite complex X, where $\mathrm{Map}_*(Y, X)$ denotes the space of pointed maps from Y to X. Therefore the homotopy fibration

$$\mathrm{Map}_*\big(BG, K(\mathbb{Z}, 2n)\big) \to \mathrm{Map}_*(BG, S^{2n+1}\langle 2n+1\rangle) \to \mathrm{Map}_*(BG, S^{2n+1})$$

which comes from the definition of $S^{2n+1}\langle 2n+1\rangle$ implies that

$$\mathrm{Map}_*\big(BG, K(\mathbb{Z}, 2n)\big) \approx \mathrm{Map}_*(BG, S^{2n+1}\langle 2n+1\rangle).$$

Applying π_{2n-2} gives

$$[BG, K(\mathbb{Z}, 2)] \cong [BG, \Omega^{2n-2} S^{2n+1}\langle 2n+1\rangle].$$

If $\Omega^{2n-2} S^{2n+1}\langle 2n+1\rangle$ had H-exponent M, then $[Y, \Omega^{2n-2} S^{2n+1}\langle 2n+1\rangle]$ would have exponent M for any space Y and in particular $[BG, K(\mathbb{Z}, 2)]$ would have a uniform exponent M independent of the group G. However $[BG, K(\mathbb{Z}, 2)] \cong H^2(BG)$ clearly does not have this property. Thus we have proved

Theorem 2.1 (Neisendorfer-Selick) *If $k \leq 2n-2$ then $\Omega^k S^{2n+1}\langle 2n+1\rangle$ has no H-space exponent.*

References

[1] Anick, D. J. [1993], *Differential algebras in topology*, Research Notes in Mathematics, A. K. Peters, Vol. 3.

[2] Anick, D. J. and Gray, B. [1995], *Small H-spaces related to Moore spaces*, Topology **34**, 859–881.

[3] Cohen, F. [1987], *A course in some aspects of classical homotopy theory*, Algebraic Topology, Proceedings, Seattle 1985, Springer-Verlag.

[4] Cohen, F., Moore, J. and Neisendorfer, J. [1979], *Torsion in homotopy groups*, Ann. of Math **109**, 121–168.

[5] Cohen, F., Moore, J. and Neisendorfer, J. [1979], *The double suspension and exponents of the homotopy groups of spheres*, Ann. of Math **110**, 549–565.

[6] Cohen, F., Moore, J. and Neisendorfer, J. [1987], *Exponents in homotopy theory*, Algebraic Topology and Algebraic K-theory, Ann. of Math. Studies **113**, 3–87.

[7] James, I. [1955], *Reduced product spaces*, Ann. of Math. **62**, 170–197.

[8] Miller, H. [1984], *The Sullivan conjecture on maps from classifying space*, Ann. of Math. **120**, 39–87.

[9] Neisendorfer, J. [1981], *3-primary exponents*, Math. Proc. Camb. Phil. Soc. **90**, 63–83.

[10] Neisendorfer, J. and Selick, P. [1981], *Some examples of spaces with or without exponents*, Can. Math. Soc. Proc. 2, Part 1, 343–357.

[11] Richter, B., *The H-space squaring map on $\Omega^3 S^{4N+1}$ factors through the double suspension*, to appear.

[12] Selick, P. [1978], *Odd primary torsion in $\pi_*(S^3)$*, Topology **17**, 407–412.

[13] Selick, P. [1981], *A decomposition of* $\pi_*(S^{2p+1}; \mathbb{Z}/p\mathbb{Z})$, Topology **20**, 175–177.

[14] Selick, P. [1983], *A reformulation of the Arf invariant* 1 *mod* p *problem and applications to atomic spaces*, Pac. J. Math. **108**, 431–449.

[15] Selick, P. [1984], *Mod* 2 *exponents for the homotopy groups of spheres*, Topology **23**, 97–99.

[16] Theriault, S. [1997], *A reconstruction of Anick's fibration* $S^{2n-1} \to T^{2n-1} \to \Omega S^{2n+1}$, thesis, University of Toronto.

Fields Institute Communications
Volume **19**, 1998

On f-Localization Functors and Connectivity

Jin-yen Tai
Department of Mathematics
Rutgers University
New Brunswick, NJ USA 08903
Current address: Department of Mathematics
Dartmouth College
Hanover, NH USA 03755
`jtai@math.rutgers.edu`

Abstract. Using the pointed mapping space, Dror Farjoun defined an f-localization functor L_f on the pointed homotopy category of CW-complexes for any map f. We show that nontrivial functors are those represented by maps f with $\pi_0(f)$ a bijection. We also study when connectivity is preserved by an f-localization functor. Finally, we study when an f-localization functor commutes with Postnikov sections and connective covers, and by applying these results to complex K-theory, we recover some results of Meier and Mislin.

1 Introduction

In [**5**], E. Dror Farjoun defined for any map f, an f-localization functor L_f on the category of unpointed or pointed simplicial sets. It is because f-localization functor can be constructed on the category of unpointed simplicial sets, many useful constructions, for example, fiberwise f-localization, can be implemented. In this paper, we restrict our attention to f-localization functor on pointed simplicial sets. The functor L_f on pointed simplicial sets induces an f-localization functor on the homotopy category of pointed CW-complexes, denoted by Ho_*. This type of localization generalizes all known localization functors. Postnikov sections, p-localization, Anderson localization, Bendersky localization and Bousfield homological localization are a few examples. It is also conjectured by E. Dror Farjoun that every monad in Ho_* is an f-localization functor. For simplicity, we work in Ho_*. We treat space (resp., connected) and CW-complex (resp., path-connected) as synonyms.

In this paper, we observe that (Theorem A) in order to have a nontrivial f-localization functor L_f, it is necessary and sufficient that $\pi_0(f)$ be a bijection. Moreover, for any nontrivial L_f, we can find a map g between connected spaces such that L_f and L_g agree on connected spaces. To prove these results, we introduce an equivalence relation on maps, called similarity. When the map is $W \to *$ or

---

1991 *Mathematics Subject Classification.* Primary 55P60.

$* \to W$, this notion was defined and used by A. K. Bousfield and E. Dror Farjoun (cf. [**3**] and [**6**]).

We go on to study when connectivity is preserved by an f-localization functor. The motivation for this problem comes from J. A. Neisendorfer who first proved in [**9**] that if f is a map from a connected space to a simply connected space, then L_f preserves simple connectivity. We show that (Theorem B and C) if the homotopy cofiber (mapping cone) C_f of f is n-connected and $j \leq n$, then for any j-connected space Y, $L_f Y$ is also j-connected. Moreover, if the map f induces isomorphisms on integral homology groups $\widetilde{H}_j(f)$ for all $j \leq n$, then we show that (cf. Sections 7 and 8) L_f commutes with Postnikov sections and connective covers to a certain extent.

Finally, we apply our results to homological localization with respect to complex K-theory. We recover some results of W. Meier [**7**] and G. Mislin [**8**].

For the sake of completeness, let us recall the definition of f-localization. Let a map $f\colon A \to B$ between spaces be fixed throughout. A space Y is called *f-local* if

$$f^* = \mathrm{map}_*(f, Y)\colon \mathrm{map}_*(B, Y) \to \mathrm{map}_*(A, Y)$$

is a weak equivalence. A map $g\colon C \to D$ is called an *L_f-equivalence* if for every f-local T, $g^*\colon \mathrm{map}_*(D, T) \to \mathrm{map}_*(C, T)$ is a weak equivalence. The *f-localization* of a space Y is a map $\eta\colon Y \to L_f Y$ in Ho_* which is an L_f-equivalence and for which $L_f Y$ is f-local. The f-localization η is the appropriate initial and terminal arrows in Ho_* in the following sense: (1) η is initial among all maps going out of Y with target an f-local space; (2) η is terminal among all L_f-equivalences going out of Y. If the map f is $W \to *$ or $* \to W$, we call f-local spaces *W-null.* We write P_W for L_f and call $\eta\colon X \to P_W X$ the *W-nullification* of X.

The organization of this paper is as follows: In Section 2, we list our three main results, Theorems A, B and C. In Section 3, we recall some basic properties of similarity classes and we define a numerical invariant $c(g)$ for each similarity class $\langle g \rangle$. We give a characterization of the condition $c(g) \geq n \geq 0$ in terms of the behavior of the localization functor L_g. We also prove Proposition 3.12. In Section 4, we prove Theorem A. In Section 5, we prove Theorem B and Theorem C. In Section 6, we study the special case when f is a map between low dimensional CW-complexes. In Sections 7 and 8, we show that, with some restrictions, the f-localization functor L_f commutes with Postnikov sections and connective covers. Finally, in Section 9, we give some applications.

2 Summary of results

In this section, we summarize the main results of this paper. The details, notations and proofs will be presented later.

We first begin by analyzing trivial f-localization functors L_f. Our result simply says that for any nontrivial localization functor L_f, over the category of connected spaces, we can always assume f is a map between connected spaces.

Theorem A Let f be a map between spaces. Then

(1) L_f is nontrivial if and only if $\pi_0(f)$ is a bijection.
(2) If $\pi_0(f)$ is a bijection, then there is a map g between connected spaces such that L_f and L_g agree on connected spaces.

Our next two results address the following question: When does L_f preserve connectivity? Let C_f denote the homotopy cofiber (mapping cone) of f. First, we prove

Theorem B Suppose a map f between connected spaces and a space Y satisfy one of the following two conditions:

(1) $\pi_1 C_f = 0$ and the space Y is arbitrary.

(2) $\widetilde{H}_1(C_f) = 0$ and $\pi_1 L_f Y$ is solvable.

Then $L_f Y$ is simply connected whenever Y is simply connected.

This generalizes the result of Neisendorfer we mentioned in the introduction. We can push Theorem B a little bit further to show:

Theorem C Let $n \geq 0$ and $j \leq n$. For any map f between connected spaces with n-connected cofiber C_f, if Y is j-connected, then $L_f Y$ is also j-connected.

Using Theorem C, with further assumptions, we show in Sections 7 and 8 that L_f commutes with the Postnikov sections and connective covers to a certain extent. We apply these results to the homological localization functor with respect to complex K-theory.

In Example 3.7, we define an invariant $c(E)$ for every generalized homology theory $E_*(-)$. We show that:

Proposition 3.12 $c(E) = 0$ for any nontrivial generalized homology theory $E_*(-)$.

This result gives a reason why homological localization functors are hard to study.

3 Similarity classes and the invariant $c(g)$

In this section, we first define the similarity class of a map g and give a partial ordering on similarity classes. Second, we define an invariant $c(g)$ of the similarity class $\langle g \rangle$ for any map g. We give a characterization of those L_g with $c(g) \geq n \geq 0$. We also show that $c(E) = 0$ for every nontrivial generalized homology theory $E_*(-)$ where $c(E)$ is defined in Example 3.7 below.

Definition 3.1 Two maps g and h are called *similar* if g and h have the same local spaces, i.e., $L_g = L_h$. We write $\langle g \rangle$ for the *similarity class* containing g.

Definition 3.2 Being given a map g, a space Y is called *g-acyclic* if $L_g Y \simeq *$.

For any map $g\colon C \to D$ and any map h, the universal property of the h-localization map $\eta_C\colon C \to L_h C$ implies that there is a map unique up to homotopy, denoted by $L_h(g)$, such that $L_h(g) \circ \eta_C \simeq \eta_D \circ g$ where η_D is the h-localization map of D.

Lemma 3.3 *Given any two maps g and h, the following five statements* (1) *to* (5) *are equivalent. We define $\langle g \rangle \leq \langle h \rangle$ to be any of these five equivalent statements.*

(1) *Every h-local space is g-local.*

(2) *Every L_g-equivalence is an L_h-equivalence.*

(3) *g is an L_h-equivalence.*

(4) *$L_h(g)$ is a homotopy equivalence.*

(5) *$L_g L_h = L_h = L_h L_g$.*

Moreover, any one of the above implies

(6) *Every g-acyclic space is h-acyclic.*

Proof It is clear from the definitions that (1) $\Rightarrow$ (2) $\Rightarrow$ (3) $\Rightarrow$ (1). (1) $\Rightarrow$ $L_gL_h = L_h$, $L_gL_h = L_h \Rightarrow$ (1) and $L_h = L_hL_g \Rightarrow$ (6) are also clear.

(2) $\Rightarrow L_h = L_hL_g$: For any space Y, $Y \to L_gY$ is an L_g-equivalence and hence an L_h-equivalence. $L_gY \to L_hL_gY$ is by definition an L_h-equivalence with target an h-local space. Hence, the composition $Y \to L_gY \to L_hL_gY$ is the h-localization map of Y. That is $L_h = L_hL_g$.

(1) $\Leftrightarrow$ (4): Consider the following commutative diagram, for any space T:

$$\begin{array}{ccc} \mathrm{map}_*(L_hD, T) & \xrightarrow{L_h(g)^*} & \mathrm{map}_*(L_hC, T) \\ \eta_D^* \downarrow & & \eta_C^* \downarrow \\ \mathrm{map}_*(D, T) & \xrightarrow{g^*} & \mathrm{map}_*(C, T). \end{array}$$

If T is an h-local space, the two vertical maps are weak equivalences. Suppose (1) is true. Since T is g-local, the bottom horizontal map is a weak equivalence and so is the top horizontal map. So $L_hC \to L_hD$ is an L_h-equivalence and L_hD is h-local. By the universal property of the h-localization functor, $L_hC \to L_hD$ is a homotopy equivalence. If (4) holds, then g^* is a weak equivalence for all h-local T, so such T are g-local, i.e., (1) is true. □

The relation '$\leq$' is a partial ordering on the similarity classes of maps between spaces. In this partial ordering '$\leq$', there is a supremum and an infimum, namely for any map g, $\langle * \to * \rangle \leq \langle g \rangle \leq \langle S^0 \to * \rangle$.

For the nullification functor P_U, we write $\langle U \rangle$ for $\langle U \to * \rangle$ ($= \langle * \to U \rangle$).

Definition 3.4 For any map g, we define $c(g)$ to be the maximum of those integers n such that $\pi_j(g)$ is an isomorphism for each $j \leq n$. If $\pi_0(g)$ is not an isomorphism, we set $c(g) = -1$. When g is the map $A \to *$ or $* \to A$, we write $c(g)$ as $\mathrm{conn}(A)$ in agreement with the notation of Bousfield in [**3**].

Theorem 3.5 *If $\langle h \rangle \leq \langle g \rangle$, then $c(h) \geq c(g)$.*

Proof Suppose $c(g) = n$. If $n = -1$, the result is clear. If $n \geq 0$, then $\langle g \rangle \leq \langle S^{n+1} \rangle$ by Lemma 3.3 since $P_{S^{n+1}}(g)$ is a homotopy equivalence. Hence, $\langle h \rangle \leq \langle S^{n+1} \rangle$, or $P_{S^{n+1}}(h)$ is a homotopy equivalence. That is, $c(h) \geq n = c(g)$. □

Theorem 3.5 implies that $c(g)$ is a well defined function of the similarity class $\langle g \rangle$. It is also clear that for any integer $n \geq -1$, there is a canonical representative $\langle g \rangle$ such that $c(g) = n$, namely $\langle S^{n+1} \rangle$.

Notice that the proof of Theorem 3.5 also gives the statement '$c(g) \geq n \geq 0$ if and only if $\langle g \rangle \leq \langle S^{n+1} \rangle$'. Hence, using spaces S^{n+1}, we can filter the class of maps by $F_n = \{g : \langle g \rangle \leq \langle S^{n+1} \rangle\}$. Then, the invariant $c(g)$ is actually the integer such that $g \in F_{c(g)} \backslash F_{c(g)+1}$.

Recall from [**10**] that for each map g and for each prime p, there is an integer $d_p(g)$, called the *p-transitional dimension of* g, which is defined to be the smallest positive integer n so that $L_gK(\mathbb{Z}/p, n) \simeq *$. The p-transitional dimension of g puts some constraints on $c(g)$.

Corollary 3.6 *For any map g and any prime p, if $d_p(g) = n$, then $c(g) \leq n-1$.*

Proof By definition, $d_p(g) = n$ implies $L_gK(\mathbb{Z}/p, n) \simeq *$. This is equivalent to $\langle K(\mathbb{Z}/p, n) \rangle \leq \langle g \rangle$ by Lemma 3.3. By Theorem 3.5, $c(g) \leq n-1 = \mathrm{conn}(K(\mathbb{Z}/p, n))$. □

Example 3.7 For any generalized homology theory $E_*(-)$, there is a map g representing the localization functor L_E, i.e. $L_g = L_E$ (cf. [**2**]). The map g is $\bigvee g_\alpha$ where $\{g_\alpha\colon A_\alpha \to B_\alpha\}$ contains each E_*-equivalence up to homotopy with cardinality of A_α and B_α less than or equal to that of $E_*(pt)$. In this case, we write $c(E)$ for $c(g)$. Let K and $K\mathbb{Z}/p$ be the complex K-theory and mod p complex K-theory respectively. Then by [**7**] and [**8**], $L_K K(\mathbb{Z}/p, 2) \simeq *$ and $L_{K\mathbb{Z}/p} K(\mathbb{Z}/q, 1) \simeq *$ where q is a prime number different from p. Hence, $c(K) \leq 1$ and $c(K\mathbb{Z}/p) \leq 0$.

We are going to give a characterization of those localization functor L_g for which $c(g) \geq n$. To simplify the statement of the next theorem, we make the following definition.

Definition 3.8 For any map g and any integer $n \geq 0$, we say that L_g has *property* C_n if for any space X, $\pi_j(\eta)$ is an isomorphism for $j \leq n$ where $\eta\colon X \to L_g X$ is the localization map of X. We say that L_g has *property* C'_n if L_g has property C_n and $\pi_{n+1}(\eta)$ is a surjection for every space X. In other words, L_g has *property* C'_n if the homotopy fiber of the localization map $X \to L_g X$ is n-connected for any space X.

Theorem 3.9 *For any map g and any space W, the following statements hold:*

(a) $\mathrm{conn}(W) \geq n$ *if and only if* P_W *has property* C'_n.
(b) $c(g) \geq n$ *if and only if* L_g *has property* C_n.

Proof (a) If W is n-connected, P_W has property C'_n since by the construction of $P_W X$ given by Bousfield we attach cells of dimension bigger than $n+1$ to W. Conversely, suppose W is not n-connected, then $\pi_j W \neq 0$ for some $j \leq n$. Since $P_W W \simeq *$ by Lemma 3.3 (see also Theorem A), $\pi_j(\eta)$ is not an isomorphism for the space W.

(b) If $c(g) \geq n$, then $P^n = P^n L_g$ by Lemma 3.3, where $P^n = P_{S^{n+1}}$ is the nth Postnikov section functor. That is, for any space Y, $P^n(\eta)\colon P^n Y \to P^n L_g Y$ is a natural equivalence. This shows that L_g has property C_n. For the converse, consider the following commutative diagram when $j \leq n$:

$$\begin{array}{ccc} \pi_j A & \xrightarrow{\pi_j(g)} & \pi_j B \\ \downarrow & & \downarrow \\ \pi_j L_g A & \xrightarrow{\pi_j L_g(g)} & \pi_j L_g B. \end{array}$$

The vertical maps are induced by the g-localization maps of A and B correspondingly. So by assumption, the vertical maps are isomorphisms. Observe that $L_g(g)$ is a homotopy equivalence by Lemma 3.3. Hence, the upper horizontal map $\pi_j(g)$ is an isomorphism. □

Remark 3.10 If we replace homotopy groups in the definitions of $c(f)$, property C_n and property C'_n by integral homology groups, Theorem 3.9 still holds in this case. We refer to this result as the homology version of Theorem 3.9 in the sequel.

Example 3.11 This is a continuation of Example 3.7. It is not hard to see that L_g has property $C_0 \Leftrightarrow K(\mathbb{Z}, 0)$ is g-local $\Leftrightarrow K(\mathbb{Z}/p, 0)$ is g-local $\Leftrightarrow K(\mathbb{Q}, 0)$ is g-local. Since $K(\mathbb{Z}/p, 0)$ is both K_*-local and $K\mathbb{Z}/p_*$-local, $0 \leq c(K) \leq 1$ and $c(K\mathbb{Z}/p) = 0$.

Similarly, $c(H\mathbb{Q}) = 0$. If $E\mathbb{Q}_*(pt) \neq 0$ where $E_*(-)$ is a generalized homology theory, then since $\langle E\mathbb{Q}\rangle = \langle H\mathbb{Q}\rangle$ by [**8**], $c(E\mathbb{Q}) = 0$.

Actually more is true. Let $L_{H\mathbb{Z}}$ denote the localization functors with respect to the ordinary integral homology theory $H_*(-;\mathbb{Z}) = H\mathbb{Z}_*(-)$ for spaces and also for groups [**2**]. Recall [**1**] that for each group G, one can define $L_{H\mathbb{Z}}G$ to be $\pi_1 L_{H\mathbb{Z}}K(G,1)$.

Proposition 3.12 *$c(E) = 0$ for any nontrivial generalized homology theory $E_*(-)$.*

Proof We first show that $c(H\mathbb{Z}) = 0$. Since not every group is $H\mathbb{Z}_*$-local, we can pick a non-$H\mathbb{Z}_*$-local group G. Then pick a space Y such that $\pi_1 Y = G$. Suppose $c(H\mathbb{Z}) \geq 1$. By Theorem 3.9, $\pi_1(\eta)\colon \pi_1 Y \to \pi_1 L_{H\mathbb{Z}} Y$ is an isomorphism for any space Y. Since $L_{H\mathbb{Z}}\pi_1 Y \cong \pi_1 L_{H\mathbb{Z}} Y$ by ([**2**], Lemma 7.2), $\pi_1 Y$ is isomorphic to $L_{H\mathbb{Z}}\pi_1 Y$ or $\pi_1 Y$ is $H\mathbb{Z}_*$-local, a contradiction. Thus, $c(H\mathbb{Z}) < 1$. From the observation in Example 3.11 or by Theorem A, $c(H\mathbb{Z}) \geq 0$ and $c(E) \geq 0$. Hence, $c(H\mathbb{Z}) = 0$.

By the Atiyah-Hirzebruch spectral sequence, we know every $H\mathbb{Z}_*$-equivalence is an E_*-equivalence. Hence, $\langle H\mathbb{Z}\rangle \leq \langle E\rangle$. By Theorem 3.5, $c(E) = 0$. □

4 Trivial localization functors

The following result shows that trivial localization functors are those represented by a map f for which $\pi_0(f)$ is not a bijection.

Theorem 4.1 *For any map f, L_f is trivial if and only if $\pi_0(f)$ is not a bijection.*

Proof If L_f is trivial, then $L_f S^0 \simeq *$. This means $\langle f\rangle = \langle S^0\rangle = \langle S^0 \to *\rangle$. Hence, $c(f) = c(S^0 \to *) = -1$, or $\pi_0(f)$ is not a bijection. Conversely, since $\pi_0(f)$ is not a bijection, either $S^0 \to *$ or $* \to S^0$ is a retract of f. It is not hard to see that L_f-equivalence is preserved by retracts. We get that $S^0 \to *$ or $* \to S^0$ is also an L_f-equivalence, and hence $\langle S^0\rangle \leq \langle f\rangle$. □

In terms of $c(f)$, Theorem 4.1 says that $c(f) \geq 0$ for any nontrivial localization functor L_f. Moreover, in the partial ordering '$\leq$' of similarity classes, there is nothing in between $\langle S^0\rangle$ and $\langle S^1\rangle$.

Combining Theorem 3.9 and Theorem 4.1, we have:

Corollary 4.2 *For any map f and any space W, L_f preserves connectivity and P_W preserves simple connectivity.*

We conclude this section by proving part (2) of Theorem A.

Proof [Proof of Theorem A(2)] Write the map $f\colon A \to B$ as a disjoint union of maps $f_\alpha\colon A_\alpha \to B_\alpha$ where A_α and B_α are the connected components of A and B respectively. Observe that a connected space is f-local if and only if it is f_α-local for every α. By Corollary 4.4, if Y is connected, then $L_f Y$ is also connected. If we take g to be the wedge of all f_α (with base points chosen for A_α and B_α), then a connected space is g-local if and only if it is f_α-local for every α. Hence, $L_f Y \simeq L_g Y$. □

5 Theorem B and Theorem C

For the rest of this paper, all maps are assumed to be maps between connected spaces. Given a map f and a space Y, let C_f denote the homotopy cofiber (mapping cone) of f and let $Y\langle n\rangle$ denote the n-connective cover of Y.

Lemma 5.1 *For any map $f\colon A \to B$ and any fibration $F \to E \to K$, if K is C_f-null and E is f-local, then F is also f-local.*

Proof Consider the following commutative diagram:

$$\begin{array}{ccccc} \mathrm{map}_*(B,F) & \longrightarrow & \overline{\mathrm{map}}_*(B,E) & \longrightarrow & \mathrm{map}_*(B,K)_0 \\ {\scriptstyle\gamma}\downarrow & & {\scriptstyle\beta}\downarrow & & {\scriptstyle\alpha}\downarrow \\ \mathrm{map}_*(A,F) & \longrightarrow & \overline{\mathrm{map}}_*(A,E) & \longrightarrow & \mathrm{map}_*(A,K)_0 \end{array}$$

where $\mathrm{map}_*(X,E)_0$ ($X = A$ or B) is the connected component of $\mathrm{map}_*(X,E)$ containing the trivial map, and $\overline{\mathrm{map}}_*(X,E)$ is the space of all those maps $X \to E$ such that the composition $X \to E \to K$ is null-homotopic, in other words, $\overline{\mathrm{map}}_*(X,E)$ consists of all those components of $\mathrm{map}_*(X,E)$ sitting over the null component $\mathrm{map}_*(X,K)_0$ of the mapping space $\mathrm{map}_*(X,K)$. Since $F \to E \to K$ is a fibration, the two rows in the diagram are also fibrations.

Since $\mathrm{map}_*(B,K)_0$ and $\mathrm{map}_*(A,K)_0$ are connected and K is C_f-null, the map $\alpha\colon \mathrm{map}_*(B,K)_0 \to \mathrm{map}_*(A,K)_0$ is a weak equivalence. If we can prove that the map $\beta\colon \overline{\mathrm{map}}_*(B,E) \to \overline{\mathrm{map}}_*(A,E)$ is a weak equivalence, then it follows that the map γ is also a weak equivalence, and hence F is f-local.

We now show that the map $\beta\colon \overline{\mathrm{map}}_*(B,E) \to \overline{\mathrm{map}}_*(A,E)$ is a weak equivalence. As E is f-local, $\mathrm{map}_*(B,E) \to \mathrm{map}_*(A,E)$ is a weak equivalence. Therefore, it is enough to show that β induces a bijection on the set of components.

Since $A \to B \to C_f$ is a cofibration, we have a fibration

$$\mathrm{map}_*(C_f,K) \to \mathrm{map}_*(B,K) \to \mathrm{map}_*(A,K)$$

where $\mathrm{map}_*(C_f,K)$ is the homotopy fiber over the null component $\mathrm{map}_*(A,K)_0$ of $\mathrm{map}_*(A,K)$. Since K is C_f-null, $\mathrm{map}_*(C_f,K)$ is connected and therefore the only component of $\mathrm{map}_*(B,K)$ that maps into the null component $\mathrm{map}_*(A,K)_0$ of $\mathrm{map}_*(A,K)$ is the null component $\mathrm{map}_*(B,K)_0$.

Now consider the following commutative diagram:

$$\begin{array}{ccc} \mathrm{map}_*(B,E) & \longrightarrow & \mathrm{map}_*(B,K) \\ \downarrow & & \downarrow \\ \mathrm{map}_*(A,E) & \longrightarrow & \mathrm{map}_*(A,K). \end{array}$$

To prove that β induces a bijection on the sets of components, we take an element $a \in \pi_0\overline{\mathrm{map}}_*(A,E)$. Since E is f-local, there is a unique $b \in \pi_0\,\mathrm{map}_*(B,E)$ which maps to a. We need to show that $b \in \pi_0\overline{\mathrm{map}}_*(B,E)$. Let $b' \in \pi_0\,\mathrm{map}_*(B,K)$ be the image of b under the map $\mathrm{map}_*(B,E) \to \mathrm{map}_*(B,K)$. By what we proved in the previous paragraph, b' has to represent the null component of $\mathrm{map}_*(B,K)$ since it maps to the image of a in $\mathrm{map}_*(A,K)$ which represents the null component of $\mathrm{map}_*(A,K)$. This shows that $b \in \pi_0\overline{\mathrm{map}}_*(B,E)$, and the required map β induces a bijection on the sets of components. This finishes the proof of the lemma. □

Theorem 5.2 *For any map f, the following statements are equivalent:*

(1) *L_f preserves simple connectivity.*

(2) $(L_fY)\langle 1\rangle$ *is* f*-local for every simply connected space* Y.
(3) $\mathrm{Hom}(\pi_1 C_f, \pi_1 L_f Y) = 0$ *for every simply connected space* Y.
(4) *the kernel of the map* $\Phi\colon \mathrm{Hom}(\pi_1 B, \pi_1 L_f Y) \to \mathrm{Hom}(\pi_1 A, \pi_1 L_f Y)$ *is trivial for every simply connected space* Y.

Proof (1) $\Rightarrow$ (2), (3) and (4) are clear. (3) $\Leftrightarrow$ (4): Since $\mathrm{map}_*(X, K(G,1)) \simeq \mathrm{Hom}(\pi_1 X, G)$ for every space X and every group G, it is not hard to see from the fibration

$$\mathrm{map}_*(C_f, K(\pi_1 L_f Y, 1)) \to \mathrm{map}_*(B, K(\pi_1 L_f Y, 1)) \to \mathrm{map}_*(A, K(\pi_1 L_f Y, 1))$$

that the set $\mathrm{Hom}(\pi_1 C_f, \pi_1 L_f Y)$ is the kernel of the map Φ. (3) $\Rightarrow$ (2): Since $\mathrm{Hom}(\pi_1 C_f, \pi_1 L_f Y) = 0$ is equivalent to $K(\pi_1 L_f Y, 1)$ being C_f-null, we can apply Lemma 5.1 to the fibration $(L_fY)\langle 1\rangle \to L_fY \to K(\pi_1 L_f Y, 1)$ to conclude that $(L_fY)\langle 1\rangle$ is f-local. (2) $\Rightarrow$ (1): Let $\pi\colon (L_fY)\langle 1\rangle \to L_fY$ be the universal covering of L_fY. Since Y is simply connected, the localization map $\eta\colon Y \to L_fY$ lifts uniquely to its universal cover to get $\tilde{\eta}$. Since $(L_fY)\langle 1\rangle$ is f-local and $\eta\colon Y \to L_fY$ is an L_f-equivalence, the induced map

$$\mathrm{map}_*(L_fY, (L_fY)\langle 1\rangle) \to \mathrm{map}_*(Y, (L_fY)\langle 1\rangle)$$

is a weak equivalence. Hence, there is a map $s\colon L_fY \to (L_fY)\langle 1\rangle$ such that $s \circ \eta = \tilde{\eta}$. By the univeral property of L_fY, $\pi \circ s$ is homotopic to the identity on L_fY. This means that L_fY is simply connected. □

Corollary 5.3 (Theorem B) *Suppose a map* f *and a space* Y *satisfy one of the following two conditions:*

(1) $\pi_1 C_f = 0$ *and the space* Y *is arbitrary.*
(2) $\widetilde{H}_1(C_f) = 0$ *and* $\pi_1 L_f Y$ *is solvable.*

Then L_fY *is simply connected whenever* Y *is simply connected.*

Proof By Corollary 4.2, L_fY is connected. By Theorem 5.2, L_fY is simply connected if $\pi_1 C_f = 0$. If $\widetilde{H}_1(C_f) = 0$, then $\pi_1 C_f$ is a perfect group. So $\mathrm{Hom}(\pi_1 C_f, G)$ is trivial for any abelian group G. Since $\pi_1 L_f Y$ is solvable, an induction argument shows that $\mathrm{Hom}(\pi_1 C_f, \pi_1 L_f Y) = 0$. By Theorem 5.2 again, L_fY is simply connected. □

Remarks 5.4 1. For any connected space Y, the space $(L_fY)\langle 1\rangle$ is always C_f-null. This indicates that L_f is close to preserve simple connectivity for any map f.

2. By the Van Kampen Theorem, if the normal closure of the image of $\pi_1 A$ in $\pi_1 B$ is the whole group $\pi_1 B$, then $\pi_1 C_f = 0$.

3. Since f is a map between connected spaces, we know from ([**10**], Section 3.2) that L_f preserves finite products in this situation. This implies L_fY is an H-space whenever Y is. Thus, part (2) of Corollary 5.3 says that if $\widetilde{H}_1(C_f) = 0$, then L_f preserves simple connectivity for H-spaces. Dror Farjoun has asked whether L_f turns nilpotent spaces into nilpotent spaces. If this is true, $\widetilde{H}_1(C_f) = 0$ is sufficient to guarantee that L_f preserves simple connectivity.

4. For any group H, let $H^{(0)} = H$. For any nonnegative integer n, define $H^{(n+1)}$ to be the commutator subgroup $[H^{(n)}, H^{(n)}]$. Consider $\mathcal{H} = \{H : \cap_{n\geq 0} H^{(n)} = 0\}$. Then the following four conditions are equivalent:

(1) G is a perfect group.
(2) $\mathrm{Hom}(G, A) = 0$ for every abelian group A.

(3) $\operatorname{Hom}(G, S) = 0$ for every solvable group S.
(4) $\operatorname{Hom}(G, H) = 0$ for every group H in $\mathcal{H}$.

We now proceed to the proof of Theorem C.

Lemma 5.5 *For any map f and any connected space Y, the homotopy cofiber C_η of the f-localization map $\eta\colon Y \to L_f Y$ is C_f-acyclic. Consequently,* $\operatorname{conn}(C_\eta) \geq \operatorname{conn}(C_f)$.

Proof The f-localization map $\eta\colon Y \to L_f Y$ is constructed by attaching to Y maps of the form $f\colon A \to B$ and $\operatorname{hocolim}(B \xleftarrow{f} A \xrightarrow{f} B) \to B$ (For more details, see [**5**]). The homotopy cofibers of these two maps are C_f and ΣC_f respectively, which are C_f-acyclic. Therefore, C_η is C_f-acyclic by ([**6**], Theorem 1.D.3). The last statement follows from Lemma 3.3 and Theorem 3.5. □

Theorem 5.6 *Let $n \geq 2$ and $j \leq n$. Suppose f is a map with n-connected cofiber C_f, then $L_f Y$ is j-connected whenever Y is j-connected.*

Moreover, for $1 \leq j \leq n-2$, if Y is j-connected, then $\pi_l(\eta)\colon \pi_l Y \to \pi_l L_f Y$ is an isomorphism for $j+1 \leq l \leq n-1$.

Proof Let Y be j-connected. By Corollary 5.3, we can assume that $j \geq 2$. Hence we have a cofibration $Y \xrightarrow{\eta} L_f Y \to C_\eta$ of simply connected spaces. It is not hard to see that $\operatorname{conn}(L_f Y) \geq \min(\operatorname{conn}(Y), \operatorname{conn}(C_\eta))$ with equality except possibly $\operatorname{conn}(Y) = \operatorname{conn}(C_\eta) - 1$. By Lemma 5.5, we know that $\operatorname{conn}(C_\eta) \geq \operatorname{conn}(C_f)$. Hence, $L_f Y$ is j-connected if Y is j-connected.

By the homology version of Theorem 3.9, $\widetilde{H}_i(\eta)\colon \widetilde{H}_i(Y;\mathbb{Z}) \to \widetilde{H}_i(L_f Y;\mathbb{Z})$ is an isomorphism for $0 \leq i \leq n-1$. Since $L_f Y$ is simply connected, Hurewicz Isomorphism Theorem implies the last assertion of the theorem. □

Corollary 5.7 (Theorem C) *Let $n \geq 0$ and $j \leq n$. For any map f with n-connected cofiber C_f, L_f preserves connectivity up to n. That is, $L_f Y$ is j-connected whenever Y is j-connected.*

Moreover, L_f preserves connectivity up to n if and only if $(L_f Y)\langle j\rangle$ is f-local for every j-connected space Y and for every $j \leq n$.

The first assertion follows from Corollary 5.3 and Theorem 5.6. The proof of last assertion is similar to the proof of ((2) ⇒ (1)) of Theorem 5.2.

Example 5.8 Let $\phi\colon G \to H$ be a surjective homomorphism of abelian groups and let $f\colon M(G,n) \to M(H,n)$ be a map between Moore spaces induced by ϕ. Then f satisfies the condition in Corollary 5.7, hence L_f preserves connectivity up to n.

For future reference, we conclude this section by giving a recognition principle for an abelian Eilenberg-Mac Lane space to be f-local.

Lemma 5.9 *Let $f\colon A \to B$ be a map (A and B are not necessarily connected). Then for any abelian group G,*

$$K(G,m) \text{ is } f\text{-local} \Leftrightarrow f^*\colon \widetilde{H}^j(B;G) \xrightarrow{\cong} \widetilde{H}^j(A;G) \text{ for all } j \leq m.$$

In particular, $K(G,1)$ is f-local if and only if $\operatorname{Hom}(\widetilde{H}_1(f), G)$ *is an isomorphism.*

Proof Since G is abelian, $K(G,m)$ is an infinite loop space. Thus the connected components of the mapping space $\operatorname{map}_*(X, K(G,m))$ are weakly equivalent to each other. Now, as $\Sigma^s A$ and $\Sigma^s B$ are at least $(s-1)$-connected, the lemma follows from the following four equivalent statements:

(1) $K(G,m)$ is f-local

(2) $f^*\colon \operatorname{map}_*(B, K(G,m)) \xrightarrow{\cong} \operatorname{map}_*(A, K(G,m))$

(3) $f^*\colon \widetilde{H}^m(\Sigma^s B; G) \xrightarrow{\cong} \widetilde{H}^m(\Sigma^s A; G)\ \forall\ s \geq 0$

(4) $f^*\colon \widetilde{H}^j(B; G) \xrightarrow{\cong} \widetilde{H}^j(A; G)\ \forall\ 0 \leq j \leq m$.

□

6 Special Cases

Suppose given a map $f\colon A \to B$. We will apply the results in the previous sections to the case when the dimensions of A and B are small.

Proposition 6.1 *For any map f, if $\dim C_f \leq 2$, then L_f preserves simple connectivity and $K(\pi_1 L_f Y, 1)$ is f-local for any connected space Y.*

Proof If $\dim C_f \leq 2$, then the map $[C_f, L_f Y] \to [C_f, K(\pi_1 L_f Y, 1)]$ is surjective. Since $L_f Y$ is f-local and hence C_f-null, so $[C_f, K(\pi_1 L_f Y, 1)] = 0$. Hence, $\operatorname{Hom}(\pi_1 C_f, \pi_1 L_f Y) = 0$. By Theorem 5.2, L_f preserves simple connectivity. For the second assertion, consider the commutative diagram:

$$\begin{array}{ccc} [B, L_f Y] & \longrightarrow & [B, K(\pi_1 L_f Y, 1)] \\ \cong\big\downarrow & & \beta\big\downarrow \\ [A, L_f Y] & \longrightarrow & [A, K(\pi_1 L_f Y, 1)]. \end{array}$$

By our assumption, the upper horizontal map is surjective and the lower horizontal map is an isomorphism. Hence, β is an isomorphism, since the mapping space $\operatorname{map}_*(X, K(G,1))$ is discrete for every space X and for every group G, $K(\pi_1 L_f Y, 1)$ is f-local. □

Remark 6.2 Using the terminology of Casacuberta and Rodríguez [**4**], if $\dim C_f \leq 2$, then L_f is π_1-compatible.

Similar to the proof of Proposition 6.1 but using Corollary 5.7 instead of Theorem 5.2, we have

Proposition 6.3 *Let $n \geq 2$. For any map f, if C_f is $(n-1)$-connected and $\dim C_f \leq n+1$, then L_f preserves connectivity up to n. Moreover, $K(\pi_n L_f Y, n)$ is f-local for any $(n-1)$-connected space Y.*

Since the homotopy cofiber of $W \to *$ is ΣW, we have

Corollary 6.4 *Let $n \geq 1$. For any $(n-1)$-connected space W, if $\dim W \leq n+1$, then P_W preserves connectivity up to $n+1$. Moreover, $K(\pi_{n+1} P_W Y, n+1)$ is W-null for any n-connected space Y.*

7 Commutation with Postnikov Sections

For W-nullification functor P_W, we have

Proposition 7.1 *Let W be a connected space and $n \geq 1$. If $\widetilde{H}_j(W) = 0$ for $j \leq n$, then for any connected space Y with solvable fundamental group, the W-nullification map $\eta\colon Y \to P_W Y$ induces an isomorphism on π_j for $j \leq n$. Moreover, the spaces $K(\pi_j Y, j)$ and $P^j Y$ are W-null for $j \leq n$ where P^j is the jth Postnikov section.*

Proof $n = 1$: Since $\widetilde{H}_1(W) = 0$, $\pi_1 W$ is perfect. So $\mathrm{map}_*(W, K(\pi_1 Y, 1)) \simeq \mathrm{Hom}(\pi_1 W, \pi_1 Y) = 0$ and hence $P^1 Y = K(\pi_1 Y, 1)$ is W-null. Thus we have a commutative diagram by fiberwise W-nullification (cf. [**3**], 4.3 and 4.5)

$$\begin{array}{ccccc} Y\langle 1\rangle & \longrightarrow & Y & \longrightarrow & K(\pi_1 Y, 1) \\ \downarrow & & \eta\downarrow & & 1\downarrow \\ P_W(Y\langle 1\rangle) & \longrightarrow & P_W Y & \xrightarrow{g} & K(\pi_1 Y, 1). \end{array}$$

Since $P_W(Y\langle 1\rangle)$ is simply connected, $\pi_1(g)$ is an isomorphism. Hence η induces an isomorphism on fundamental groups and $P^1 P_W Y = K(\pi_1 P_W Y, 1)$ is W-null.

$n \geq 2$: By Lemma 5.9, all abelian Eilenberg-Mac Lane spaces $K(G, j)$ are W-null for $j \leq n$. By considering the fibration $K(\pi_j Y, j) \to P^j Y \to P^{j-1} Y$ and using induction, $P^j Y$ is W-null for $j \leq n$. Then, by fiberwise W-nullification, we have a commutative diagram

$$\begin{array}{ccccc} Y\langle j\rangle & \longrightarrow & Y & \longrightarrow & P^j Y \\ \downarrow & & \eta\downarrow & & 1\downarrow \\ P_W(Y\langle j\rangle) & \longrightarrow & P_W Y & \longrightarrow & P^j Y. \end{array}$$

As the homotopy cofiber of $W \to *$ is ΣW, ΣW is $(n+1)$-connected by our assumption on W. Hence $P_W(Y\langle j\rangle)$ is j-connected by Theorem 5.6. Thus, η induces an isomorphism on π_i for $i \leq j$. □

Corollary 7.2 *With the same assumptions in Proposition 7.1, we have a weak equivalence* $P_W P^j Y \simeq P^j P_W Y$ *for* $j \leq n$.

Proof Since $\eta\colon Y \to P_W Y$ induces an isomorphism on π_j for $j \leq n$, $P^j Y \to P^j P_W Y$ is a weak equivalence. Since $P^j Y$ is also W-null, $P_W P^j Y \simeq P^j P_W Y$. □

For f-localization functor L_f, we have

Proposition 7.3 *Let* $n \geq 1$. *Suppose* $f\colon A \to B$ *is a map which induces isomorphisms* $\widetilde{H}_j(f)$ *for* $j \leq n$. *Then for any connected nilpotent space* Y, *the* f*-localization map* $\eta\colon Y \to L_f Y$ *induces monomorphism on* π_j *for* $j < n$.

If, in addition Y *is an* H*-space,* $\pi_j(\eta)$ *is an isomorphism for* $j < n$, *and an epimorphism for* $j = n$.

Proof When $n > 1$, consider the following commutative diagram:

$$\begin{array}{ccc} \pi_j Y & \longrightarrow & \pi_j L_f Y \\ \downarrow & & \downarrow \\ \pi_j(L_{H\mathbb{Z}} Y) & \longrightarrow & \pi_j(L_{H\mathbb{Z}} L_f Y). \end{array}$$

By the homology version of Theorem 3.9, $\widetilde{H}_j(\eta)\colon \widetilde{H}_j(Y;\mathbb{Z}) \to \widetilde{H}_j(L_f Y;\mathbb{Z})$ is an isomorphism. Now the Whitehead theorem of Bousfield [**1**] implies that the bottom horizontal map is an isomorphism for $j < n$ and an epimorphism for $j = n$. Since Y is nilpotent, the left vertical map is an isomorphism (cf. [**1**]). Hence, $\pi_j(\eta)$ is a monomorphism for $j < n$.

Suppose now that Y is an H-space. Since the localization functor preserves finite products, $L_f Y$ is also an H-space. Since both $\pi_1 Y$ and $\pi_1 L_f Y$ are abelian, they are actually isomorphic by the Hurewicz Isomorphism Theorem. Let M_η be the mapping cylinder of η. Then $\pi_1(M_\eta, Y) = 0$. Since M_η is homotopy equivalent to L_fY, M_η is also an H-space. Thus, the pair (M_η, Y) is r-simple for all $r \geq 1$. Now using the relative Hurewicz Isomorphism Theorem and induction, we have $\pi_j(M_\eta, Y) = 0$ for $j \leq n$. The result follows from the long exact homotopy sequence of (M_η, Y).

When $n = 1$, the first assertion is obvious. If Y is an H-space, $\pi_1(\eta)$ is actually an isomorphism. □

Corollary 7.4 *Suppose f is as in Proposition 7.3 and $n > 1$. If Y is a connected space with abelian fundamental group, then P^jY is f-local for $j < n$.*

If, in addition Y is an H-space, there are natural equivalences $P^j(\eta)\colon P^jY \to P^jL_fY$ for $j < n$.

Proof By considering the fiber sequence $P^jY \to P^{j-1}Y \to K(\pi_j(Y), j+1)$, the first assertion follows by induction. The last assertion follows from Proposition 7.3. □

8 Commutation with Connective Covers

In the last section, we showed (with some restrictions on f) that the functors L_f and P^j commute in a certain range. In this section, we show with further restrictions, that the functors L_f and j-connective cover also commute.

Theorem 8.1 *Let f be a map and Y be a connected space. If $\widetilde{H}_j(C_f) = 0$ for $j \leq n$, $\pi_1 Y$ is solvable and $L_fY \simeq P_{C_f}Y$, then $L_f(Y\langle j\rangle) \simeq (L_fY)\langle j\rangle$ for $j \leq n$.*

Proof By Proposition 7.1, the map $Y \to P_{C_f}Y \simeq L_fY$ induces an isomorphism on π_j, $j \leq n$. Hence, the map $P^jY \to P^jL_fY$ is a weak equivalence. Consider the following commutative diagram

$$\begin{array}{ccccc} Y\langle j\rangle & \longrightarrow & Y & \longrightarrow & P^jY \\ {\scriptstyle\mu}\downarrow & & {\scriptstyle\eta}\downarrow & & \downarrow \\ (L_fY)\langle j\rangle & \longrightarrow & L_fY & \longrightarrow & P^jL_fY. \end{array}$$

We see that the homotopy fibers of the maps η and μ are the same. Since $P_{C_f}Y \simeq L_fY$, the homotopy fiber of η is C_f-acyclic by ([**6**], Theorem 1.H.2). So the homotopy fiber of μ is C_f-acyclic and hence f-acyclic by Lemma 3.3. By Zabrodsky's lemma (cf. [**3**], Theorem 4.6 or [**11**]), μ is an L_f-equivalence. Since $P^jL_fY \simeq P^jP_{C_f}Y$ is C_f-null by Proposition 7.1, $(L_fY)\langle j\rangle$ is f-local by Lemma 5.1. Hence μ is the f-localization map of $Y\langle j\rangle$, and so $L_f(Y\langle j\rangle) \simeq (L_fY)\langle j\rangle$. □

Remark 8.2 The condition $L_fY \simeq P_{C_f}Y$ is always satisfied if Y is f-local. See Example 9.4 below for an application.

The proof of the following theorem is basically the same as that of Theorem 8.1.

Theorem 8.3 *Suppose $c(f) \geq n$. For any connected space Y and $1 \leq j \leq n$, if $L_fY \simeq P_{C_f}Y$, then $L_f(Y\langle j\rangle) \simeq (L_fY)\langle j\rangle$.*

9 Application to complex K-theory

Let $g\colon C \to D$ be a map representing the localization functor of complex K-theory, i.e., $L_g = L_K$. Since we are only going to apply L_g to connected spaces, we can assume both C and D are connected. Then we have

Proposition 9.1 *For any connected space Y with $H_*(-;\mathbb{Z})$-local fundamental group, $\pi_1(\eta)\colon \pi_1 Y \to \pi_1(L_K Y)$ is an epimorphism; and is an isomorphism provided that $\pi_1 Y$ is abelian. In particular, L_K preserves simple connectivity.*

Proof Since $K(G,1)$ is K_*-local for any abelian group G by [**8**], $\widetilde{H}_1(g)$ is an isomorphism. From the proof of Proposition 7.3, we know that $\pi_1 L_{H\mathbb{Z}} Y \to \pi_1 L_{H\mathbb{Z}} L_g Y$ is surjective. From the proof of Proposition 3.12, we know that $L_g = L_{H\mathbb{Z}} L_g$. Hence, $\pi_1 L_{H\mathbb{Z}} Y \to \pi_1 L_K Y$ is surjective. If $\pi_1 Y$ is $H_*(-;\mathbb{Z})$-local, then $\pi_1 Y = \pi_1 L_{H\mathbb{Z}} Y$ and so $\pi_1(\eta)$ is surjective. Furthermore, if $\pi_1 Y$ is abelian, the Hurewicz Isomorphism Theorem shows that $\pi_1(\eta)$ is an isomorphism. □

Proposition 9.2 *For any connected space Y with solvable fundamental group, if $P_{C_g} Y \simeq L_K Y$, then there is a natural equivalence*

$$L_K(Y\langle 1\rangle) \simeq (L_K Y)\langle 1\rangle.$$

Proof Since $\widetilde{H}_1(g)$ is an isomorphism, $\widetilde{H}_1(C_g) = 0$. Then this proposition follows from Theorem 8.1. □

Remarks 9.3 1. The above result on the commutation between L_K and the n-connective cover is the best we can get, as the example $Y = \mathrm{BSO}$ shows. Indeed, by ([**7**], Proposition 2.2), $L_K\,\mathrm{BSpin} = \mathrm{BSO}$. Moreover, this example also shows that L_K does not preserve connectivity up to 2.

2. The nullification functor P_{C_g} is also called the complex K-theory plus-construction.

Example 9.4 Since BU and BO are K_*-local spaces, the spaces U, O, SO, Spin and BSO are also K_*-local by using Proposition 9.2.

Acknowledgements

This paper is part of the author's thesis written under the supervision of Peter Landweber. The author would like to thank Peter for his patience and encouragement. The author would also like to thank Martin Bendersky, Myles Tierney and Chuck Weibel for many helpful discussions. My thanks also go to the referee whose suggestions improve this paper a lot.

References

[1] Bousfield, A. K. [1974], *Homological localization of spaces, groups and π-modules*, Lecture Notes in Math. **304**, Springer-Verlag, Berlin and New York, 22–30.

[2] Bousfield, A. K. [1975], *The localization of spaces with respect to homology*, Topology **14**, 133–150.

[3] Bousfield, A. K. [1994], *Localization and periodicity in unstable homotopy theory*, J. of AMS **7**, 831–873.

[4] Casacuberta, C and Rodrígues, J. L., *On tower approximating homological localizations*, J. of London Math. Soc., to appear.

[5] Dror Farjoun, E. [1992], *Homotopy localization and v_1-periodic spaces*, Lecture Notes in Math. **1509**, Springer-Verlag, Berlin and New York, 104–113.

[6] Dror Farjoun, E. [1996], *Cellular Spaces, Null Spaces and Homotopy Localization*, Lecture Notes in Math. **1622**, Springer-Verlag, Berlin and New York.

[7] Meier, W. [1979], *Complex and real K-theory and localization*, J. of Pure and Applied Alg. **14**, 59–71.

[8] Mislin, G. [1977], *Localization with respect to K-theory*, J. of Pure and Applied Alg. **10**, 201–213.

[9] Neisendorfer, J. A. [1995], *Localization and connected covers of finite complexes*, Contemporary Math., vol. 181, AMS, 385–390.

[10] Tai, J.-Y. [1996], *f-localization and connectivity*, Ph.D. Thesis, Rutgers University.

[11] Tai, J.-Y., *A version of Zabrodsky's lemma*, PAMS, to appear.

Fields Institute Communications
Volume **19**, 1998

On Characterizations of Rational Homotopy Types with some Rational Cyclic Cohomologies

Toshihiro Yamaguchi
Graduate School of Science and Technology
Okayama University
Okayama 700, Japan
t_yama@math.okayama-u.ac.jp

Abstract. Let X be a simply connected CW complex such that $H^*(X;\mathbf{Q})$ is finitely generated as a $\mathbf{Q}$-algebra and X is not rationally contractible. Then we show that the homotopy orbit space of the free loop space $EX^1 \times_{S^1} X^{S^1}$ is formal if and only if X has the rational homotopy type of an odd dimensional sphere. Also we characterize the rational homotopy types of X from some graded structures of rational cyclic cohomologies $HC^*(X;\mathbf{Q})$.

1 Introduction

D. Sullivan's theory of minimal model says that the rational homotopy type of a nilpotent CW complex X is determined by its minimal model $\mathcal{M}(X)$ [**2**]. Let $\mathbf{Q}$ be the rational number field. If X is *formal* (see Definition 2.2 below), $\mathcal{M}(X)$ is a purely algebraic construct closely related to a resolution of $H^*(X;\mathbf{Q})$ by free commutative $\mathbf{Q}$-algebras [**8**]. It is known, for example, that spheres, complex projective spaces, and Lie groups are formal. Symmetric spaces and compact Kähler manifolds are also formal ([**5**, X] and [**2**, Section 6], respectively). The most basic necessary and sufficient condition of formal spaces is given in [**2**, Theorem (4.1)] by P. Deligne et al. (see (2.3) below).

In the following, we suppose that X is a connected and simply connected CW complex such that $H^*(X;\mathbf{Q})$ is finitely generated as a $\mathbf{Q}$-algebra and suppose that X is not rationally contractible, unless we explicitly say otherwise. Then we propose a question: can we characterize the rational homotopy types of X when its rational cyclic cohomology $HC^*(X;\mathbf{Q})$ has a certain graded structure? To seek the answer, we make use of the facts that the minimal models of the free loop space X^{S^1} and its homotopy orbit space by the natural free S^1-action $ES^1 \times_{S^1} X^{S^1}$ are given in terms of the minimal model of X ([**15**] and [**14**], respectively) and that $HC^*(X;\mathbf{Q})$ is isomorphic to $H^*(ES^1 \times_{S^1} X^{S^1};\mathbf{Q})$ [**1**].

1991 *Mathematics Subject Classification.* 55P62, 19D55.

Let BS^1 be the classifying space of S^1. Then we know that $H^*(BS^1;\mathbf{Q}) \cong \mathbf{Q}[\alpha]$; the polynomial algebra generated by an element α with deg $\alpha = 2$. M. Vigué-Poirrier proves in [**12**, Theorem 1] that there is a decomposition of $HC^*(X;\mathbf{Q})$ into a direct sum of graded $\mathbf{Q}[\alpha]$-modules:

$$HC^*(X;\mathbf{Q}) \cong \mathbf{Q}[\alpha] \oplus T^*$$

with T^* a graded $\mathbf{Q}[\alpha]$-module given in terms of the minimal model of X^{S^1} (see (4.1) below). The main purpose of this paper is to characterize the rational homotopy type of X when $ES^1 \times_{S^1} X^{S^1}$ is formal as follows:

Theorem 1.1 *The following are equivalent:*

(1) $ES^1 \times_{S^1} X^{S^1}$ *is formal,*
(2) X *has the rational homotopy type of an odd dimensional sphere, and*
(3) $T^* \cong \mathbf{Q}\{z_i\}_{i>0}$ *as graded* $\mathbf{Q}$*-modules with deg* z_i *even and* deg $z_i = i$ deg z_1 *for any* i.

If a space is formal, the minimal model is calculated as a formal consequence of the graded $\mathbf{Q}$-algebra structure of its rational cohomology and we can see from the equivalence of (1) and (2) in Theorem 1.1 that the spaces $ES^1 \times_{S^1} X^{S^1}$ are rarely formal. But even when $ES^1 \times_{S^1} X^{S^1}$ is not formal in general, we have the following theorems, which partially answer the question mentioned above:

Theorem 1.2 T^* *is concentrated in even degrees if and only if* X *has the rational homotopy type of a bouquet of a finite number of odd dimensional spheres.*

Theorem 1.3 T^* *is concentrated in odd degrees if and only if* X *has the rational homotopy type of the James' reduced product space* $(S^{2m})_\infty$ *of* S^{2m} *for some* m *or the* $2mn$ *skeleton* $(S^{2m})_n$ *of it for some* n.

M. Vigué-Poirrier proves in [**13**, Theorem B] that if X is formal, then the structure of $\mathbf{Q}[\alpha]$-module on T^* is trivial, that is, $\alpha \cdot T^* = 0$. Since we know that a bouquet of a finite number of odd dimensional spheres, the James' reduced product space of S^{2m} for some m, and the $2mn$ skeleton of it for some n are formal ([**8**, Lemma 1.6] and [**3**, p.576], respectively), we obtain the following from Theorems 1.2 and 1.3:

Corollary 1.4 *If* T^* *is concentrated in even degrees or odd degrees, then* $\alpha \cdot T^* = 0$.

This paper is organized as follows: in Section 2, we give the definitions of a minimal model, a formal space and a normal minimal model, and give a useful condition (see Lemma 2.8 below) for the formality of $ES^1 \times_{S^1} X^{S^1}$ corresponding Lemma 2.5 below, which gives it the formality of X. We prove Theorem 1.1 in Section 3, and Theorems 1.2 and 1.3 in Section 4. Throughout this paper, an algebra, a vector space and a cohomology are defined over $\mathbf{Q}$. Also we often omit the symbol "$\cdot$" of the product.

2 Some preliminaries

A graded algebra $A = \oplus_{i\geq 0} A^i$ is said to be *commutative* if $xy = (-1)^{ij} yx$ for $x \in A^i$ and $y \in A^j$. Here A^i is the subspace of A whose elements are homogeneous of degree i. A commutative graded algebra A is called a *commutative differential graded algebra* (CDGA for short) if it has a differential d_A of degree 1. We denote it by (A, d_A). An algebra homomorphism of degree zero $\phi : (A, d_A) \to (B, d_B)$ is called a *CDGA-morphism* if it commutes with differentials: $\phi \circ d_A = d_B \circ \phi$.

Definition 2.1 [**2**, Section 1] *A CDGA $\mathcal{M} = (M, d)$ is the* minimal model *of a CDGA $\mathcal{A} = (A, d_A)$ if*

(1) *M is a free graded commutative algebra,*
(2) *the differential d is decomposable, and*
(3) *there exists a CDGA-morphism $\rho : \mathcal{M} \to \mathcal{A}$ which induces an algebra isomorphism $\rho^* : H^*(\mathcal{M}) \cong H^*(\mathcal{A})$.*

The minimal model exists, for any CDGA $\mathcal{A}$, uniquely up to CDGA-isomorphism [**7**, Chapter 6]. So the minimal model of $\mathcal{A}$ is denoted by $\mathcal{M}(\mathcal{A})$. Let $\mathcal{A}(X)$ be the rational de Rham complex [**6**, p.183] of a path connected space X, namely, the CDGA of the $\mathbf{Q}$-polynomial forms on X with exterior differentials. There is an isomorphism $H^*(\mathcal{A}(X)) \cong H^*(X)$ induced by the integration chain map from $\mathcal{A}(X)$ to the singular cochain complex of X with $\mathbf{Q}$-coefficient [**6**, p.183]. The minimal model of the CDGA $\mathcal{A}(X)$ is called the *minimal model of* X, denoted by $\mathcal{M}(X)$.

By Definition 2.1 (1), $\mathcal{M}$ is given by $(\wedge Z, d)$, where $\wedge Z$ is the free algebra over a graded vector space $Z = \oplus_{i \geq 1} Z^i$ (Z^i is the subspace of Z whose elements are of degree i), that is,

$$\wedge Z = (\text{the polynomial algebra over } Z^{even}) \otimes (\text{the exterior algebra over } Z^{odd}),$$

where $Z^{even} = \oplus_{i \geq 1} Z^{2i}$ and $Z^{odd} = \oplus_{i \geq 0} Z^{2i+1}$. Let $\wedge^i Z$ denote the subspace of $\wedge Z$ generated by the products of i elements of Z. Put $\wedge^+ Z = \oplus_{i \geq 1} \wedge^i Z$; this is the ideal of $\wedge Z$ generated by a basis of Z. Then we can express the decomposability of d in Definition 2.1 (2) as $d(z) \in \wedge^+ Z \cdot \wedge^+ Z$ for any $z \in Z$.

Suppose X is a nilpotent CW complex with $\mathcal{M}(X)$ given by $(\wedge Z, d)$. Then it is known [**2**, (3.3)] that $\mathcal{M}(X)$ determines the rational homotopy type of X as follows:

$$Z^i \cong \mathrm{Hom}(\pi_i(X), \mathbf{Q}) \tag{2.1}$$

for any i if X is simply connected. Also for nilpotent spaces X and Y, it is known [**2**, (3.3)] that

$$X_{(0)} \simeq Y_{(0)} \quad \textit{if and only if} \quad \mathcal{M}(X) \cong \mathcal{M}(Y) \quad \textit{as} \quad \textit{CDGAs}, \tag{2.2}$$

where "$\simeq$" means a homotopy equivalence, and $X_{(0)}$ is the $\mathbf{Q}$-localization of X.

Let $(H^*(X), 0)$ be the CDGA equipped with the differential $d = 0$ on the commutative graded algebra $H^*(X)$. In general, $\mathcal{M}(X)$ is not isomorphic to the minimal model of $(H^*(X), 0)$. For example, one can see in [**5**, p.486] that the minimal model of the homogeneous space $SU(6)/SU(3) \times SU(3)$ is different from that of $(H^*(SU(6)/SU(3) \times SU(3)), 0)$.

Definition 2.2 [**2**, Section 4] *A CDGA $\mathcal{A}$ is called* formal *if $\mathcal{M}(\mathcal{A})$ is CDGA-isomorphic to the minimal model of $(H^*(\mathcal{A}), 0)$.*

A nilpotent space X is called *formal* if $\mathcal{A}(X)$ is formal. If $\mathcal{A}$ is formal, there is a CDGA-morphism $\psi : \mathcal{M}(\mathcal{A}) = (\wedge Z, d) \to (H^*(\mathcal{A}), 0)$ such that $\psi^* : H^*(\wedge Z, d) \cong H^*(H^*(\mathcal{A}), 0)$. Then ψ is called a *quasi-isomorphism.* An element w of $\wedge Z$ is called a *d-cocycle* if $d(w) = 0$ or is said to be *closed* and is said to be *d-exact* if there exists an element v of $\wedge Z$ such that $d(v) = w$. Let C denote the subspace of d-cocycles in Z, that is, $C = \mathrm{Ker}(d|_Z)$ and let $I(S)$ denote the ideal in the algebra $\wedge Z$ generated by a basis of a subspace S in Z. When $\mathcal{A}$ is formal, we can choose a minimal model $\mathcal{M} = (\wedge Z, d)$ of $\mathcal{A}$ such that $Z = C \oplus \mathrm{Ker}\,(\psi|_Z)$ for a quasi-isomorphism $\psi : \mathcal{M} \to H^*(\mathcal{A})$. Therefore, following [**2**, Theorem (4.1)],

$$\text{$\mathcal{A}$ is formal if and only if there is a complement N to C, $Z = C \oplus N$, such that any d-cocycle of $I(N)$ is d-exact.} \tag{2.3}$$

Here we state about the construction of a minimal model and some properties of it. In the following, we use the symbol $[w]$ for the element which is represented by a cocycle w in a cohomology. Let $\mathcal{A}$ be a CDGA. According to [**2**, p.251], [**6**, IX,D] or [**11**, p.172], the minimal model $\mathcal{M}$ of $\mathcal{A}$ and $\rho : \mathcal{M} \to \mathcal{A}$ satisfying the condition (3) of Definition 2.1 is inductively constructed as $\mathcal{M} = \cup_n \mathcal{M}(n)$ and $\rho = \cup_n \rho_n$ with $\rho_n : \mathcal{M}(n) \to \mathcal{A}$, where $\mathcal{M}(n)$ is minimal and generated by elements in degrees $\leq n$, ρ_n^* is an isomorphism in degrees $\leq n$, and ρ_n^* is an injection in degree $n+1$. Here $\mathcal{M}(n+1)$ is defined by an elementary extension [**6**, IX,D] $\mathcal{M}(n+1) = \mathcal{M}(n) \otimes_d \wedge Z^{n+1}$ where $Z^{n+1} = C^{n+1} \oplus K^{n+1}$,

$$C^{n+1} = \mathrm{Coker}(\rho^* : H^{n+1}(\mathcal{M}(n)) \to H^{n+1}(\mathcal{A})),$$

$$K^{n+1} = \mathrm{Ker}(\rho^* : H^{n+2}(\mathcal{M}(n)) \to H^{n+2}(\mathcal{A})),$$

$d|_{C^{n+1}} = 0$, and passing to cohomology $[d|_{K^{n+1}}]$ is an injection. In the following, we put $C = \oplus_n C^n$, and $K = \oplus_n K^n$.

Definition 2.3 [**11**, Definition 1.2] *A minimal model $(\wedge Z, d)$ is called* normal *if* $\mathrm{Ker}[d|_Z] = \mathrm{Ker}(d|_Z)$.

For an element z of Z^n, $z \in Ker[d|_Z]$ means that $dz = du$ for an element u of $\wedge Z^{<n}$. So it is clear that $\mathrm{Ker}[d|_Z] \supset \mathrm{Ker}(d|_Z)$ in general. Further we shall say in this paper, when we put $Z^{\leq n} = \oplus_{i \leq n} Z^i$, that

Definition 2.4 *A minimal model $(\wedge Z, d)$ is* (n)-normal *if* $\mathrm{Ker}[d|_{Z^{\leq n}}] = \mathrm{Ker}(d|_{Z^{\leq n}})$.

If $(\wedge Z, d)$ is (n)-normal, then it is (i)-normal for $i \leq n$. Also if $(\wedge Z, d)$ is (n)-normal and $\mathrm{Ker}[d|_{Z^{n+1}}] = \mathrm{Ker}(d|_{Z^{n+1}})$, then it is $(n+1)$-normal.

It is known [**11**, p.172] that the minimal model constructed above is normal and [**11**, Lemma 1.8] that

$$\text{If a minimal model $\mathcal{M} = (\wedge Z, d)$ is normal and formal, then $H^*(\mathcal{M})$ is generated by $[\mathrm{Ker}(d|_Z)]$ as an algebra.} \tag{2.4}$$

Under the construction above, we have $C = \mathrm{Ker}(d|_Z)$ and $Z = C \oplus K$ for $\mathcal{M} = (\wedge Z, d)$. Then we see that the condition (2.3) for the formality of $\mathcal{A}$ is modified as

Lemma 2.5 *$\mathcal{A}$ is formal if and only if any d-cocycle of $I(K)$ is d-exact.*

Proof The "if" part is clear from (2.3) since K is a complement to C, $Z = C \oplus K$. The "only if" part is proved since there exists a quasi-isomorphism $\rho : \mathcal{M} \to (H^*(\mathcal{M}), 0)$ such that $\rho|_C = [id_C]$, $\rho|_K = 0$ and $\rho|_{d(K)} = [0]$ from (2.4). If an element w of $I(K)$ is closed, then $[w] = \rho(w) = 0$, and w is d-exact in $\mathcal{M}$ since ρ^* is an isomorphism on cohomology. □

In the following, we assume that X is a simply connected CW complex and that $(\wedge Z, d)$ is the minimal model of X constructed above, which is normal. We know that if X is a simply connected, then X^{S^1} and $ES^1 \times_{S^1} X^{S^1}$ are nilpotent ([**9**, Theorem 2.5] and [**12**, p.41] respectively). Following [**15**], we define a CDGA

$(\wedge\overline{Z}\otimes\wedge Z,\delta)$ as follows: Let $\overline{Z}^i = Z^{i+1}$ and $\overline{Z} = \oplus_{i\geq 1}\overline{Z}^i$. Define $\beta : \wedge\overline{Z}\otimes\wedge Z \to \wedge\overline{Z}\otimes\wedge Z$ to be the unique derivation of degree -1 with the property:

$$\beta(z) = \overline{z} \text{ for all } z \in Z \text{ and } \beta(\overline{z}) = 0 \text{ for all } \overline{z} \in \overline{Z}.$$

(Here a linear map β is called a *derivation* if $\beta(uv) = \beta(u)v + (-1)^{deg(u)}u\beta(v)$ for any $u, v \in \wedge\overline{Z}\otimes\wedge Z$.) Then it is easy to check that $\beta\circ\beta = 0$. Here recall [**14**, p. 251-252] that

$$(\wedge\overline{Z}\otimes\wedge Z,\beta) \textit{ is acyclic and especially } \beta\mid_{\wedge Z} \textit{ is an injection.} \tag{2.5}$$

Now define $\delta : \wedge\overline{Z}\otimes\wedge Z \to \wedge\overline{Z}\otimes\wedge Z$ as the unique derivation of degree $+1$ with the property:

$$\delta(z) = d(z) \text{ for all } z \in Z \text{ and } \delta(\overline{z}) = -\beta(d(z)) \text{ for all } \overline{z} \in \overline{Z}.$$

Then it is also easy to check that $\delta\circ\delta = 0$ and $\delta\circ\beta + \beta\circ\delta = 0$. According to Theorem of [**15**], the minimal model of X^{S^1} can be expressed in terms of that of X as follows:

$$\textit{The minimal model of } X^{S^1} \textit{ is given by } (\wedge\overline{Z}\otimes\wedge Z,\delta). \tag{2.6}$$

Let $C_1 = C\oplus\beta(C)$, $K_1 = K\oplus\beta(K)$, and $I_1(S)$ the ideal in the algebra $\wedge\overline{Z}\otimes\wedge Z$ generated by a basis of a subspace S of $\overline{Z}\oplus Z$. Then we have $\mathrm{Ker}(\delta|_{\overline{Z}\oplus Z}) = C_1$ and $\overline{Z}\oplus Z = C_1\oplus K_1$.

Following [**14**], we define a CDGA $(\wedge\alpha\otimes\wedge\overline{Z}\otimes\wedge Z, D)$ as follows: Let α be an element of degree 2, $\mathbf{Q}\{\alpha\}$ the vector space with a basis α and $\wedge\alpha = \wedge\mathbf{Q}\{\alpha\}$, the polynomial algebra generated by α. Then define $D : \wedge\alpha\otimes\wedge\overline{Z}\otimes\wedge Z \to \wedge\alpha\otimes\wedge\overline{Z}\otimes\wedge Z$ as the unique derivation of degree $+1$ with the property:

$$D(\alpha) = 0 \text{ and } D(u) = \delta(u) + \alpha\cdot\beta(u) \text{ for all } u \in \wedge\overline{Z}\otimes\wedge Z.$$

Then it holds that $D\circ D = 0$. According to Theorem A of [**14**], the minimal model of $ES^1\times_{S^1}X^{S^1}$ can be expressed in terms of that of X as follows:

$$\textit{The minimal model of } ES^1\times_{S^1}X^{S^1} \textit{ is given by } (\wedge\alpha\otimes\wedge\overline{Z}\otimes\wedge Z, D). \tag{2.7}$$

Let $C_2 = \mathbf{Q}\{\alpha\}\oplus\beta(C)$, $K_2 = C\oplus K_1$, and $I_2(S)$ the ideal in the algebra $\wedge\alpha\otimes\wedge\overline{Z}\otimes\wedge Z$ generated by a basis of a subspace S of $\mathbf{Q}\{\alpha\}\oplus\overline{Z}\oplus Z$. Then we have $\mathrm{Ker}(D|_{\mathbf{Q}\{\alpha\}\oplus\overline{Z}\oplus Z}) = C_2$ and $\mathbf{Q}\{\alpha\}\oplus\overline{Z}\oplus Z = C_2\oplus K_2$.

Lemma 2.6 *If $(\wedge\alpha\otimes\wedge\overline{Z}\otimes\wedge Z, D)$ is (n)-normal and formal, then any D-cocycle of $I_2({K_2}^{\leq n})$ is D-exact.*

Proof There exists a quasi-isomorphism $\rho : (\wedge\alpha\otimes\wedge\overline{Z}\otimes\wedge Z, D) \to (H^*(\wedge\alpha\otimes\wedge\overline{Z}\otimes\wedge Z, D), 0)$ such that $\rho|_{C_2} = [id_{C_2}]$ since $(\wedge\alpha\otimes\wedge\overline{Z}\otimes\wedge Z, D)$ is formal. Also since $(\wedge\alpha\otimes\wedge\overline{Z}\otimes\wedge Z, D)$ is (n)-normal, we have the following commutative diagram of CDGA-morphisms from Definition 2.4:

$$\begin{array}{ccc}
(\wedge(\mathbf{Q}\{\alpha\}\oplus\overline{Z}\oplus Z)^{\leq n}, D) & \xrightarrow{\ \pi\ } & (\wedge {C_2}^{\leq n}/(D(I_2({K_2}^{\leq n}))\cap \wedge {C_2}^{\leq n}), 0) \\
i \downarrow & & \downarrow \phi \\
(\wedge\alpha\otimes\wedge\overline{Z}\otimes\wedge Z, D) & \xrightarrow[\ \rho\ ]{} & (H^*(\wedge\alpha\otimes\wedge\overline{Z}\otimes\wedge Z, D), 0),
\end{array}$$

where π is the projection with $\pi|_{{C_2}^{\leq n}} = [id_{{C_2}^{\leq n}}]$, $\pi|_{{K_2}^{\leq n}} = 0$ and $\pi|_{d({K_2}^{\leq n})} = [0]$, ϕ is defined by $[\phi|_{{C_2}^{\leq n}}] = [id_{{C_2}^{\leq n}}]$, and i is the inclusion induced by $(\mathbf{Q}\{\alpha\}\oplus\overline{Z}\oplus Z)^{\leq n} \hookrightarrow \mathbf{Q}\{\alpha\}\oplus\overline{Z}\oplus Z$. Then we have $\rho|_{{K_2}^{\leq n}} = \rho\circ i|_{{K_2}^{\leq n}} = \phi\circ\pi|_{{K_2}^{\leq n}} = 0$. If an element w of $I_2({K_2}^{\leq n})$ is a D-cocycle, then $[w] = \rho(w) = 0$, and w is D-exact in $(\wedge\alpha\otimes\wedge\overline{Z}\otimes\wedge Z, D)$ since ρ^* is an isomorphism on cohomology. □

Lemma 2.7 *If $(\wedge\alpha\otimes\wedge\overline{Z}\otimes\wedge Z, D)$ is formal, then it is normal.*

Proof We show this lemma inductively. It is clear that $(\wedge\alpha\otimes\wedge\overline{Z}\otimes\wedge Z, D)$ is (1)-normal since

$$\mathrm{Ker}[D|_{(\mathbf{Q}\{\alpha\}\oplus\overline{Z}\oplus Z)^1}] = \mathrm{Ker}[D|_{\overline{Z}^1}] = \overline{Z}^1 = \mathrm{Ker}(D|_{\overline{Z}^1}) = \mathrm{Ker}(D|_{(\mathbf{Q}\{\alpha\}\oplus\overline{Z}\oplus Z)^1}).$$

Suppose that $(\wedge\alpha\otimes\wedge\overline{Z}\otimes\wedge Z, D)$ is (n)-normal. If we can show that $\mathrm{Ker}[\delta|_{\overline{Z}^{n+1}}] = \beta(\mathrm{Ker}[d|_{Z^{n+2}}])$, then we see that

$$\begin{aligned}
\mathrm{Ker}[D|_{(\mathbf{Q}\{\alpha\}\oplus\overline{Z}\oplus Z)^{n+1}}] &= \mathbf{Q}\{\delta_n\cdot\alpha\}\oplus\mathrm{Ker}[\delta|_{\overline{Z}^{n+1}}] \\
&= \mathbf{Q}\{\delta_n\cdot\alpha\}\oplus\beta(\mathrm{Ker}[d|_{Z^{n+2}}]) \\
&= \mathbf{Q}\{\delta_n\cdot\alpha\}\oplus\beta(\mathrm{Ker}(d|_{Z^{n+2}})) \\
&= \mathbf{Q}\{\delta_n\cdot\alpha\}\oplus\mathrm{Ker}(\delta|_{\overline{Z}^{n+1}}) \\
&= \mathrm{Ker}(D|_{(\mathbf{Q}\{\alpha\}\oplus\overline{Z}\oplus Z)^{n+1}}),
\end{aligned}$$

where δ_n is 1 if $n = 1$ and 0 if $n > 1$, that is, $(\wedge\alpha\otimes\wedge\overline{Z}\otimes\wedge Z, D)$ is $(n+1)$-normal. The inclusion $\mathrm{Ker}[\delta|_{\overline{Z}^{n+1}}] \supset \beta(\mathrm{Ker}[d|_{Z^{n+2}}])$ is clear since $\delta(\overline{z}) = \delta(\beta(u))$ if $d(z) = d(u)$ for elements z of Z^{n+2} and u of $(\wedge Z)^{n+2}$. Therefore we show that $\mathrm{Ker}[\delta|_{\overline{Z}^{n+1}}] \subset \beta(\mathrm{Ker}[d|_{Z^{n+2}}])$.

Suppose that $\overline{z}\in\mathrm{Ker}[\delta|_{\overline{Z}^{n+1}}]$, that is, there is an element w_1 in $(\wedge^1\overline{Z}\otimes\wedge Z)^{n+1}\cap I_1(K_1)$ such that

$$\delta(\overline{z}) = \delta(w_1)$$

for an element $\overline{z}$ of $\overline{Z}^{n+1}$. If $\beta(w_1) = 0$, we have $d(z) = d(u)$ for an element u of $(\wedge Z)^{n+2}$ such that $\beta(u) = w_1$, from (2.5). This means that $z\in\mathrm{Ker}[d|_{Z^{n+2}}]$. Let $\beta(w_1)\neq 0$. Then we have $\beta(w_1)\in I_2(K_2^{\leq n})$ and $D(\beta(w_1)) = \delta(\beta(w_1)) = -\beta(\delta(w_1)) = -\beta(\delta(\overline{z})) = \beta(\beta(d(z))) = 0$. From Lemma 2.6, there is an element $w_2 = \sum_{i\geq 2}\alpha^{i-2}\cdot w_{2,i}$ in $\wedge\alpha\otimes\wedge^{\geq 2}\overline{Z}\otimes\wedge Z$, where $w_{2,i}\in\wedge^i\overline{Z}\otimes\wedge Z$, such that

$$D(w_2) = \beta(w_1),\ \text{especially}\ \delta(w_{2,2}) = \beta(w_1).$$

Then we have $\beta(w_{2,2})\in I_2({K_2}^{\leq n})$ and $D(\beta(w_{2,2})) = \delta(\beta(w_{2,2})) = -\beta(\delta(w_{2,2})) = -\beta(\beta(w_1)) = 0$. From Lemma 2.6, there is an element $w_3 = \sum_{i\geq 3}\alpha^{i-3}\cdot w_{3,i}$ in $\wedge\alpha\otimes\wedge^{\geq 3}\overline{Z}\otimes\wedge Z$, where $w_{3,i}\in\wedge^i\overline{Z}\otimes\wedge Z$, such that

$$D(w_3) = \beta(w_{2,2}),\ \text{especially}\ \delta(w_{3,3}) = \beta(w_{2,2}).$$

Iterating this argument yields an element $w_m = \sum_{i\geq m} \alpha^{i-m} \cdot w_{m,i}$ in $\wedge\alpha\otimes\wedge^{\geq m}\overline{Z}\otimes\wedge Z$, where $w_{m,i}\in\wedge^i\overline{Z}\otimes\wedge Z$, such that

$D(w_m) = \beta(w_{m-1,m-1})$, especially $\delta(w_{m,m}) = \beta(w_{m-1,m-1})$ and $\beta(w_{m,m}) = 0$.

From (2.5), there is an element u_{m-1} in $\wedge^{m-1}\overline{Z}\otimes\wedge Z$ such that

$$\beta(u_{m-1}) = w_{m,m}$$

and there is an element u_{m-2} in $\wedge^{m-2}\overline{Z}\otimes\wedge Z$ such that

$$\beta(u_{m-2}) = \delta(u_{m-1}) + w_{m-1,m-1}$$

since $\beta(\delta(u_{m-1}) + w_{m-1,m-1}) = -\delta(\beta(u_{m-1})) + \beta(w_{m-1,m-1}) = -\delta(w_{m,m}) + \beta(w_{m-1,m-1}) = -\beta(w_{m-1,m-1}) + \beta(w_{m-1,m-1}) = 0$. Iterating this argument for $i = m-3,..,0$, there is an element u_0 in $\wedge Z$ such that

$$\beta(u_0) = \delta(u_1) + w_1.$$

Then $\beta(d(u_0)) = -\delta(\beta(u_0)) = -\delta(w_1) = -\delta(\overline{z}) = \beta(d(z))$. Thus we have $d(z) = d(u_0)$ from (2.5), that is, $z\in\mathrm{Ker}[d|_{Z^{n+2}}]$. Hence $\mathrm{Ker}[\delta|_{\overline{Z}^{n+1}}]\subset\beta(\mathrm{Ker}[d|_{Z^{n+2}}])$. □

Lemma 2.8 *$ES^1\times_{S^1}X^{S^1}$ is formal if and only if any D-cocycle of $I_2(K_2)$ is D-exact.*

Proof The "if" part follows from (2.3) since C_2 is the subspace of D-cocycles in $\mathbf{Q}\{\alpha\}\oplus\overline{Z}\oplus Z$ and K_2 is a complement to C_2: $\mathbf{Q}\{\alpha\}\oplus\overline{Z}\oplus Z = C_2\oplus K_2$.

The "only if" part follows as in the proof of Lemma 2.5 since $(\wedge\alpha\otimes\wedge\overline{Z}\otimes\wedge Z, D)$ is normal from Lemma 2.7. □

3 Proof of Theorem 1.1

Proof of the equivalence of (1) and (2) of Theorem 1.1 The "if" part is proved as follows: Suppose $X_{(0)}\simeq S^{2n+1}{}_{(0)}$ for a positive integer n. Then we have $\mathcal{M}(X) = (\wedge(x_{2n+1}), 0)$ with x_{2n+1} a generator of the algebra $H^*(S^{2n+1})$. By (2.7), there is a CDGA-isomorphism:

$$\mathcal{M}(ES^1\times_{S^1}X^{S^1})\cong(\wedge(\alpha,\overline{x}_{2n+1},x_{2n+1}),D), \tag{3.1}$$

where $D(\overline{x}_{2n+1}) = 0$ and $D(x_{2n+1}) = \alpha\overline{x}_{2n+1}$. Then we have $I_2(K_2) = I_2(x_{2n+1})$. Since every nonzero element of $I_2(x_{2n+1})$ is not a D-cocycle, we see that $ES^1\times_{S^1}(S^{2n+1})^{S^1}$ is formal by Lemma 2.8.

The "only if" part is proved by using Lemma 2.8 and the following lemma:

Lemma 3.1 *If $ES^1\times_{S^1}X^{S^1}$ is formal, then*

(1) *X is formal,*
(2) *there are elements y_{ij} of K such that $d(y_{ij}) = x_ix_j$ for any i, j where $\{x_1,\cdots,x_l\}$ is a basis of C, and*
(3) *$C^{even} = 0$.*

By (1) of Lemma 3.1, X is formal. Then we see that

$$H^*(X)\cong\wedge C/(d(I(K))\cap\wedge C) \tag{3.2}$$

as algebras from (2.4) . By (2) of Lemma 3.1, we see that (3.2) is given as follows:

$$H^*(X)\cong\wedge(x_1,x_2,\cdots,x_l)/(x_ix_j\ ;1\leq i<j\leq l),$$

where $\{x_1, \cdots, x_l\}$ is a basis of C. By (3) of Lemma 3.1, each x_i is of odd degree. We put deg $x_i = 2n_i + 1$ for $i = 1, \cdots, l$.

Let $l > 1$. Then we can see, from a formal consequence of $H^*(X)$ (Section 3 of [**8**]), that the minimal model of it is given as follows:

$$\mathcal{M}(X) \cong (\wedge(x_1, x_2, \cdots, x_l, y_{12}, \cdots, y_{l-1\,l}, v_1, v_2, \cdots),\ d\ ),$$

where deg $x_i = 2n_i+1$ $(1 \leq i \leq l)$, deg $y_{ij} = 2n_i+2n_j+1$ $(1 \leq i < j \leq l)$, deg $v_1 = 4n_1 + 2n_2 + 1$, deg $v_2 = 2n_1 + 4n_2 + 1$, and the differential d is given as

$$d(x_1) = 0, \cdots, d(x_l) = 0,\ d(y_{ij}) = x_i x_j,\ d(v_1) = y_{12}x_1,\ d(v_2) = y_{12}x_2$$

for some elements of low degrees. Define an element w of $I_2(K_2)$ as

$$w = 2\ \overline{v}_1\overline{x}_2 - 2\ \overline{v}_2\overline{x}_1 + \overline{y}_{12}{}^2.$$

The element w is a D-cocycle since

$$\begin{aligned} D(w) &= \delta(w) \\ &= 2\delta(\overline{v}_1)\overline{x}_2 - 2\delta(\overline{v}_2)\overline{x}_1 + \delta(\overline{y}_{12}{}^2) \\ &= 2\ (-\overline{y}_{12}x_1 + y_{12}\overline{x}_1)\overline{x}_2 - 2\ (-\overline{y}_{12}x_2 + y_{12}\overline{x}_2)\overline{x}_1 + 2\ \overline{y}_{12}(-\overline{x}_1 x_2 + x_1\overline{x}_2) \\ &= 0. \end{aligned}$$

Since $\delta(\wedge\overline{Z}\otimes\wedge Z) \subset \wedge\overline{Z}\otimes\wedge^+ Z$, we have Im D = Im $(\delta+\alpha\cdot\beta) \subset \wedge\overline{Z}\otimes\wedge^+ Z \oplus I_2(\alpha)$. Thus we see that

$$\textit{any element of } \wedge\overline{Z} \textit{ is neither } \delta\textit{-exact nor } D\textit{-exact.} \qquad (3.3)$$

From (3.3), w is not D-exact. Thus if $l > 1$, we see that $ES^1 \times_{S^1} X^{S^1}$ is not formal from Lemma 2.8. Hence we have $l = 1$, that is, $X_{(0)} \simeq S^{2n+1}{}_{(0)}$ for a positive integer n. □

Proof of the equivalence of (2) **and** (3) **of Theorem 1.1** The "only if" part is proved as follows. Suppose $X_{(0)} \simeq S^{2n+1}{}_{(0)}$. By (3.1), we have

$$H^*(ES^1 \times_{S^1} X^{S^1}) \cong \mathbf{Q}[\alpha, \overline{x}_{2n+1}]/(\alpha\cdot\overline{x}_{2n+1}) \cong \mathbf{Q}[\alpha] \oplus \mathbf{Q}\{\overline{x}^i_{2n+1};\ i > 0\}.$$

Thus $T^* \cong \mathbf{Q}\{\overline{x}^i_{2n+1};\ i > 0\}$.

The "if" part is proved as follows. Suppose $C^{even} \neq 0$. Then for a nonzero element x of C^{even}, the element $\overline{x}$ is a generator of $H^*(ES^1 \times_{S^1} X^{S^1})$ such that its degree is odd. So we have

$$C^{even} = 0.$$

Also suppose $\dim_{\mathbf{Q}} C^{odd} \geq 2$. If a basis of C^{odd} is given as $\{x_1, \cdots, x_n\}$, then $\overline{x}_1, \cdots, \overline{x}_n$ are the generators of even degrees of T^*. Let deg $\overline{x}_1 \leq$ deg $\overline{x}_2$. Even if deg $\overline{x}_2 = a\cdot$ deg $\overline{x}_1$ for some positive integer a, it contradicts to (3). For there are two elements of same degrees $\overline{x}_1^a$, $\overline{x}_2$ in T^*. So we have

$$\dim_{\mathbf{Q}} C^{odd} = 1.$$

Since the differential d is decomposable, we have

$$K = 0.$$

Thus we see that $H^*(X) \cong \wedge(x_{2n+1})$ with the deg$x_{2n+1} = 2n+1$. This means that $X_{(0)} \simeq S^{2n+1}_{(0)}$. □

Proof of (1) of Lemma 3.1 Let an element w of $I(K)$ be a d-cocycle. Then we see that $\beta(w)$ is a D-cocycle of $I_2(K_2)$ since $D(\beta(w)) = -\beta(d(w)) = 0$. From

Lemma 2.8, there is an element $u = \sum_{i=1}^{n} \alpha^{i-1} u_i$ in $\wedge\alpha \otimes \wedge\overline{Z} \otimes \wedge Z$ for some n, where $u_i \in \wedge^i \overline{Z} \otimes \wedge Z$, such that

$$D(u) = \delta(u_1) = \beta(w).$$

This means that

$$\beta(u_i) = -\delta(u_{i+1}) \text{ for } i = 1, .., n-1 \text{ and } \beta(u_n) = 0.$$

From (2.5), there is an element v_{n-1} in $\wedge^{n-1}\overline{Z} \otimes \wedge Z$ such that

$$\beta(v_{n-1}) = u_n.$$

Then we have $\beta(\delta(v_{n-1})) = -\delta(\beta(v_{n-1})) = -\delta(u_n) = \beta(u_{n-1})$, that is, $\beta(\delta(v_{n-1}) - u_{n-1}) = 0$. From (2.5), there is an element v_{n-2} in $\wedge^{n-2}\overline{Z} \otimes \wedge Z$ such that

$$\beta(v_{n-2}) = \delta(v_{n-1}) - u_{n-1}.$$

Then we have $\beta(\delta(v_{n-2})) = -\delta(\beta(v_{n-2})) = -\delta(\delta(v_{n-1})) + \delta(u_{n-1}) = \delta(u_{n-1}) = -\beta(u_{n-2})$, that is, $\beta(\delta(v_{n-2}) + u_{n-2}) = 0$. From (2.5), there is an element v_{n-3} in $\wedge^{n-3}\overline{Z} \otimes \wedge Z$ such that

$$\beta(v_{n-3}) = \delta(v_{n-2}) + u_{n-2}.$$

Iterating this argument for $i = n-4, .., 0$, we see that there is an element $v_0 \in \wedge Z$ such that

$$\beta(v_0) = \delta(v_1) + (-1)^{n+1} u_1.$$

Then we have $\beta(d(v_0)) = -\delta(\beta(v_0)) = (-1)^n \delta(\delta(v_1)) + (-1)^n \delta(u_1) = (-1)^n \delta(u_1) = (-1)^n \beta(w)$. From (2.5), we have

$$d((-1)^n v_0) = (-1)^n \delta(v_0) = w.$$

Thus X is formal from Lemma 2.5. □

Proof of (2) of Lemma 3.1 For any i, j where $\{x_1, \cdots, x_m\}$ is a basis of C, we see that

$$\beta(x_i x_j) = \overline{x}_i x_j + (-1)^{deg(x_i)} x_i \overline{x}_j \in I_2(K_2).$$

This element $\beta(x_i x_j)$ is a D-cocycle. By the same argument of the proof of (1) of Lemma 3.1, there is an element u in $I(K)$ such that $d(u) = x_i x_j \in \wedge^2 Z$. From the decomposability of d, we see that u is not decomposable, but an element of K. □

If $ES^1 \times_{S^1} X^{S^1}$ is formal, then we see

there is no element y of $K - \{0\}$ such that $d(y) \in \wedge C^{even}$. (3.4)

In fact, (3.4) is shown as follows: Suppose that there are an element y in K and an element f in $\wedge C^{even}$ such that $d(y) = f$. If a basis of C^{even} is $\{x_{i_1}, \cdots, x_{i_n}\}$, we have

$$\overline{y}\, \overline{x}_{i_1} \cdots \overline{x}_{i_n} \in I_2(K_2).$$

This element $\overline{y}\, \overline{x}_{i_1} \cdots \overline{x}_{i_n}$ is a D-cocycle since

$$D(\overline{y}\, \overline{x}_{i_1} \cdots \overline{x}_{i_n}) = \delta(\overline{y})\overline{x}_{i_1} \cdots \overline{x}_{i_n} = -\beta(f)\overline{x}_{i_1} \cdots \overline{x}_{i_n} = 0.$$

By (3.3), we see that $\overline{y}\, \overline{x}_{i_1} \cdots \overline{x}_{i_n}$ is not D-exact. Then $ES^1 \times_{S^1} X^{S^1}$ is not formal from Lemma 2.8. Thus we have (3.4).

Proof of (3) of Lemma 3.1 Suppose $C^{even} \neq 0$. For a generator x of C^{even},

$$\beta(x^2) = 2\overline{x}x \in I_2(K_2).$$

This element $\beta(x^2) = 2\overline{x}x$ is a D-cocycle and not D-exact. This contradicts Lemma 2.8. Thus we have $C^{even} = 0$.

In fact, if it is D-exact, then there is an element $\overline{y}$ in $\overline{Z}$ such that $\delta(\overline{y}) = D(\overline{y}) = \beta(x^2)$ from the decomposability of D. Then there is an element y in K such that $d(y) = -x^2$ from (2.5). But we can not have such an element from (3.4). □

4 Proofs of Theorems 1.2 and 1.3

Let $\mathcal{A} = (A, d)$ be a CDGA. To show Lemma 4.1 below, we define the normalized cyclic bar complex $(\mathbf{N}(\mathcal{A})[\alpha], b + \alpha \cdot B)$ of $\mathcal{A}$ and the cyclic homology of $\mathcal{A}$ following [**4**, §2] :

$$\mathbf{N}(\mathcal{A}) = \sum_{k=0}^{\infty} A \otimes \bar{A}^{\otimes k},$$

the derivation $b = b_0 + b_1$ and B have the properties such that

$$b_0([z_0, .., z_k]) = -\sum_{i=0}^{k} (-1)^{\varepsilon_{i-1}} [z_0, .., z_{i-1}, d(z_i), z_{i+1}, .., z_k],$$

$$b_1([z_0, .., z_k]) = -\sum_{i=0}^{k-1} (-1)^{\varepsilon_i} [z_0, .., z_{i-1}, z_i z_{i+1}, z_{i+2}, .., z_k] + (-1)^{(degz_k - 1)\varepsilon_{k-1}} [z_k z_0, .., z_{k-1}],$$

$$B([z_0, .., z_k]) = \sum_{i=0}^{k} (-1)^{(\varepsilon_{i-1}+1)(\varepsilon_k - \varepsilon_{i-1})} [1, z_i, .., z_k, z_0, .., z_{i-1}],$$

and $b_0(\alpha) = b_1(\alpha) = B(\alpha) = 0$, where $\bar{A} = \oplus_{i>0} A^i$, $\{\deg[z_0, .., z_k] = \deg z_0 + \cdots + \deg z_k - k\}$, for $[z_0, .., z_k]$ in $\mathbf{N}(\mathcal{A})$, and $\{\varepsilon_i = \deg z_0 + \cdots + \deg z_i - i\}$. Since the formulas $b \circ B + B \circ b = 0$ and $b^2 = B^2 = 0$ hold, the derivation $b + \alpha \cdot B$ is a differential with degree 1. Its homology is denoted $HC^*(\mathcal{A})$, which is called the cyclic homology of $\mathcal{A}$. We denote that of rational de Rham complex $\mathcal{A}(X)$ of X by $HC^*(X; \mathbf{Q})$ or simply by $HC^*(X)$ it is called the *(rational) cyclic cohomology of* X. If $\mathcal{A}$ is formal, we see that $HC^*(\mathcal{A}) \cong HC^*(\mathcal{M}(\mathcal{A})) \cong HC^*(H^*(\mathcal{A}))$ which are induced from quasi-isomorphisms $\mathcal{A} \leftarrow \mathcal{M}(\mathcal{A}) \rightarrow H^*(\mathcal{A})$ and that $b = b_1$ in $\mathbf{N}(H^*(\mathcal{A}))$ since the differential of $H^*(\mathcal{A})$ is trivial. Under these notations, we show

Lemma 4.1 *If* $X_{(0)} \simeq \vee_{i=1}^{l} {S^{2n_i+1}}_{(0)}$, *then* $HC^*(X)$ *is concentrated in even degrees.*

Proof Since X is formal, it is sufficient to show that $HC^*(H^*(X))$ is evenly graded. Let

$$H^*(X) = \wedge(x_1, .., x_l)/(x_i x_j;\ 1 \leq i < j \leq l),$$

where deg x_i is odd for any i. Then an element w with degree odd of $\mathbf{N}(H^*(X))[\alpha]$ is uniquely expressed as

$$w = \sum_{n \geq 0} w_n \ \text{ with } w_n = \sum_i \left(\sum_{(i_0,..,i_n) \in I_{N_i}} a_{(i_0,..,i_n)} [x_{i_0}, .., x_{i_n}] \right) \cdot \alpha^{N_i},$$

where $a_{(i_0,..,i_n)} \in \mathbf{Q} - \{0\}$ for any $(i_0, .., i_n) \in I_{N_i}$ and $N_i \neq N_j$ if $i \neq j$.

Suppose that w is a $(b + \alpha \cdot B)$-cocycle. Since Im $b \cap$ Im $\alpha \cdot B = 0$ in this case, we have

$$B(w_n) = 0 \text{ for any } n.$$

The equation $B(w_0) = 0$ means $w_0 = 0$. Let $n > 0$. The cyclic group $\mathbf{Z}/(n+1)\mathbf{Z}$ action on $\overline{H^*(X)}^{\otimes n+1}$ is given by letting its generator t act by

$$t \cdot [x_{i_0}, .., x_{i_{n-1}}, x_{i_n}] = [x_{i_n}, x_{i_0}, .., x_{i_{n-1}}],$$

which is equivalent to the action by $t \cdot (i_0, .., i_{n-1}, i_n) = (i_n, i_0, .., i_{n-1})$ on $(\mathbf{Z}/(l+1)\mathbf{Z} - \{0\})^{\times n+1}$. Then the index sets $\{I_{N_i}\}_i$ of w_n are represented as the following disjoint unions:

$$I_{N_i} = \cup_\gamma I_{N_i}^\gamma \quad ; \; I_{N_i}^\gamma = \{\kappa \in I_{N_i} \mid \kappa = t^j \cdot \gamma \;\text{ for some } 0 \leq j \leq n\}.$$

For any index γ of $I_{N_i}^\gamma$, we can fix $k = k_\gamma$ and $0 = j_0 < j_1 < \cdots < j_k \leq n$ such that

$$I_{N_i}^\gamma = \{\gamma = (i_0,.., i_n), t^{j_1} \cdot \gamma = (i_{n-j_1+1},.., i_n, i_0, .., i_{n-j_1}), \cdots, t^{j_k} \cdot \gamma = (i_{n-j_k+1},.., i_{n-j_k})\}.$$

Put $[x]_\gamma = [x_{i_0},.., x_{i_n}]$. Then we remark

$$t^{j_s} \cdot [x]_\gamma = [x]_{t^{j_s} \cdot \gamma} = [x_{i_{n-j_1+1}}, .., x_{i_n}, x_{i_0}, .., x_{i_{n-j_1}}].$$

The equation

$$0 = B(w_n) = B\left(\sum_i \left(\sum_{\gamma \in I_{N_i}} \sum_{s=0}^{k} a_{\gamma_s} t^{j_s} \cdot [x]_\gamma\right) \cdot \alpha^{N_i}\right)$$

$$= \sum_i \left(\sum_{\gamma \in I_{N_i}} \sum_{s=0}^{k} a_{\gamma_s} B(t^{j_s} \cdot [x]_\gamma)\right) \cdot \alpha^{N_i}$$

means that

$$\sum_{s=0}^{k} a_{\gamma_s} = 0 \text{ for any } \gamma \in I_{N_i} \text{ and } i$$

since $B(t^j \cdot [x]_\gamma) = B([x]_\gamma)$ for any $0 \leq j \leq n$ and $\{B([x]_\gamma)\}_\gamma$ are linearly independent. If we put

$$[x]_b^a = [1, x_{i_{n-j_b+2}}, .., x_{i_n}, x_{i_0}, .., x_{i_{n-j_b+1}}] + [1, x_{i_{n-j_b+3}}, .., x_{i_{n-j_b+2}}]$$

$$+ \cdots + [1, x_{i_{n-j_a+1}}, .., x_{i_{n-j_a}}]$$

for $0 \leq a < b \leq k$ and

$$[x]_0^k = [1, x_{i_1}, .., x_{i_n}, x_{i_0}] + [1, x_{i_2}, .., x_{i_1}] + \cdots + [1, x_{i_{n-j_k+1}}, .., x_{i_{n-j_k}}],$$

then we have

$$b([x]_b^a) = [x_{i_{n-j_b+1}}, .., x_{i_{n-j_b}}] - [x_{i_{n-j_a+1}}, .., x_{i_{n-j_a}}] = t^{j_b} \cdot [x]_\gamma - t^{j_a} \cdot [x]_\gamma \text{ and}$$

$$b([x]_0^k) = [x_{i_0}, .., x_{i_n}] - [x_{i_{n-j_k+1}}, .., x_{i_{n-j_k}}] = [x]_\gamma - t^{j_k} \cdot [x]_\gamma,$$

respectively. From these equations, if we put

$$\begin{aligned}
u_n^\gamma = &(k a_{\gamma_0} + \;0\; + a_{\gamma_2} + 2a_{\gamma_3} + \cdots + (k-1)a_{\gamma_k})[x]_1^0 \\
&+ ((k-1)a_{\gamma_0} + k a_{\gamma_1} + \;0\; + a_{\gamma_3} + \cdots + (k-2)a_{\gamma_k})[x]_2^1 \\
&+ ((k-2)a_{\gamma_0} + (k-1)a_{\gamma_1} + k a_{\gamma_2} + \;0\; + \cdots + (k-3)a_{\gamma_k})[x]_3^2 \\
&\;\vdots \\
&+ (a_{\gamma_0} + 2a_{\gamma_1} + 3a_{\gamma_2} + \cdots + k a_{\gamma_{k-1}} + \;0\;)[x]_k^{k-1} \\
&+ (\;0\; + a_{\gamma_1} + 2a_{\gamma_2} + 3a_{\gamma_3} + \cdots + k a_{\gamma_k})[x]_0^k,
\end{aligned}$$

then we have its image by differential b as

$$b(u_n^\gamma) = (ka_{\gamma_0}+0+a_{\gamma_2}+2a_{\gamma_3}+\cdots)t^{j_1}[x]_\gamma - (ka_{\gamma_0}+0+a_{\gamma_2}+2a_{\gamma_3}+\cdots)[x]_\gamma$$
$$+((k-1)a_{\gamma_0}+ka_{\gamma_1}+0+a_{\gamma_3}+\cdots)t^{j_2}[x]_\gamma$$
$$-((k-1)a_{\gamma_0}+ka_{\gamma_1}+0+a_{\gamma_3}+\cdots)t^{j_1}[x]_\gamma$$
$$+((k-2)a_{\gamma_0}+(k-1)a_{\gamma_1}+ka_{\gamma_2}+\cdots)t^{j_3}[x]_\gamma$$
$$-((k-2)a_{\gamma_0}+(k-1)a_{\gamma_1}+ka_{\gamma_2}+\cdots)t^{j_2}[x]_\gamma$$
$$\vdots$$
$$+(a_{\gamma_0}+2a_{\gamma_1}\cdots+ka_{\gamma_{k-1}}+0)t^{j_k}[x]_\gamma - (a_{\gamma_0}+2a_{\gamma_1}\cdots+ka_{\gamma_{k-1}}+0)t^{j_{k-1}}[x]_\gamma$$
$$+(0+a_{\gamma_1}+2a_{\gamma_2}+3a_{\gamma_3}+\cdots)[x]_\gamma - (0+a_{\gamma_1}+2a_{\gamma_2}+3a_{\gamma_3}+\cdots)t^{j_k}[x]_\gamma$$
$$= -(k+1)(a_{\gamma_0}[x]_\gamma + a_{\gamma_1}t^{j_1}\cdot[x]_\gamma + \cdots a_{\gamma_k}t^{j_k}\cdot[x]_\gamma)$$
$$= -(k+1)\sum_{s=0}^{k}(a_{\gamma_s}t^{j_s}\cdot[x]_\gamma).$$

Thus we have

$$b\left(\sum_{n>0}\sum_i\left(\sum_{\gamma\in I_{N_i}}\frac{-1}{k_\gamma+1}u_n^\gamma\right)\cdot\alpha^{N_i}\right)$$
$$=\sum_{n>0}\sum_i\left(\sum_{\gamma\in I_{N_i}}\sum_{s=0}^{k_\gamma}a_{\gamma_s}t^{j_s}\cdot[x]_\gamma\right)\cdot\alpha^{N_i}$$
$$=\sum_{n>0}w_n = w,$$

that is, w is b-exact. Further we see that w is $(b+\alpha\cdot B)$-exact since $B(u_n^\gamma)=0$ for any $n>0$ and γ. Hence there is no element with degree odd in $HC^*(H^*(X))$. □

Let us return to the arguments of minimal models in Section 2. We know $HC^*(X)\cong H^*(ES^1\times_{S^1}X^{S^1})\cong H^*(\wedge\alpha\otimes\wedge\overline{Z}\otimes\wedge Z, D)$ from [**1**] and (2.7), respectively. Let

$$T^* = H^*(\wedge\alpha\otimes\wedge^+(\overline{Z}\oplus Z), D).$$

Then we can decompose $H^*(\wedge\alpha\otimes\wedge\overline{Z}\otimes\wedge Z, D)$ into a direct sum of graded $\mathbf{Q}[\alpha]$-modules $\mathbf{Q}[\alpha]\oplus T^*$ since $\mathrm{Im}\, D\subset\wedge\alpha\otimes\wedge^+(\overline{Z}\oplus Z)$. M. Vigué-Poirrier proves in Lemma 2 of [**12**] that

$$T^*\cong\frac{\beta(\wedge\overline{Z}\otimes\wedge Z)\cap\mathrm{Ker}\delta}{\delta(\beta(\wedge\overline{Z}\otimes\wedge Z))} \tag{4.1}$$

as graded $\mathbf{Q}$-modules.

Lemma 4.2 *If $T^*=T^{even}$, then $H^*(X)=H^{odd}(X)$.*

Proof Suppose $C^{even} \neq 0$. Then for a nonzero element x of C^{even}, the element $\overline{x}$ is a generator of T^* of odd degree. This is a contradiction. So we have

$$C^{even} = 0.$$

Also suppose that w is a decomposable d-cocycle of $\wedge Z$ of even degree. From (2.5), $\beta(w) \neq 0$. Then we see that $\beta(w)$ is a δ-cocycle of odd degree. Since $T^* = T^{even}$, there is an element v of $\wedge \overline{Z} \otimes \wedge Z$ such that $(\delta \circ \beta)(v) = \beta(w)$ from (4.1), that is, $d(-v) = w$ from (2.5). So we have $w = 0$ in $H^*(X)$. Thus we see that $H^*(X) = H^{odd}(X)$. □

Proof of Theorem 1.2 The "if" part follows from Lemma 4.1. The "only if" part follows by combining Lemma 4.2 and Theorem 1.5 of [**8**]. □

Lemma 4.3 (1) *If $H^*(X) \cong \mathbf{Q}[x_{2n}]$ and* $\deg x_{2n} = 2n$, *then*

$\dim_{\mathbf{Q}} T^s = 1$ *for $s = 2n(i+1) - 1$ with $i \geq 0$ and*

$\dim_{\mathbf{Q}} T^s = 0$ *for other s.*

(2) *If $H^*(X) \cong \mathbf{Q}[x_{2n}]/(x_{2n}^k)$ and $k > 1$, then*

$\dim_{\mathbf{Q}} T^s = 1$ *for $s = 2n(jk+i+1) - 2j - 1$ with $j \geq 0$, $0 \leq i \leq k-2$ and*

$\dim_{\mathbf{Q}} T^s = 0$ *for other s.*

Proof Let $\mathbf{Q}\{S\}$ be the vector space generated by S. (1) is shown as follows: If $H^*(X) \cong \mathbf{Q}[x_{2n}]$, we have $\mathcal{M}(X) \cong (\wedge(x), 0)$ with $\deg x = 2n$. Then we see

$$T^* \cong \mathbf{Q}\{\overline{x}x^i;\ i \geq 0\}$$

as vector spaces. Note that $\deg(\overline{x}x^i) = 2n(i+1) - 1$.

(2) is shown as follows: If $H^*(X) \cong \mathbf{Q}[x_{2n}]/(x_{2n}^k)$, we have $\mathcal{M}(X) \cong (\wedge(x,y), d)$ with $\deg x = 2n$, $\deg y = 2kn - 1$, $d(x) = 0$, and $d(y) = x^k$. Then, following [**14**],

$$T^* \cong \mathbf{Q}\{\overline{y}^j\overline{x}x^i;\ j \geq 0,\ 0 \leq i \leq k-2\}$$

as vector spaces. Note that $\deg(\overline{y}^j\overline{x}x^i) = 2n(jk+i+1) - 2j - 1$. □

Lemma 4.4 *If $T^* = T^{odd}$, then the algebra $H^*(X)$ is generated by a single element of even degree.*

Proof Suppose $C^{odd} \neq 0$. Then for a nonzero element x of C^{odd}, the element $\overline{x}$ is a generator of T^* of even degree. So we have

$$C^{odd} = 0.$$

Suppose $\dim_{\mathbf{Q}} C^{even} \geq 2$. If a basis of C^{even} is given as $\{x_1, \cdots, x_n\}$, then any $\overline{x}_i\overline{x}_j$ for $i \neq j$ is also a generator of T^* of even degree by (3.3). So we have

$$\dim_{\mathbf{Q}} C^{even} = 1.$$

If $\dim_{\mathbf{Q}} K > 1$, we can choose two elements y, z in K such that $d(y) = x^m$ and $d(z) = x^n$ with $2 \leq m \leq n$ for the only generator x of C and that they are linearly independent. Here we can assume $m < n$ from the construction of a minimal model in Section 2. Then we have

$$\mathrm{Ker}[d|_Z] \supset \mathbf{Q}\{x,\ z\} \underset{\neq}{\supset} \mathbf{Q}\{x\} = C = \mathrm{Ker}(d|_Z)$$

since $z \in \mathrm{Ker}[d|_Z]$ by $d(z) = x^n = d(x^{n-m} \cdot y)$. This contradicts the normality (see Definition 2.3). So we have

$$\dim_{\mathbf{Q}} K \leq 1.$$

Thus we see that $(\wedge Z, d) = (\wedge(x), 0)$ or $(\wedge Z, d) = (\wedge(x, y), d)$ with $d(x) = 0$ and $d(y) = x^n$ for some n. Since every nonzero element of $I(y)$ is not a d-cocycle, $(\wedge Z, d)$ is formal from Lemma 2.5. Thus we have this lemma from (2.4). □

Proof of Theorem 1.3 Let $(S^{2m})_\infty$ be the James' reduced product space of S^{2m} for some m and $(S^{2m})_n$ the $2mn$ skeleton of it for some n. Then we recall that

$$H^*(X) \cong \mathbf{Q}[x]/(x^{n+1}) \text{ if and only if } X \cong (S^{2m})_n \text{ with } \deg x = 2m \tag{4.2}$$

as a special case of Corollary 2.6 of [**10**]. The "if" part follows from (4.2) and Lemma 4.3. The "only if" part follows from Lemma 4.4 and (4.2). □

Acknowledgements

The author wishes to thank the referee for his attention to some mistakes in the methods of proofs in the first version and Professors M. Mimura and K. Shimakawa and Mr. T. Nishimoto for their helpful comments.

References

[1] Burghelea, D. and Fiedorowicz, Z. [1986], *Cyclic homology and algebraic K-theory of spaces II*, Topology **25**, 303-317.

[2] Deligne, P., Griffiths, P., Morgan, J. and Sullivan, D. [1975], *The real homotopy theory of Kaehler manifolds*, Invent. Math. **29**, 245-274.

[3] Felix, Y. and Halperin, S. [1982], *Formal spaces with finite-dimensional rational homotopy*, American Math. Soc. **270**, 575-588.

[4] Getzler, E., Jones, J. and Petrack, S. [1991], *Differential forms on loop spaces and the cyclic bar complex*, Topology **30**, 339-371.

[5] Greub, W., Halperin, S. and Vanstone, R. [1976], *Connection, curvature and cohomology*, Vol. III, Academic Press.

[6] Griffiths, P. and Morgan, J. [1981], *Rational homotopy theory and differential forms*, Birkhaüser, **16**.

[7] Halperin, S. [1983], *Lectures on minimal models*, Mem. Soc. Math. France, Vol.**9/10**.

[8] Halperin, S. and Stasheff, J. [1979], *Obstructions to homotopy equivalences*, Advance in Math. **32**, 233-279.

[9] Hilton, P., Mislin, G. and Roitberg, J. [1975], *Localization of nilpotent groups and spaces*, Math. Studies **15**, North Holland.

[10] Mimura, M. and Toda, H. [1971], *On p-equivalences and p-universal spaces*, Comment. Math. Helv. **46**, 87-97.

[11] Papadima, S. [1985], *The cellular structure of formal homotopy types,* Journal of Pure and Applied Algebra, **35**, 171-184.

[12] Vigué-Poirrier, M. [1988], *Sur l'algebre de cohomologie cyclique d'un espace 1-connexe applications a la geometrie des varietes*, Illinois J. of Math. **32** No. 1, 40-52.

[13] Vigué-Poirrier, M. [1994], *Homologie cyclique des espaces formels,* Journal of Pure and Applied Algebra **91**, 347-354.

[14] Vigué-Poirrier, M. and Burghelea, D. [1985], *A model for cyclic homology and algebraic K-theory of 1-connected topological spaces*, J. Differential Geometry **22**, 243-253.

[15] Vigué-Poirrier, M. and Sullivan, D. [1976], *The homology theory of the closed geodesic problem*, J. Differential Geometry **11**, 633-644.

Appendix

List of Participants

Adem, Alejandro, University of Wisconsin, Madison, Wisconsin, USA
Aguade, Jaume, Barcelona Univ., Barcelona, Spain
Andersen, Kasper, The Fields Institute, Toronto, Ontario, Canada
Ando, Matthew, University of Virginia, Charlottesville, Virginia, USA
Arone, Greg, University of Chicago, Chicago, Illinois, USA
Badzioch, Bernard, University of Notre Dame, South Bend, Indiana, USA
Bahri, Tony, Ryder College, Princeton, New Jersey, USA
Baues, Hans, University of Bonn, Bonn, Germany
Benson, David, University of Georgia, Athens, Georgia, USA
Bisson, Terry, SUNY at Buffalo, Buffalo, New York, USA
Blanc, David, University of Haifa, Haifa, Israel
Boardman, Michael, Johns Hopkins University, Baltimore, Maryland, USA
Botvinnik, Boris, University of Oregon, Eugene, Oregon, USA
Bria, Frank, Purdue University, West Lafayette, Indiana, USA
Campbell, Eddy, Queen's University, Kingston, Ontario, Canada
Casacuberta, Carles, University of Barcelona, Barcelona, Spain
Chacholski, Wojcieck, The Fields Institute, Toronto, Ontario, Canada
Christensen, Dan, MIT, Cambridge, Massachusetts, USA
Cohen, Fred, University of Rochester, Rochester, New York, USA
Cornea, Octavian, Université de Lille, Lille, France
Crawford, Arleigh, The Fields Institute, Toronto, Ontario, Canada
Cruickshank, James, University of Alberta, Edmonton, Alberta, Canada
Cuvilliez, Max, University de Louvain, Louvain-la-Neuve, Belgium
Davis, Don, Lehigh University, Bethlehem, Pennsylvania, USA
Devinatz, Ethan, University of Chicago, Chicago, Illinois, USA
Dumey, Luc, Université de Lille, Lille, France
Dwyer, Bill, University of Notre Dame, South Bend, Indiana, USA
Elliott, George, The Fields Institute, Toronto, Ontario, Canada
Falk, Jeffrey, University of Rochester, Rochester, New York, USA
Felix, Yves, Université de Louvain, Louvain-la-Neuve, Belgium
Fernandez-Suarez, Lucia, Université de Lille, Lille, France
Florin, Marion, University of Notre Dame, South Bend, Indiana, USA
Franjou, Vincent, Université de Paris-Sud, Orsay, France
Gammelin, Herve, Université de Lille, Lille, France
Ghorbal, Sonia, Université de Louvain, Louvain-la-Neuve, Belgium
Gitler, Sam, University of Rochester, Rochester, New York USA
Gorbunov, Vassily, University of Kentucky, Lexington, Kentucky, USA
Gray, Brayton, University of Illinois at Chicago Circle, Chicago, Illinois, USA
Greenlees, John, University of Sheffield, Sheffield, England
Grodal, Jesper, The Fields Institute, Toronto, Ontario, Canada
Halperin, Steve, University of Toronto, Toronto, Ontario, Canada

Harper, John, University of Rochester, Rochester, New York, USA
Henn, Hans-Verner, Heidelberg University, Heidelberg, Germany
Hess, Kathryn, Ecole Polytechnique de Lausanne, Lausanne, Switzerland
Hessenholt, Lars, MIT, Cambridge, Massachusetts, USA
Hopkins, Mike, MIT, Cambridge, Massachusetts, USA
Hovey, Mark, University of Kentucky, Lexington, Kentucky, USA
Hubbuck, John, University of Aberdeen, Aberdeen, Scotland
Hunter, David, University of Virginia, Charlottesville, Virginia, USA
Jackowski, Stefan, University of Warsaw, Warsaw, Poland
Jardine, Rick, The University of Western Ontario, London, Ontario, Canada
Jeffrey, Lisa, McGill University, Montreal, Quebec, Canada
Jessup, Barry, University of Ottawa, Ottawa, Ontario, Canada
Johnson, Brenda, Union College, Schenectady, New York, USA
Johnson, David, University of Kentucky, Lexington, Kentucky, USA
Johnson, Keith, Dalhousie University, Halifax, Nova Scotia, Canada
Kahl, Thomas, Université de Louvain, Louvain-la-Neuve, Belgium
Kallel, Sadok, Université de Montreal, Montreal, Quebec, Canada
Kane, Richard, The University of Western Ontario, London, Ontario, Canada
Kechegias, Nondas, University of Iannina, Iannina, Greece
Kitchloo, Nitya, MIT, Cambridge, Massachusetts, USA
Kochman, Stan, York University, Downsview, Ontario, Canada
Kuhn, Nick, University of Virginia, Charlottesville, Virginia, USA
Lam, Kee, University of British Columbia, Vancouver, B.C., Canada
Lambrechts, Pascal, Université de Louvain, Louvain-la-Neuve, Belgium
Langsetmo, Lisa, University of Ottawa, Ottawa, Ontario, Canada
Laures, Gerd, MIT, Cambridge, Massachusetts, USA
Lazarev, Andrey, University of Pennsylvania, Philadelphia, Pennsylvania, USA
Lemaire, Jacques, Université de Nice, Nice, France
Lesh, Kathryn, University of Toledo, Toledo, Ohio, USA
Levi, Ran, Northwestern University, Evanston, Illinois, USA
Libine, Matvei, University of Toronto, Toronto, Ontario, Canada
Lin, Jim, University of San Diego, La Jolla, California, USA
Lin, W.-H., National Tsiang Hua University, Hsinchu, Taiwan
Madsen, Ib, Aarhus University, Aarhus, Denmark
Mahowald, Mark, Northwestern University, Evanston, Illinois, USA
Majewski, Martin, Universität of Berlin, Berlin, Germany
Mather, Michael, The Fields Institute, Toronto, Ontario, Canada
May, Peter, University of Chicago, Chicago, Illinois, USA
McCleary, John, Vassar College, Poughkeepsie, New York, USA
McClure, Jim, Purdue University, West Lafayette, Indiana, USA
McGibbon, Chuck, Wayne State University, Detroit, Michigan, USA
Menichi, Luc, Université de Lille, Lille, France
Miller, Haynes, MIT, Cambridge, Massachusetts, USA
Minami, Norihiko, University of Alabama, Birmingham, Alabama USA
Møller, Jesper, University of Copenhagen, Copenhagen, Denmark
Moore, J. C., University of Rochester, Rochester, New York, USA
Morava, Jack, Johns Hopkins University, Baltimore, Maryland, USA
Morrison, Wendy, Trent University, Peterborough, Ontario, Canada
Murillo, Aniceto, University of Malaga, Malaga, Spain
Ndombol, Bitjong, Université de Lille, Lille, France

Neisendorfer, Joe, University of Rochester, Rochester, New York, USA
Notbohm, Dietrich, Göttingen Universität, Göttingen, Germany
Palmieri, John, MIT, Cambridge, Massachusetts, USA
Parent, Paul, University of Ottawa, Ottawa, Ontario, Canada
Pemmaraju, Satya, Northwestern University, Evanston, Illinois, USA
Pengelley, David, New Mexico State University, Las Cruces, New Mexico, USA
Peschke, George, University of Alberta, Edmonton, Alberta, Canada
Peterson, Frank, MIT, Cambridge, Massachusetts, USA
Priddy, Stuart, Northwestern University, Evanston, Illinois, USA
Przezdzicki, Adam, University of Chicago, Chicago, Illinois, USA
Randall, Duane, Loyola University, Chicago, Illinois, USA
Ravenel, Doug, University of Rochester, Rochester, New York, USA
Ray, Nigel, University of Manchester, Manchester, England
Rezk, Charles, MIT, Cambridge, Massachusetts, USA
Richter, Bill, Northwestern University, Evanston, Illinois, USA
Rodriguez, Jose, University of Barcelona, Barcelona, Spain
Sadofsky, Hal, University of Oregon, Eugene, Oregon, USA
Santos, Marta, University of Barcelona, Barcelona, Spain
Sato, Takeshi, University of Rochester, Rochester, New York, USA
Schuster, Bjorn, The Fields Institute, Toronto, Ontario, Canada
Scott, Jonathan, University of Toronto, Toronto, Ontario, Canada
Selick, Paul, University of Toronto, Toronto, Ontario, Canada
Shank, Jim, Queen's University, Kingston, Ontario, Canada
Sharpe, Rick, University of Toronto, Toronto, Ontario, Canada
Sheerer, Jerome, Ecole Polytechnique de Lausanne, Lausanne, Switzerland
Shick, Paul, John Carroll Univ., Cleveland, Ohio, USA
Shipley, Brooke, University of Notre Dame, South Bend, Indiana, USA
Silverman, Judith, University of Michigan, Ann Arbor, Michigan, USA
Singer, Bill, Fordham University, New York, New York, USA
Slack, Michael, Western Michigan University, Kalamazoo, Michigan, USA
Smith, Jeff, Purdue University, West Lafayette, Indiana, USA
Smith, Larry, Universität of Göttingen, Göttingen, Germany
Snaith, Vic, McMaster University, Hamilton, Ontario, Canada
Sottile, Frank, University of Toronto, Toronto, Ontario, Canada
Stanley, Don, University of Toronto, Toronto, Ontario, Canada
Strickland, Neil, MIT, Cambridge, Massachusetts, USA
Strounine, Alex, University of Notre Dame, South Bend, Indiana, USA
Tai, Jin-Yen, Rutgers University, New Brunswick, New Jersey, USA
Tanre, Daniel, Université de Lille, Lille, France
Theriault, Steve, University of Toronto, Toronto, Ontario, Canada
Thiercelin, J. M., Université de Lille, Lille, France
Thomas, Jean-Claude, Université de Lille, Lille, France
Thompson, Rob, Hunter University, CUNY, New York, New York, USA
Tse, Kung-Kuen, University of Rochester, Rochester, New York, USA
Vandembroucq, Lucile, Université de Lille, Lille, France
Vershinin, Vladimir, University of Novosibirsk, Novosibirsk, Russia
Wang, Xiangjun, University of Nankai, Nankai, China
Wehlau, David, Royal Military College, Kingston, Ontario, Canada
Wilkerson, Clarence, Purdue University, West Lafayette, Indiana, USA
Wilson, Steve, Johns Hopkins University, Baltimore, Maryland, USA

Witherspoon, Sarah, University of Toronto, Toronto, Ontario, Canada
Wolbert, Jerome, University of Chicago, Chicago, Illinois, USA
Wu, Jie, Mathematical Sciences Research Institute, Berkeley, California, USA
Xicotenatl, Miguel, University of Rochester, Rochester, New York, USA
Yagita, Nbuaki, Ibaraki University, Ibaraki, Japan
Yang, Huajian, Lehigh University, Bethlehem, Pennsylvania, USA
Yamaguchi, Toshihiro, Okayama University, Okayama, Japan
Zelov, Vitaly, Lehigh University, Bethlehem, Pennsylvania, USA